Pyramid Algorithms

A Dynamic Programming Approach to Curves and Surfaces for Geometric Modeling

The Morgan Kaufmann Series in Computer Graphics and Geometric Modeling

Series Editor: Brian A. Barsky, University of California, Berkeley

Pyramid Algorithms

A Dynamic Programming Approach to Curves and Surfaces for Geometric Modeling

Ron Goldman
Rice University

MORGAN KAUFMANN PUBLISHERS

AN IMPRINT OF ELSEVIER SCIENCE

AMSTERDAM   BOSTON   LONDON   NEW YORK
OXFORD   PARIS   SAN DIEGO   SAN FRANCISCO
SINGAPORE   SYDNEY   TOKYO

Publishing Director   Diane Cerra
Assistant Publishing Services Manager   Edward Wade
Senior Production Editor   Elisabeth Beller
Developmental Editor   Marilyn Alan
Cover Design   Ross Carron Design
Cover Image   Getty/Mark Downey
Text Design   Rebecca Evans & Associates
Technical Illustration   Dartmouth Publishing, Inc.
Composition   Nancy Logan
Copyeditor   Ken DellaPenta
Proofreader   Jennifer McClain
Indexer   Steve Rath
Printer   Courier Corporation

Designations used by companies to distinguish their products are often claimed as trademarks or registered trademarks. In all instances in which Morgan Kaufmann Publishers is aware of a claim, the product names appear in initial capital or all capital letters. Readers, however, should contact the appropriate companies for more complete information regarding trademarks and registration.

Morgan Kaufmann Publishers
An Imprint of Elsevier Science
340 Pine Street, Sixth Floor, San Francisco, CA 94104-3205, USA
www.mkp.com

Printed and bound by CPI Group (UK) Ltd, Croydon, CR0 4YY

Transferred to Digital Print 2011

Library of Congress Control Number: 2002107240

ISBN 1-55860-354-9

This book is printed on acid-free paper.

For my graduate students—past, present, and future

This writer . . . would have to prepare his book with meticulous care, perpetually regrouping his forces like a general conducting an offensive, and he would have also to endure his book like a form of fatigue, to accept it like a discipline, build it up like a church, follow it like a medical regime, vanquish it like an obstacle, win it like a friendship, cosset it like a little child, create it like a new world without neglecting those mysteries whose explanation is to be found probably only in worlds other than our own and the presentiment of which is the thing that moves us most deeply in life and in art.

Marcel Proust, *Remembrance of Things Past*

Contents

Foreword

About forty years ago, the development of computer-aided design and manufacturing created the strong need for new ways to mathematically represent curves and surfaces: the new representations should possess enough flexibility to describe almost arbitrary geometric shapes; be compatible with efficient algorithms; and be readily accessible to designers who could manipulate them simply and intuitively. Although these new requirements presented a difficult challenge, the search for appropriate mathematical tools has been very successful within a relatively short period of time. Curves and surfaces with a piecewise polynomial or rational parametric representation have become the favorites, in particular if they are represented in so-called Bézier or B-spline form. A new field, called computer-aided geometric design (CAGD) emerged. Deeply rooted in approximation theory and numerical analysis, CAGD greatly benefited from results in the classical geometric disciplines such as differential, projective, and algebraic geometry.

Today, CAGD is a mature field that branches into various areas of mathematics, computer science, and engineering. Its boundaries have become less defined, but its kernel still consists of algorithms for interpolation and approximation with piecewise polynomial or rational curves and surfaces.

Pyramid Algorithms presents this kernel in a unique way. A few celebrated examples of pyramid algorithms are known to many people: I think of the de Casteljau algorithm and de Boor's algorithm for evaluation and subdividing a Bézier or B-spline curve, respectively. However, as Ron Goldman tells us in this fascinating book, pyramid algorithms occur almost everywhere in CAGD: they are used for polynomial interpolation, approximation, and change of basis procedures; they are even dualizable. Dr. Goldman discusses pyramid algorithms for polynomial curves, piecewise polynomial curves, tensor product surfaces, and triangular and multisided surface patches.

Though the book focuses mainly on topics well known in CAGD, there are many parts with unconventional approaches, interesting new views, and new insights. Surprises already appear in Chapter 1 on foundations: I had always thought that projective geometry was the ideal framework for a deeper study of rational curves and surfaces, but I must admit that Ron Goldman's preference of Grassmann space over projective space has its distinct advantages for the topics discussed in this book. Surprises continue to occur throughout the book and culminate in the last chapter on multisided patches. Here we find a brand new exposition of very recent results, which eloquently connects the well-established theory of CAGD to ongoing research in the field.

I am convinced that reading this book will be a pleasure for everyone interested in the mathematical and algorithmic aspects of CAGD. Ron Goldman is a leading expert who knows the fundamental concepts and their interconnectedness as well as the small details. He skillfully guides the reader through subtle subjects without getting lost in pure formalism. The elegance of the writing and of the methods used to present the material allows us to get a deep understanding of the central concepts of CAGD. The presentation is clear and precise but never stiff or too abstract. This is a mathematically substantial book that lets the reader enjoy the beauty of the subject. It achieves its goal even without illustrating the creative shape potential of free-form curves and surfaces and without espousing the many important applications this field has in numerous branches of science and technology. In its simplicity and pure beauty, the theory indeed resembles the pyramids.

Helmut Pottmann
Vienna University of Technology, Austria

Preface

Every mature technical subject has its own distinct point of view—favorite methods, cherished formulas, standard examples, preferred algorithms, characteristic projects, common folklore, pet principles and paradigms. Initially, computer-aided geometric design (CAGD) grew out of approximation theory and numerical analysis, adapting the tools of these disciplines for its own devices. Differential, algebraic, and projective geometry also contributed to the development of CAGD, which borrowed heavily from each of these fields.

Outside of mathematics, CAGD was strongly influenced by computer science and mechanical engineering. Indeed, it was the ability to solve computational problems in mechanical design and manufacture that gave CAGD its initial impetus, its original reason to exist. It is no accident that many of the founders of the field—Bezier, de Casteljau, Coons, and Gordon—worked in some capacity for automotive companies.

Today CAGD is a distinct science with its own unique criteria and prerequisites, themes and leitmotifs, tactics and strategies, goals and aspirations, models and representations, problems and procedures, challenges and requirements. The purpose of this book is to present this fresh point of view by reinvestigating polynomial, rational, and piecewise polynomial interpolation and approximation from this contemporary computational perspective.

There is a unity to CAGD—dynamic programming procedures, pyramid algorithms, up and down recurrences, basis functions, dual functionals, rational schemes, tensor product and triangular patches—these themes recur again and again in different guises throughout the subject. One deliberate goal of this book is to capture this unity by presenting different topics exercising these same basic techniques.

To achieve this goal, this book begins with an introductory chapter, followed by two main parts: Part I covers Interpolation (Chapters 2–4) and Part II, Approximation (Chapters 5–8).

Foundations are presented in Chapter 1. These root topics are the underlying geometric ideas—the essential, often unwritten, fundamentals of the field.

Geometry is the bedrock of CAGD. Numerical analysis and approximation theory investigate functions defined over the fields of real or complex numbers; classical algebraic geometry focuses on polynomials defined on real or complex projective spaces. In contrast, the natural geometric domains for CAGD are affine spaces and Grassmann spaces. Polynomial curves and surfaces, along with their control points, control polygons, and control polyhedra, typically lie in affine spaces; rational curves and surfaces are projections of polynomial curves and surfaces from Grassmann spaces to affine or projective spaces. Thus the control structures of rational curves and surfaces consist of mass-points in Grassmann space, not of ordinary points in affine space or homogeneous points in projective space. Blossoming too requires the formalisms of affine spaces and Grassmann spaces. The domain of the standard blossom is affine space; the domain of the homogeneous blossom is Grassmann space.

Although affine spaces and Grassmann spaces are known in classical mathematics, they are somewhat obscure and rarely treated in standard texts. Students are almost never familiar with these geometric spaces. We begin, then, with an overview of different ambient spaces—vector spaces, affine spaces, Grassmann spaces, and projective spaces—the spaces that support the geometry of CAGD. Definitions, examples, distinguishing features, embeddings, projections, and other relationships between these four distinct spaces are discussed in considerable detail to clarify the underlying algebraic structures of these geometric spaces.

We stress as well in Chapter 1 a coordinate-free approach to geometry. This style has at least two distinct advantages over coordinate methods: First, this technique allows us to distinguish clearly between points and vectors in affine space, as well as between mass-points in Grassmann space and homogeneous points in projective space. Coordinates obscure these distinctions, confusing rather than clarifying these issues for the student. Second, this coordinate-free notation is very concise; it is a great deal more convenient to write one formula for points or for vectors rather than two or three formulas for their coordinates. Algorithms, identities, and computations are much cleaner to express and a good deal easier to understand in this notation.

The main objects of study in CAGD are smooth curves and surfaces. Many representations for curves and surfaces are available; smooth shapes can be defined by functions that are explicit, implicit, parametric, or procedural. In Chapter 1 we fix our attention once and for all on parametric curves and surfaces. Differential geometry also studies smooth parametric curves and surfaces, but with this difference: in differential geometry the parametrizations are typically implicit; in CAGD they are always explicit. In differential geometry it is enough to know that a smooth parametrization exists—one often works implicitly with the arc length parametrization of a differentiable curve; in CAGD the parametrization must be explicit—typically a polynomial or rational function—for computational considerations.

Last, but not least, in Chapter 1 we also discuss barycentric coordinates. These coordinates provide the natural domain parameters for the triangular patches that occur in various guises throughout the text, including triangular Lagrange interpolation, triangular Bezier approximation, and triangular B-patches. Generalized barycentric coordinates are also central to the construction in Chapter 8 of a general theory of multisided Bezier patches.

Part I of this text focuses on polynomial and rational interpolation. This investigation of interpolation is divided into three chapters: Lagrange (Chapter 2), Hermite (Chapter 3), and Newton (Chapter 4).

We introduce Lagrange and Hermite interpolation via Neville's algorithm. This approach allows us to discuss many of the classical themes of CAGD—dynamic programming procedures, pyramid algorithms, up and down recurrences, tensor product and triangular patches, existence and uniqueness theorems—in the context of interpolation, which many beginning students find more intuitive than approximation. One innovation here is that we often derive properties and formulas for Lagrange and Hermite interpolation directly from the dynamic programming diagram for Neville's algorithm, an approach that students typically find more appealing than manipulating formulas in which many distracting indices may be present. This method of reasoning from the structure of the diagram for a dynamic programming procedure will be pursued throughout the text; several of the basic properties, fundamental formulas, and principal algorithms for Bezier and B-spline curves and surfaces are also derived in this manner.

Our choice of topics in interpolation is not always standard. For example, our interest in triangular grids for bivariate Lagrange interpolation is a bit unconventional but not at all whimsical; we intentionally emphasize this topic here to pave the way later on for the study of triangular Bezier patches. We also examine rational Lagrange and rational Hermite interpolation—topics not typically covered in texts on CAGD—as a prelude to the investigation of rational Bezier and rational B-spline curves and surfaces. Finally, we present one additional innovation: as an application of Lagrange interpolation, we examine the Fast Fourier Transform. Strictly speaking, Fast Fourier Transform is not a concern of CAGD, but this topic provides an excellent application of Lagrange interpolation in computational science outside of strict data interpolation, so we include a short discussion of this subject here as well, to broaden the outlook of the student.

Newton interpolation and the divided difference round out our discussion of polynomial interpolation. Although these subjects are classical topics in approximation theory and numerical analysis, they are not as well known as they should be to the CAGD community. Therefore we take the time to provide a thorough presentation of these topics in Chapter 4.

The divided difference provides the Newton coefficients of the polynomial interpolant. This observation allows us to introduce the notion of dual functionals, anticipating our investigation of blossoming in Chapter 6. To prepare the way for blossoming, we provide, in addition to the classical definitions, an axiomatic characterization of the divided difference. These divided difference axioms are akin to the blossoming axioms, so some of the analysis techniques developed here can be reprised in Chapter 6 when blossoming is investigated.

Numerous identities for the divided difference are developed in the text and in the exercises. Since there are so many divided difference identities, we list these formulas for easy reference at the end of Chapter 4. These identities can be applied in several areas: for example, B-splines are often presented from the perspective of divided differences. Therefore we shall have occasion to return to several of these identities when we study B-splines in Chapter 7. Also, because the blossoming axioms are so similar to the divided difference axioms, some blossoming identities have analogues in divided difference identities. We shall follow up some of these connections to divided difference when we come to study blossoming in Chapter 6.

Part II of this text is devoted to polynomial, rational, and piecewise polynomial approximation. Bezier curves and surfaces, blossoming, and B-splines are the most successful areas of CAGD, and in Part II of this text a chapter is devoted to each of these topics.

We begin our investigation of polynomial approximation with the study of Bezier curves and surfaces. Free-form curves and surfaces are shapes with no name—hard to describe in precise words or explicit formulas but conspicuous in aesthetic and practical design. The key geometric feature of Bezier curves and surfaces from the perspective of CAGD is that they approximate—in a way that is intuitively natural and can be made mathematically precise—the contour described by their control points. Thus they lend themselves readily to the design of free-form shapes.

What makes Bezier curves and surfaces so attractive analytically is that they possess straightforward algorithms for evaluation, subdivision, differentiation, and degree elevation. This suite of algorithms is what permits exhaustive computer analysis of free-form shapes represented in Bezier form. For example, recursive subdivision leads to simple divide-and-conquer procedures for rendering and intersecting Bezier curves and surfaces.

Bernstein/Bezier approximation is an extremely rich theory, and we purposely approach this topic from as many different analytic perspectives as possible, including dynamic programming procedures, Bernstein polynomials, generating functions, the binomial theorem, the binomial distribution, and discrete convolution. Although any one of these techniques may be powerful enough to develop the entire Bernstein/ Bezier canon, we have intentionally avoided consistently adopting any one particular method in order not to impoverish the theory. At first, this very richness of the theory may seem daunting to the novice, but the student should keep in mind when faced with new problems that a variety of approaches are possible. There are many weapons in the Bernstein/Bezier arsenal.

In contrast to standard texts that treat Bezier curves and surfaces, we have incorporated the following innovations in our approach to this subject:

- Reasoning directly from the dynamic programming diagram for the de Casteljau algorithm to provide easy derivations for some elementary properties of Bezier curves and surfaces.

- Developing algorithms for differentiating and blossoming Bezier curves and surfaces by differentiating and blossoming the diagram for the de Casteljau algorithm.

- Introducing general principles of duality to simplify the study of change of basis procedures.

- Providing an elementary proof of the Weierstrass Approximation Theorem, which is then applied to establish the convergence of the degree-elevation algorithm for Bezier curves.

- Presenting Wang's formula to avoid flatness testing and speed up algorithms for rendering and intersection based on recursive subdivision.

- Using discrete convolution to derive differentiation formulas for the Bernstein polynomials. This approach not only simplifies the study of derivative algorithms for Bezier curves and surfaces, but also prepares the way for understanding the symmetry property when in Chapter 6 we study how to blossom the de Casteljau algorithm. It also anticipates the central role that discrete convolution plays in Chapter 8, where we study multisided Bezier schemes.

- Treating the subject of integration for the Bernstein polynomials. Definite integrals provide the most direct way to prove that the arc length of a Bezier curve is bounded by the perimeter of its control polygon. In addition, integration formulas for the Bernstein basis functions prepare the way for developing integration formulas for the B-splines.

- Comparing and contrasting pyramid algorithms with the de Casteljau approach to evaluation and differentiation for tensor product and triangular Bezier surfaces.

In addition, at the end of the chapter, we provide a comprehensive list of identities for the univariate and bivariate Bernstein basis functions for easy reference.

Blossoming is an elegant and potent tool for analyzing Bezier and B-spline curves and surfaces. Nevertheless, we resolutely postpone blossoming till Chapter 6, even though blossoming could effectively be applied in Chapter 5 to derive algorithms for subdivision, degree elevation, differentiation, and change of basis for Bezier curves and surfaces.

There are two problems with introducing blossoming too early in the text. First, blossoming is too powerful. If students come to believe that they can do everything with blossoming, why should they learn any other approach? We delay blossoming so that students are forced to learn a variety of techniques that they may then use in extensions of the Bezier setting where blossoming no longer applies. Second, students do not appreciate the real power of blossoming unless they get to see how many disparate techniques blossoming can be used to replace. In Chapter 5 we derive subdivision from the binomial distribution, degree elevation from polynomial identities, differentiation from discrete convolution, and change of basis from monomial to Bezier form by invoking the binomial theorem and generating functions. Each of these approaches is quite elegant when viewed in isolation, but altogether this variety of approaches can be quite overwhelming. For Bezier curves and surfaces, blossoming can be used to replace all of these methods. By deferring blossoming till after our initial investigation of Bezier curves and surfaces and then reinvestigating topics such as subdivision, differentiation, degree elevation, and change of basis in light of this new tool, students come to appreciate the full power of blossoming.

We highlight both the affine and the homogeneous blossom. The affine blossom is appropriate for studying points, function values, and change of basis procedures; the homogeneous blossom is the natural way to investigate derivatives. The study of derivatives via blossoming is applied in Chapter 7 in our investigation of splines since blossoming can be used to determine when two polynomials meet smoothly at their join.

Blossoming prepares the way for B-splines. Some authors begin the study of B-splines by writing down the de Boor recurrence without any motivation. Students may then follow the development of the theory, but they are unable to fathom the inspiration for this recurrence. Blossoming provides the motivation for the de Boor algorithm since the de Boor recurrence is identical with the blossoming recurrence for computing values along the diagonal, and this blossoming recurrence is, in turn, a straightforward generalization of the de Casteljau algorithm.

B-spline curves and surfaces have two advantages over Bezier curves and surfaces. For a large collection of control points, a Bezier curve or surface approximates the data with a single polynomial of high degree. But high-degree polynomials take a long time to compute and are numerically unstable. B-splines provide low-degree approximations, which are faster to compute and numerically more secure. For these reasons B-splines have become extremely popular in large-scale industrial applications.

We begin the study of B-splines by analyzing the dynamic programming diagram for the de Boor algorithm. Reasoning from the diagram, we can derive many of the elementary properties of B-spline curves such as the local convex hull property. By overlapping these dynamic programming diagrams for adjacent polynomial segments and then using blossoming to differentiate these diagrams, we provide a simple proof that adjacent polynomial segments meet smoothly at their join. This proof from overlapping de Boor diagrams is much more natural and easier for students to grasp than proofs by induction or by divided difference. We also develop algorithms for differentiating and blossoming B-spline curves and surfaces by showing how to differentiate and blossom the diagram for the de Boor algorithm.

Knot insertion is one of the main innovations of CAGD. Nested knot vectors generate nested spline spaces. Given a knot sequence and a control polygon, knot insertion algorithms construct a new control polygon that generates the same B-spline curve as the original control polygon by inserting control points corresponding to the new knots. The motivation is to create a control polygon with additional control points that more closely approximates the curve than the original control polygon.

Knot insertion is to B-splines what subdivision is to Bezier schemes. The new control polygons generated by knot insertion can be used for rendering and intersecting B-spline curves and surfaces, as well as for providing additional control over the shape of a B-spline curve or surface. Differentiation, too, can be viewed as a knot insertion procedure. Both the standard derivative algorithm and Boehm's derivative algorithm can be understood in terms of knot insertion. The variation diminishing property for B-spline curves also follows from knot insertion, an insight unique to the geometric spirit of CAGD.

Many knot insertion algorithms are now available: Boehm's algorithm, the Oslo algorithm, factored knot insertion, Sablonniere's algorithm, and the Lane-Riesenfeld algorithm for uniform B-splines. Blossoming provides a unified approach to knot insertion as well as insight into the connections between different knot insertion procedures, so we use blossoming to derive most of these algorithms. We study each of these knot insertion algorithms in turn, and we compare and contrast their relative benefits and limitations.

Midway through Chapter 7, we shift the focus to the B-spline basis functions, and we explain the links between B-splines and divided differences. Divided difference is the classical way to introduce B-splines, so students need to learn this approach if only to be able to understand many of the standard tracts on B-splines. In addition, divided differences allow us to derive those properties of B-splines that do not follow readily from blossoming. For example, we use the connection between B-splines and divided differences together to develop a geometric characterization of the univariate B-splines. This geometric approach is often taken as the starting point for the development of the theory of multivariate B-splines, so it is important for students to see this formula first in the univariate setting.

The Bernstein basis functions can be generated from discrete convolution; uniform B-splines can be constructed from continuous convolution. We derive this convolution formula and then use this convolution technique to derive the Lane-Riesenfeld knot insertion algorithm for uniform B-spline curves.

NURBS is an acronym for *non-uniform rational B-splines*. At this late stage in the text, students are well prepared for the study of rational B-splines since they have already encountered rational schemes in the Lagrange and Bezier settings. NURBS are the projection from Grassmann space to affine or projective space of integral B-spline curves and surfaces. Therefore NURBS inherit most of the standard properties and algorithms of ordinary B-spline curves and surfaces. Thus once we have explained B-splines thoroughly, NURBS are quite easy to understand.

Catmull-Rom splines are interpolating splines constructed by combining Lagrange interpolation with B-spline approximation, fusing Neville's algorithm together with the de Boor algorithm. Studying Catmull-Rom splines near the end of Chapter 7 allows us to reprise some of the high points of interpolation and approximation in the context of interpolating spline curves.

We close Chapter 7 with the study of B-spline surfaces. Tensor product surfaces are introduced in the standard way by repeated application of the univariate de Boor algorithm. But there is another approach to tensor product B-spline surfaces, less well known than the de Boor algorithm, but highly in keeping with one of the major themes of this book, pyramid algorithms. We derive the pyramid algorithm for tensor product B-spline surfaces from blossoming and then show that this algorithm can be extended to a kind of local triangular B-spline surface—the B-patch. Unlike the pyramid construction for the tensor product B-spline surface, there is no easy way to piece together polynomial B-patches to form a spline surface over a triangular grid. There is, however, a construction of multivariate B-splines from B-patches, but, unfortunately, this construction is a bit beyond the scope of this text.

At the end of Chapter 7—as we did at the end of Chapter 4 for the divided difference, at the end of Chapter 5 for the Bernstein polynomials, and at the end of Chapter 6 for blossoming—we gather, for easy reference, a comprehensive list of identities for the B-spline basis functions.

Chapter 8 is devoted to multisided Bezier patches, including S-patches, C-patches, and toric Bezier patches. Each of these schemes has a pyramid evaluation algorithm that generalizes the de Casteljau evaluation algorithm for triangular and tensor product Bezier patches. These pyramid algorithms can also be blossomed to provide the dual functionals for these multisided Bezier schemes. Three key ideas link together and unify these different constructions of multisided Bezier patches: discrete convolution, Minkowski sum, and the general pyramid algorithm. A vital role is also played by different approaches to indexing multisided arrays and different ways to construct generalized barycentric coordinate functions. These concepts and techniques extend many of the salient ideas and insights encountered in earlier chapters, so this topic, which is still ongoing research, makes a fitting final chapter for this book.

Unlike the rest of this book, much of the material in Chapter 8 is new and is presented here in a coherent and unified fashion for the first time. Although for the most part I have based this chapter on what has come before it in the text, a higher level of mathematical sophistication is required here on the part of the reader. In contrast to the other chapters, this chapter is written for experts rather than for neophytes.

My goal has been to write a book that can serve both as a reference and as a text. As a text for a one-semester 15-week course, I envision that the class could cover the first seven chapters, devoting the first week to Chapter 1 and roughly two weeks apiece to Chapters 2–7. Obviously, there is far too much material to read everything in each of these chapters, so instructors must pick and choose according to their tastes. The remaining material can serve either as a future reference or as material for a second semester course.

Many exercises are included at the end of each section, and many should be assigned; it is not possible to learn this material to any depth without working through many, many exercises. The exercises are intended both to complement and to illuminate the text. Alternative approaches, as well as additional examples, algorithms, identities, theorems, and proofs, are included in the exercises. The relative difficulty of these exercises varies widely, ranging from simple illustrative examples to straightforward algorithms to complicated proofs. I have provided hints for some of the more challenging and more interesting problems.

This book is intended for engineers and computer scientists, as well as for applied mathematicians. To accommodate the engineers, I have tried to include enough detail to make the subject fully intelligible while not drowning in rigor. I have also tried to keep notation to a minimum, erring if necessary on the side of naivete rather than pedantry. I trust that applied mathematicians will also benefit from this presentation.

Prerequisites include only a standard freshman calculus course along with a limited amount of linear algebra. Students should have at least a passing acquaintance, for example, with vector spaces and linear transformations. To provide some familiar models of affine spaces for engineers, Chapter 1 refers briefly to matrix algebra and to ordinary differential equations. Readers not versed in matrices or differential equations can simply skip these examples with little loss of content. The remainder of the book is self-contained.

A word about some of the choices in this book: Inevitably I have had to leave out subjects that others would consider vital to the field. I have not included material on rates of convergence, algebraic curves and surfaces, Coons patches and Gordon surfaces, Pythagorean hodographs, or geometric continuity. Partly these omissions reflect my own unconscious biases and interests, and partly they reflect a conscious decision on my part to stick to certain major themes and not to stray too far or too often from this path. Others I am sure would make different choices, just as valid, equally compelling.

Finally, this point is the place in the preface where I get to thank all those people who inspired me and helped me to write this book. Unfortunately, this list is way too long for publication. Family and friends, teachers and students, colleagues and confidants, collaborators and competitors, predecessors and contemporaries, Americans, Europeans, Asians, Africans, and Australians have all contributed to this effort. I have borrowed ideas from almost everyone I know who works in the field—scientists in academia, engineers in industry, and even aspiring undergraduate and graduate students in colleges and universities. Conspicuously, this book was written on three different continents, where I was hosted by various genial colleagues and supported by several generous grants. CAGD, like most large-scale human endeavors, is a collaborative effort. This book is the work of many, many people. I am only their conduit. Do not confuse the dancer with the dance.

CHAPTER 1

Introduction: Foundations

We begin with some background material that will be assumed throughout the remainder of this book. Although we shall discuss several generic types of curve and surface representations, our main focus here is on the ambient mathematical spaces in which these shapes reside. We will also review barycentric coordinates, a topic that is central to the construction of conventional triangular surface patches.

1.1 Ambient Spaces

Four different kinds of mathematical spaces support the representation and analysis of free-form curves and surfaces: *vector spaces, affine spaces, Grassmann spaces,* and *projective spaces.* When first reading this chapter, you should focus your attention on vector spaces and affine spaces. Not only are these spaces more familiar, but they are fundamental both for the construction of polynomial curves and surfaces and for the development of the more complicated Grassmann spaces and projective spaces. Grassmann spaces and projective spaces are discussed here as well, but you can defer reading about these mathematical spaces till later in the book when we shall need these tools to clarify some of the properties of rational curves and surfaces. We adopt a coordinate-free approach to geometry. Try to get used to coordinate-free methods now because we plan to employ this approach throughout the text.

1.1.1 Vector Spaces

You should already be familiar with vector spaces from linear algebra. Informally a *vector space* is a collection of objects called *vectors* that can be added and subtracted as well as multiplied by constants. Vectors are often represented geometrically by arrows. These arrows are added and subtracted by the familiar triangle rules; multiplication by constants is represented by stretching or shrinking the arrows, reversing the orientation when the constant is negative (Figure 1.1).

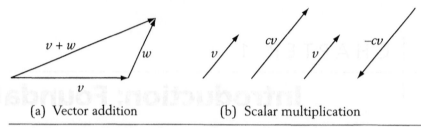

(a) Vector addition   (b) Scalar multiplication

Figure 1.1   Addition and scalar multiplication of vectors.

Examples of vectors include the standard elements of mechanics—velocity, acceleration, and force. However, there are many other important models of abstract vector spaces. For example, the space of all polynomials of degree less than or equal to n is a vector space because addition, subtraction, and scalar multiplication are all well-defined operations for polynomials of degree less than or equal to n.

Solutions to systems of homogeneous linear equations also form a vector space. Consider the system of m linear equations in n unknowns:

$$a_{11}x_1 + a_{12}x_2 + \cdots + a_{1n}x_n = 0 \tag{1.1}$$
$$\vdots$$
$$a_{m1}x_1 + a_{m2}x_2 + \cdots + a_{mn}x_n = 0 \ .$$

Collecting the coefficients $\{a_{ij}\}$ into a matrix A and the variables $\{x_k\}$ into a column array X, we can rewrite (1.1) in matrix notation as

$$AX = 0. \tag{1.2}$$

Now if S_1 and S_2 are solutions of (1.2), then by the linearity of matrix multiplication so are $S_1 \pm S_2$ and cS_1, so the arrays S that represent solutions of (1.2) form a vector space.

Similarly, solutions to linear homogeneous ordinary differential equations form a vector space. Consider the ordinary differential equation

$$a_n(x)y^{(n)} + a_{n-1}(x)y^{(n-1)} + \cdots + a_1(x)y' + a_0(x)y = 0. \tag{1.3}$$

If $y = f_1(x)$ and $y = f_2(x)$ are solutions of (1.3), then by the linearity of differentiation so are $y = f_1(x) \pm f_2(x)$ and $y = cf_1(x)$, so the functions $y = f(x)$ that represent solutions of (1.3) also form a vector space.

Thus vector spaces are ubiquitous in science and mathematics. It is for this reason that you have encountered vectors and vector spaces before and studied them formally in courses on linear algebra.

1.1.2   Affine Spaces

Despite their familiarity, vector spaces are not the appropriate mathematical setting for studying the geometry of curves and surfaces. Curves and surfaces are collections of points, and points are not vectors. Vectors add and scale; points do not. Vectors have size; points do not. Vectors are represented geometrically by arrows; points by dots. Arrows have direction and length, but no fixed position; dots have a fixed position, but no direction or length. Points are moved by translation; vectors are unaffected by translation. To emphasize these distinctions, we shall typically use lowercase letters $u, v, w, \ldots$ from the end of the alphabet to represent vectors and uppercase letters $P, Q, R, \ldots$ from the middle of the alphabet to represent points. Scalars will typically be represented either by lowercase letters $a, b, c, \ldots$ from the beginning of the alphabet or by lowercase Greek letters $\alpha, \beta, \gamma \ldots$.

Although points do not add, we would still like to have an algebra for points. If we were to introduce a rectangular coordinate system, then we could add two points P and Q by adding their coordinates. Unfortunately, the resulting point would depend on our choice of coordinate system (see Figure 1.2). In disciplines such as computer graphics, robotics, and geometric design, there may be several local coordinate systems in any particular model. It would be extremely confusing if our notion of addition were to depend on our choice of coordinate system. Notice that the definition of vector addition in Figure 1.1 is independent of any coordinate system. We seek a similar coordinate-free algebra for points.

Consider the expression $(P+Q)/2$. Even though the term $P+Q$ is indeterminate, the full expression $(P+Q)/2$ does have a clear coordinate-free meaning: it denotes the midpoint of the line segment joining P and Q. Are there any other unambiguous expressions involving points?

If $v_0, \ldots, v_n$ is a collection of vectors and $c_0, \ldots, c_n$ is a collection of constants, then we can form the linear combination

$$v = \sum_{k=0}^{n} c_k v_k,$$

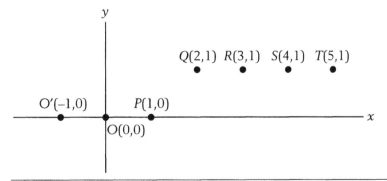

Figure 1.2 Where is $P + Q$? If the origin is at O, then $P = (1,0)$ and $Q = (2,1)$ so $P + Q = (3,1) = R$. But if the origin is moved to $O' = (-1,0)$, then the coordinates of P and Q change. Now $P = (2,0)$ and $Q = (3,1)$, so $P + Q = (5,1)$ and this point is located at $S = (4,1)$ in the original coordinate system. Confused? You should be. Adding points by adding their coordinates is a bad idea and leads both to general bewilderment and incorrect programs.

and since addition and scalar multiplication are defined for vectors, v is a well-defined vector. Similarly if $P_0,...,P_n$ is a collection of points, then we would like to form the combination

$$P = \sum_{k=0}^{n} c_k P_k .$$

Unfortunately, this expression generally has no fixed, coordinate-free meaning, so P is not really a point. What is to be done?

Let us begin again by thinking geometrically about points and vectors. Points are represented by dots, vectors by arrows. Two points P,Q determine a vector v—the arrow joining the dots P and Q. We shall denote this vector by $Q - P$. Subtraction is then defined for points. Similarly, a point P and a vector v determine another point Q by placing the tail of the vector v at the point P and letting Q be the point at which the head of v rests. We shall denote this point Q at the head of v by the sum $P + v$ (see Figure 1.3).

To summarize: so far we have defined $Q - P$ to be a vector and $P + v$ to be a point. Moreover, these definitions are consistent with the usual cancellation rules for addition and subtraction. For example, it follows from our geometric interpretation that

   i. $P + (Q - P) = Q$

   ii. $(R - Q) + (Q - P) = R - P$.

The first rule is just the definition of subtraction for points; the second rule is the triangle rule of vector addition (redraw Figure 1.1(a) with P,Q,R at the vertices of the triangle).

Now we are ready to define the notion of an *affine combination* of points. Earlier we saw that we could not take arbitrary linear combinations of points, but some combinations like $(P + Q)/2$ do make sense. This particular expression can be rewritten using the formal identity

$$\frac{(P + Q)}{2} = P + \frac{(Q - P)}{2} .$$

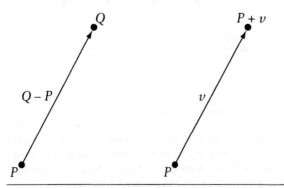

Figure 1.3 Subtraction of points and addition of points and vectors.

The right-hand side of this equation defines the left-hand side. Moreover, the right-hand side now has a clear meaning, since it is the sum of a point P and a vector $(Q - P)/2$. We make this definition because formally it obeys the usual rules of arithmetic; that is, if we want the standard rules of addition, subtraction, and scalar multiplication to apply, then this identity must hold.

Taking our cue from this example, we see that we want to define

$$\sum_{k=0}^{n} c_k P_k = (\sum_{k=0}^{n} c_k) P_0 + \sum_{k=1}^{n} c_k (P_k - P_0) \tag{1.4}$$

since formally all the terms involving P_0 on the right-hand side cancel except $c_0 P_0$. The second summation on the right-hand side makes good sense since $\sum_k c_k (P_k - P_0)$ is a linear combination of vectors, but what meaning can we assign to $(\sum_k c_k) P_0$? In general, we cannot multiply a point by a scalar in a coordinate-free manner, but there are two exceptions. We can define

$$1 \cdot P = P$$
$$0 \cdot P = 0$$
$$cP = \text{undefined} \qquad c \neq 0, 1.$$

By the way, notice that in the second equation the zero on the left-hand side is the zero scalar, but the zero on the right-hand side denotes the zero vector.

Now Equation (1.4) suggests that

$$\sum_{k=0}^{n} c_k P_k = P_0 + \sum_{k=1}^{n} c_k (P_k - P_0) \text{ provided that } \sum_{k=0}^{n} c_k = 1$$

$$= \sum_{k=1}^{n} c_k (P_k - P_0) \qquad \text{provided that } \sum_{k=0}^{n} c_k = 0$$

$$= \text{undefined} \qquad \text{otherwise.}$$

We shall adopt these definitions because formally they obey the usual rules of arithmetic.

Combinations of the form $\sum_k c_k P_k$ where $\sum_k c_k = 1$ are called *affine combinations*. Each difference $P_k - P_0$ is a vector, so the affine combination $\sum_k c_k P_k$ is the sum of a point P_0 and a vector $v = \sum_k c_k (P_k - P_0)$; thus an affine combination of a collection of points generates a new point. For example,

$$\frac{P + Q}{2} \qquad (1 - c)P + cQ \qquad \frac{P + 2Q + R}{4}$$

are affine combinations of points and thus are well-defined points, but

$$\frac{P}{2} + \frac{Q}{3} \qquad \frac{P + Q}{3} \qquad \frac{2P + 5Q + 2R}{10}$$

are meaningless expressions. Notice, in particular, that the points on the line determined by P and Q can be represented by taking the affine combinations $(1-t)P + tQ$ for all values of t.

There are other combinations of points that have meaning, even though the coefficients do not sum to one. For example, $Q - P$ is a meaningful expression, but the sum of the coefficients is zero. In general, $\sum_k c_k P_k$ is a vector, not a point, when $\sum_k c_k = 0$. Again this definition is consistent with the usual rules of arithmetic as well as with the geometry we have constructed for points and vectors.

An *affine space* is a collection of elements called *points* for which affine combinations are defined. Associated with every affine space is a vector space whose vectors are generated by differencing the points. Vectors can be added to the points in the affine space, and the sum of a point and a vector is a point. The usual rules of addition, subtraction, and scalar multiplication apply, but the only combinations of points allowed are those where the scalar coefficients sum to either zero or one.

Since we can subtract points to produce vectors, for every pair of points P and Q, there is a unique vector v such that $Q = P + v$; indeed $v = Q - P$. Thus most of the information contained in an affine space is stored in the vectors; you really need to know only one point. In geometry you often simply pick an origin and then represent the points as vectors emanating from the origin. It is for this reason that vector spaces are studied so intensely and affine spaces may seem so unfamiliar.

Nevertheless, in this book we shall insist on the framework of affine spaces. One reason is that to avoid confusion we need to do all our work independent of the choice of any coordinate system. We want to study the intrinsic properties of curves and surfaces, not their relationships to coordinate systems. Moreover, it is actually often simpler to work directly with the curves and surfaces without referring to any specific coordinates. For this purpose, affine spaces are often the most appropriate setting.

There are many important models of affine spaces. In computer graphics, the points on the graphics screen form an affine space, not a vector space. Points can be translated by adding vectors, but only the points, not the vectors, are actually visible on the graphics terminal.

As with vector spaces, there are also abstract, nongeometric models of affine spaces. The space of all monic polynomials of degree n—that is, all polynomials of degree n with leading coefficient 1—is an affine space. Affine combinations of monic polynomials generate monic polynomials. The difference of two monic polynomials of degree n is a polynomial of degree at most $n - 1$. Thus the associated vector space is the space of all polynomials of degree less than or equal to $n - 1$. The sum of a monic polynomial of degree n and an arbitrary polynomial of degree at most $n - 1$ is again a monic polynomial of degree n. Thus, the sum of a point and a vector is a point as it should be.

Solutions to systems of nonhomogeneous linear equations also form an affine space. Consider the system of m linear equations in n unknowns:

$$a_{11}x_1 + a_{12}x_2 + \cdots + a_{1n}x_n = b_1 \tag{1.5}$$
$$\vdots$$
$$a_{m1}x_1 + a_{m2}x_2 + \cdots + a_{mn}x_n = b_m$$

As before, we can rewrite (1.5) in matrix notation as

$$AX = B. \tag{1.6}$$

If $S_0,...,S_p$ are solutions of (1.6), then by the linearity of matrix multiplication so is $\sum_k c_k S_k$ provided that $\sum_k c_k = 1$, so the arrays S that represent solutions of (1.6) form an affine space. The difference of two solutions of (1.6) is a solution of the associated homogeneous system (1.2). Thus the associated vector space consists of the solutions to the associated system of homogeneous linear equations. Again by the linearity of matrix multiplication, if S is a solution of (1.6) and v is a solution of (1.2), then $S + v$ is a solution of (1.6)—that is, the sum of a point and a vector is a point as required.

Similarly, solutions to nonhomogeneous linear ordinary differential equations form an affine space. Consider the ordinary differential equation

$$a_n(x)y^{(n)} + a_{n-1}(x)y^{(n-1)} + \cdots + a_1(x)y' + a_0(x)y = b(x). \tag{1.7}$$

If $F_1(x),...,F_p(x)$ are solutions of (1.7), then by the linearity of differentiation so is $\sum_k c_k F_k(x)$ provided that $\sum_k c_k = 1$, so the functions $y = F(x)$ that are solutions of (1.7) form an affine space. The difference of two solutions of (1.7) is a solution of the associated homogeneous ordinary differential equation (1.3). Thus the associated vector space consists of the solutions to the associated homogeneous ordinary differential equation. Again by the linearity of differentiation, if $F(x)$ is a solution of (1.7) and $v(x)$ is a solution of (1.3), then $F(x) + v(x)$ is a solution of (1.7)—that is, the sum of a point and a vector is a point.

Thus, like vector spaces, affine spaces are really omnipresent in computational science and engineering.

When we study geometry, we often need to know the dimension of the ambient space. For vector spaces and affine spaces the notion of dimension is tied to the concept of independence. Recall that a collection of vectors $v_1,...,v_n$ is said to be *linearly independent* if we cannot write any vector in the set as a linear combination of the remaining vectors. A maximal linearly independent set of vectors is called a *basis*. The dimension of a vector space is the maximum number of linearly independent vectors in the space—that is, the number of vectors in a basis.

Similarly, a collection of points $P_0,...,P_n$ is said to be *affinely independent* if we cannot write any point in the set as an affine combination of the remaining points. A maximal affinely independent set of points is said to be an *affine basis* for the affine space. The dimension of an affine space is one less than the maximum number of affinely independent points in the space.

For example, the dimension of a single point is zero, the dimension of the affine line is one, and the dimension of the affine plane is two. Thus our notion of dimension is consistent with the standard dimensions in geometry. Moreover, it is not hard to show that the dimension of an affine space is the same as the dimension of its associated vector space (see Exercise 6).

The natural transformations on vector spaces are the transformations that preserve linear combinations. We say then that a transformation L is a *linear transformation* if

$$L(\textstyle\sum_k c_k v_k) = \textstyle\sum_k c_k L(v_k).$$

Similarly, the natural transformations on affine spaces are the transformations that preserve affine combinations. Thus we say that a transformation A is an *affine transformation* if

$$A(\textstyle\sum_k c_k P_k) = \textstyle\sum_k c_k A(P_k) \text{ whenever } \textstyle\sum_k c_k = 1.$$

Many familiar transformations of space such as translation, rotation, scaling, and shearing are affine transformations. To learn more about affine transformations, see Exercises 15–17.

We mentioned at the start of this chapter that there are two other kinds of spaces that arise in the study of free-form curves and surfaces: projective spaces and Grassmann spaces. Projective spaces introduce points at infinity that are convenient for investigating intersections and poles. Projective spaces are also related to the homogeneous coordinates that you may already have encountered in computer graphics. In Grassmann spaces, points have mass as well as location. Assigning mass to points is a venerable technique for studying geometry by applying mechanical principles, an idea first introduced by Archimedes and later refined by Grassmann. Mass-points permit us to complete the definition of $\sum_k c_k P_k$ by allowing us to define combinations where $\sum_k c_k \neq 0,1$. Projective spaces and Grassmann spaces both come into play during the construction of rational curves and surfaces. We shall discuss these spaces in the next two sections, but if you like you can postpone reading these sections for now and come back to them later when we introduce rational curves and surfaces in subsequent chapters.

Exercises

1. Prove that if $\sum_k c_k = 1$, then for any two points R, S

$$R + \textstyle\sum_k c_k (P_k - R) = S + \textstyle\sum_k c_k (P_k - S).$$

   Interpret this result geometrically.

2. Archimedes' law of the lever asserts that the center of mass P of two masses m_1 and m_2 situated respectively at the points P_1 and P_2 is located at the point along the line segment $P_1 P_2$ characterized by the property that the first moment of m_1 around P balances the first moment of m_2 around P. Thus $m_1 \mid P_1 - P \mid = m_2 \mid P_2 - P \mid$.

   a. Use Archimedes' law of the lever to verify that the center of mass of two masses m_1 and m_2 situated respectively at the affine points P_1 and P_2 is located at the affine point $P = (m_1 P_1 + m_2 P_2)/(m_1 + m_2)$.

   b. Suppose that the masses m_k are located at the affine points P_k, $k = 1,...,n$. Use induction to show that the center of mass of the masses m_k, $k = 1,...,n$, is located at the affine point

$$P = \sum_{k=1}^{n} m_k P_k \,/\, \sum_{k=1}^{n} m_k \,.$$

    c. Show that the center of mass of three mass-points $(m_1P_1, m_1), (m_2P_2, m_2)$, (m_3P_3, m_3) can be computed by first computing the center of mass $(m_{12}P_{12}, m_{12})$ of (m_1P_1, m_1) and (m_2P_2, m_2) and then computing the center of mass of $(m_{12}P_{12}, m_{12})$ and (m_3P_3, m_3).

3. Prove that the affine dimension of the affine line is one and that the affine dimension of the affine plane is two.

4. Prove that every vector can be written in a *unique* way as a linear combination of a fixed basis.

5. Prove that every point can be written in a *unique* way as an affine combination of a fixed affine basis.

6. Prove that the dimension of an affine space is the same as the dimension of its associated vector space.

7. What is the dimension of the affine space of monic polynomials of degree n? Justify your answer.

8. Consider the collection of all arrays of real numbers $(a_0, ..., a_n)$ for which $\sum_k a_k = 1$.

    a. Show that these arrays form an affine space under coordinate addition and scalar multiplication.

    b. What is the dimension of this affine space?

    c. Describe the associated vector space.

9. Consider the collection of all arrays of real numbers $(a_0, ..., a_n)$ for which $a_0 = 1$.

    a. Show that these arrays form an affine space under coordinate addition and scalar multiplication.

    b. What is the dimension of this affine space?

    c. Describe the associated vector space.

    (When $n = 3$, these affine coordinates correspond to the standard affine coordinates used in computer graphics and robotics.)

10. Suppose that $A, B, C \neq 0$. Show that the points (x, y) on the line $Ax + By + C = 0$ form an affine space. What is the associated vector space?

11. Show that under the usual operations of addition and scalar multiplication on functions the collection of all real-valued functions on a set S such that $f(a) = 1$ for all $a \in A$, where $A \subset S$, forms an affine space. Describe the associated vector space.

12. Consider the collection of all sequences $\{a_n\}$ that satisfy the linear recurrence relation

$$a_n = c_1 a_{n-1} + c_2 a_{n-2} + \cdots + c_k a_{n-k} + d,$$

where $c_1, ..., c_k, d$ are fixed nonzero constants.

a. Show that these sequences form an affine space under the usual operations of addition and scalar multiplication for sequences.

b. What is the dimension of this affine space?

c. Describe the associated vector space.

13. Let $v_1,...,v_n$ be a basis for a vector space V.

a. Show that $A = \left\{ \sum_k c_k v_k \mid \sum_k c_k = 1 \right\}$ is an affine space.

b. What is the associated vector space?

c. How is the dimension of A related to the dimension of V?

14. In the text we showed that with every affine space there is associated a corresponding vector space. Here we show that the converse is also true—that with every vector space there is associated a corresponding affine space. Let V be a vector space and let $A = \{(v,1) \mid v \in V\}$.

a. Show that A is an affine space under coordinate addition and scalar multiplication.

b. Show that the vector space associated with A is isomorphic to V.

15. Prove that the translation map $T(P) = P + v$ is an affine transformation.

16. Prove that an affine transformation is completely defined by its action on an affine basis.

17. Let A be an affine transformation.

a. Show that if $Q - P = T - R$, then $A(Q) - A(P) = A(T) - A(R)$.

b. Let $v = Q - P$, and define $A(v) = A(Q) - A(P)$. Using part (a), show that A induces a well-defined map on vectors. That is, $A(v)$ is independent of the choice of P and Q.

c. Show that $A(P + v) = A(P) + A(v)$.

d. Show that A induces a linear transformation on vectors—that is,

$$A(u + v) = A(u) + A(v)$$

$$A(cu) = cA(u).$$

e. Conclude from parts (c) and (d) that every affine transformation A is determined by a linear transformation L together with the value of A at a single point.

f. Prove that an affine transformation maps each line in an affine space to either a point or a line.

1.1.3 Grassmann Spaces and Mass-Points

The algebra of points in affine space is incomplete. For example, scalar multiplication of points is defined only for the scalars $0,1$; that is,

$$
\begin{aligned}
cP &= P & c &= 1 \\
&= 0 & c &= 0 \\
&= \text{undefined} & c &\neq 0,1 \ .
\end{aligned}
$$

More generally, we cannot take arbitrary linear combinations of points; only affine combinations—combinations where the scalars sum to one—yield new points. Grassmann spaces extend affine spaces so that all the usual operations of arithmetic are valid.

How is this done? Since Grassmann spaces are not so familiar as vector spaces or even affine spaces, we shall provide three distinct models for Grassmann space: physical, algebraic, and geometric. We shall then combine all three models into a single diagram (Figure 1.6).

We take our initial inspiration from physics. In classical mechanics there are points (locations) and vectors (forces), but in addition there are also objects (masses) on which the forces act. The masses reside at points, so it is natural to combine a mass and a point into a single entity called simply a *mass-point*. In this framework masses are allowed to be negative, so perhaps we should call them charges instead of masses, but the term *mass-point* is fairly standard so we shall stick to it here. Vectors are incorporated into this scheme as entities with zero mass.

To develop an algebra for mass-points, we need suitable notation. It might seem reasonable, at first, to denote mass-points by pairs (P,m), where P is a point in affine space and m is a nonzero scalar mass. Unfortunately, the algebra of mass-points is not at all natural in this notation. For example, the sum of two mass-points is not simply the sum of the points juxtaposed with the sum of the masses. Indeed, the sum of two points in affine space is not even well defined.

By introducing a slight abuse of notation, however, we can generate a simple algebraic formalism for mass-points consistent both with the mathematics of affine spaces and with the physics of classical mechanics. We shall denote the mass-point with the nonzero mass m located at the affine point P by the pair (mP,m). Of course, strictly speaking, the expression mP by itself is meaningless, since mP is not a well-defined, coordinate-free expression in affine space. (If we were to introduce rectangular coordinates, however, then the expression mP would represent the first moment of the mass m around each of the coordinate planes. The pair (mP,m) is then called the *Grassmann coordinates* of the mass-point—see Section 1.2.2.) Nevertheless, if we adopt the convention of writing our mass-points in this way, we can certainly recover the affine point P by formally dividing the expression mP by the mass m. Since, by convention, vectors have zero mass, vectors v are written as $(v,0)$.

Scalar multiplication of mass-points is defined by multiplying the mass by the scalar and leaving the point unchanged. If the scalar is zero, then its product with the mass would be zero, so we set the result to the zero vector. Thus, we define

$$
\begin{aligned}
c(mP,m) &= (cmP,cm) & c &\neq 0 & \text{(1.8)} \\
&= (0,0) & c &= 0 \ .
\end{aligned}
$$

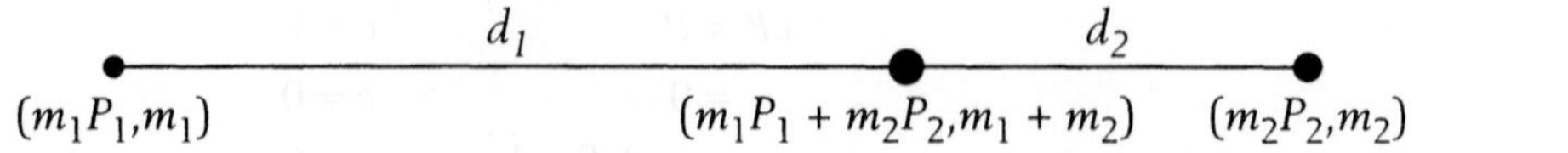

Figure 1.4 The sum of two mass-points (m_1P_1, m_1) and (m_2P_2, m_2)—represented here by dots at the corresponding points of sizes proportional to their masses is located at their center of mass, where $m_1d_1 = m_2d_2$.

To add two mass-points, we need to specify both the position and the mass of the sum. We define the position to be the center of mass (see Exercise 2 of Section 1.1.2) of the two mass-points and the mass to be the sum of the two masses (see Figure 1.4). Formally this means that

$$(m_1P_1, m_1) + (m_2P_2, m_2) = \left(m_1P_1 + m_2P_2, m_1 + m_2\right) \qquad m_1 + m_2 \neq 0. \tag{1.9}$$

Our inspiration for this definition comes from classical mechanics, where typically we can replace the physical effects of two masses by a single mass that is the sum of the two masses located at their center of mass. Since in this framework masses can be negative, we also need to worry about what happens when $m_1 + m_2 = 0$. In this case we define the sum to be the vector from P_1 to P_2 scaled by the mass at P_2. That is,

$$(-mP_1, -m) + (mP_2, m) = \left(m(P_2 - P_1), 0\right) . \tag{1.10}$$

Notice that with these definitions addition of mass-points is associative and commutative. Moreover, scalar multiplication distributes through addition, since

$$c\{(m_1P_1, m_1) + (m_2P_2, m_2)\} = c\left(m_1P_1 + m_2P_2, m_1 + m_2\right)$$
$$= \left(cm_1P_1 + cm_2P_2, c(m_1 + m_2)\right)$$
$$= c(m_1P_1, m_1) + c(m_2P_2, m_2) .$$

Equations (1.8), (1.9), and (1.10) define a complete arithmetic for mass-points. But what about vectors? Addition and scalar multiplication are already defined for vectors, so we can just carry over these definitions in the obvious manner. That is, we set

$$(v, 0) + (w, 0) = (v + w, 0) \tag{1.11}$$
$$c(v, 0) = (cv, 0) .$$

To complete our algebra of mass-points and vectors, we need to define how to add a vector to a mass-point. Again we take our inspiration from mechanics. Think of the vectors as forces acting on the masses. The forces try to pull the masses in the directions of the force vectors. But mass has inertia. The more mass there is at a point, the harder it is to move the mass. A convenient convention is that a force v relocates a mass-point (mP, m) to the new position $P + v/m$. Thus, the larger the mass m, the smaller the net effect of the force v. Therefore we define

$$(mP,m) + (v,0) = (mP + v,m) \ . \tag{1.12}$$

Notice that if a unit mass is located at P, then the vector v moves the mass-point $(P,1)$ to the location $P + v$, which is the location of the standard sum of a point and a vector in affine space. Multiplication distributes through addition for mass-points and vectors, since

$$\begin{aligned} c\{(mP,m) + (v,0)\} &= c(mP + v,m) \\ &= (cmP + cv, cm) \\ &= c(mP,m) + c(v,0) \ . \end{aligned}$$

Thus we have a complete algebra for mass-points and vectors that extends the limited algebra of points and vectors in affine space. Indeed we see now that the algebra of points and vectors in affine space is the algebra of mass-points and vectors, where the point masses are restricted to unit masses.

We have a complete algebra of mass-points and vectors because addition, subtraction, and scalar multiplication are well-defined operations that satisfy the usual rules of arithmetic. But whenever these operations satisfy the standard rules, we have a vector space. So the mass-points and vectors form a vector space that incorporates both the original affine space and its associated vector space. This new vector space is called *Grassmann space*. The dimension of the Grassmann space of mass-points and vectors is one higher than the dimension of the original affine space. This new dimension arises from the masses, which at any location form a one-dimensional subspace.

Adopting the convention of writing mass-points in the form (mP,m) makes addition and scalar multiplication quite natural computationally; it also allows us to avoid, or at least to postpone, division by storing denominators as masses. Equations (1.8)–(1.12) provide as well a purely formal algebraic model for Grassmann space. This algebraic model also guarantees that Grassmann space is a vector space, since the arithmetic operations are performed independently on the coordinates pairs.

There is also a simple geometric model for Grassmann space. Affine space actually consists of two disjoint components represented by the points (*mass* = 1) and the vectors (*mass* = 0)—see Figure 1.5. We can embed these two distinct models of n-dimensional space as two isolated components inside a vector space of dimension $n + 1$. This higher-dimensional vector space is the geometric model of Grassmann space.

In this geometric model for Grassmann space, the notation (mP,m), $m \neq 0$, represents the point on the line $L(t) = (1 - t)(0,0) + t(P,1)$—the line from the zero vector through the affine point P—located at the parameter $t = m$. Equivalently, (mP,m) is equal to the vector (arrow) from the zero vector $(0,0)$ to the affine point $(P,1)$ scaled by the mass m. Thus, geometrically, Grassmann space is the vector space consisting of all affine vectors, together with the points on the lines connecting the zero vector with the points in affine space (see Figure 1.5). In this purely geometric model, mass encodes distance rather than an inertial property of matter.

We can tie together our physical, algebraic, and geometric models of Grassmann space within a single diagram (see Figure 1.6). In the physical model, we start with

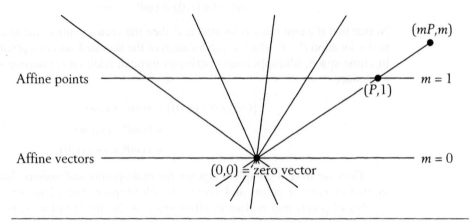

Figure 1.5 A geometric model for Grassmann space as the space of all affine vectors, together with the points on the lines connecting the zero vector with the points in affine space.

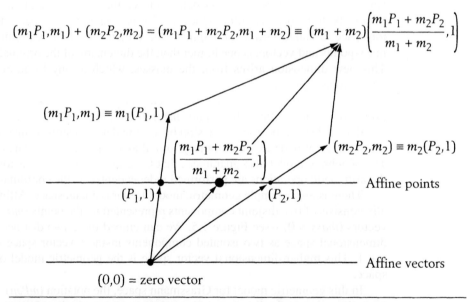

Figure 1.6 Two characterizations for addition in Grassmann space: physical and geometric. Addition is represented by adding mass-points (dots of different sizes) in an affine space or by adding vectors (arrows of different lengths) in a higher-dimensional vector space.

two points P_1, P_2 in affine space with which we associate the scalar masses m_1, m_2. As mass-points, their sum is given by the point $(m_1 P_1 + m_2 P_2)/(m_1 + m_2)$ in affine space associated with the mass $m_1 + m_2$. In this physical model, mass-points (mP,m) are represented by dots of different sizes in affine space (see Figure 1.4). Alternatively, in the geometric model, we encode a mass-point (mP,m) by the arrow from the zero vector $(0,0)$ to the affine point $(P,1)$ scaled by the mass m; that is,

$$(mP,m) = m((P,1) - (0,0)) \equiv m(P,1).$$

Now the sum $(m_1 P_1, m_1) + (m_2 P_2, m_2)$ is given by adding the corresponding arrows $m_1(P_1,1) + m_2(P_2,1)$ using the standard triangle rule for vector addition (see Figure 1.1).

To demonstrate that our physical and geometric models are consistent, we need to show that

$$m_1(P_1,1) + m_2(P_2,1) = (m_1 + m_2)((m_1 P_1 + m_2 P_2)/(m_1 + m_2),1).$$

We proceed in the following manner. Let P be the projection into the affine space $m = 1$ along the line to the origin of the sum $m_1(P_1,1) + m_2(P_2,1)$. Then there are constants α, λ such that

$$m_1(P_1,1) + m_2(P_2,1) = \lambda(P,1)$$
$$(1-\alpha)(P_1,1) + \alpha(P_2,1) = (P,1) .$$

The first equation holds from the definition of P as a projection, and the second equation must hold because $(P,1)$ lies on the line in affine space joining the points $(P_1,1)$ and $(P_2,1)$. Multiplying the second equation by λ and subtracting it from the first equation yields

$$(m_1 - (1-\alpha)\lambda)(P_1,1) + (m_2 - \alpha\lambda)(P_2,1) = 0 .$$

Since P_1 and P_2 are distinct points in affine space, the vectors $(P_1,1)$ and $(P_2,1)$ in Grassmann space are linearly independent, so

$$(1-\alpha)\lambda = m_1$$
$$\alpha\lambda = m_2 .$$

Adding these equations and solving first for λ and then for α yields

$$\lambda = m_1 + m_2$$
$$\alpha = m_2/(m_1 + m_2) .$$

It follows that the sum of the arrows generates the vector $(P,1) = ((m_1 P_1 + m_2 P_2)/(m_1 + m_2),1)$ scaled by the mass $\lambda = m_1 + m_2$. Thus the projection of the arrow $m_1(P_1,1) + m_2(P_2,1)$ back into affine space gives the point in affine space corresponding to the addition of the original mass-points (center of mass), and the scale factor is the sum of the original masses. So in the geometric model of Grassmann space, affine points encode direction, mass encodes scale, and these encodings are consistent with the standard addition of mass-points from classical mechanics.

One final observation about our notation for mass-points in Grassmann space: consider what happens in the limit to a mass-point (mP,m) as $m \to 0$. From the geometric interpretation of Grassmann space (Figures 1.5 and 1.6), we observe that

$\lim_{m \to 0}(mP,m) = (0,0) =$ zero vector. This limit is the same for all affine points P, so in Grassmann space the zero vector is arbitrarily close to every point when the mass is small. This phenomenon occurs because the mass-points (mP,m), P fixed, $m \neq 0$, all lie on the line through $(0,0)$ and $(P,1)$, rather than on the vertical line over $(P,1)$. Thus, from the geometric perspective as well as from the algebraic point of view, the natural representation for mass-points in Grassmann space is indeed (mP,m) and not (P,m) (see also Section 1.2.2 on Grassmann coordinates).

Exercises

1. (Ceva's Theorem) Let P_1, P_2, P_3 be the vertices of a triangle, and let M_i be a point on side $P_j P_k$, $i \neq j,k$. Define

$$r_i = \frac{\left|P_j - M_i\right|}{\left|P_k - M_i\right|} \quad i = 1,2,3.$$

Prove that the lines $P_1 M_1, P_2 M_2, P_3 M_3$ are concurrent if and only if $r_1 r_2 r_3 = 1$.

(Hint: Place appropriate masses at the vertices of the triangle.)

2. The space of all monic polynomials of degree n—all polynomials of degree n with leading coefficient 1—is an affine space.

    a. What is the associated Grassmann space?

    b. Where are the masses stored?

    c. What are the elements of this Grassmann space with zero mass?

    d. Describe addition and scalar multiplication on this space.

3. Show that a Grassmann space can always be embedded in an affine space of the same dimension. (Hint: See Exercise 14 of Section 1.1.2.)

4. Consider a system of homogenized linear equations:

$$a_{11}x_1 + a_{12}x_2 + \cdots + a_{1n}x_n = b_1 w$$

$$\vdots$$

$$a_{m1}x_1 + a_{m2}x_2 + \cdots + a_{mn}x_n = b_m w \ .$$

We can rewrite these equations in matrix notation as $AX = Bw$.

    a. Show that the solutions (X,w) of this homogenized system of linear equations form a Grassmann space.

    b. What are the points with unit mass?

    c. What are the vectors?

5. Define the product of two elements of Grassmann space by setting

$$(m_1 P_1, m_1) \bullet (m_2 P_2, m_2) = m_1 m_2 (P_2 - P_1) \qquad m_1 m_2 \neq 0$$

$$(mP,m) \bullet (v,0) = mv \qquad m \neq 0$$

$$(v,0) \bullet (mP,m) = -mv \qquad m \neq 0$$

$$(v,0) \bullet (w,0) = 0 \ .$$

Show that this product

a. is not associative:

$$\{(m_1P_1,m_1)\bullet(m_2P_2,m_2)\}\bullet(m_3P_3,m_3) \neq (m_1P_1,m_1)\bullet\{(m_2P_2,m_2)\bullet(m_3P_3,m_3)\}.$$

b. is anti-commutative:

$$(m_1P_1,m_1)\bullet(m_2P_2,m_2) = -(m_2P_2,m_2)\bullet(m_1P_1,m_1).$$

c. commutes with scalar multiplication:

$$c\{(m_1P_1,m_1)\bullet(m_2P_2,m_2)\} = \{c(m_1P_1,m_1)\}\bullet(m_2P_2,m_2) = (m_1P_1,m_1)\bullet\{c(m_2P_2,m_2)\}.$$

d. distributes through addition:

$$(wQ,w)\bullet\{(m_1P_1,m_1)+(m_2P_2,m_2)\} = (wQ,w)\bullet(m_1P_1,m_1)+(wQ,w)\bullet(m_2P_2,m_2).$$

1.1.4 Projective Spaces and Points at Infinity

Affine space is flawed in two ways: both its algebra and its geometry are incomplete. Grassmann space completes the algebra; projective space completes the geometry.

The geometry of affine space is incomplete because there are no points at infinity. Typically two lines in the affine plane intersect at a point. But where do two parallel lines intersect? We need points at infinity to complete the geometry of the affine plane.

Points at infinity are needed as well to complete the definition of perspective projection. Given an eye point E and an affine plane S not containing E, we can map points P in affine 3-space onto S by perspective projection—that is, by taking the intersection of the line EP with the plane S (see Figure 1.7). For most points P, we get a well-defined intersection point, but what does perspective projection do to the points P' on the plane through E parallel to S? Where do lines parallel to the plane S intersect S? Again we need points at infinity to complete our geometry.

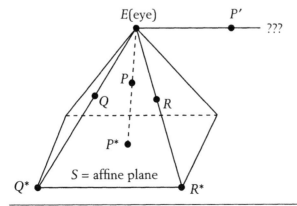

Figure 1.7 Perspective projection from the eye point E to the affine plane S. The points P,Q,R have well-defined images $P*,Q*,R*$ in the plane S. But what is the image of the point P'?

Projective space is, by definition, the collection of the points in affine space together with the points at infinity. The points at infinity are constructed by asserting that in each direction there lies a unique point at infinity (see Figure 1.8). By convention, directions that are 180° apart define the same point at infinity; otherwise parallel lines would intersect at two points instead of just one.

A point at infinity can be represented by a direction, and a direction can be described by a vector. So we shall use vectors to represent points at infinity. But there is a slight problem with this approach because the vectors incorporate length as well as direction. That is, the point at infinity represented by the vector v is the same as the point at infinity represented by the vector cv, $c \neq 0$. To overcome this difficulty, we shall simply identify v and cv as the same point at infinity; that is, we shall ignore nonzero scalar multiples. Notice too that the zero vector does not represent a point at infinity, since the zero vector does not correspond to a fixed direction.

We can adapt the notation of mass-points to represent points in projective space. For vectors we ignore length; for points we ignore mass. That is, we simply identify two mass-points with different masses if they are located at the same affine point. Thus in projective space,

$$[v,0] = [cv,0] \qquad c \neq 0,$$

$$[P,1] = [cP,c] \qquad c \neq 0 \quad .$$

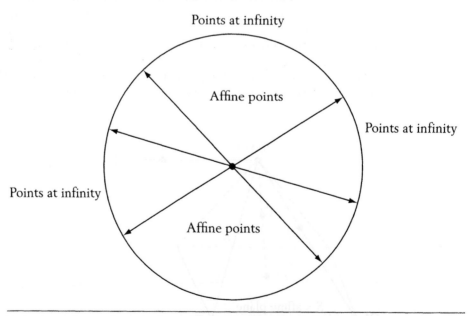

Figure 1.8 Projective space consists of affine points and points at infinity. Intuitively, the vectors are pasted onto the affine points as points at infinity, so projective space consists of a single connected component.

The pairs $[cP,c]$, where $c \neq 0$, represent affine points; the pairs $[v,0]$ represent points at infinity. Now every pair $[X,w] \neq [0,0]$ has a well-defined meaning in projective space, either as a point in affine space or as a point at infinity. The parameter w is called a *homogeneous coordinate;* these are the standard homogeneous coordinates used in computer graphics in order to represent projective transformations such as perspective projection by 4×4 matrices. Points at infinity have homogeneous coordinate zero, reminding us that they are often introduced to account for division by zero.

The homogeneous coordinates w in this representation for projective space are sometimes confounded with the masses m of Grassmann space, and the points $[X,w]$ of projective space are sometimes confused with the mass-points (X,w) of Grassmann space. These identifications are not correct. In Grassmann space $(mP,m) \neq (P,1)$ because even though these mass-points are located at the same affine point P they have different masses. On the other hand, in projective space there is no distinction whatsoever between the affine points $[P,1]$ and $[cP,c]$; mass is not a constituent of projective space. Notice too that $(0,0)$ is a well-defined object in Grassmann space, namely, the zero vector, but the pair $[0,0]$ is meaningless in projective space.

Nevertheless, Grassmann space and projective space are intimately related. *The points in projective space are made up of equivalence classes of points in Grassmann space.* Indeed each line through the origin in Grassmann space corresponds to a distinct point in projective space (see Figure 1.5). But the topology of Grassmann space and the topology of projective space are quite different. Grassmann space connects the two disjoint components (points and vectors) of affine space by embedding them in a vector space of one higher dimension. Projective space connects these two components by pasting the vectors onto the affine points at infinity. As a consequence there are no longer any affine vectors in projective space, so there are no notions of direction or length in projective space. Also, there is no concept of orientation in projective space, since v and $-v$ are identified.

The algebra of projective space is also quite different from the algebra of Grassmann space. We have already seen that in projective space scalar multiplication has no effect. What about addition? We cannot simply define

$$[c_1P_1,c_1] + [c_2P_2,c_2] = [c_1P_1 + c_2P_2, c_1 + c_2] \tag{1.13}$$

as we did for mass-points in Grassmann space because the projective point on the right-hand side of (1.13) would depend on our choices for the representatives of the projective points $[c_1P_1,c_1]$ and $[c_2P_2,c_2]$. Indeed evidently

$$[P_1,1] + [P_2,1] = [P_1 + P_2, 2] \neq [c_1P_1 + c_2P_2, c_1 + c_2] = [c_1P_1,c_1] + [c_2P,c_2].$$

What works instead is to define the sum of $[c_1P_1,c_1]$ and $[c_2P_2,c_2]$ to be the projective line joining these two projective points. Then the right-hand side of Equation (1.13) represents one point on this line, the affine point $(c_1P_1 + c_2P_2)/(c_1 + c_2)$. By taking all the representatives of $[c_1P_1,c_1]$ and $[c_2P_2,c_2]$ and adding their coordinates, we can generate all the points along the projective line joining these two projective points. This convention works equally well for two affine points, for one affine point and one point at infinity, or for two points at infinity. Thus we add two projective points by taking the projective line determined by these two points; we sum three projective points by taking the projective plane determined by the three points; and so on for more and more points. With this definition, addition of projective points is associative and commutative. But projective space is not a vector space because the sum of two projective points is not a projective point. Moreover, there is no nontrivial notion of scalar multiplication in projective space.

Thus the Grassmann space of mass-points and vectors is a vector space, but the projective space of affine points and points at infinity is not a vector space. The vector space algebra of Grassmann space is much more powerful than the limited algebra of projective space. Consequently, to construct free-form curves and surfaces algebraically, we shall prefer to work primarily in Grassmann space. Only when we need to complete our geometry with points at infinity shall we appeal to projective space.

Exercises

1. Consider ordered pairs of integers (p,q) . We shall say that two such pairs (p,q) and (r,s) are equivalent if $(mp,mq) = (nr,ns)$ for some nonzero integers m,n. Denote by $[p,q]$ the equivalence class of the pair (p,q). We can identify an ordered pair (p,q) with the *fraction p/q* and the equivalence class $[p,q]$ with the *rational number p/q* (or with ∞ if $q = 0$).

   a. Show that the operations of addition, subtraction, and scalar multiplication defined by

   $$\frac{p}{q} \pm \frac{r}{s} = \frac{p \pm r}{q \pm s}$$

   $$n \times \frac{p}{q} = \underbrace{\frac{p}{q} + \cdots + \frac{p}{q}}_{n} = \frac{np}{nq}$$

   are well defined on fractions (ordered pairs), but not on rational numbers (equivalence classes of ordered pairs).

   b. What is the identity for this addition operation?

   c. Which set is more like a projective space: the set of fractions or the set of rational numbers? Which set is more like a Grassmann space?

2. The projective plane is not oriented because the vectors v and $-v$ are identified with the same point at infinity. We can, however, define an oriented version of the projective plane by setting

   $$[cv,0] = [dv,0] \qquad cd > 0,$$
   $$[cP,c] = [dP,d] \qquad cd > 0.$$

   a. Explain the geometric relationship between Grassmann space and the oriented projective plane.

b. How should Figure 1.8 be changed to model the oriented projective plane?

3. Show that by choosing different representatives for the points $[c_1 P_1, c_1]$ and $[c_2 P_2, c_2]$ Equation (1.13) can be used to generate all the points on the projective line joining $[c_1 P_1, c_1]$ and $[c_2 P_2, c_2]$.

4. What is the analogue to Equation (1.13) for generating points along the line joining two points at infinity? Show that by choosing different representatives for the same two points at infinity this formula generates all the points along the projective line joining these two projective points.

1.1.5   Mappings between Ambient Spaces

We have constructed four kinds of ambient spaces: vector spaces, affine spaces, Grassmann spaces, and projective spaces. These four spaces are intimately related: the affine points live inside of Grassmann space as the points with unit mass and the vectors reside there as well as objects with zero mass, while in projective space all mass-points located at the same affine point but with different mass are identified to the same projective point and vectors are replaced by points at infinity. These observations lead to the following four natural maps between these ambient spaces:

affine space → *Grassmann space*   *affine space* → *projective space*

$$P \to (P,1)$$ $$P \to [P,1]$$
$$v \to (v,0)$$ $$v \to [v,0]$$

Grassmann space → *affine space*   *Grassmann space* → *projective space*

$$(mP,m) \to \frac{mP}{m} = P$$ $$(mP,m) \to [mP,m] \equiv [P,1]$$
$$(v,0) \to v$$ $$(v,0) \to [v,0] \equiv \left[\frac{v}{|v|}, 0\right]$$

Notice that the projections from Grassmann space and affine space onto projective space are well defined everywhere, except at the zero vector.

The coordinates $(P,1)$ for affine points and $(v,0)$ for affine vectors introduced by the embedding of affine space into Grassmann space are called *affine coordinates* and are familiar both in computer graphics and in robotics, where the additional coordinate is used to distinguish between points and vectors in affine space (see Section 1.2.2). The embedding from affine space to Grassmann space via affine coordinates captures the algebraic structure of affine space by preserving affine combinations. Indeed, if $\sum_k c_k = 0, 1$, then

$$\sum_{k=0}^{n} c_k P_k \to \sum_{k=0}^{n} c_k (P_k, 1).$$

The natural projection from Grassmann space onto affine space is the left-sided inverse of the natural embedding of affine space into Grassmann space. Notice, however, that this projection is not a continuous map because Grassmann space is connected whereas the space of affine points and affine vectors consists of two disjoint components.

The projections from affine space and Grassmann space onto projective space are continuous maps. But these projections do not preserve the algebraic structures on their domain spaces because, as we have seen in Section 1.1.4, projective space is not a vector space—addition and scalar multiplication are not well-defined operations in projective space.

Exercises

1. Show that the following diagram commutes:

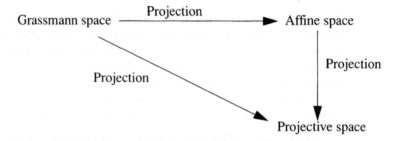

Conclude that the projection from Grassmann space onto projective space factors through the projection from Grassmann space onto affine space, even though the projection onto projective space is continuous while the projection onto affine space is discontinuous.

2. Show that the affine points $P_0, ..., P_n$ form an affine basis on an affine space if and only if the mass-points $(P_0, 1), ..., (P_n, 1)$ form a vector space basis for the associated Grassmann space.

3. A transformation A on Grassmann space is said to be *mass preserving* if the mass of a transformed mass-point is the same as the mass of the corresponding untransformed mass-point. Let A be an affine transformation on affine space. Define a transformation A^* on Grassmann space by setting: $A^*(mP, m) = (mA(P), m)$ and $A^*(v, 0) = (A(v), 0)$.

    a. Show that A^* is a mass-preserving linear transformation on Grassmann space.

    Conversely, let A^* be a mass-preserving linear transformation on Grassmann space. Define a transformation A on affine space by setting $A(P) = A^*(P, 1)$.

    b. Show that A is an affine transformation on affine space.

    c. Show that the following diagram commutes:

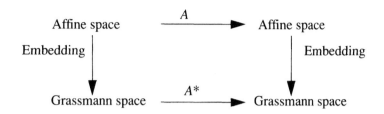

d. Conclude that the affine transformations on affine space are equivalent to the mass-preserving linear transformation on Grassmann space.

4. Let L be a linear transformation on Grassmann space.

a. Show that if L is nonsingular, then L induces a unique, well-defined transformation L^* on projective space so that the following diagram commutes:

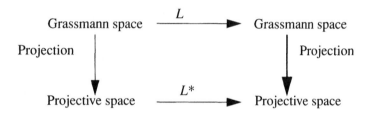

b. Show that if L is singular, then L still induces a unique, well-defined transformation

$$L^*: \text{Projective space} - [ker(L)] \rightarrow \text{Projective space},$$

where $ker(L) = \{P \mid L(P) = 0\}$, so that the following diagram commutes:

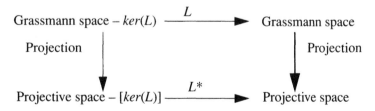

A transformation L^* on projective space induced in this fashion by a linear transformation L on Grassmann space is called a *projective transformation*.

5. Let L: Grassmann space $\rightarrow \mathbf{R}$ be a linear map between vector spaces. Then $H = \{P \mid L(P) = 0\}$ is called the *hyperplane* defined by L, and $H^* = \{[P] \mid L(P) = 0\}$ is called the *projective hyperplane* induced by L. Fix a point E in Grassmann space such that E is not an element of H, and define the map

Perps: Grassmann space $\rightarrow$ Hyperplane H

by setting

$$Persp(P) = L(P)E - L(E)P.$$

Show that

a. *Persp(P)* lies on the hyperplane *H*.

b. *Persp(P)* = 0 if and only if *P* = *cE*.

c. *Persp* induces a unique projective transformation

  Persp*: Projective space – [*E*] → Projective hyperplane *H**

such that *Persp**[*P*] = [*Persp(P)*]. (Hint: See Exercise 4.)

d. if *P* and *E* are points in affine space, then

  i.   *Persp**[*P*] lies on the intersection of the line [*EP*] and the hyperplane *H**;

  ii. *Persp**[*P*] is a point at infinity in projective space if and only if the vector *P* – *E* lies in the hyperplane *H*.

In three dimensions, the map *Persp** is the standard perspective projection from an eye point [*E*} onto a projective plane *H**.

1.1.6   Polynomial and Rational Curves and Surfaces

For the study of polynomial curves and surfaces, affine spaces usually suffice. Typically we define a polynomial curve $P(t)$ by choosing a sequence of affine control points $P_0,\ldots,P_n$ and a collection of polynomial blending functions $B_0(t),\ldots,B_n(t)$, and setting $P(t)$ to be the set of points in affine space determined by the equation

$$P(t) = \sum_{k=0}^{n} B_k(t)P_k . \tag{1.14}$$

Similarly, we define a polynomial surface by choosing an array of affine control points $\{P_{ij}\}$ and a collection of bivariate polynomial blending functions $\{B_{ij}(s,t)\}$, and setting $P(s,t)$ to be the set of points in affine space determined by the equation

$$P(s,t) = \sum_{ij} B_{ij}(s,t)P_{ij} . \tag{1.15}$$

For this to work in affine space, the blending functions $\{B_k(t)$ for curves and $\{B_{ij}(s,t)\}$ for surfaces must form a partition of unity; that is, we must have

$$\sum_{k=0}^{n} B_k(t) \equiv 1 \tag{1.16}$$

$$\sum_{ij} B_{ij}(s,t) \equiv 1 .$$

We shall see in subsequent chapters that Lagrange polynomials, as well as Bézier and B-spline curves and surfaces, are defined in precisely this fashion.

Curves and surfaces generated by Equations (1.14) and (1.15) are said to be *translation invariant*. Translation invariance means that to translate the curve or surface by a vector v, we need only translate each control point by v. This result follows from the fact that the blending functions form a partition of unity, since for such blending functions

$$P(t) + v = \sum_{k=0}^{n} B_k(t)P_k + \sum_{k=0}^{n} B_k(t)v = \sum_{k=0}^{n} B_k(t)(P_k + v)$$

$$P(s,t) + v = \sum_{ij} B_{ij}(s,t)P_{ij} + \sum_{ij} B_{ij}(s,t)v = \sum_{ij} B_{ij}(s,t)(P_{ij} + v) \ .$$

More generally, these curves and surfaces are *affine invariant*. That is, if A is any affine transformation, then since A preserves affine combinations

$$A\big(P(t)\big) = \sum_{k=0}^{n} B_k(t)A(P_k)$$

$$A\big(P(s,t)\big) = \sum_{ij} B_{ij}(s,t)A(P_{ij}) \ .$$

Thus to perform an affine transformation such as rotation or scaling on such a curve or surface, we need only apply the affine transformation to each control point.

The Grassmann space of mass-points is used for the construction of rational curves and surfaces. To define a rational curve $R(t)$, we start with a sequence of mass-points and vectors $(m_0 P_0, m_0), \ldots, (m_n P_n, m_n)$. (If $m_k = 0$, we replace the mass-point $(m_k P_k, m_k)$ by a vector $(v_k, 0)$, and we do not insist that $m_k P_k = 0$.) Now given an arbitrary collection of polynomial blending functions $B_0(t), \ldots, B_n(t)$, we define $R(t)$ to be the set of points in *affine space* determined by the curve $P(t)$ in *Grassmann space* given by the equation

$$P(t) = \sum_{k=0}^{n} B_k(t)(m_k P_k, m_k) = \left(\sum_{k=0}^{n} B_k(t)m_k P_k, \ \sum_{k=0}^{n} B_k(t)m_k \right). \tag{1.17}$$

To project $P(t)$ into affine space, we divide by the mass to generate the curve

$$R(t) = \frac{\displaystyle\sum_{j=0}^{n} m_j P_j B_j(t)}{\displaystyle\sum_{k=0}^{n} m_k B_k(t)} = \sum_{j=0}^{n} \frac{m_j B_j(t)}{\displaystyle\sum_{k=0}^{n} m_k B_k(t)} P_j, \tag{1.18}$$

which is indeed a rational curve in affine space. Rational surfaces are defined in an analogous fashion. Observe that to define $P(t)$, we need to work in a vector space so that we can perform both addition and scalar multiplication. Thus $P(t)$ must be constructed in Grassmann space, not in projective space. Notice too that changing the mass of one of the mass-points $(m_k P_k, m_k)$ alters both $P(t)$ and $R(t)$, so again the control points we are dealing with here are mass-points, not points in projective space.

Projective space comes into the picture when the denominator of $R(t)$ vanishes—that is, when

$$\sum_{k=0}^{n} m_k B_k(t_0) = 0$$

for some parameter $t = t_0$. Division by zero is not well defined in affine space, and therefore in affine space the rational curve $R(t)$ would have a discontinuity. Moreover, the natural projection from Grassmann space to affine space would send

$$\left(\sum_{k=0}^{n} B_k(t_0)m_k P_k, 0 \right) \rightarrow \sum_{k=0}^{n} B_k(t_0)m_k P_k,$$

which is a vector, not a point, in affine space. But curves and surfaces are collections of points, not points and vectors. To avoid these problems, we typically map $P(t)$ from Grassmann space, not to affine space, but to projective space using the canonical projection. That is, we simply set

$$R(t) = \left[\sum_{k=0}^{n} B_k(t)(m_k P_k, m_k) \right] = \left[\sum_{k=0}^{n} B_k(t)m_k P_k, \sum_{k=0}^{n} B_k(t)m_k \right]. \tag{1.19}$$

Notice that the affine points on this curve are the same as in Equation (1.18), but now $R(t_0)$ is a point at infinity in projective space rather than a vector in affine space. Moreover, the curve $R(t)$ is a continuous curve in projective space. Thus for rational curves, the control points lie in Grassmann space, but the curves reside in projective space!

Exercises

1. Let $P(t)$ be a curve in affine space.

    a. Using the definition of $P'(t)$ as the limit of a difference quotient, show that $P'(t)$ is a vector field—that is, a one-parameter family of vectors—in affine space. Interpret this vector field geometrically.

    b. Suppose that $P(t) = \sum_k B_k(t)P_k$ is a polynomial curve, where $\sum_k B_k(t) \equiv 1$. Without appealing to part (a), show that the derivative $P'(t) = \sum_k B_k'(t)P_k$ is a vector field.

2. Show that for curves in Grassmann space differentiation and projection do not commute. That is, show that the following diagram does not commute:

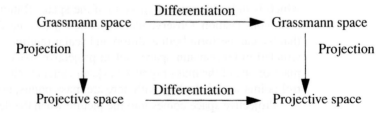

3. Recall that when we project a mass-point from Grassmann space to projective space, the result is a point in affine space—only the vectors project to points at infinity.

a. Show that for mass-points translation and projection to affine space do not commute. That is, for mass-points the following diagram does not commute:

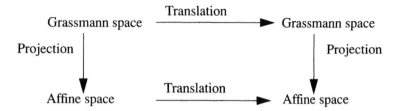

b. What operation does translation of mass-points in Grassmann space correspond to in affine space?

c. Consider a polynomial curve in Grassmann space that projects to a rational curve in affine space. Show that we generate two different curves in affine space, depending on whether we apply translation before or after projection.

4. Show that rational curves are invariant under projective transformations (see Exercise 4 of Section 1.1.5). That is, show that if L^* is a projective transformation induced by the transformation L on Grassmann space and if

$$R(t) = \left[\sum_{k=0}^{n} B_k(t)(m_k P_k, m_k) \right]$$

is a rational curve in projective space, then

$$L^*(R(t)) = \left[\sum_{k=0}^{n} B_k(t)L(m_k P_k, m_k) \right].$$

1.2 Coordinates

In most of this book, we shall adopt a coordinate-free approach to geometry. We have already illustrated this technique in the preceding section in our discussion of ambient spaces. This coordinate-free style works well for the range space where our curves and surfaces reside, but we are going to study parametric curves and surfaces, so we need as well a way to represent the parametric domain. In Section 1.1.6 we implicitly resorted to rectangular coordinates for our parameters. For certain types of surfaces, however, in particular for triangular patches, rectangular coordinates are not the most convenient way to represent the parameter domain. Here we shall introduce another kind of coordinates, called *barycentric coordinates,* which are more suitable for representing the domain of a triangular surface patch.

We begin with a brief review of rectangular coordinates and go on to provide a short sketch of affine, Grassmann, and homogeneous coordinates—that is, rectangular coordinates adapted to affine, Grassmann, and projective spaces. We then present a more thorough exposition of barycentric coordinates for affine spaces. You can skip this section for now if you like and return to it later when we study triangular patches in subsequent chapters.

1.2.1   Rectangular Coordinates

In Euclidian space it is often convenient to introduce rectangular (Cartesian) coordinates. This can be done by selecting an orthonormal basis $v_1,...,v_n$—a basis whose vectors are mutually orthogonal unit vectors—and representing any vector v by a unique linear combination of these basis vectors. If

$$v = \sum_{k=1}^{n} c_k v_k \, ,$$

then we say that $(c_1,...,c_n)$ are the *rectangular coordinates* of v. When the basis is fixed, often we abuse notation and write $v = (c_1,...,c_n)$.

We can proceed in a similar fashion in affine space. Here we need a fixed point O in the affine space as well as an orthonormal basis $v_1,...,v_n$ for the associated vector space. Now any point P in the affine space can be written uniquely as

$$P = O + \sum_{k=1}^{n} c_k v_k .$$

Again we say that $(c_1,...,c_n)$ are the *rectangular coordinates* of P. The point O plays the role of the origin, and the vectors $v_1,...,v_n$ are parallel to the coordinate axes (see Figure 1.10(a) in Section 1.2.3). Once more when the origin and axes are fixed, we often abuse notation and write $P = (c_1,...,c_n)$.

1.2.2   Affine Coordinates, Grassmann Coordinates, and Homogeneous Coordinates

Rectangular coordinates do not permit us to distinguish between points and vectors. In an *n*-dimensional affine space both points and vectors are represented by n rectangular coordinates. But points and vectors convey different information, and the rules of linear algebra are different for points and for vectors. Therefore it is important for us to differentiate somehow between coordinates that represent points and coordinates that represent vectors.

The natural embedding from affine space to Grassmann space presented in Section 1.1.5 provides a simple way to discriminate the points from the vectors. This embedding assigns an additional mass coordinate to both points and vectors: points are assigned a mass equal to one, vectors a mass equal to zero. Thus we write

$$P = (c_1,...,c_n,1)$$
$$v = (c_1,...,c_n,0) .$$

These rectangular coordinates followed by a zero or a one are called *affine coordinates*, and they are the coordinates most commonly adopted for affine space.

Grassmann space extends affine space by incorporating mass-points with arbitrary masses. The mass-points are combinations of affine points P and scalar masses m. If we were to use rectangular coordinates $(c_1,...,c_n)$ to represent the affine point P and one additional coordinate to represent the scalar mass m, then a mass-point would be written in terms of coordinates as

$$(P,m) = (c_1,...,c_n,m).$$

But we observed in Section 1.1.3 that with this representation the rules for addition and scalar multiplication for mass-points would not correspond to the natural rules of addition and scalar multiplication on coordinates. To adapt our coordinates to the algebra of mass-points, we instead represent a mass-point with the notation

$$(mP,m) = (mc_1,...,mc_n,m).$$

We call these coordinates the *Grassmann coordinates* of a mass-point. Note that we can recover the rectangular coordinates $(c_1,...,c_n)$ of the affine point from the Grassmann coordinates $(mc_1,...,mc_n,m)$ of the corresponding mass-point by dividing the first n coordinates $(mc_1,...,mc_n)$ by the $(n + 1)$st coordinate m. For points with unit mass, Grassmann coordinates coincide with affine coordinates.

Thus the first n coordinates of a mass-point are the rectangular coordinates of the affine point P scaled by the mass m. For vectors, however, the mass is zero, so it would not be prudent to scale the rectangular coordinates $(c_1,...,c_n)$ of a vector v by its mass. Instead, the Grassmann coordinates of a vector are just its rectangular coordinates followed by a zero mass, just like in affine space—that is, for vectors we write the Grassmann coordinates as

$$(v,0) = (c_1,...,c_n,0).$$

Points in projective space are equivalence classes of points in Grassmann space. Thus we can adapt Grassmann coordinates to represent points in projective space by writing

$$[mP,m] = [mc_1,...,mc_n,m]$$
$$[v,0] = [c_1,...,c_n,0] \ .$$

These coordinates for points in projective space are called *homogeneous coordinates*. Note that, unlike in Grassmann space, in projective space

$$[mc_1,...,mc_n,m] = [c_1,...,c_n,1]$$
$$[\mu c_1,...,\mu c_n,0] = [c_1,...,c_n,0] \ ,$$

since in projective space we are dealing with equivalence classes of points in Grassmann space. A point in projective space that corresponds to a point in affine space has a nonzero final coordinate. Thus, just as in Grassmann space, we can recover the rectangular coordinates $(c_1,...,c_n)$ of an affine point from the homogeneous coordinates $[mc_1,...,mc_n,m]$ ot the corresponding projective point by dividing the first n homogeneous coordinates $(mc_1,...,mc_n)$ by the $(n + 1)$st homogeneous coordinate m.

Exercises

1. Let L be a linear transformation on an n-dimensional vector space with a fixed basis $v_1,...,v_n$ and let $v = (c_1,...,c_n)$ be an arbitrary vector. Suppose that

$$L(v_k) = (c_{k1},...,c_{kn}), \ k = 1,...,n.$$

   Show that

$$L(v) = (c_1,...,c_n)\begin{pmatrix} L(v_1) \\ \vdots \\ L(v_n) \end{pmatrix} = (c_1,...,c_n)\begin{pmatrix} c_{11} & \cdots & c_{1n} \\ \vdots & \vdots & \vdots \\ c_{n1} & \cdots & c_{nn} \end{pmatrix}.$$

   Thus linear transformations on vectors can be computed by matrix multiplication on their coordinates.

2. Let A be an affine transformation on an n-dimensional affine space. Let $v_1,...,v_n$ be a fixed orthonormal basis of the associated vector space, and let O be a fixed point in the affine space. With respect to this origin and axes, suppose that

$$A(v_k) = (c_{k1},...,c_{kn},0), \ k = 1,...,n$$
$$A(O) = (d_1,...,d_n,1) \ .$$

   Suppose further that P is an arbitrary affine point, that v is an arbitrary affine vector, and that, with respect to the same origin and axes

$$P = (p_1,...,p_n,1)$$
$$v = (c_1,...,c_n,0) \ .$$

   Show that the affine coordinates of $A(P), A(v)$ are given by

$$A(P) = (p_1,...,p_n,1)\begin{pmatrix} A(v_1) & 0 \\ \vdots & 0 \\ A(v_n) & 0 \\ A(O) & 1 \end{pmatrix} = (p_1,...,p_n,1)\begin{pmatrix} c_{11} & \cdots & c_{1n} & 0 \\ \vdots & \vdots & \vdots & \vdots \\ c_{n1} & \cdots & c_{nn} & 0 \\ d_1 & \cdots & d_n & 1 \end{pmatrix}.$$

$$A(v) = (c_1,...,c_n,0)\begin{pmatrix} A(v_1) & 0 \\ \vdots & 0 \\ A(v_n) & 0 \\ A(O) & 1 \end{pmatrix} = (c_1,...,c_n,0)\begin{pmatrix} c_{11} & \cdots & c_{1n} & 0 \\ \vdots & \vdots & \vdots & \vdots \\ c_{n1} & \cdots & c_{nn} & 0 \\ d_1 & \cdots & d_n & 1 \end{pmatrix}$$

Thus affine transformations on points and vectors can be computed by matrix multiplication on their affine coordinates.

3. Show that in Grassmann space, linear transformations can be computed by matrix multiplication on Grassmann coordinates. Given a fixed origin O and axis vectors $v_1,...,v_n$ for the naturally embedded affine space, what are the entries of the matrix M associated to the linear transformation T?

4. Show that in projective space, projective transformations (see Exercise 4 of Section 1.1.5) can be computed by matrix multiplication on homogeneous coordinates. Given a fixed origin O and axis vectors $v_1,...,v_n$ for the naturally embedded affine space, what are the entries of the matrix M^* associated to the projective transformation T^*?

1.2.3 Barycentric Coordinates

Rectangular coordinates and affine coordinates are not always the most convenient way to represent points in affine space. We introduced affine spaces because we plan to work directly with points rather than vectors, so rectangular or affine coordinates are often unnatural for our purposes. Therefore, especially when representing triangular surfaces, we shall routinely replace rectangular coordinates for the domain parameters with another type of coordinates called *barycentric coordinates*.

Let $P_0,...,P_n$ be an affine basis. Then any point P can be represented by a unique affine combination of the points $P_0,...,P_n$. Thus we can write

$$P = \sum_{k=0}^{n} \beta_k P_k \text{ where } \sum_{k=0}^{n} \beta_k = 1.$$

We call $(\beta_0,...,\beta_n)$ the *barycentric coordinates* of P with respect to the affine basis $P_0,...,P_n$. Just like rectangular coordinates, barycentric coordinates depend on the choice of basis: the same point P will have different barycentric coordinates with respect to different affine bases. Notice, however, that in an affine space of n dimensions each point has $n+1$ barycentric coordinates but only n rectangular coordinates, since an affine basis has $n+1$ elements whereas a basis for the associated vector space has only n elements. Still, barycentric coordinates represent only n degrees of freedom because, unlike rectangular coordinates, barycentric coordinates satisfy a relation—they sum to one.

Barycentric coordinates in affine space are related to masses in Grassmann space. If $P_0,...,P_n$ is an affine basis, then for any affine point P and any mass $m \neq 0$ there is a collection of masses $m_0,...,m_n$ such that (mP,m) is the center of mass of the mass-points $(m_0P_0,m_0),...,(m_nP_n,m_n)$ (see Exercise 2 of Section 1.1.5); that is,

$$P = \frac{\sum_{k=0}^{n} m_k P_k}{\sum_{j=0}^{n} m_j}.$$

The barycentric coordinates of P are given by

$$\beta_k = \frac{m_k}{\sum\limits_{j=0}^{n} m_j}.$$

It is easy to check that indeed

$$P = \sum_{k=0}^{n} \beta_k P_k.$$

Notice, however, that the barycentric coordinates β_k are unique, but the masses m_k are defined only up to constant multiples.

We shall most often apply barycentric coordinates in one and two dimensions, so let us now get a feel for these new coordinates by computing explicit formulas for them in low dimensions.

The one-dimensional case is easy. An affine basis for the affine line is given by two distinct points T_0 and T_1. To compute the barycentric coordinates (β_0, β_1) of an arbitrary point T in terms of T_0 and T_1, we write

$$T = \beta_0 T_0 + \beta_1 T_1.$$

Since $\beta_0 + \beta_1 = 1$,

$$T = (1 - \beta_1)T_0 + \beta_1 T_1 = T_0 + \beta_1(T_1 - T_0). \tag{1.20}$$

Let $|T_1 - T_0|$ denote the distance between T_0 and T_1. Then subtracting T_0 from both sides of (1.20) and solving for β_1, we get

$$|\beta_1| = \frac{|T - T_0|}{|T_1 - T_0|}. \tag{1.21}$$

Similarly, we also find that

$$|\beta_0| = \frac{|T_1 - T|}{|T_1 - T_0|}. \tag{1.22}$$

The signs of β_0 and β_1 depend upon the relative ordering of T_0, T_1, T along the affine line: by Equation (1.20) β_1 is positive if and only if T is on the same side of T_0 as T_1; a similar analysis shows that β_0 is positive if and only if T is on the same side of T_1 as T_0. Thus β_0 and β_1 are both positive along the line segment $T_0 T_1$. Moreover,

$$
\begin{array}{llcll}
\beta_0 = 1 & T = T_0 & & \beta_1 = 0 & T = T_0 \\
 = 0 & T = T_1 & \text{and} & = 1 & T = T_1.
\end{array}
$$

Equations (1.21) and (1.22) represent the barycentric coordinates of T in terms of distances between T and the affine basis T_0, T_1. We can also apply these equations to convert from rectangular to barycentric coordinates. Let t_0 and t_1 be the rectangular coordinates of T_0 and T_1, and let t be the rectangular coordinate of T along the affine line. Then by (1.21) and (1.22)

$$\beta_1 = \frac{t - t_0}{t_1 - t_0} \quad \text{and} \quad \beta_0 = \frac{t_1 - t}{t_1 - t_0}.$$

(See Figure 1.9.) Notice that in these two equations the signs of the barycentric coordinates are generated automatically from the signs of the rectangular coordinates.

Let us look now at the affine plane. Given any three noncollinear points P_1, P_2, P_3, we can represent any other point Q in the affine plane as an affine combination

$$Q = \beta_1 P_1 + \beta_2 P_2 + \beta_3 P_3 \text{ where } \beta_1 + \beta_2 + \beta_3 = 1. \tag{1.23}$$

We can solve for the barycentric coordinates $\beta_1, \beta_2, \beta_3$ explicitly using determinants. Substituting $\beta_1 = 1 - \beta_2 - \beta_3$ and rearranging the terms in (1.23), we find that

$$Q - P_1 = \beta_2 (P_2 - P_1) + \beta_3 (P_3 - P_1).$$

Taking the determinant of both sides with $P_2 - P_1$ and recalling that the determinant is multilinear and that $\det(v, v) = 0$, we obtain

$$\det(Q - P_1, P_2 - P_1) = \beta_3 \det(P_3 - P_1, P_2 - P_1).$$

So, solving for β_3, we arrive at

$$\beta_3 = \frac{\det(Q - P_1, P_2 - P_1)}{\det(P_3 - P_1, P_2 - P_1)}. \tag{1.24}$$

Similarly, we find that

$$\beta_2 = \frac{\det(Q - P_3, P_1 - P_3)}{\det(P_2 - P_3, P_1 - P_3)} \tag{1.25}$$

$$\beta_1 = \frac{\det(Q - P_2, P_3 - P_2)}{\det(P_1 - P_2, P_3 - P_2)}. \tag{1.26}$$

When we want to specify that β_k is the barycentric coordinate of a point Q, we shall write $\beta_k(Q)$. Notice, in particular, from Equations (1.24)–(1.26) we can conclude that

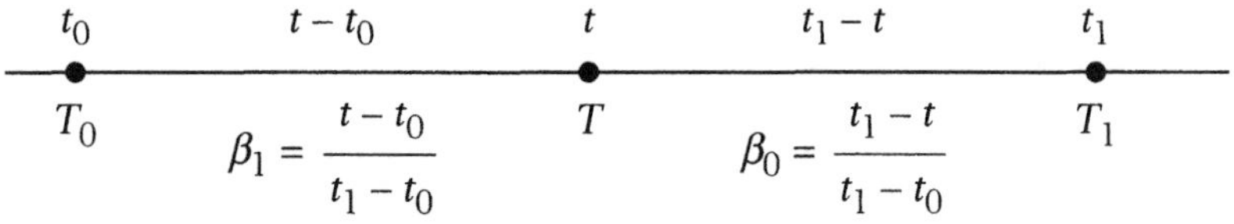

Figure 1.9 Rectangular and barycentric coordinates along the affine line.

$$\beta_k(P_j) = 0 \qquad j \neq k \tag{1.27}$$
$$= 1 \qquad j = k \ .$$

More generally, if Q is any point on the line joining P_i and P_j, then $Q = P_i + t(P_j - P_i)$; hence again from Equations (1.24)–(1.26) and the multilinearity of the determinant function, $\beta_k(Q) = 0$.

Since, up to sign, determinants represent areas (see Exercise 1), barycentric coordinates in the plane have a geometric interpretation. Equations (1.24)–(1.26) yield

$$\beta_3(Q) = \pm \frac{area(\Delta Q P_1 P_2)}{area(\Delta P_1 P_2 P_3)},$$

$$\beta_2(Q) = \pm \frac{area(\Delta Q P_1 P_3)}{area(\Delta P_1 P_2 P_3)},$$

$$\beta_1(Q) = \pm \frac{area(\Delta Q P_2 P_3)}{area(\Delta P_1 P_2 P_3)},$$

where the sign of $\beta_i(Q)$ is positive if Q lies inside $\Delta P_1 P_2 P_3$ and negative when Q crosses the line $P_j P_k, j,k \neq i$. These area formulas are illustrated in Figure 1.10(b).

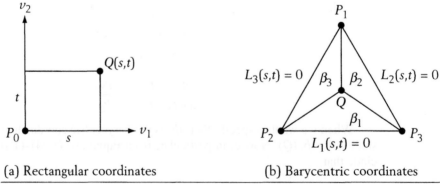

(a) Rectangular coordinates   (b) Barycentric coordinates

Figure 1.10 Rectangular and barycentric coordinates in the affine plane. Rectangular coordinates are represented by signed ratios of lengths; barycentric coordinates, by signed ratios of areas. The rectangular coordinate of Q relative to the axis v_k is given by the projection of Q on the v_k axis (divided by the unit length). The barycentric coordinate of Q relative to the point P_k is represented by the area of the triangle opposite to P_k divided by the area of $\Delta P_1 P_2 P_3$. Since Q splits $\Delta P_1 P_2 P_3$ into three subtriangles, it follows immediately from this normalization that the barycentric coordinates of Q sum to one. The barycentric coordinate of Q relative to P_k can also be represented by the value at Q of the linear expression $L_k(s,t)$ for the line $P_i P_j$ when properly normalized.

Just as on the affine line, Equations (1.24)–(1.26) can be used to convert from rectangular coordinates to barycentric coordinates in the affine plane. By writing P_1, P_2, P_3, Q in terms of rectangular coordinates and expanding the determinants on the right-hand side of these equations, it is easy to see that the barycentric coordinates $\beta_1(Q), \beta_2(Q), \beta_3(Q)$ are linear functions in the rectangular coordinates s,t of Q. This observation leads to another interesting way to generate barycentric coordinates, which we will have occasion to use in subsequent chapters.

Consider the lines joining the points P_1, P_2, P_3. Let $l_k(s,t) = 0$ denote the equation in rectangular coordinates of the line in the st-plane determined by P_i and P_j, $i, j \neq k$. We are going to show that the barycentric coordinates relative to the affine basis P_1, P_2, P_3 are given by the functions $l_k(s,t)$ when properly normalized.

First observe that l_k and β_k agree on the line determined by P_i and P_j, $i, j \neq k$. In particular, by construction, $l_k(P_\alpha) = 0$, and by Equation (1.27), $\beta_k(P_\alpha) = 0$, $\alpha \neq k$, so we are off to a good start. However, we still need to be sure that $l_k(P_k) = \beta_k(P_k)$. In fact, this need not be the case. In rectangular coordinates

$$l_k(s,t) = a_k s + b_k t + c_k$$

for some constant coefficients a_k, b_k, c_k, but the coefficients a_k, b_k, c_k are uniquely determined only up to constant multiples because multiplying $l_k(s,t)$ by a constant does not alter the line $l_k(s,t) = 0$. Thus, without some normalization, we cannot know the value of $l_k(s,t)$ off the line. Let (s_k, t_k) be the rectangular coordinates of P_k. Since the points P_1, P_2, P_3 form an affine basis, the point P_k cannot lie on the line l_k; hence $l_k(s_k, t_k) \neq 0$. Therefore we can normalize the coefficients a_k, b_k, c_k by setting

$$L_k(s,t) = \frac{l_k(s,t)}{l_k(s_k, t_k)} \ .$$

After this normalization, we can be sure by Equation (1.27) that $L_k(P_k) = \beta_k(P_k) = 1$. Since L_k and β_k are both linear functions of the rectangular coordinates s,t and since they also agree at the points P_i, P_j, $i, j \neq k$, it follows that $\beta_k(Q) = L_k(Q)$ for all Q, since linear functions in the plane that agree at three noncollinear points are identical (see part (a) of Exercise 4).

The area formula and the line formula for barycentric coordinates in the affine plane can both be extended to higher-dimensional affine spaces. However, our focus here is on the affine line and the affine plane because these spaces serve as the parameter spaces for curves and surfaces. Hence we shall leave the extension of these formulas for barycentric coordinates in higher dimensions to the exercises (see Exercises 6 and 7). Their derivations are much the same as in the planar case.

We close this section with a theorem summarizing for future reference the main properties we have just derived of barycentric coordinates in the affine plane.

THEOREM
1.1

Properties of Barycentric Coordinates in the Affine Plane

Let $\beta_1, \beta_2, \beta_3$ be barycentric coordinates relative to an affine basis P_1, P_2, P_3. Then

1. $\sum_{k=1}^{3} \beta_k = 1$.

2. $\beta_k > 0$ in the interior of $\Delta P_1 P_2 P_3$.

3. $\beta_k = 0$ on the line $P_i P_j$, $k \neq i, j$.

4. $\beta_k(P_j) = 0 \qquad j \neq k$

   $\qquad\qquad = 1 \qquad j = k$.

5. $\beta_1, \beta_2, \beta_3$ are linear functions in the rectangular coordinates s, t.

Exercises

1. Let P_0, P_1, P_2 be the vertices of a triangle. Show that the following are equivalent:

   a. $2 \times area(\Delta P_1 P_2 P_3)$

   b. $|P_2 - P_1| \, |P_3 - P_1| \sin\theta$, where θ is the angle between $P_2 - P_1$ and $P_3 - P_1$

   c. $|(P_2 - P_1) \times (P_3 - P_1)|$

   d. $|\det(P_2 - P_1, P_3 - P_1)|$

2. Let $\beta_1, \beta_2, \beta_3$ be barycentric coordinates for the affine plane relative to the affine basis P_1, P_2, P_3, and let γ_i, γ_j be the barycentric coordinates for the affine line $P_i P_j$ relative to the points P_i, P_j. Suppose that Q is a point on the line $P_i P_j$. Show that $\beta_i(Q) = \gamma_i(Q)$ and $\beta_j(Q) = \gamma_j(Q)$.

3. Let β_0, β_1 be barycentric coordinates for the affine line relative to the affine basis T_0, T_1, and let L, L_1, L_2 be linear functions on the affine line. Show that

   a. If $L_1(t)$ and $L_2(t)$ agree at two distinct values of t, then $L_1(t) = L_2(t)$ for all t.

   b. $L(T) = L(T_0)\beta_0(T) + L(T_1)\beta_1(T)$ for all points T on the affine line.

4. Let $\beta_1, \beta_2, \beta_3$ be barycentric coordinates for the affine plane relative to the affine basis P_1, P_2, P_3, and let L, L_1, L_2 be linear functions on the affine plane. Show that

   a. If $L_1(s,t)$ and $L_2(s,t)$ agree at three noncollinear points, then $L_1(s,t) = L_2(s,t)$ for all (s,t).

   b. $L(Q) = \sum_{k=1}^{3} \beta_k(Q) L(P_k)$ for all points Q in the affine plane.

c. $Q = \sum_{k=1}^{3} \beta_k(Q)P_k$ for all points Q in the affine plane.

5. Prove that the barycentric coordinate functions $\beta_1, \beta_2, \beta_3$ are the only functions satisfying the five properties listed in Theorem 1.1. (Hint: Use part (a) of Exercise 4).

6. Let $\beta_0(Q), \ldots, \beta_n(Q)$ be the barycentric coordinates of Q relative to an affine basis $P_0, \ldots, P_n$.

   a. Prove that

   $$\beta_k(Q) = \frac{\det(Q - P_j, P_0 - P_j, \ldots, P_n - P_j)}{\det(P_k - P_j, P_0 - P_j, \ldots, P_n - P_j)} \quad j \neq k$$

   where the terms $P_k - P_j$, $P_j - P_j$ are omitted from the sequences $P_0 - P_j, \ldots, P_n - P_j$ in the numerator and denominator.

   b. Conclude that

   $$\beta_k(P_j) = 0 \quad j \neq k$$
   $$= 1 \quad j = k \ .$$

   c. Interpret the result in part (a) geometrically when $n = 3$.

7. Let $\beta_0(Q), \ldots, \beta_n(Q)$ be the barycentric coordinates of Q relative to an affine basis $P_0, \ldots, P_n$. Introduce rectangular coordinates $(t_1, \ldots, t_n)$ and call a function $L(Q)$ linear if it is linear in $(t_1, \ldots, t_n)$. Prove that

   a. If $L_1(P)$ and $L_2(P)$ are two linear functions that agree at the $n+1$ points $P_0, \ldots, P_n$, then they agree everywhere.

   b. For each k there is a linear equation $L_k(P) = 0$ satisfied by all the points in the affine basis except for P_k.

   c. If the function $L_k(P)$ in part (b) is normalized so that $L_k(P_k) = 1$, then $\beta_k(Q) = L_k(Q)$.

   d. If L is a linear function, then $L(Q) = \sum_{k=0}^{n} \beta_k(Q)L(P_k)$ for all points Q in affine n space.

   e. $Q = \sum_{k=0}^{n} \beta_k(Q)P_k$ for all points Q in affine n space.

8. Consider the rectangle in Figure 1.11.

   a. Show that the functions

   $$\beta_1(s,t) = (1-s)(1-t)$$
   $$\beta_2(s,t) = (1-s)t$$
   $$\beta_3(s,t) = s(1-t)$$
   $$\beta_4(s,t) = st$$

   behave like barycentric coordinates for the rectangle $P_1P_2P_3P_4$.

In particular, show that these functions satisfy all five conditions of Theorem 1.1, except for condition 5, which must be replaced by 5*. $\beta_1, \beta_2, \beta_3, \beta_4$ are bilinear functions.

b. Give a geometric interpretation for these barycentric coordinates.

c. Generalize the results in parts (a) and (b) to arbitrary rectangles with sides parallel to the coordinate axes.

1.3   Curve and Surface Representations

This is a book about curves and surfaces. So far, however, we have discussed mostly ambient spaces and coordinate systems. Be patient. We still must address one more preliminary issue before we can proceed to our main theme. We need to decide how we shall represent curves and surfaces inside our ambient spaces.

Four types of representations for curves and surfaces are common in computer graphics and geometric design: *explicit, implicit, parametric,* and *procedural.* Here we shall look briefly at each of these alternatives and then settle on one particular form to use throughout this text.

When you first studied analytic geometry, you used rectangular coordinates and considered equations of the form $y = f(x)$. The graphs $(x, f(x))$ of these functions are curves in the plane. For example, $y = 3x + 1$ represents a straight line, and $y = x^2$ represents a parabola (see Figure 1.11). Similarly, you could generate surfaces by considering equations of the form $z = f(x,y)$: the equation $z = 2x + 5y - 7$ represents a plane in 3-space, and $z = x^2 - y^2$ represents a hyperbolic paraboloid. Expressions of the form $y = f(x)$ or $z = f(x,y)$ are called *explicit representations* because they express one variable explicitly in terms of the other variables.

Not all curves and surfaces can be captured readily by a single explicit expression. For example, the unit circle centered at the origin is represented implicitly by all solutions to the equation $x^2 + y^2 - 1 = 0$. If we try to solve explicitly for y in terms of x, we obtain

$$y = \sqrt{1 - x^2},$$

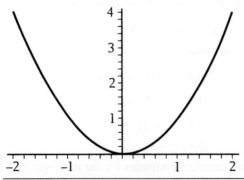

Figure 1.11   Graph of the parabola given by the explicit function $y = x^2$.

which represents only the upper half circle. We must use two explicit formulas

$$y = \pm\sqrt{1 - x^2}$$

to capture the entire circle. Often it is easier just to stick with the original implicit equation rather than to solve explicitly for one of the variables. Thus $x^2 + y^2 - 1 = 0$ represents a circle, and $x^2 + y^2 + z^2 - 1 = 0$ represents a sphere. Equations of the form $f(x,y) = 0$ or $f(x,y,z) = 0$ are called *implicit representations* because they represent the curve or surface implicitly without explicitly solving for one of the variables.

Implicit representations are more general than explicit representations. The explicit curve $y = f(x)$ is the same as the implicit curve $y - f(x) = 0$, but as we have seen it is not always a simple matter to convert an implicit curve into a single explicit formula. Moreover, implicit equations can be used to define closed curves and surfaces or curves and surfaces that self-intersect, shapes that are impossible to represent with explicit functions (see Figure 1.12).

For closed curves and surfaces, the implicit equation can also be used to distinguish the inside from the outside by looking at the sign of the implicit expression. For example, for points inside the unit circle $x^2 + y^2 - 1 < 0$, and for points outside the unit circle $x^2 + y^2 - 1 > 0$. This ability to distinguish easily between the inside and the outside of a closed curve or surface is often important in solid modeling applications.

Nevertheless, implicit representations also have their drawbacks. Given an explicit representation $y = f(x)$, we can easily find lots of points on the curve $(x, f(x))$ by selecting values for x and computing $f(x)$. If our functions $f(x)$ are restricted to elementary functions like polynomials, then for each x there is a unique,

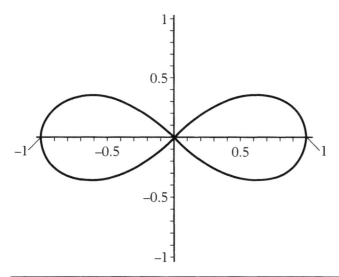

Figure 1.12 The lemniscate of Bernoulli: $(x^2 + y^2)^2 - (x^2 - y^2) = 0$. Notice that unlike explicit functions, the graphs of implicit equations can self-intersect.

easily computable y. Thus it is a simple matter to graph the curve $y = f(x)$. On the other hand, it may not be so easy to find points on the curve $f(x,y) = 0$. For many values of x there may be no y at all, or there may be several values of y, even if we restrict our functions $f(x,y)$ to polynomials in x and y. Finding points on implicit surfaces $f(x,y,z) = 0$ can be even more formidable. Thus it can be difficult to render implicitly defined curves and surfaces.

There is another standard way to represent curves and surfaces that is more general than the explicit form and yet is still easy to render. We can express curves and surfaces parametrically by representing each coordinate with an explicit equation in a new set of parameters. For planar curves we set $x = x(t)$ and $y = y(t)$; for surfaces in 3-space we set $x = x(s,t)$, $y = y(s,t)$, and $z = z(s,t)$. For example, the parametric equations

$$x(t) = \frac{2t}{1+t^2} \qquad y(t) = \frac{1-t^2}{1+t^2}$$

represent the unit circle centered at the origin because by simple substitution we can readily verify that $x^2(t) + y^2(t) - 1 = 0$. Similarly, the parametric equations

$$x(s,t) = \frac{2s}{1+s^2+t^2} \qquad y(s,t) = \frac{2t}{1+s^2+t^2} \qquad z(s,t) = \frac{1-s^2-t^2}{1+s^2+t^2}$$

represent a unit sphere, since $x^2(s,t) + y^2(s,t) + z^2(s,t) - 1 = 0$. Often we shall restrict the parameter domain. Thus a parametric curve is typically the image of a line segment; a parametric surface, the image of a region—usually rectangular or triangular—of the plane.

The parametric representation has several advantages. Like the explicit representation, the parametric representation is easy to render: simply evaluate the coordinate functions at various values of the parameters. Like implicit equations, parametric equations can also be used to represent closed curves and surfaces as well as curves and surfaces that self-intersect. In addition, the parametric representation has another advantage: it is easy to extend to higher dimensions. To illustrate: if we want to represent a curve in 3-space, all we need do is introduce an additional equation $z = z(t)$. Thus the parametric equations

$$x(t) = 2t - 5 \qquad y(t) = 3t + 7 \qquad z(t) = 4t + 1$$

represent a line in 3-space. Figure 1.13 illustrates a more complicated parametric curve in 3-space.

The parametric representation has its own idiosyncrasies. The explicit representation of a curve is unique: the graph of $y = g(x)$ is the same curve as the graph of $y = f(x)$ if and only if $g(x) = f(x)$. Similarly, if we restrict to polynomial functions, then the implicit representation $f(x,y) = 0$ is essentially unique. Indeed if $f(x,y)$ and $g(x,y)$ are polynomials, then $g(x,y) = 0$ represents the same curve as $f(x,y) = 0$ over the complex numbers if and only if $g(x,y)$ is a constant times a power of $f(x,y)$. However, the parametric representation of a curve is not unique. For example, the equations

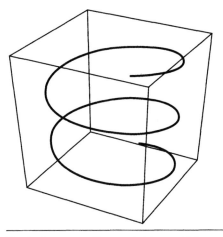

Figure 1.13 The helix: $x = \cos(t)$, $y = \sin(t)$, $z = t / 5$.

$$x(t) = \frac{2t}{1+t^2} \qquad y(t) = \frac{1-t^2}{1+t^2}$$

$$x(t) = \sin(t) \qquad y(t) = \cos(t)$$

are two very different parametric representations for the unit circle $x^2 + y^2 = 1$. Moreover, if we restrict our attention, as we shall in most of this text, to polynomial or rational parametrizations, then it is known that every such parametric curve or surface lies on an implicit polynomial curve or surface. The converse, however, is not true. There exist implicit polynomial curves and surfaces that have no polynomial or rational parametrization. Thus, the implicit polynomial form is more general than the rational parametric form.

Nevertheless, because of their power, simplicity, and ease of use, we shall choose to represent all the curves and surfaces in this book using parametric representations. Moreover, our curves and surfaces will lie in an unspecified number of dimensions, since the parametric representation works equally well in an arbitrary number of dimensions. Note that in the one-dimensional case the parametric representation is the same as the explicit representation, so we cover explicit representations automatically as a special case. Sometimes it will be helpful to think about the special case of explicit representations, but more often than not this can confuse the issue because parametric curves exhibit geometric properties such as self-intersection that can never occur in explicit representations. Planar parametric curves $(x(t), y(t))$ are much more flexible than the planar graphs $\big(t, x(t)\big)$ of explicit functions.

It remains to say what kinds of functions we shall allow in our parametric representations. *Most of the remainder of this book is about how to choose the parametric functions in order to generate suitable curves and surfaces.* Generally our functions shall be variants of polynomials: either simple polynomials or rational functions (ratios of polynomials) or piecewise polynomials (splines) or piecewise rational functions.

Polynomials have many advantages, especially when used in conjunction with a computer. Polynomials are easy to evaluate. Furthermore, more complicated functions are generally evaluated by computing some polynomial approximation, so nothing is really lost by restricting to polynomials in the first place. In addition, there is a well-developed theory of polynomials in numerical analysis and approximation theory; computer graphics and geometric modeling borrow extensively from this theory.

We have yet to mention procedurally defined curves or surfaces. In geometric design, offsets, blends, and fillets are often specified by procedures rather than by formulas. In solid modeling, geometry is often constructed procedurally using Boolean operations such as union, intersection, and difference. Most fractal surfaces and space-filling curves are defined by recursive algorithms rather than with explicit formulas. We shall not discuss any of these kinds of procedures in this text. Sculpting or subdivision is another paradigm for defining curves and surfaces by exploiting recursive procedures. Since certain subdivision techniques are closely related to parametric curves and surfaces, we will have a good deal more to say about these methods later in this book.

One final point. Although we are going to resort to parametric representations, we want to get away almost entirely from using coordinate systems and coordinate functions. In Section 1.1 we spoke extensively about ambient spaces and coordinate-free operations, and we want to take advantage of these notions. How then shall we proceed?

We will write $P(t)$ to represent a parametric curve and $P(s,t)$ to represent a parametric surface. Applying the algebra of affine space or Grassmann space, we will provide explicit formulas or recursive procedures for computing $P(t)$ and $P(s,t)$ directly without resorting to coordinates in the range. We have already encountered such formulas in Section 1.1.6, where we wrote

$$P(t) = \sum_{k=0}^{n} B_k(t) P_k$$

$$P(s,t) = \sum_{ij} B_{ij}(s,t) P_{ij} \ .$$

In terms of rectangular coordinates $P_k = (x_k, y_k, z_k)$, $P_{ij} = (x_{ij}, y_{ij}, z_{ij})$, $P(t) = (x(t), y(t), z(t))$, and $P(s,t) = (x(s,t), y(s,t), z(s,t))$, but we shall almost never write such explicit coordinate formulas in this text. There are several reasons for adopting a coordinate-independent approach. First, we do not want to carry around coordinates all the time; it is simpler and cleaner to deal with one equation for P rather than with three equations for x, y, z. Also, as we stressed in Section 1.1, all our algebraic operations are going to be coordinate free. Thus not only do we not need coordinates, they would actually get in our way by obscuring the geometric *meaning* of our algorithms. Coordinate techniques are for computation; coordinate-free methods are for comprehension. It may take a little getting used to, but in the end we expect the coordinate-free approach to pay off in much better geometric understanding and far more comprehensible programming.

Exercises

1. Let P,Q,R be the vertices of an isosceles right triangle with P opposite to the hypotenuse. Show that the parametric equation

$$P(t) = P + \frac{2t}{1+t^2}(Q-P) + \frac{1-t^2}{1+t^2}(R-P)$$

   represents a circle centered at P.

2. Let a,b be fixed constants.

   a. How is the graph of $y = f(ax+b)$ related to the graph of $y = f(x)$?

   b. How is the implicit curve $f(ax+b, y) = 0$ related to the implicit curve $f(x,y) = 0$?

   c. How is the parametric curve $P(at+b)$ related to the parametric curve $P(t)$?

1.4 Summary

In this chapter, we have discussed ambient spaces (vector spaces, affine spaces, Grassmann spaces, and projective spaces), coordinate systems (rectangular, affine, Grassmann, homogeneous, and barycentric), and curve and surface representations (explicit, implicit, parametric, and procedural). We fixed on the parametric representation for curves and surfaces, and settled upon affine spaces for modeling such polynomial schemes. A coordinate-free approach was adopted for the range, but barycentric coordinates were chosen for representing the domain. Grassmann spaces and projective spaces were studied in order to prepare the way for investigating rational curves and surfaces.

If you have understood all the tools in this chapter, you will have a solid foundation for reading the rest of this book. If not, you may want to return to this chapter from time to time to refresh your understanding of this material. But you should not proceed any further until you have a firm grasp at least of affine spaces and coordinate-free methods. These techniques are assumed almost everywhere throughout the text, so you need to be comfortable with them before you proceed to subsequent chapters where we begin in earnest the study of free-form curves and surfaces.

Part I

Interpolation

CHAPTER 2

Lagrange Interpolation and Neville's Algorithm

Perhaps the easiest way to describe a shape is to select some points on the shape. Given enough data points, the eye has a natural tendency to interpolate smoothly between the data. Here we are going to study this problem mathematically. Given a finite collection of points in affine space, we shall investigate methods for generating polynomial curves and surfaces to go through the points. We begin with schemes for curves and later extend these techniques to surfaces.

2.1 Linear Interpolation

Two points determine a line. Suppose we want the equation of the line $P(t)$ passing through the two points P and Q in affine space. Then we can write

$$P(t) = P + t(Q - P). \tag{2.1}$$

The curve $P(t)$ passes through P at $t = 0$ and Q at $t = 1$. Moreover, as t varies, the points on $P(t)$ extend in the direction along the vector from P to Q; thus, these points lie along the line in affine space generated by P and Q. Rearranging terms, we can rewrite (2.1) as

$$P(t) = (1 - t)P + tQ. \tag{2.2}$$

Equation (2.2) is called *linear interpolation;* this equation is the foundation of all we plan to accomplish in this chapter.

Notice that the formula for linear interpolation is given by an affine combination, so the right-hand side of (2.2) represents a well-defined collection of points in affine space.

One subtle issue. We saw that in (2.2) $P(t)$ passes through P at $t = 0$ and through Q at $t = 1$. We did not specify this requirement in the original problem. All we wanted was a line passing through the two points P and Q; the parameters t at which the line was to pass through these points were not mentioned. Suppose, however, that we do wish to specify these parameters as well. That is, now we require a line $P_{01}(t)$ to pass through P_0 at $t = t_0$ and through P_1 at $t = t_1$. Mimicking (2.2), we expect to write an equation of the form

$$P_{01}(t) = (1 - f(t))P_0 + f(t)P_1. \tag{2.3}$$

Moreover, still emulating (2.2), we want $f(t)$ to be linear and to satisfy

$$f(t_0) = 0 \text{ and } f(t_1) = 1.$$

These equations for $u = f(t)$ represent another linear interpolation problem; this time in the tu-plane. That is, now we need to find the line in the coordinate plane interpolating the data $(t_0, 0)$ and $(t_1, 1)$. Of course you learned long ago, when you first studied analytic geometry, how to solve such problems. This line is given by the equation

$$f(t) = \frac{(t - t_0)}{(t_1 - t_0)}, \tag{2.4}$$

as you can readily verify by evaluating $f(t)$ at $t = t_0$ and $t = t_1$. Substituting (2.4) into (2.3), we obtain

$$P_{01}(t) = \frac{t_1 - t}{t_1 - t_0}P_0 + \frac{t - t_0}{t_1 - t_0}P_1, \tag{2.5}$$

where we have used the identity $1 - f(t) = (t_1 - t)/(t_1 - t_0)$. Notice, by the way, that the coefficients of P_0 and P_1 are precisely the barycentric coordinates (see Section 1.2.3) of the point $P_{01}(t)$ with respect to the points P_0 and P_1, so linear interpolation is just another way of deriving barycentric coordinates along a line.

Equation (2.5) is so fundamental that we are going to represent it graphically with a simple diagram. In Figure 2.1(a) the value at the apex of the triangle is computed by multiplying the points at the base by the values along the arrows and then adding the results. The end product is just Equation (2.5). Figure 2.1(b) represents exactly the same computation as Figure 2.1(a). Here, however, we have removed the normalization in the denominator to simplify the diagram. The denominator can be retrieved by summing the numerators, since in affine space the functions multiplying the points must sum to one. The advantage of Figure 2.1(b) is that it is much less

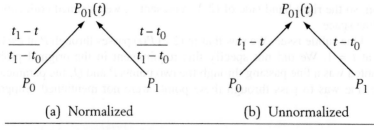

(a) Normalized   (b) Unnormalized

Figure 2.1 Graphical representations of Equation (2.5).

cluttered than Figure 2.1(a), so in the future we shall usually draw these graphs in this unnormalized form. You should get used to this simple diagram now because you are going to see many more like it throughout this book.

Exercises

1. Describe the curve represented by the equation $P_{01}(t) = (1 - f(t))P_0 + f(t)P_1$ when $f(t) = t^2, t^3, \cos(t), e^t$.

2. A table of sines states that $\sin(24°) = 0.40674$ and $\sin(25°) = 0.42262$. Use linear interpolation to estimate $\sin(24.3°)$.

3. Let $\beta_1, \beta_2, \beta_3$ be barycentric coordinate functions relative to $\Delta P_1 P_2 P_3$. Let Q be an arbitrary point in $\Delta P_1 P_2 P_3$ and let $R_k = P_i P_j \cap P_k Q$. Suppose that $\alpha_k(Q)$ is the coefficient of P_k computed by first performing linear interpolation along $P_i P_j$ to find R_k, and then performing linear interpolation along $P_k R_k$ to find Q. Show that $\beta_k(Q) = \alpha_k(Q)$, $k = 1, 2, 3$.

2.2 Neville's Algorithm

Let's try a slightly harder problem. Suppose we now have three points P_0, P_1, P_2 in affine space that we wish to interpolate at the parameters t_0, t_1, t_2. How shall we proceed?

We already have a way to interpolate P_0, P_1 at t_0, t_1; we can join these points with the straight line

$$P_{01}(t) = \frac{t_1 - t}{t_1 - t_0} P_0 + \frac{t - t_0}{t_1 - t_0} P_1.$$

Similarly, by reindexing, we can interpolate P_1, P_2 at t_1, t_2 with the straight line

$$P_{12}(t) = \frac{t_2 - t}{t_2 - t_1} P_1 + \frac{t - t_1}{t_2 - t_1} P_2.$$

The piecewise linear curve given by

$$P(t) = P_{01}(t) \qquad t \le t_1$$
$$= P_{12}(t) \qquad t \ge t_1$$

certainly interpolates the points P_0, P_1, P_2 at the parameters t_0, t_1, t_2. However, this curve is not smooth; it has a sharp point at P_1. Sharp points are potentially dangerous and hence undesirable in objects designed for human consumption. We seek a smooth curve that does the job.

To generate a smooth curve, apply linear interpolation to the two curves $P_{01}(t)$ and $P_{12}(t)$:

$$P_{012}(t) = \frac{t_2 - t}{t_2 - t_0} P_{01}(t) + \frac{t - t_0}{t_2 - t_0} P_{12}(t). \qquad (2.6)$$

By substitution it is easy to verify that $P_{012}(t)$ interpolates P_0 and P_2 at t_0 and t_2, since by (2.6)

$$P_{012}(t_0) = P_{01}(t_0) = P_0$$
$$P_{012}(t_2) = P_{12}(t_2) = P_2.$$

To verify that $P_{012}(t)$ also interpolates P_1 at t_1, observe that $P_{01}(t)$ and $P_{12}(t)$ both interpolate P_1 at t_1. Therefore

$$P_{012}(t_1) = \frac{t_2 - t_1}{t_2 - t_0} P_{01}(t_1) + \frac{t_1 - t_0}{t_2 - t_0} P_{12}(t_1)$$

$$= \frac{t_2 - t_1}{t_2 - t_0} P_1 + \frac{t_1 - t_0}{t_2 - t_0} P_1$$

$$= P_1 \ .$$

If we were to expand the right-hand side of (2.6), we would find that $P_{012}(t)$ is a quadratic polynomial in t, since $P_{01}(t)$ and $P_{12}(t)$ are both linear in t. Thus we have constructed a smooth curve that interpolates the given points at the specified parameter values (see Figure 2.2). Figure 2.3 is a graphical representation of Equation (2.6).

What if we want to interpolate four points P_0, P_1, P_2, P_3 at parameter values t_0, t_1, t_2, t_3? We already know how to build quadratic curves to interpolate portions of this data. We can construct $P_{012}(t)$ to interpolate P_0, P_1, P_2 at t_0, t_1, t_2 and $P_{123}(t)$ to interpolate P_1, P_2, P_3 at t_1, t_2, t_3. Diagramming $P_{123}(t)$ yields Figure 2.4.

Figures 2.3 and 2.4 share the little subtriangle with vertex $P_{12}(t)$. Overlapping these two figures and joining $P_{012}(t)$ and $P_{123}(t)$ by yet another linear interpolation step

$$P_{0123}(t) = \frac{t_3 - t}{t_3 - t_0} P_{012}(t) + \frac{t - t_0}{t_3 - t_0} P_{123}(t),$$

we arrive at Figure 2.5.

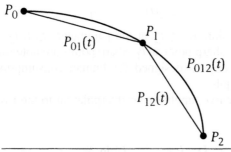

Figure 2.2 The two lines $P_{01}(t)$ and $P_{12}(t)$, and the quadratic interpolant $P_{012}(t)$.

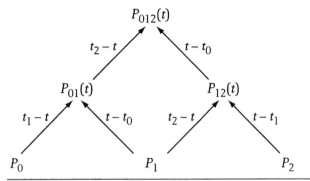

Figure 2.3 A graphical representation of Equation (2.6). The first level is just two juxtaposed copies of Figure 2.1, one for $P_{01}(t)$ and one for $P_{12}(t)$. The second level represents the linear interpolation step joining $P_{01}(t)$ and $P_{12}(t)$. Here we have adopted our convention of leaving off the denominators to avoid cluttering the diagram.

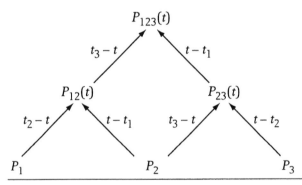

Figure 2.4 A graphical representation for the curve $P_{123}(t)$.

Now it is easy to verify directly from the figure that $P_{0123}(t)$ interpolates the given data at the specified parameter values. By substitution, we see that

$$P_{0123}(t_0) = P_{012}(t_0) = P_0$$
$$P_{0123}(t_3) = P_{123}(t_3) = P_3 \ .$$

Moreover we already know that

$$P_{012}(t_k) = P_{123}(t_k) = P_k \qquad\qquad k = 1,2$$

and since the labels on the arrows exiting $P_{012}(t)$ and $P_{123}(t)$ sum to one (remember the normalization), it follows that

$$P_{0123}(t_k) = P_k \qquad\qquad k = 1,2 \ .$$

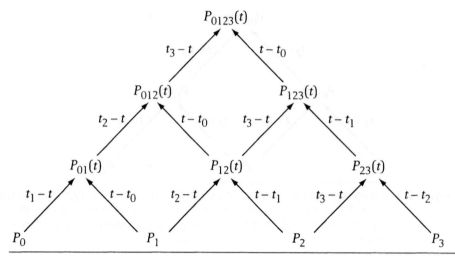

Figure 2.5 Neville's algorithm for cubic interpolation.

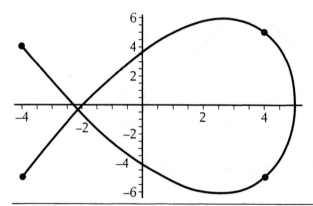

Figure 2.6 Figure 2.6: The cubic Lagrange polynomial for the control points $P_0 = (-4,4)$, $P_1 = (4,-5)$, $P_2 = (4,5)$, $P_3 = (-4,-5)$ (dots), interpolated at the nodes $t_k = k$, $k = 0,...,3$.

The algorithm for computing $P_{0123}(t)$ represented by Figure 2.5 is called *Neville's algorithm*. We shall have a lot more to say about this algorithm shortly. The curves generated by Neville's algorithm are called *Lagrange interpolating polynomials*. We illustrate an example of a Lagrange interpolating polynomial in Figure 2.6.

We could go on introducing more and more data points and constructing higher- and higher-order curves, but by now it should be clear how to proceed. Instead, let's summarize what we expect to be true in the following theorem.

THEOREM
2.1

Given affine points $P_0,...,P_n$ and distinct parameters $t_0,...,t_n$, there is a polynomial curve $P_{0...n}(t)$ of degree n that interpolates the given points at the specified parameters. That is, $P_{0...n}(t_k) = P_k$, $k = 0,...,n$.

Proof The proof is by induction on n. We have already established this result by construction for $n = 0,1,2,3$. Suppose this result is true for $n - 1$. Then by the inductive hypothesis, there are polynomial curves $P_{0...n-1}(t)$ and $P_{1...n}(t)$ of degree $n - 1$ that interpolate the points $P_0,...,P_{n-1}$ at the parameters $t_0,...,t_{n-1}$ and the points $P_1,...,P_n$ at the parameters $t_1,...,t_n$. Define

$$P_{0...n}(t) = \frac{t_n - t}{t_n - t_0} P_{0...n-1}(t) + \frac{t - t_0}{t_n - t_0} P_{1...n}(t). \qquad (2.7)$$

Then applying the same arguments we used in the quadratic and cubic cases, you can easily verify that

$$P_{0...n}(t_k) = P_k \qquad\qquad k = 0,...,n.$$

Moreover since $P_{0...n-1}(t)$ and $P_{1...n}(t)$ are polynomials of degree $n - 1$, it follows from (2.7) that $P_{0...n}(t)$ is a polynomial of degree n.

The parameter values $t_0,...,t_n$ at which the interpolation occurs are called *nodes*, and the points $P_0,...,P_n$ that are interpolated are called *control points* (see Figure 2.6). In general, if we change the nodes, then the interpolating curve $P_{0...n}(t)$ changes even if we leave the control points fixed (see Exercise 3).

Exercises

1. Complete the proof of Theorem 2.1 by showing that $P_{0...n}(t_k) = P_k$.

2. Let $P_{0...p,m}(t)$ denote a polynomial curve of degree $p + 1$ that interpolates the points $P_0,...,P_p,P_m$ at the parameters $t_0,...,t_p,t_m$. Prove that $P_{0...p,m}(t)$ can be generated from the recurrence

$$P_{0...p,m}(t) = \frac{t_m - t}{t_m - t_p} P_{0...p}(t) + \frac{t - t_p}{t_m - t_p} P_{0...p-1,m}(t).$$

3. Give an example to show that changing the nodes alters the interpolating curve $P_{0...n}(t)$ even if we leave the control points fixed.

4. Let $P(t)$ be the Lagrange interpolating polynomial for the control points $P_0,...,P_n$ and nodes $t_0,...,t_n$. Form a new Lagrange interpolating curve $Q(t)$ by replacing each node t_k by the node $\tau_k = at_k + b$ for some fixed constants $a > 0$ and b. Show that changing all the nodes in this way has no affect on the shape of the interpolating curve. In particular, using Neville's algorithm, show that $Q(at + b) = P(t)$. What happens if we choose $a < 0$?

2.3 **The Structure of Neville's Algorithm**

Equation (2.7) is a recursive formula for $P_{0...n}(t)$. It asserts that we can compute $P_{0...n}(t)$ by calculating $P_{0...n-1}(t)$ and $P_{1...n}(t)$ and then taking a specific affine combination of the results. Continuing in this manner, we can also compute $P_{0...n-1}(t)$ and $P_{1...n}(t)$ recursively. This recursion bottoms out at the constant functions $P_k(t) = P_k$.

If we proceed in this manner, we arrive at an algorithm with the structure of a binary tree as illustrated in Figure 2.7. This algorithm is very inefficient because it requires an exponential number of procedure calls. Moreover, all the interior nodes—that is, all the nodes not lying along the periphery of the diagram—are computed twice; for example, $P_{12}(t)$ is computed once during the computation of $P_{012}(t)$ and once again during the computation of $P_{123}(t)$. Thus, implementing (2.7) recursively is not a good idea.

There is a better way: apply dynamic programming. In dynamic programming, we first compute all the constant interpolants, then all the linear interpolants, then all the quadratic interpolants, continuing to build higher- and higher-order interpolants as we go. In this fashion, each interpolant is computed only once. This approach leads to an $O(n^2)$ algorithm—there are n linear interpolants, $n-1$ quadratic interpolants, $n-2$ cubic interpolants, and so on, so altogether there are $n + (n-1) + ... + 1$ $= n(n+1)/2 = O(n^2)$ interpolants—rather than an exponential algorithm. It is precisely this technique that is illustrated for cubic curves in Figure 2.5.

Moreover, while the time complexity of this dynamic programming algorithm is $O(n^2)$, the space complexity is only $O(n)$. Indeed once we have computed the interpolants of order $k + 1$, we can discard the interpolants of order k, since they are no longer needed to compute the higher-order interpolants. This space efficiency is another advantage of dynamic programming.

This dynamic programming approach to interpolation is called Neville's algorithm. This algorithm and algorithms like it are at the heart of what we plan to study throughout this text. Get accustomed to it now because it will be fundamental to all our work later on. In particular, be sure you understand the difference between the

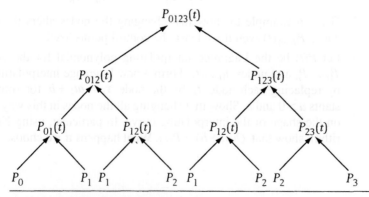

Figure 2.7   Performing interpolation by recursive calls: the cubic case.

dynamic programming algorithm illustrated in Figure 2.5 and the recursive procedure illustrated in Figure 2.7.

Neville's algorithm has an interesting structure. Call the base of the diagram the zeroth level and the apex the nth level. Then the kth level of the algorithm represents kth-order interpolants because, by construction, $P_j,...,P_{j+k}(t)$ interpolates the control points $P_j,...,P_{j+k}$ at the nodes $t_j,...,t_{j+k}$. Notice that in the figure the points $P_j,...,P_{j+k}$ lie in the span of the curve $P_{j...j+k}(t)$; that is, the points $P_j,...,P_{j+k}$ form the base of the triangle with apex $P_{j...j+k}(t)$. Thus each subtriangle reproduces in the small the structure of the entire triangle in the large.

The diagram for Neville's algorithm is easy to remember. Start with $P_{0...n}(t)$ at the apex. Strip off the index n and place $P_{0...n-1}(t)$ below it to the left; strip off the index 0 and place $P_{1...n}(t)$ below it to the right. Since the index n was removed to the left, label the left arrow with $t_n - t$; since the index 0 was removed to the right, label the right arrow with $t - t_0$. Now proceed recursively stripping off labels from $P_{0...n-1}(t)$ and $P_{1...n}(t)$ and labeling the arrows accordingly. Remember to join $P_{1...n-1}(t)$ to both $P_{0...n-1}(t)$ and $P_{1...n}(t)$ to generate a dynamic programming algorithm instead of a recursive procedure. Refer to Figure 2.5 for an illustration of the cubic case.

There is another important structural property of Neville's algorithm that is an artifact of this construction. Look at Figure 2.5. Pick any direction and consider the labels along parallel arrows as you ascend the triangle in that direction. Notice that these labels appear to be identical; this observation holds for any degree. In fact, these labels are not really identical because we have suppressed the denominators. Only the numerators match; the denominators differ from level to level. Nevertheless, this *parallel property* of matching numerators along parallel arrows is fairly important, and we shall return to it again in subsequent sections.

Neville's algorithm has one additional significant property. Suppose we have already interpolated the control points $P_0,...,P_n$ at the nodes $t_0,...,t_n$ and later we discover that we need to interpolate one additional point P_{n+1} at the parameter t_{n+1}. We need not restart our computations from the beginning. If we have saved the original triangular computation for $P_{0...n}(t)$, then we need only add the edge computing $P_{n,n+1}(t),...,P_{0...n+1}(t)$. That is, in the dynamic programming algorithm for $P_{0...n+1}(t)$, we need only add the computation of one curve of each degree. Thus, at the cost of increasing our storage from $O(n)$ to $O(n^2)$, to complete our calculation, we need only add a computation of $O(n)$ instead of redoing work of $O(n^2)$. This savings is yet another advantage of the dynamic programming approach to interpolation.

Exercises

1. Aitken's algorithm is very similar to Neville's algorithm except that it is based on the recurrence

$$P_{0...p,m}(t) = \frac{t_m - t}{t_m - t_p} P_{0...p}(t) + \frac{t - t_p}{t_m - t_p} P_{0...p-1,m}(t) \text{ for all } m > p.$$

a. Use this recurrence to give an alternative proof of Theorem 2.1.

  b. Explain how to generate a dynamic programming algorithm for interpolation based on Aitken's recurrence.

  c. Illustrate Aitken's algorithm with a diagram for the cubic case.

2. Implement Neville's algorithm. Experiment with curves of different degrees.

  a. How does changing the order of the control points without altering the order of the nodes affect the shape of the curve?

  b. How does changing the values of the nodes affect the shape of the curve?

  c. Place the nodes at the integers $0,1,...,n$, and graph the curves with control points at $P_k = (k,0)$, $k \neq j$, and $P_j = (j,1)$.

2.4 Uniqueness of Polynomial Interpolants and Taylor's Theorem

Theorem 2.1 asserts that given any arbitrary sequence of points $P_0,...,P_n$ and any collection of distinct parameters $t_0,...,t_n$, there *exists* a polynomial curve $P_{0...n}(t)$ of degree n that interpolates the given points at the specified parameters. Here we are going to show that this polynomial curve is *unique*, extending the result that two points determine a unique line. Notice, however, that uniqueness requires us to specify the nodes as well as the control points. We begin by recalling some simple facts about polynomials.

THEOREM 2.2

Taylor's Theorem

Let $P(t)$ be a polynomial of degree n, and let r be a real number. Then

$$P(t) = P(r) + P'(r)(t-r) + P''(r)\frac{(t-r)^2}{2!} + \cdots + P^{(n)}(r)\frac{(t-r)^n}{n!}.$$

Proof Since $P(t)$ is a polynomial of degree n, there must be constants $c_0,...,c_n$ such that

$$P(t) = c_0 + c_1 t + \cdots + c_n t^n.$$

Let $Q(t) = P(t+r)$. Then

$$Q(t) = c_0 + c_1(t+r) + \cdots + c_n(t+r)^n.$$

Expanding the powers of $(t+r)$ and collecting the coefficients of the powers of t, we see that $Q(t)$ is also a polynomial of degree n in t, so there must be constants $d_0,...,d_n$ such that

$$Q(t) = d_0 + d_1 t + \cdots + d_n t^n.$$

But $P(t) = Q(t-r)$, so by substitution

$$P(t) = d_0 + d_1(t-r) + \cdots + d_n(t-r)^n.$$

Differentiating both sides k times and evaluating at $t = r$ yields $d_k = P^{(k)}(r)/k!$.

COROLLARY 2.3

Let $P(t)$ be a polynomial of degree n. Then r is a root of $P(t)$ if and only if $t - r$ is a factor of $P(t)$.

Proof Let $P(t)$ be a polynomial of degree n. Then by Taylor's Theorem

$$P(t) = P(r) + P'(r)(t - r) + P''(r)\frac{(t - r)^2}{2!} + \cdots + P^{(n)}(r)\frac{(t - r)^n}{n!} \; .$$

Therefore, by inspection, $P(r) = 0$ if and only if $t - r$ is a factor of $P(t)$.

COROLLARY 2.4

Every nonzero polynomial of degree n has at most n roots.

Proof This result is an immediate consequence of Corollary 2.3, since a polynomial of degree n can have at most n linear factors.

COROLLARY 2.5

Let $P(t)$ and $Q(t)$ be two polynomials of degree n that agree at $n + 1$ parameter values. Then $P(t) = Q(t)$.

Proof Let $R(t) = Q(t) - P(t)$. Then $R(t)$ is a polynomial of degree n. Moreover, since $P(t)$ and $Q(t)$ agree at $n + 1$ parameter values, $R(t)$ has $n + 1$ roots. Therefore, by Corollary 2.4, $R(t)$ must be the zero polynomial, so $P(t) = Q(t)$.

THEOREM 2.6

Given affine points $P_0, \ldots, P_n$ and distinct parameters $t_0, \ldots, t_n$, there exists a unique polynomial curve of degree n that interpolates the given points at the specified parameters.

Proof Existence has already been established in Theorem 2.1; it remains to demonstrate uniqueness. Suppose that $P(t)$ and $Q(t)$ are two polynomial curves of degree n that interpolate the given control points at the specified nodes. Then $P(t)$ and $Q(t)$ are polynomials of degree n that agree at the $n + 1$ parameter values $t_0, \ldots, t_n$. Hence by Corollary 2.5, $Q(t) = P(t)$, so the interpolating polynomial is unique.

Exercises

1. Prove that the polynomials $1, (t - r), \ldots, (t - r)^n$ are linearly independent. Conclude that the polynomials $1, (t - r), \ldots, (t - r)^n$ form a basis for the polynomials of degree n and use this fact to provide an alternative proof of Taylor's Theorem.

2. A polynomial $P(t)$ is said to have a root of multiplicity m at the parameter r if $P^{(k)}(r) = 0$, $k = 0,...,m-1$.

   a. Show that a polynomial $P(t)$ has a root of multiplicity m at r if and only if $(t-r)^m$ is a factor of $P(t)$.

   b. Show that every nonzero polynomial of degree n can have no more than n roots counting multiplicities.

3. Let $P(t)$ be a polynomial of degree n, and let $P_{0\ldots n}(t)$ be the polynomial that interpolates the control points $P(t_0),...,P(t_n)$ at the nodes $t_0,...,t_n$. Prove that $P_{0\ldots n}(t) = P(t)$.

4. Let $f(t)$ be a polynomial of degree n and let r be an arbitrary constant.

   a. Using long division of polynomials, show that there is a polynomial $g(t)$ of degree $n-1$ such that $f(t) = (t-r)g(t) + f(r)$.

   b. Using part (a), conclude that $f(r) = 0 \Leftrightarrow t - r$ is a factor of $f(t)$.

5. Let $P(t) = a_n t^n + \cdots + a_1 t + a_0$. Then $P(t)$ interpolates the control points $P_0,...,P_n$ at parameters $t_0,...,t_n$ if and only if

$$a_n t_0{}^n + \cdots + a_1 t_0 + a_0 = P_0$$
$$\vdots \qquad\qquad \vdots$$
$$a_n t_n{}^n + \cdots + a_1 t_n + a_0 = P_n \ .$$

   a. Prove that this system of linear equations in the unknowns $a_0,...,a_n$ has a unique solution by showing that the determinant of the coefficients

$$\begin{vmatrix} t_0^n & \cdots & t_0 & 1 \\ \vdots & \vdots & \vdots & \vdots \\ t_n^n & \cdots & t_n & 1 \end{vmatrix} \neq 0 \ .$$

   (Hint: Replace t_n by t. Show that this determinant is a polynomial of degree n in t by proving that the coefficient of t^n is not zero. Then using the properties of determinants, show that $t_0,...,t_{n-1}$ are n roots of this polynomial. It follows by Corollary 2.4 that t_n cannot also be a root of this polynomial, so the determinant cannot be zero.)

   b. Conclude that $P_{0\ldots n}(t)$ exists and is unique.

2.5   **Lagrange Basis Functions**

So far we have developed a recursive formula and a dynamic programming algorithm for computing the polynomial interpolant $P_{0\ldots n}(t)$. Here we shall develop an explicit formula for this interpolant.

We begin by observing that there must exist polynomials $L_0^n(t \mid t_0,...,t_n),...,L_n^n(t \mid t_0,...,t_n)$—polynomials of degree n in the variable t whose coefficients depend on the nodes $t_0,...,t_n$—such that

$$P_{0...n}(t) = \sum_{k=0}^{n} L_k^n(t \mid t_0,...,t_n)P_k \ .$$

This is clearly true for $n = 1$; in fact (2.5) gives explicit formulas for $L_0^1(t \mid t_0,t_1)$ and $L_1^1(t \mid t_0,t_1)$. Now we proceed by induction on n. Suppose that

$$P_{0...n-1}(t) = \sum_{k=0}^{n-1} L_k^{n-1}(t \mid t_0,...,t_{n-1})P_k$$

$$P_{1...n}(t) = \sum_{k=0}^{n-1} L_k^{n-1}(t \mid t_1,...,t_n)P_{k+1} \ .$$

Then by (2.7)

$$P_{0...n}(t) = \frac{t_n - t}{t_n - t_0} P_{0...n-1}(t) + \frac{t - t_0}{t_n - t_0} P_{1...n}(t) \ ,$$

so substituting the preceding formulas for $P_{0...n-1}(t)$ and $P_{1...n}(t)$ we obtain

$$\sum_{k=0}^{n} L_k^n(t \mid t_0,...,t_n)P_k = \frac{t_n - t}{t_n - t_0} \sum_{k=0}^{n-1} L_k^{n-1}(t \mid t_0,...,t_{n-1})P_k + \frac{t - t_0}{t_n - t_0} \sum_{k=0}^{n-1} L_k^{n-1}(t \mid t_1,...,t_n)P_{k+1}.$$

Equating the coefficients of P_k on both sides of this equation yields the recurrence

$$L_k^n(t \mid t_0,...,t_n) = \frac{t_n - t}{t_n - t_0} L_k^{n-1}(t \mid t_0,...,t_{n-1}) + \frac{t - t_0}{t_n - t_0} L_{k-1}^{n-1}(t \mid t_1,...,t_n) \quad k = 0,...,n, \quad (2.8)$$

where $L_k^n(t \mid t_0,...,t_n)$ is defined to be zero whenever $k < 0$ or $k > n$. Since by the inductive hypothesis $L_k^{n-1}(t \mid t_0,...,t_{n-1})$ and $L_{k-1}^{n-1}(t \mid t_1,...,t_n)$ are both polynomials of degree $n - 1$ in t, it follows from (2.8) that $L_k^n(t \mid t_0,...,t_n)$ must be a polynomial of degree n in t. Our goal is to find explicit formulas for these polynomials and to study their properties.

Let's begin with the cubic case. Consider Figure 2.5, and let's try to calculate, for example, $L_1^3(t \mid t_0,...,t_3)$. The contribution of P_1 to $P_{0123}(t)$ is the sum over all paths from P_1 to $P_{0123}(t)$, where a path is the product of the labels along the arrows. But notice that because of the parallel property all paths from P_1 to $P_{0123}(t)$ produce the same product. In fact, disregarding signs, this product is just $(t - t_0)(t - t_2)(t - t_3)$. Since we have omitted the normalization in the denominator, it follows that $L_1^3(t \mid t_0,...,t_3)$ is actually some constant multiple of this product—that is,

$$L_1^3(t \mid t_0,...,t_3) = c_1(t - t_0)(t - t_2)(t - t_3) \ .$$

Similarly, we find that

$$L_0^3(t \mid t_0,...,t_3) = c_0(t - t_1)(t - t_2)(t - t_3)$$

$$L_2^3(t \mid t_0,...,t_3) = c_2(t - t_0)(t - t_1)(t - t_3)$$

$$L_3^3(t \mid t_0,...,t_3) = c_3(t - t_0)(t - t_1)(t - t_2).$$

There is an obvious pattern here. Each polynomial $L_k^3(t \mid t_0,...,t_3)$ contains three of the four factors $(t - t_0), (t - t_1), (t - t_2), (t - t_3)$, and the missing factor is $(t - t_k)$.

It remains to determine the values of the constant coefficients c_j. This is easy to do because

$$P_j = P_{0123}(t_j) = \sum_{k=0}^{3} L_k^3(t_j \mid t_0,...,t_3) P_k.$$

But we have seen that if $j \neq k$, then $L_k^3(t \mid t_0,...,t_3)$ contains the factor $(t - t_j)$; hence

$$L_k^3(t_j \mid t_0,...,t_3) = 0 \qquad\qquad j \neq k.$$

This leaves us with

$$P_j = L_j^3(t_j \mid t_0,...,t_3) P_j,$$

so we must have

$$L_j^3(t_j \mid t_0,...,t_3) = 1.$$

Now we have one equation with one unknown, so we can easily solve for c_j. For example, if $j = 1$ we have

$$c_1(t_1 - t_0)(t_1 - t_2)(t_1 - t_3) = 1,$$

so we obtain

$$c_1 = \frac{1}{(t_1 - t_0)(t_1 - t_2)(t_1 - t_3)}.$$

Thus

$$L_1^3(t \mid t_0,...,t_3) = \frac{(t - t_0)(t - t_2)(t - t_3)}{(t_1 - t_0)(t_1 - t_2)(t_1 - t_3)}.$$

Notice that the denominator is just the numerator evaluated at $t = t_1$. Using this trick of evaluating the numerator at $t = t_k$ to find the denominator and recalling that the numerator is missing the factor $(t - t_k)$, we obtain the general formula

$$L_k^3(t \mid t_0,...,t_3) = \frac{\prod_{j \neq k}(t - t_j)}{\prod_{j \neq k}(t_k - t_j)}.$$

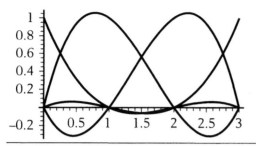

Figure 2.8 The four cubic Lagrange basis functions for the nodes $t_k = k$, $k = 0,...,3$.

These functions are called the *cubic Lagrange basis functions*. We illustrate the cubic Lagrange basis functions for the nodes $t_k = k$, $k = 0,...,3$, in Figure 2.8.

We shall now show that this same analysis and a similar formula are valid for any degree, not just for $n = 3$. Let's generalize what we have discovered so far. We begin with two important observations connecting the polynomials $L_k^n(t \mid t_0,...,t_n)$ with the structure of the triangle diagram for Neville's algorithm.

1. $L_k^n(t \mid t_0,...,t_n) =$ the sum over all paths from P_k to $P_{0...n}(t)$, where a path is the product of the labels along the arrows.

2. All paths from P_k to $P_{0...n}(t)$ are identical, up to constant multiples.

Statement 1 is just the observation that the contribution of P_k to $P_{0...n}(t)$ is the sum of all paths from P_k to $P_{0...n}(t)$, and since $L_k^n(t \mid t_0,...,t_n)$ is the coefficient of P_k in $P_{0...n}(t)$, it must represent the sum of all these paths. Statement 2 is a simple consequence of the parallel property of Neville's algorithm, which was discussed in Section 2.3.

Any path from P_k to $P_{0...n}(t)$ must take exactly k left turns and $n - k$ right turns. By the parallel property the labels on the k left turns are identical to the first k labels—counting down from $P_{0...n}(t)$—on the right edge of Neville's triangle, and by construction these labels are $(t - t_0),(t - t_1),...,(t - t_{k-1})$. Similarly, the labels on the $n - k$ right turns are identical to the first $n - k$ labels—counting down from $P_{0...n}(t)$—on the left edge of Neville's triangle, and by construction these labels are $(t - t_n),(t - t_{n-1}),...,(t - t_{k+1})$. Multiplying all these labels together, we find that any path from P_k to $P_{0...n}(t)$ is a constant multiple of $\prod_{j \neq k}(t - t_j)$. Thus

$$L_k^n(t \mid t_0,...,t_n) = c_k \prod_{j \neq k}(t - t_j) \qquad k = 0,...,n. \tag{2.9}$$

It remains to find the constants c_k.

We proceed again exactly as in the cubic case. We know that

$$P_k = P_{0...n}(t_k) = \sum_{j=0}^{n} L_j^n(t_k \mid t_0,...,t_n)P_j. \tag{2.10}$$

But by (2.9) if $j \neq k$, then $L_j^n(t_k \mid t_0,...,t_n)$ contains the factor $(t - t_k)$; hence

$$L_j^n(t_k \mid t_0,...,t_n) = 0 \qquad j \neq k,$$

so by (2.10) we must have

$$L_k^n(t_k \mid t_0,...,t_n) = 1.$$

Thus from (2.9) we obtain

$$c_k = \frac{1}{\prod_{j \neq k}(t_k - t_j)}.$$

Therefore, as in the cubic case,

$$L_k^n(t \mid t_0,...,t_n) = \frac{\prod_{j \neq k}(t - t_j)}{\prod_{j \neq k}(t_k - t_j)} \qquad k = 0,...,n.$$

Notice again that the denominator is simply the numerator evaluated at $t = t_k$.

The polynomials

$$L_0^n(t \mid t_0,...,t_n),...,L_n^n(t \mid t_0,...,t_n)$$

are called the *Lagrange basis functions* for the nodes $t_0,...,t_n$. These functions play a fundamental role in the theory of polynomial interpolation. The following theorem and corollary summarize their principal properties.

THEOREM 2.7

Properties of Lagrange Basis Functions

1. $L_k^n(t \mid t_0,...,t_n) = \dfrac{\prod_{j \neq k}(t - t_j)}{\prod_{j \neq k}(t_k - t_j)} \qquad k = 0,...,n$. (2.11)

2. $L_k^n(t \mid t_0,...,t_n)$ is a polynomial of degree n.

3. $L_k^n(t_j \mid t_0,...,t_n) = 0 \qquad j \neq k$ (2.12)
   $\qquad\qquad\qquad\quad = 1 \qquad j = k.$

4. If $P(t)$ is a polynomial of degree n, then

$$P(t) = \sum_{k=0}^{n} P(t_k) L_k^n(t \mid t_0,...,t_n).$$ (2.13)

5. $P_{0 \cdots n}(t) = \sum_{k=0}^{n} L_k^n(t \mid t_0,...,t_n) P_k.$ (2.14)

6. $\sum_{k=0}^{n} L_k^n(t \mid t_0,...,t_n) \equiv 1.$ (2.15)

Proof Property 1 is the definition of $L_k^n(t \mid t_0,...,t_n)$, and Property 2 is immediate from this definition since the numerator has n linear factors. Property 3 follows by substituting $t = t_j$ into the definition of $L_k^n(t \mid t_0,...,t_n)$ and observing that

    i.   the numerator of $L_k^n(t \mid t_0,...,t_n)$ has a factor of $(t - t_j)$ if $j \neq k$

    ii. the denominator of $L_j^n(t \mid t_0,...,t_n)$ is its numerator evaluated at $t = t_j$

To prove Property 4, let $P(t)$ be an arbitrary polynomial of degree n and define

$$Q(t) = \sum_{k=0}^{n} P(t_k) L_k^n(t \mid t_0,...,t_n).$$

Then by Property 3, $Q(t_j) = P(t_j)$, $j = 0,...,n$. Hence by the uniqueness of the polynomial interpolant (Theorem 2.6), $P(t) = Q(t)$. Property 5 is an immediate consequence of Property 4, since by Property 4

$$P_{0 \cdots n}(t) = \sum_{k=0}^{n} L_k^n(t \mid t_0,...,t_n) P_{0 \cdots n}(t_k) = \sum_{k=0}^{n} L_k^n(t \mid t_0,...,t_n) P_k.$$

Finally, Property 6 is the special case of Property 4 where $P(t) \equiv 1$.

COROLLARY 2.8 The polynomials $L_0^n(t \mid t_0,...,t_n),...,L_n^n(t \mid t_0,...,t_n)$ form a basis for all polynomials of degree n.

Proof We need to show that the polynomials $L_0^n(t \mid t_0,...,t_n),...,L_n^n(t \mid t_0,...,t_n)$ are linearly independent and that they span the space of all polynomials of degree n. By (2.13) they span the space of polynomials of degree n, so it remains only to verify that these functions are linearly independent. Suppose then that

$$\sum_{k=0}^{n} c_k L_k^n(t \mid t_0,...,t_n) = 0.$$

Substituting $t = t_j$ on the left-hand side and applying (2.12) yields $c_j = 0$. Hence the polynomials $L_0^n(t \mid t_0,...,t_n),...,L_n^n(t \mid t_0,...,t_n)$ are linearly independent.

Several parts of Theorem 2.7 stand out as exceptionally important. Equations (2.11) and (2.14) assert that we have indeed succeeded in finding an explicit formula for the polynomial interpolant. Equation (2.15) is critical because otherwise Equation (2.14) would not make sense in affine space. Finally, notice that the proof of Corollary 2.8 relies heavily upon (2.12). The conditions in Equation (2.12) are called the *cardinal conditions*. These equations are fundamental to many interpolation schemes (see Exercise 4).

Exercises

1. Prove the identity $(x - t)^n = \sum_{k=0}^{n} L_k^n(t \mid t_0,...,t_n)(x - t_k)^n.$

2. Prove that

$$\sum_{k=0}^{n} L_k^n(t \mid t_0,...,t_n) \equiv 1$$

by choosing $P_k = 1$, $k = 0,...,n$, as the control points in Neville's algorithm. (Hint: By (2.14),

$$P_{0\cdots n}(t) = \sum_{k=0}^{n} L_k^n(t \mid t_0,...,t_n),$$

so it is enough to prove that $P_{0\cdots n}(t) \equiv 1$. Now observe by induction from the bottom level up that when all the control points are set to one, the value in each node of Neville's algorithm is identically one (see Figure 2.9). Hence the value at the apex must be one.)

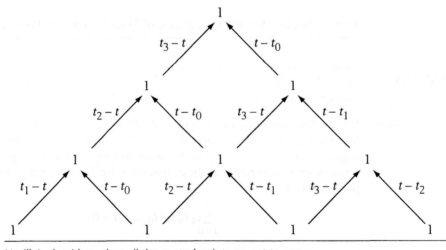

Figure 2.9 Neville's algorithm when all the control points are set to one.

3. Prove that

$$\sum_{k=0}^{n} L_k^n(t \mid t_0,...,t_n) \equiv 1$$

by using the cardinal conditions (Equation (2.12)) and invoking Corollary 2.5.

4. Suppose that $D_0^n(t),...,D_n^n(t)$ are a collection of functions, not necessarily polynomials, that satisfy the cardinal conditions at $t_0,...,t_n$—that is,

$$D_k^n(t_j) = 0 \qquad j \neq k$$
$$= 1 \qquad j = k \; .$$

Let

$$Q(t) = \sum_{k=0}^{n} D_k^n(t) P_k.$$

Show that $Q(t)$ interpolates the points $P_0,...,P_n$ at the parameters $t_0,...,t_n$.

5. Give necessary and sufficient conditions on the control points for the curve $P_{0...n}(t)$ to collapse to a single point. Justify your answer.

6. Give necessary and sufficient conditions on the control points for the curve $P_{0...n}(t)$ to collapse to a straight line. Justify your answer.

7. Consider the curve $P(t) = \sum_k L_k^n(t \mid t_0,...,t_n) P_k$. What are the control points of this curve relative to

   a. the degree $n + 1$ Lagrange basis with respect to the nodes $t_0,...,t_n,t_{n+1}$;

   b. the degree n Lagrange basis with respect to the nodes $s_0,...,s_n$.

8. Let $\tau_k = at_k + b$, $k = 0,...,n$, for some fixed constants $a > 0$ and b. Show that

$$L_k^n(at + b \mid \tau_0,...,\tau_n) = L_k^n(t \mid t_0,...,t_n).$$

Compare this result to Exercise 4 of Section 2.2.

9. Prove that

   a. $t = \sum_k t_k L_k^n(t \mid t_0,...,t_n)$

   b. $t^p = \sum_k t_k^p L_k^n(t \mid t_0,...,t_n)$, $p = 0,...,n$.

10. Let $P(t)$ be a polynomial curve. Define $Graph(P) = (t, P(t))$. Show that if $P_0,...,P_n$ are the control points for $P(t)$ relative to the Lagrange basis $L_0^n(t \mid t_0,...,t_n),...,L_n^n(t \mid t_0,...,t_n)$, then $(t_0,P_0),...,(t_n,P_n)$ are the control points for $Graph(P)$ relative to the same Lagrange basis.

2.6 Computational Techniques for Lagrange Interpolation

Equation (2.11) is an explicit formula for the Lagrange basis functions, but these basis functions can also be computed by a dynamic programming algorithm. Since by (2.14)

$$P_{0...n}(t) = \sum_{k=0}^{n} L_k^n(t \mid t_0,...,t_n) P_k,$$

if we choose the control points

$$P_k = 0 \qquad k \neq j$$
$$= 1 \qquad k = j$$

then we get

$$P_{0\cdots n}(t) = L_j^n(t \mid t_0,\ldots,t_n).$$

Thus the dynamic programming algorithm for $P_{0\cdots n}(t)$ becomes a dynamic programming algorithm for $L_j^n(t \mid t_0,\ldots,t_n)$ when the jth control point is one and all the remaining control points are zero. We illustrate this *up recurrence* for $L_2^3(t \mid t_0,\ldots,t_3)$ in Figure 2.10. This up recurrence is yet another manifestation of the observation we made in the previous section that $L_k^n(t \mid t_0,\ldots,t_n)$ is the sum over all paths from P_k at the base of the triangle to $P_{0\cdots n}(t)$ at the apex.

This insight about paths leads to yet another recurrence for the polynomials $L_0^n(t \mid t_0,\ldots,t_n),\ldots,L_n^n(t \mid t_0,\ldots,t_n)$. Paths from the base to the apex of the triangle are identical to paths from the apex to the base. Thus we can compute all the Lagrange basis functions simultaneously by starting with a one at the apex, reversing all the arrows in Neville's algorithm, and collecting the results at the base of the triangle. The Lagrange basis functions emerge at the base because the functions at the base represent all paths from the apex to the base. We illustrate this *down recurrence* for the cubic Lagrange basis functions in Figure 2.11. Notice that although the Lagrange basis functions of degree n lie at the base of the triangle, the functions at nodes in intermediate levels are not Lagrange basis functions of lower degree.

Both the up recurrence and the down recurrence are $O(n^2)$ algorithms. However, we need to run the up recurrence once for each basis function, so naively the up recurrence is $O(n^3)$ if we use it to compute all the basis functions. If we trim away

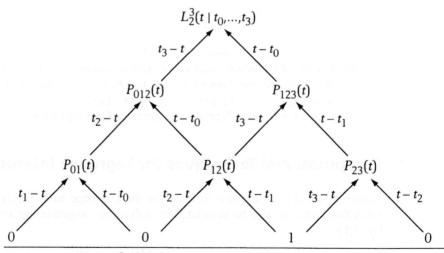

Figure 2.10 The up recurrence for $L_2^3(t \mid t_0,\ldots,t_3)$. Arrows emerging from a zero may be trimmed away to simplify the algorithm.

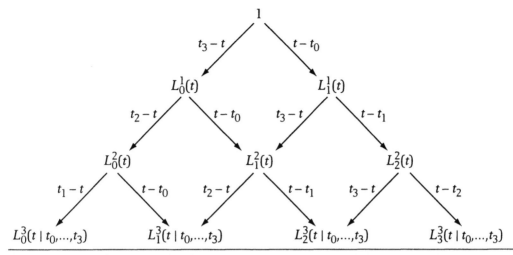

Figure 2.11 The down recurrence for the cubic Lagrange basis functions. Notice, for example, that the intermediate function $L_1^2(t)$ is not a degree 2 Lagrange basis function.

the zeros, the up recurrence is $O(n)$ for each basis function and hence only $O(n^2)$ for all the basis functions. Nevertheless, the up recurrence is still less efficient than the down recurrence (see Exercise 4).

If the control points lie in an affine space of dimension greater than one, Neville's algorithm for $P_{0\ldots n}(t)$ is somewhat inefficient because it must be applied once for each coordinate function. It is often faster to use the down recurrence once to compute all the Lagrange basis functions and then to multiply $L_k^n(t \mid t_0,\ldots,t_n)$ and P_k directly to compute

$$P_{0\ldots n}(t) = \sum_{k=0}^{n} L_k^n(t \mid t_0,\ldots,t_n)P_k.$$

This approach uses a single $O(n^2)$ algorithm to calculate the basis functions and $O(n)$ multiplications to generate each coordinate, rather than $O(n^2)$ computations for each coordinate.

Utilizing the down recurrence in this fashion still generates an $O(n^2)$ evaluation algorithm for $P_{0\ldots n}(t)$. There is, however, an $O(n)$ evaluation algorithm due to Joe Warren that takes advantage of the special structure of the Lagrange basis functions. Whereas Neville's algorithm lies on a triangle, Warren's algorithm lies on a ladder, so we shall call it the *ladder algorithm*.

Label the arrows on the left side of the ladder with the functions $t - t_0,\ldots,t - t_{n-1}$, and on the right side with the functions $t - t_1,\ldots,t - t_n$. Label the rungs with the constant values

$$Q_k = \frac{P_k}{\prod_{j \neq k}(t_k - t_j)} \qquad k = 0,\ldots,n,$$

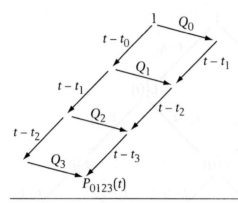

Figure 2.12 The ladder algorithm for $P_{0123}(t)$.

and place a one at the top of the ladder. Values at any node in the ladder are computed in the usual way by multiplying the arrows that enter the node by the values at the nodes from which they emerge and adding the results. Tracing through the paths in the ladder, it is easy to see that

$$P_{0\cdots n}(t) = \sum_{k=0}^{n} L_k^n(t \mid t_0,\ldots,t_n) P_k$$

indeed emerges at the bottom of the ladder (see Figure 2.12). This ladder algorithm uses only $3n + 1$ multiplications for each coordinate function—n along each side of the ladder and $n + 1$ along the rungs.

Exercises

1. Give an example to show that the functions at intermediate levels of the down recurrence are not Lagrange basis functions.

2. a. Show that the functions on every level of the down recurrence sum to one.

   b. Conclude that $\sum_{k=0}^{n} L_k^n(t \mid t_0,\ldots,t_n) \equiv 1.$

3. Define the polynomials $p_1(t),\ldots,p_n(t)$ and $q_0(t),\ldots,q_{n-1}(t)$ recursively by

$$p_n(t) = t - t_n$$
$$p_k(t) = (t - t_k) p_{k+1}(t) \qquad 1 \le k \le n-1$$

$$q_0(t) = t - t_0$$
$$q_k(t) = (t - t_k) q_{k-1}(t) \qquad 1 \le k \le n-1 \ .$$

a. Show that the numerators of the Lagrange basis functions $L_k^n(t \mid t_0,...,t_n)$ are given by

$$\Lambda_k^n(t \mid t_0,...,t_n) = p_{k+1}(t)q_{k-1}(t) \qquad k = 0,...,n\,.$$

b. As in the ladder algorithm, let

$$Q_k = \frac{P_k}{\underset{j \neq k}{\prod}(t_k - t_j)} \qquad\qquad k = 0,...,n.$$

Show that

$$P_{0\cdots n}(t) = \sum_{k=0}^{n} \Lambda_k^n(t \mid t_0,...,t_n)Q_k\,.$$

c. Explain why the preceding evaluation algorithm for $P_{0\cdots n}(t)$ is $O(n)$.

d. Under what circumstances is this evaluation algorithm faster than the ladder algorithm?

4. Show that if we trim away zeros, the up recurrence is identical to the ladder recurrence for $L_k^n(t \mid t_0,...,t_n)$. Conclude that the up recurrence is $O(n)$ for one basis function and $O(n^2)$ for all the basis functions. Explain why the up recurrence is still less efficient than the down recurrence.

5. Generate the dynamic programming algorithm for the recurrence for $L_k^n(t \mid t_0,...,t_n)$ given in (2.8). Is this algorithm the same as either the up recurrence or the down recurrence or is it yet another recurrence?

6. Implement the ladder algorithm. Run test cases comparing the speed of the ladder algorithm to the speed of Neville's algorithm.

7. Let $t_1,...,t_{2n}$ be fixed parameters, and define functions $\Psi_k^n(t)$ by setting

$$\Psi_k^n(t) = (t - t_{k+1})\cdots(t - t_{k+n}) \qquad k = 0,...,n\cdot$$

Develop a ladder algorithm for evaluating expressions of the form

$$E(t) = \sum_{k=0}^{n} c_k \Psi_k^n(t)\,.$$

8. Develop an algorithm for differentiating
   a. the ladder algorithm
   b. Neville's algorithm

2.7 Rational Lagrange Curves

Lagrange interpolation generates polynomial curves. But many simple curves in computer graphics and computer-aided geometric design cannot be represented exactly by parametric polynomials. Indeed, even the circle has no polynomial parametrization. We can, however, try to approximate a circle with polynomials by interpolating points along the circle. Figure 2.13 shows polynomial approximations of degree 2 and degree 4 to a semicircle, and Figure 2.14 shows a polynomial approximation of degree 6.

So one way to proceed for a curve that is not a polynomial is to approximate the curve with a polynomial by interpolating more and more points along the curve. In many, but not all, cases this strategy works well, but the degree of the approximating polynomial may need to be quite high. For example, we see from Figures 2.13 and 2.14 that we need to use a degree 6 polynomial to achieve a really good Lagrange

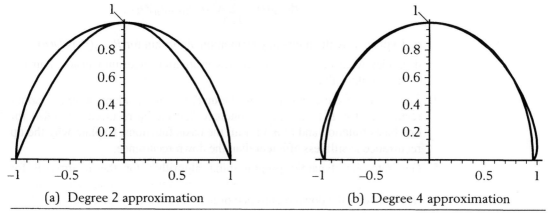

(a) Degree 2 approximation   (b) Degree 4 approximation

Figure 2.13 Polynomial approximations to the semicircle: (a) a degree 2 approximation interpolating three evenly spaced points, and (b) a degree 4 approximation interpolating five evenly spaced points. The degree 2 approximation is a parabola, which undershoots the semicircle, while the degree 4 approximation overshoots the semicircle near the end points.

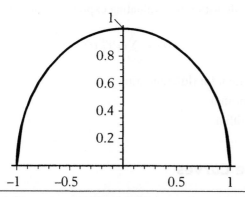

Figure 2.14 A degree 6 approximation to the semicircle using a Lagrange polynomial that interpolates seven evenly spaced points. Here it is difficult to see any difference between the semicircle and the approximating polynomial.

approximation for the semicircle. If we should want to approximate more than half the circle, we would have to use an even higher-degree polynomial. But high-degree polynomials are unwieldy to compute. Worse yet, in many cases high-degree interpolatory polynomials introduce unwanted oscillations (see, for example, Exercise 1).

For the circle and for many other curves, there is a better solution. Although the circle does not possess a polynomial parametrization, the circle does have a rational parametrization

$$x(t) = \frac{2t}{1+t^2} \qquad y(t) = \frac{1-t^2}{1+t^2},$$

since it is easy to verify that $x^2(t) + y^2(t) = 1$. Thus if we want to represent the circle exactly, we can resort to rational functions.

To use Lagrange interpolation to represent the circle exactly, lift the circle from affine space to Grassmann space by treating the denominator in affine space as a mass in Grassmann space. (Now would be a good time to review Section 1.1.3 on Grassmann space.) Lifting replaces rational curves in affine space by polynomial curves in Grassmann space. The circle lifts to the polynomial curve

$$P(t) = \left(2t, 1 - t^2, 1 + t^2\right)$$

in Grassmann space, and this polynomial curve projects to the rational curve

$$R(t) = \left(\frac{2t}{1+t^2}, \frac{1-t^2}{1+t^2}\right)$$

in affine space. So all we need to do to apply Lagrange interpolation is to find the mass-points $(m_k P_k, m_k)$ in Grassmann space that are the control points for the polynomial curve $P(t)$ and then project onto affine space to retrieve the circle $R(t)$.

To carry out this computation explicitly, observe that $P(t)$ is a curve of degree 2, so we can use Lagrange polynomials of degree 2 to represent $P(t)$. We still need to choose our nodes. Any nodes will do, but since $R(t)$ sweeps out the upper half circle for $-1 \leq t \leq 1$, we shall use the nodes $(-1, 0, 1)$. Now we need to find the mass-points $(m_k P_k, m_k)$, $k = 0, 1, 2$, so that

$$P_{012}(t) = \sum_{k=0}^{2} (m_k P_k, m_k) L_k^2(t \mid -1, 0, 1) = \left(2t, 1 - t^2, 1 + t^2\right) = P(t).$$

By (2.13) we must choose

$$\begin{aligned}
(m_0 P_0, m_0) &= P(-1) = (-2, 0, 2) \\
(m_1 P_1, m_1) &= P(0) = (0, 1, 1) \qquad\qquad (2.16) \\
(m_2 P_2, m_2) &= P(1) = (2, 0, 2) \ .
\end{aligned}$$

Thus we can read off the masses and the control points from (2.16):

$$m_0 = 2, \ m_1 = 1, \ m_2 = 2$$
$$P_0 = (-1,0), \ P_1 = (0,1), \ P_2 = (1,0) \ .$$

In Grassmann space, we obtain the polynomial curve

$$P(t) = \sum_{k=0}^{2}(m_k P_k, m_k)L_k^2(t \mid -1,0,1);$$

projecting into affine space generates the circle

$$R(t) = \frac{\sum\limits_{k=0}^{2} m_k P_k L_k^2(t \mid -1,0,1)}{\sum\limits_{k=0}^{2} m_k L_k^2(t \mid -1,0,1)} \ .$$

Figure 2.15 illustrates the semicircle with its three control points using rational Lagrange interpolation.

In general, we define a *rational Lagrange curve* in affine space to be the projection of a Lagrange polynomial curve

$$P(t) = \sum_{k=0}^{n}(m_k P_k, m_k)L_k^n(t \mid t_0,\ldots,t_n)$$

in Grassmann space. In affine space $P(t)$ projects to the rational curve

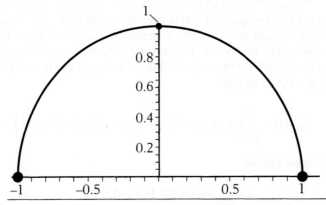

Figure 2.15 The semicircle as a degree 2 rational Lagrange curve. The three mass-points that serve as the control points are shown. The size of the dots indicates the relative masses of the points: the mass is two at $(-1,0)$ and $(1,0)$, while the mass is one at $(0,1)$.

$$R_{0\cdots n}(t) = \frac{\sum_{k=0}^{n} m_k P_k L_k^n(t \mid t_0,\ldots,t_n)}{\sum_{k=0}^{n} m_k L_k^n(t \mid t_0,\ldots,t_n)} .$$

Thus $R_{0\cdots n}(t)$ is called a rational Lagrange curve. Just like Lagrange interpolating polynomials, we have the following interpolation result for rational Lagrange curves.

THEOREM $R_{0\cdots n}(t_j) = P_j, \quad j = 0,\ldots,n.$
2.9

Proof This result follows because the Lagrange curve

$$P(t) = \sum_{k=0}^{n}(m_k P_k, m_k) L_k^n(t \mid t_0,\ldots,t_n)$$

in Grassmann space interpolates the mass-point $(m_j P_j, m_j)$ at the parameter t_j—that is, $P(t_j) = (m_j P_j, m_j)$. Projecting into affine space, we get $R_{0\cdots n}(t_j) = P_j$. We can also obtain this result directly by observing that

$$R_{0\cdots n}(t_j) = \frac{\sum_{k=0}^{n} m_k P_k L_k^n(t_j \mid t_0,\ldots,t_n)}{\sum_{k=0}^{n} m_k L_k^n(t_j \mid t_0,\ldots,t_n)} = \frac{m_j P_j}{m_j} = P_j.$$

Mass does not affect interpolation at the control points, but mass does alter the shape of the interpolating curve. Thus the masses in the rational Lagrange formulation serve as shape parameters (see, for example, Figure 2.16). These shape parameters are not always benign; even moderate changes to a single mass can produce

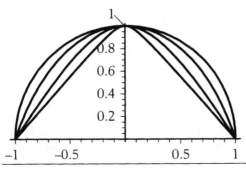

Figure 2.16 The effect of mass on the shape of a rational Lagrange curve. The semicircle has a unit mass at (0,1). Here we enlarge this mass to the values 1.5,3,10. The effect on the curve is to tense it towards the lines joining its control points and to increase the curvature at (0,1).

extreme changes in the shape of a curve and may even introduce singularities. Large changes to a single mass may create cusps and higher-order irregularities (see Exercise 8). Thus care must be taken when altering the masses of a rational Lagrange curve.

Indeed, although a rational Lagrange curve is continuous everywhere except at the parameter values where the denominator vanishes, the limit curve as any single mass m_j approaches infinity is not a continuous curve. In fact, the limit curve collapses to the control points, since

$$\lim_{m_j \to \infty} R_{0\cdots n}(t_k) = P_k, \ k = 0,...,n$$

$$\lim_{m_j \to \infty} R_{0\cdots n}(t) = P_j \text{ for all } t \neq t_k, \ k = 0,...,n \, .$$

The first limit is valid because $R_{0...n}(t_k) = P_k$ independent of the value of the mass m_j; the second limit holds because

$$\lim_{m_j \to \infty} R_{0...n}(t) = \frac{\lim_{m_j \to \infty} \sum_{k=0}^n \frac{m_k}{m_j} P_k L_k^n(t \mid t_0,...,t_n)}{\lim_{m_j \to \infty} \sum_{k=0}^n \frac{m_k}{m_j} L_k^n(t \mid t_0,...,t_n)} = \frac{P_j L_j^n(t \mid t_0,...,t_n)}{L_j^n(t \mid t_0,...,t_n)} = P_j \, .$$

Thus we should expect a rational Lagrange curve to behave very strangely as we continue to increase the values of the masses (see Exercises 2, 3, 8, and 9).

In a rational Lagrange curve some masses m_k may be set to zero (see Exercise 6). When $m_k = 0$, the mass-point $(m_k P_k, m_k)$ is replaced by a vector $(v_k, 0)$ in Grassmann space. Thus $m_k = 0$ does not necessarily imply that $v_k = 0$. Typically, when $m_k = 0$, the rational Lagrange curve $R_{0...n}(t)$ has a singularity at $t = t_k$ because when $m_k = 0$ all the Lagrange basis functions in the denominator vanish at $t = t_k$. If, however, both $m_k = 0$ and $v_k = 0$, then there is no longer a singularity at $t = t_k$ because the factor $t - t_k$ appears in every term in both the numerator and denominator and therefore can be canceled. The resulting rational curve still interpolates the control points P_j, $j \neq k$.

Computing points on a rational Lagrange curve is no different from computing points on a Lagrange interpolating polynomial. Although our analysis of interpolating polynomials was carried out entirely in affine space, the same analysis holds as well in any vector space because all we require are affine combinations, and vector spaces permit arbitrary linear combinations. Thus to compute values on a rational Lagrange curve, we can apply Neville's algorithm or the ladder algorithm in Grassmann space and then simply divide by the mass to project to the corresponding point in affine space.

However, if the mass of a Lagrange curve in Grassmann space is ever zero, then the projection of the curve into affine space is not continuous. If the corresponding vector in Grassmann space is not zero, we can avoid these discontinuities by projecting the curve instead into projective space. Therefore, for a rational Lagrange curve, the control points reside in Grassmann space, but the curve itself may lie in projective space.

Exercises

1. Consider the Gaussian curve $G(t) = (t, e^{-t^2})$.

   a. Plot $G(t)$ on the interval $[-1,1]$.

   b. Approximate $G(t)$ on the interval $[-1,1]$ using polynomial Lagrange interpolation with 5, 7, 9, and 11 evenly spaced points, and compare your results to the original curve.

   c. Observe that Lagrange interpolation introduces oscillations that are not present in the curve $G(t)$.

2. Consider the rational quadratic parametrization of the semicircle given by the masses and control points in (2.16).

   a. Plot the point with $t = .99$ for larger and larger values of the mass at $(0,1)$.

   b. What do you observe? Explain what is happening.

3. Consider the rational quadratic parametrization of the semicircle given by the masses and control points in (2.16).

   a. Plot some curves where the mass at $(-1,0)$ is increased and the masses at $(0,1)$ and $(1,0)$ are left unchanged.

   b. Plot some curves where the masses at $(-1,0)$ and $(1,0)$ are increased and the mass at $(0,1)$ is left unchanged.

   c. Explain why these two effects are different.

4. Experiment with altering the masses in a rational Lagrange curve.

   a. What are the local and global effects of altering a single mass?

   b. What is the effect of a negative mass?

   c. What happens if all the masses are changed simultaneously?

5. Consider the ellipse with the implicit equation $x^2/a^2 + y^2/b^2 = 1$.

   a. Verify that this ellipse has the rational parametrization

   $$R(t) = \left(\frac{2at}{1+t^2}, \frac{b(1+t^2)}{1+t^2} \right).$$

   b. Find the control points and weights of this ellipse for the nodes $(-1,0,1)$.

   c. Use Neville's algorithm to draw this ellipse for $a = 1$ and $b = 3$.

6. Consider the hyperbola with the implicit equation $x^2/a^2 - y^2/b^2 = 1$.

   a. Verify that this hyperbola has the rational parametrization

   $$R(t) = \left(\frac{a(1+t^2)}{1-t^2}, \frac{2bt}{1-t^2} \right).$$

   b. Find the control points and weights of this hyperbola for the nodes

       i. $(-0.5, 0, 0.5)$

    ii. $(-4,-3,-2)$

    iii. $(-1,0,1)$

  c. Use Neville's algorithm to draw different parts of this hyperbola for $a = 1$ and $b = 3$.

7. Consider a rational Lagrange curve with nodes $t_0,\ldots,t_n$ and control points $(m_0 P_0, m_0),\ldots,(m_n P_n, m_n)$. What does the limit curve look like if two or more masses are allowed to increase simultaneously?

8. Let $R(t)$ be a rational Lagrange curve with nodes $t_0,\ldots,t_n$ and control points $(m_0 P_0, m_0),\ldots,(m_n P_n, m_n)$. Define

$$P(t) = \sum_{k=0}^{n} m_k P_k L_k^n(t \mid t_0,\ldots,t_n)$$

$$w(t) = \sum_{k=0}^{n} m_k L_k^n(t \mid t_0,\ldots,t_n) \ .$$

Then $R(t) = P(t)/w(t)$ or equivalently $w(t)R(t) = P(t)$.

  a. Show that $R'(t_j) = \dfrac{P'(t_j) - w'(t_j)P_j}{m_j}$.

  b. Conclude that $\lim_{m_j \to \infty} R'(t_j) = 0$.

  c. Prove that $\sum_{i=0}^{k} \binom{k}{i} w^{(i)}(t) R^{(k-i)}(t) = P^{(k)}(t)$.

  d. Using induction and part (c), show that $\lim_{m_j \to \infty} R^{(k)}(t_j) = 0$ for all $k \geq 0$.

  e. Conclude that as the mass m_j gets large while the other masses are held fixed, the point at $R(t_j) = P_j$ becomes highly irregular.

9. Let $R(t)$ be a rational Lagrange curve with nodes $t_0,\ldots,t_n$ and control points $(m_0 P_0, m_0),\ldots,(m_n P_n, m_n)$.

  a. Using arguments similar to those in Exercise 8 show that

$$\lim_{m_j \to \infty} R^{(k)}(t) = 0 \text{ for all } k \geq 0, \text{ whenever } t \neq t_k, \ k \neq j.$$

  b. Explain how this result is consistent with the results in the text and in Exercise 8.

  c. Explore what happens to the derivative if two or more masses are allowed to increase simultaneously.

10. Given a collection of nodes $t_0,\ldots,t_n$ and masses $m_0,\ldots,m_n$, define

$$R_k(t \mid t_0,\ldots,t_n) = \frac{m_k L_k^n(t \mid t_0,\ldots,t_n)}{\sum_{j=0}^{n} m_j L_j^n(t \mid t_0,\ldots,t_n)}, \quad k = 0,\ldots,n.$$

Show that these functions behave like rational Lagrange basis functions. In particular,

a. $\displaystyle\sum_{k=0}^{n} R_k(t \mid t_0,\ldots,t_n) \equiv 1$

b. $R_k(t_i \mid t_0,\ldots,t_n) = 0 \quad i \neq k$
$$= 1 \quad i = k$$

c. $\displaystyle R_{0\cdots n}(t) = \sum_{k=0}^{n} R_k(t \mid t_0,\ldots,t_n) P_k.$

2.8 Fast Fourier Transform

Before we move on to study the interpolation problem for surfaces, we pause briefly to consider another important application of univariate Lagrange interpolation: Fast Fourier Transform (FFT). One purpose of FFT is to perform fast multiplication of polynomials written in terms of the standard monomial basis $1,t,\ldots,t^n$. We shall now examine how this is done.

Consider two polynomials of degree n

$$f(t) = \sum_{k=0}^{n} a_k t^k \text{ and } g(t) = \sum_{k=0}^{n} b_k t^k.$$

Multiplying these polynomials in the usual fashion,

$$f(t)g(t) = \sum_{k=0}^{2n} \left(\sum_{i+j=k} a_i b_j \right) t^k,$$

would require computing $O(n^2)$ products, since we would need to compute every product $a_i b_j$, $i,j = 0,\ldots,n$. Suppose, however, that we had expressed these two polynomials with respect to a Lagrange basis of degree $2n$:

$$L_0^{2n}(t \mid t_0,\ldots,t_{2n}),\ldots,L_{2n}^{2n}(t \mid t_0,\ldots,t_{2n}).$$

Then by (2.13)

$$f(t) = \sum_{k=0}^{2n} f(t_k) L_k^{2n}(t \mid t_0,\ldots,t_{2n}), \quad g(t) = \sum_{k=0}^{2n} g(t_k) L_k^{2n}(t \mid t_0,\ldots,t_{2n}),$$

and

$$f(t)g(t) = \sum_{k=0}^{2n} f(t_k)g(t_k) L_k^{2n}(t \mid t_0,\ldots,t_{2n}).$$

Thus when polynomials are written in terms of the Lagrange basis, the product requires only $O(n)$ multiplications, since we need only compute the $2n+1$ products $f(t_k)g(t_k)$, $k = 0,...,2n$.

There are two subtleties here. First, although the original polynomials $f(t)$ and $g(t)$ are of degree n, we need to express them in terms of a Lagrange basis of degree $2n$, since their product is of degree $2n$. We can certainly do so, since every polynomial of degree n is also a polynomial of degree $2n$. Second, even though we have written $f(t)$ and $g(t)$ in terms of a Lagrange basis of degree $2n$, their product is still a polynomial of degree $2n$, so the product can be expressed in terms of the same Lagrange basis as the original polynomials. In fact, it is for this reason that we initially choose a Lagrange basis of degree $2n$ instead of a Lagrange basis of degree n. These same observations hold if we use any Lagrange basis of degree $m \geq 2n$. It will actually be more convenient to employ a basis whose degree is almost a power of two. Thus we shall use a Lagrange basis of degree $m = 2^p - 1$, where $2^{p-1} - 1 < 2n < 2^p - 1$.

Now our strategy for multiplying two polynomials is as follows.

1. Convert from the monomial basis of degree n to a Lagrange basis of degree m, where $m + 1 = 2^p > 2n$.

2. Perform fast multiplication in the Lagrange basis.

3. Convert back from the Lagrange basis to the monomial basis.

Step 2 can be performed with $O(n)$ multiplies, so if we could perform steps 1 and 3 with fewer than $O(n^2)$ multiplies, we would have a fast way to perform polynomial multiplication starting and ending with a monomial representation.

To convert from the monomial to the Lagrange basis, we need to perform polynomial evaluation. The standard way to evaluate a polynomial written in terms of the monomial basis is to apply Horner's method (see Exercise 3), which employs $O(n)$ multiplies to evaluate a polynomial of degree n at a single parameter value. To convert from monomial to Lagrange form, we need to evaluate a polynomial of degree n at $m + 1 > 2n$ nodes, so using Horner's method would require $O(n^2)$ multiplies. We need to do better.

The trick is to choose a special Lagrange basis—that is, a Lagrange basis with special nodes $t_0,...,t_m$. In this book we deal mostly with real variables, but everything we have done so far in this chapter with polynomials is valid as well for complex variables. So our Lagrange basis functions can have complex nodes. We shall choose our nodes to be complex roots of unity and show that with this choice of nodes we can convert back and forth between the monomial and Lagrange representation of a polynomial of degree n in $O(n \log(n))$ time.

Let $t_0,...,t_m$ denote the $m + 1$ distinct complex roots of unity—the complex numbers that satisfy the equation $t^{m+1} = 1$. Recall from complex analysis that

$$e^{it} = \cos(t) + i\sin(t).$$

Therefore, $e^{2\pi i} = \cos(2\pi) = 1$, so

$$t_k = e^{2ki\pi/(m+1)}, \quad k = 0,...,m.$$

Define $\omega_{m+1} = e^{2i\pi/(m+1)}$; then $t_k = \omega_{m+1}^k$, $k = 0,\ldots,m$. Moreover,

$$t_k^2 = (\omega_{m+1}^k)^2 = \omega_{(m+1)/2}^k. \tag{2.17}$$

Now comes the key observation on how to convert from the monomial coefficients $a_0,\ldots,a_n$ of a polynomial $f(t)$ to its Lagrange coefficients $f(t_0),\ldots,f(t_m)$ in $O(n\log(n))$ time. Define

$$f_{\text{even}}(t) = a_0 + a_2 t + a_4 t^2 + \cdots + a_{m-1} t^{(m-1)/2}$$

$$f_{\text{odd}}(t) = a_1 + a_3 t + a_5 t^2 + \cdots + a_m t^{(m-1)/2} \quad .$$

Then it is easy to verify that

$$f(t) = f_{\text{even}}(t^2) + t f_{\text{odd}}(t^2). \tag{2.18}$$

So evaluating one polynomial $f(t)$ of degree m at the $m+1$ values $t_0,\ldots,t_m$ is equivalent to evaluating two polynomials $f_{\text{even}}(t)$ and $f_{\text{odd}}(t)$ of degree $(m-1)/2$ at the values $t_0^2,\ldots,t_m^2$. But by (2.17), the $m+1$ values

$$t_k^2 = (\omega_{m+1}^k)^2, \quad k = 0,\ldots,m$$

are the same as the $(m+1)/2$ values

$$\tau_j = \omega_{(m+1)/2}^j, \quad j = 0,\ldots,(m-1)/2.$$

Therefore, we have reduced one problem of size $m+1$ to two equivalent problems of size $(m+1)/2$. Proceeding recursively in this manner, after $p = \log(m+1)$ steps we are reduced to solving $2^p = m+1$ problems of size one. To solve the original problem, we must then combine these solutions using (2.18). Since there are p recursive steps, the work required to recombine each solution is $p = \log(m+1)$. Hence the total amount of work to convert from monomial to Lagrange form—that is, to evaluate the polynomial at the nodes $t_0,\ldots,t_m$—is $p2^p = O(m\log(m)) = O(n\log(n))$. This recursive algorithm to convert from the standard monomial to the special Lagrange form is called the Discrete Fourier Transform (DFT).

We can use the DFT to convert from the monomial to the Lagrange basis in $O(n\log(n))$ time. We can then multiply the two polynomials in the Lagrange basis in $O(n)$ time. But after multiplying the two polynomials in the Lagrange basis, we still need to convert back from Lagrange to monomial form. It turns out that we can use the DFT to perform this operation as well. Here is how. Observe that the matrix that converts from monomial to Lagrange form is

$$M = \begin{pmatrix} 1 & 1 & \cdots & 1 \\ 1 & \omega_{m+1} & \cdots & \omega_{m+1}^m \\ \vdots & \vdots & \vdots & \vdots \\ 1 & \omega_{m+1}^m & \cdots & \omega_{m+1}^{m^2} \end{pmatrix} = \begin{pmatrix} 1 & 1 & \cdots & 1 \\ t_0 & t_1 & \cdots & t_m \\ \vdots & \vdots & \vdots & \vdots \\ t_0^m & t_1^m & \cdots & t_m^m \end{pmatrix}.$$

That is, if we were to multiply the monomial coefficients by M on the right, we would obtain the Lagrange coefficients because multiplication by M on the right is equivalent to evaluation at $t_0,...,t_m$. Discrete Fourier Transform is just a fast way to perform this matrix multiplication. To convert from Lagrange form back to monomial form, we need to apply M^{-1}.

LEMMA
2.10

$$(m+1)M^{-1} = \begin{pmatrix} 1 & 1 & \cdots & 1 \\ 1 & \omega_{m+1}^{-1} & \cdots & \omega_{m+1}^{-m} \\ \vdots & \vdots & \vdots & \vdots \\ 1 & \omega_{m+1}^{-m} & \cdots & \omega_{m+1}^{-m^2} \end{pmatrix}$$

Proof We can verify this result by multiplying the ith row of M by the jth column of $(m+1)M^{-1}$

$$\sum_{k=0}^{m}\omega_{m+1}^{ik}\omega_{m+1}^{-jk} = \sum_{k=0}^{m}\omega_{m+1}^{(i-j)k}.$$

If $i = j$, then

$$\sum_{k=0}^{m}\omega_{m+1}^{ik}\omega_{m+1}^{-jk} = \sum_{k=0}^{m}1 = m+1.$$

If $i \neq j$, then setting $h = i - j$ yields

$$\sum_{k=0}^{m}\omega_{m+1}^{ik}\omega_{m+1}^{-jk} = \sum_{k=0}^{m}\omega_{m+1}^{hk} = \frac{\left(\omega_{m+1}^{h}\right)^{m+1}-1}{\omega_{m+1}^{h}-1} = \frac{\left(\omega_{m+1}^{m+1}\right)^{h}-1}{\omega_{m+1}^{h}-1} = 0.$$

Hence $M * (m+1)M^{-1} = (m+1)I$ as required.

We get from the matrix M to the matrix $(m+1)M^{-1}$ by replacing ω_{m+1} by ω_{m+1}^{-1}. Thus to convert from Lagrange to monomial form, we can simply replace ω_{m+1} by ω_{m+1}^{-1} in the Discrete Fourier Transform (DFT^{-1}) and divide the final result by $m+1$. That is, we can proceed in the following fashion. Suppose that we have computed

$$f(t)g(t) = \sum_{k=0}^{m}f(t_k)g(t_k)L_k^m(t \mid t_0,...,t_m).$$

Let

$$h(t) = \sum_{k=0}^{m}f(t_k)g(t_k)t^k.$$

Applying DFT to $h(t)$ with ω_{m+1} replaced by ω_{m+1}^{-1} and then dividing the result by $m+1$ will yield a polynomial $h^*(t)$ in Lagrange form whose Lagrange coefficients are identical to the monomial coefficients of $f(t)g(t)$. Again this algorithm has a speed of $O(n\log(n))$.

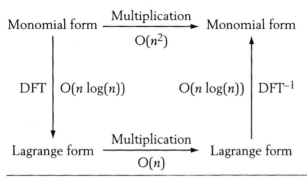

Figure 2.17 The Fast Fourier Transform (FFT) for polynomial multiplication.

Figure 2.17 summarizes our results. In the exercises we shall show how to extend these results on fast polynomial multiplication from the univariate to the bivariate setting (see Exercises 4 and 5).

Exercises

1. The convolution of two sequences $A = (a_0,...,a_n)$ and $B = (b_0,...,b_n)$ is given by the sequence $C = (c_0,...,c_{2n})$, where $c_k = \sum_{i+j=k} a_i b_j$.

   a. Show that convolution of sequences is equivalent to multiplication of polynomials.

   b. Conclude that two sequences of size $n+1$ can be convolved in time $O(n \log(n))$.

2. The elementary symmetric functions $\sigma_k(u_1,...,u_n)$, $k = 0,...,n$ are defined by setting $\sigma_k(u_1,...,u^n) = \sum u_{i_1} \cdots u_{i_k}$, where the sum is taken over all subsets of $\{1,...,n\}$ of order k.

   a. Let $f(t)$ be a polynomial of degree n with roots $u_1,...,u_n$. Show that the coefficients of $f(t)$ in the monomial basis are given by $\sigma_k(u_1,...,u_n)$, $k = 0,...,n$.

   b. Given n arbitrary values $u_1,...,u_n$, show how to use FFT to compute $\sigma_k(u_1,...,u_n)$, $k = 0,...,n$, in time $O(n\log^2(n))$.

3. Let $f(t) = a_n t^n + a_{n-1} t^{n-1} + \cdots + a_1 t + a_0$ be a polynomial of degree n. Define

   $$f_k(t) = a_n t^k + a_{n-1} t^{k-1} + \cdots + a_{n-k+1} t + a_{n-k} \quad k = 0,...,n.$$

   Show that

   a. $f_{k+1}(t) = t f_k(t) + a_{n-k-1}$

   b. $f_n(t) = f(t)$

   Computing $f(t)$ in this fashion by starting with $f_0(t) = a_n$ and recursively computing $f_k(t)$, $k = 1,...,n$, is called Horner's method.

c. Conclude from Horner's method that every polynomial of degree n can be evaluated with at most n multiplications and n additions.

d. Verify that Horner's method for cubic polynomials can be diagrammed as in Figure 2.18, and provide an analogous diagram of Horner's method for polynomials of arbitrary degree.

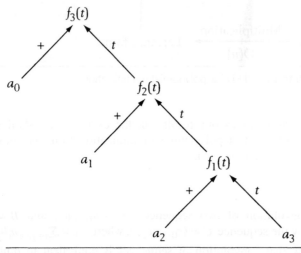

Figure 2.18 Horner's method for cubic polynomials.

4. Let $f(s,t)$ and $g(s,t)$ be two polynomials of bidegree n. That is,

$$f(s,t) = \sum_{i=0}^{n} \sum_{j=0}^{n} a_{ij}s^i t^j \text{ and } g(s,t) = \sum_{i=0}^{n} \sum_{j=0}^{n} b_{ij}s^i t^j.$$

a. Show that the naive algorithm for multiplying $f(s,t)$ and $g(s,t)$ would have a speed of $O(n^4)$.

b. Now rewrite $f(s,t)$ and $g(s,t)$ as polynomials of degree n in t with coefficients that are polynomials of degree n in s. Show that using this approach $f(s,t)$ and $g(s,t)$ can be multiplied using FFT in time $O(n^3\log(n))$.

5. Let

$$h(s,t) = \sum_{i=0}^{2n} \sum_{j=0}^{2n} c_{ij}s^i t^j$$

be a bivariate polynomial of bidegree $2n$. Define a univariate polynomial $h*(u)$ by setting

$$h*(u) = \sum_{i,j} c_{ij}u^{j(2n+1)+i}.$$

a. Show that the map $h \to h*$ sends a bivariate polynomial of bidegree $2n$ to a univariate polynomial of total degree $4n^2 + 4n$.

b. Show that the map $h \to h*$ is 1:1 and onto.

c. Show that $(h_1 + h_2)* = h_1^* + h_2^*$

d. Let $f(s,t)$ and $g(s,t)$ be two polynomials of bidegree n, show that $(fg)* = f * g *$.

e. Use parts (b) and (d) to develop an algorithm for multiplying two polynomials of bidegree n in time $O(n^2 \log(n))$.

2.9 **Recapitulation**

We have encountered three strategies for solving the interpolation problem for curves: undetermined coefficients, recursion, and cardinal basis functions.

1. *Undetermined Coefficients* (Section 2.4, Exercise 5)

   Solve the system of linear equations

   $$a_n t_0{}^n + \cdots + a_1 t_0 + a_0 = P_0$$
   $$\vdots \qquad\qquad \vdots$$
   $$a_n t_n{}^n + \cdots + a_1 t_n + a_0 = P_n$$

   for the unknown coefficients $a_0, \ldots, a_n$. This method is effective for demonstrating the existence and uniqueness of the polynomial interpolant, but it is not computationally stable.

2. *Recursion* (Section 2.2, Theorem 2.1)

   Apply the recurrence

   $$P_{0 \cdots n}(t) = \frac{t_n - t}{t_n - t_0} P_{0 \cdots n-1}(t) + \frac{t - t_0}{t_n - t_0} P_{1 \cdots n}(t)$$
   $$P_k(t) = P_k \quad .$$

   This recurrence leads directly to Neville's dynamic programming algorithm for polynomial interpolation.

3. *Cardinal Basis Functions* (Section 2.5, Theorem 2.7)

   Employ the basis functions

   $$L_k^n(t \mid t_0, \ldots, t_n) = \frac{\prod_{j \neq k}(t - t_j)}{\prod_{j \neq k}(t_k - t_j)} \qquad k = 0, \ldots, n$$

   $$P_{0 \cdots n}(t) = \sum_{k=0}^{n} L_k^n(t \mid t_0, \ldots, t_n) P_k$$

These formulas provide an explicit expression for the polynomial interpolant.

Each approach provides a different insight into the interpolation problem for curves, and each method leads to an alternative computational technique. We have focused mostly on recursion because this technique generates the most elegant computational scheme. Interpolation for surfaces is much harder than for curves, but these same fundamental approaches still apply. We shall consider each of these three basic strategies in our study of surface interpolation.

Exercises

1. Let $F_0(t),...,F_n(t)$ be a collection of blending functions. Show that there is a function

$$F(t) = \sum_{k=0}^{n} c_k F_k(t)$$

such that $F(t_k) = P_k$, $k = 0,...,n$, for any choice of $P_0,...,P_n$ if and only if

$$\begin{vmatrix} F_0(t_0) & \cdots & F_n(t_0) \\ \vdots & \vdots & \vdots \\ F_0(t_n) & \cdots & F_n(t_n) \end{vmatrix} \neq 0 .$$

2. Let $F_0(t),...,F_n(t)$ be a collection of blending functions. Show that there is a function

$$F(t) = \sum_{k=0}^{n} c_k F_k(t)$$

such that $F(t_k) = P_k$, $k = 0,...,n$ for any choice of distinct nodes $t_0,...,t_n$ and any choice of control points $P_0,...,P_n$ if and only if there are functions

$$A_j(t \mid t_0,...,t_n) = \sum_{k=0}^{n} a_{jk} F_k(t), \ j = 0,...,n ,$$

such that

$$A_j(t_k \mid t_0,...,t_n) = 0 \quad j \neq k$$
$$= 1 \quad j = k .$$

2.10 Surface Interpolation

Given $n + 1$ points $P_0,...,P_n$ and $n + 1$ distinct nodes $t_0,...,t_n$, we showed in Theorem 2.6 that there exists a unique polynomial curve $P_{0\cdots n}(t)$ of degree n that interpolates the given points at the specified parameters. That is,

$$P_{0\cdots n}(t_k) = P_k \qquad k = 0,...,n .$$

Now we would like to study the equivalent problem for surfaces.

Instead of a sequence of points and a sequence of nodes, we start with an array $\{P_{ij}\}$ of control points and an array $\{(s_{ij}, t_{ij})\}$ of parameter values. We seek a polynomial surface $P(s,t)$ that interpolates the given array of control points at the specified array of parameter values.

How many control points could we hope to interpolate with a bivariate polynomial of degree n? A univariate polynomial of degree n has $n + 1$ coefficients, and we can interpolate $n + 1$ arbitrary control points. Similarly, a bivariate polynomial of degree n has

$$\binom{n+2}{2} = (n+2)(n+1)/2$$

coefficients (see Exercise 1); therefore applying the method of undetermined coefficients, we might hope to interpolate $\binom{n+2}{2}$ arbitrary control points. Unlike for curves, however, it is not always possible to solve the general bivariate interpolation problem for surfaces.

Consider quadratic interpolation. Here we have a bivariate polynomial

$$P(s,t) = A_{20}s^2 + A_{11}st + A_{02}t^2 + A_{10}s + A_{01}t + A_{00}$$

with six undetermined coefficients $A_{20}, ..., A_{00}$, and we wish to interpolate six arbitrary points $P_1, ..., P_6$ at six arbitrary, distinct parameter pairs $(s_1, t_1), ..., (s_6, t_6)$. Thus we need to solve a system of six linear equations:

$$A_{20}s_1^2 + A_{11}s_1t_1 + A_{02}t_1^2 + A_{10}s_1 + A_{01}t_1 + A_{00} = P_1$$
$$\vdots \qquad\qquad\qquad\qquad \vdots$$
$$A_{20}s_6^2 + A_{11}s_6t_6 + A_{02}t_6^2 + A_{10}s_6 + A_{01}t_6 + A_{00} = P_6$$

for the six unknown coefficients $A_{20}, ..., A_{00}$. Rewriting these equations in matrix form, we have

$$\begin{pmatrix} s_1^2 & s_1t_1 & t_1^2 & s_1 & t_1 & 1 \\ \vdots & \vdots & \vdots & \vdots & \vdots & \vdots \\ s_6^2 & s_6t_6 & t_6^2 & s_6 & t_6 & 1 \end{pmatrix} \begin{pmatrix} A_{20} \\ \vdots \\ A_{00} \end{pmatrix} = \begin{pmatrix} P_1 \\ \vdots \\ P_6 \end{pmatrix}.$$

We can solve for the six unknowns $A_{20}, ..., A_{00}$ if and only if the 6×6 matrix of coefficients is invertible. But if the six parameter pairs lie on a conic

$$Q_{20}s^2 + Q_{11}st + Q_{02}t^2 + Q_{10}s + Q_{01}t + Q_{00} = 0$$

—that is, a second-degree curve in the st-parameter plane—then

$$\begin{pmatrix} s_1^2 & s_1t_1 & t_1^2 & s_1 & t_1 & 1 \\ \vdots & \vdots & \vdots & \vdots & \vdots & \vdots \\ s_6^2 & s_6t_6 & t_6^2 & s_6 & t_6 & 1 \end{pmatrix} \begin{pmatrix} Q_{20} \\ \vdots \\ Q_{00} \end{pmatrix} = \begin{pmatrix} 0 \\ \vdots \\ 0 \end{pmatrix}.$$

In this case the columns of the coefficient matrix are linearly dependent, so this matrix is not invertible. Thus we cannot solve the general quadratic bivariate interpolation problem when the parameters are aligned in certain positions. The same difficulty arises for arbitrary degree.

Since it is hopeless to solve the general bivariate interpolation problem using polynomials of minimal degree, we shall examine instead some important special cases, where the nodes lie in special configurations.

Exercises

1. Show that the number of terms in a bivariate polynomial of degree n is

$$\binom{n+2}{2} = \frac{(n+2)(n+1)}{2}.$$

2. What conditions must be placed on the nodes in a plane in order to be able to interpolate three arbitrary control points with a linear function in two variables?

3. Show that it is always possible to interpolate $n + 1$ control points at $n + 1$ distinct parameter values with a bivariate polynomial of degree n.

4. Consider a triangular array $\{P_{ij}\}$ of $\binom{n+2}{2}$ control points in 3-space and another triangular array $\{(s_{ij},t_{ij})\}$ of $\binom{n+2}{2}$ nodes in the parameter plane. Suppose that there are $n + 1$ lines $L_1,\ldots,L_{n+1}$ in the parameter plane such that k nodes lie on line L_k, $k = 1,\ldots,n+1$, and no node lies on two lines L_i,L_j.

   a. Prove that there exists a bivariate polynomial of degree n that interpolates the given array of control points at the specified parameter values. (Hint: Use induction on n and Exercise 3.)

   b. Bezout's Theorem states that if a polynomial curve $f(s,t) = 0$ of degree m intersects another polynomial curve $g(s,t) = 0$ of degree n in more than mn points, then $f(s,t)$ and $g(s,t)$ must have a common factor. Use Bezout's Theorem to prove that the interpolant in part (a) is unique.

2.11   Rectangular Tensor Product Lagrange Surfaces

Tensor product surfaces are some of the simplest surfaces to construct, but they are also some of the most important surfaces in computer graphics and computer-aided design. In the tensor product construction, we start with two rectangular arrays of size $(m+1) \times (n+1)$: one for the control points $\{P_{ij}\}$ and one for the nodes $\{(s_{ij},t_{ij})\}$, where $0 \le i \le m$ and $0 \le j \le n$ (see Figure 2.19).

The nodes are in special positions because they lie on a rectangular grid in the parameter plane; that is, they lie along the parameter lines $s = s_i$ and $t = t_j$. We shall assume further that $s_0 < \cdots < s_m$ and $t_0 < \cdots < t_n$. We seek a bivariate polynomial $P(s,t)$ of degree m in s and degree n in t that interpolates the given control points at the specified parameter values. That is, we seek a bivariate polynomial $P(s,t)$ of bidegree (m,n) such that $P(s_i,t_j) = P_{ij}$.

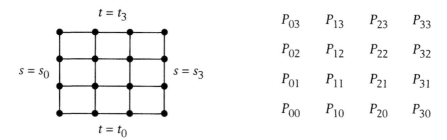

(a) Domain—rectangular grid (b) Range—rectangular array of points

Figure 2.19 Data for a tensor product bicubic interpolant: (a) represents the nodes in the domain and (b) represents the control points in the range. The nodes lie on a rectangular grid, but the control points may be in arbitrary positions. The surface $P(s,t)$ must interpolate the control points P_{ij} at the nodes (s_i, t_j)—that is, $P(s_i, t_j) = P_{ij}$.

This bivariate interpolation problem is easy to solve using univariate methods. In fact, we can simply set

$$P(s,t) = \sum_{k=0}^{m} \sum_{l=0}^{n} L_k^m(s \mid s_0,...,s_m) L_l^n(t \mid t_0,...,t_n) P_{kl}, \tag{2.19}$$

where $L_k^m(s \mid s_0,...,s_m)$ and $L_l^n(t \mid t_0,...,t_n)$ represent the Lagrange basis functions for the nodes $s_0,...,s_m$ and $t_0,...,t_n$ (see Section 2.5). By the cardinal conditions (2.12) it follows immediately that

$$P(s_i, t_j) = P_{ij},$$

so we have indeed solved this interpolation problem. The surface defined by (2.19) is called a *tensor product Lagrange surface* because the basis functions $L_k^m(s \mid s_0,...,s_m) L_l^n(t \mid t_0,...,t_n)$ that multiply the control points P_{kl} are formed from products of univariate Lagrange basis functions (see Figure 2.20).

The tensor product construction is a standard technique in geometric modeling. Become familiar with it now because you will see lots of other tensor product surfaces later in this text.

We can rewrite (2.19) in the following manner:

$$P(s,t) = \sum_{k=0}^{m} L_k^m(s \mid s_0,...,s_m) \left(\sum_{l=0}^{n} L_l^n(t \mid t_0,...,t_n) P_{kl} \right). \tag{2.20}$$

Let

$$P_k(t) = \sum_{l=0}^{n} L_l^n(t \mid t_0,...,t_n) P_{kl} \qquad k = 0,...,m. \tag{2.21}$$

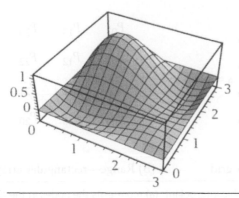

Figure 2.20 The bicubic Lagrange basis function $L_1^3(s \mid s_0,...,s_3)L_2^3(t \mid t_0,...,t_3)$. Here the nodes are at the integers—that is, $s_k = t_k = k$, $k = 0,...,3$.

Then

$$P(s,t) = \sum_{k=0}^{m} L_k^m(s \mid s_0,...,s_m)P_k(t). \qquad (2.22)$$

If we fix the value of $t = t^*$, then $P(s,t^*)$ is simply the univariate polynomial of degree m that interpolates the points $P_0(t^*),...,P_m(t^*)$ at the parameter values $s_0,...,s_m$. Similarly, each degree n univariate polynomial $P_k(t)$ interpolates the control points $P_{k0},...,P_{kn}$ at the nodes $t_0,...,t_n$. Thus the interpolation surface $P(s,t)$ interpolates the interpolation curves $P_0(t),...,P_m(t)$ (see Figure 2.21). An analogous argument shows that the surface $P(s,t)$ also interpolates the curves

$$Q_l(s) = \sum_{k=0}^{m} L_k^m(s \mid s_0,...,s_m)P_{kl} \qquad l = 0,...,n,$$

since

$$P(s,t) = \sum_{l=0}^{n} L_l^n(t \mid t_0,...,t_n)Q_l(s).$$

Thus the surface $P(s,t)$ actually interpolates the mesh of space curves $\{P_k(t),Q_l(s)\}$ (see Figure 2.22).

Notice that if we restrict to the domain $s_0 \le s \le s_m$ and $t_0 \le t \le t_n$, then we get a four-sided surface patch. Moreover, it is easy to see that the boundary curves of this rectangular patch are the Lagrange polynomial curves that interpolate the boundary control points.

Equations (2.21) and (2.22) lead to a bivariate version of Neville's algorithm. First apply Neville's algorithm $m + 1$ times to calculate points on each of the curves $P_0(t),...,P_m(t)$; then apply Neville's algorithm one more time with $P_0(t),...,P_m(t)$ as

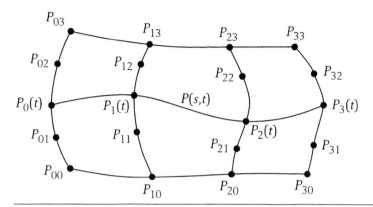

Figure 2.21 Schematic view of bicubic interpolation. The curve $P_k(t)$ interpolates the control points $P_{k0},...,P_{k3}$, and the surface $P(s,t)$ interpolates the control curves $P_0(t),...,P_3(t)$. The boundary curves are the interpolating curves of the boundary control points.

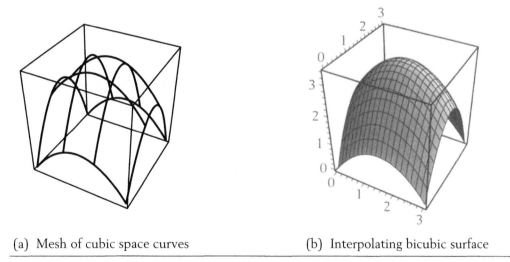

(a) Mesh of cubic space curves   (b) Interpolating bicubic surface

Figure 2.22 Bicubic interpolation.

control points to compute $P(s,t)$ (see Figure 2.23). Similarly, we could apply Neville's algorithm $n + 1$ times to compute points along the curves $Q_0(s),...,Q_n(s)$ and then apply Neville's algorithm one more time to interpolate points on these curves (see Exercise 3).

There is another dynamic programming algorithm for tensor product surfaces similar in spirit to Neville's algorithm but with a somewhat different structure. Let's assume for the sake of simplicity that the degree in s is the same as the degree in t

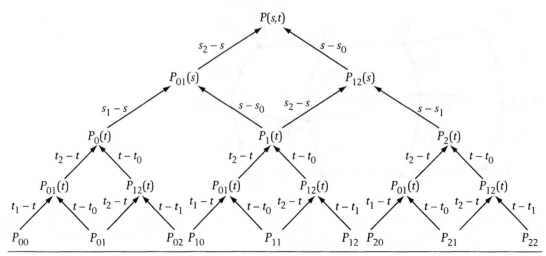

Figure 2.23 Neville's algorithm for a biquadratic patch. The three lower triangles in the two bottom tiers represent univariate interpolation in the *t* direction. The triangle at the top interpolates these results in the *s* direction.

(i.e., $m = n$). Instead of constructing the interpolating surface from a sequence of linear interpolations, we can construct the interpolating surface from a sequence of bilinear interpolations. This algorithm is based on the following bilinear generalization of linear interpolation. Let

$$B(s,t) = \frac{(s_1 - s)(t_1 - t)}{(s_1 - s_0)(t_1 - t_0)} P_{00} + \frac{(s - s_0)(t_1 - t)}{(s_1 - s_0)(t_1 - t_0)} P_{10}$$
$$+ \frac{(s_1 - s)(t - t_0)}{(s_1 - s_0)(t_1 - t_0)} P_{01} + \frac{(s - s_0)(t - t_0)}{(s_1 - s_0)(t_1 - t_0)} P_{11} \ .$$

Then it is easy to check that

$$B(s_i, t_j) = P_{ij} \qquad i, j = 0, 1.$$

Now let $P_{00}(s,t), P_{0n}(s,t), P_{n0}(s,t), P_{nn}(s,t)$ be the rectangular interpolants for the four overlapping rectangular grids of size $n \times n$ with vertices at $(s_0, t_0), (s_0, t_n), (s_n, t_0)$, (s_n, t_n), and corresponding overlapping $n \times n$ arrays of control points with corners at $P_{00}, P_{0n}, P_{n0}, P_{nn}$. That is, each surface $P_{ij}(s,t)$ has the same index as the unique corner point interpolated by the surface (see Figure 2.24).

By construction

$$P_{ij}(s_k, t_l) = P_{kl} \quad \frac{i}{n} \le k \le \frac{i}{n} + n - 1 \ \text{ and } \ \frac{j}{n} \le l \le \frac{j}{n} + n - 1 \ \text{ for } \ i, j = 0, n \ .$$

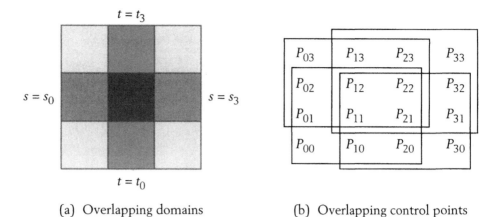

(a) Overlapping domains   (b) Overlapping control points

Figure 2.24 Overlapping data for four biquadratic rectangular interpolants: (a) represents the four over-lapping domains. Each domain contains a 3 × 3 array of nodes; interior nodes (the black square) belong to all four domains; nodes along the edges (dark gray squares) belong to two overlapping domains; and nodes at the corners belong to a single domain. (b) represents the four overlapping arrays of control points corresponding to these four overlapping domains. Interior control points belong to all four arrays; control points along the edges belong to two arrays; and control points at the corners belong to a single array.

Therefore it is easy to verify that the surface defined recursively by

$$P(s,t) = \frac{(s_n - s)(t_n - t)}{(s_n - s_0)(t_n - t_0)} P_{00}(s,t) + \frac{(s - s_0)(t_n - t)}{(s_n - s_0)(t_n - t_0)} P_{n0}(s,t)$$
$$+ \frac{(s_n - s)(t - t_0)}{(s_n - s_0)(t_n - t_0)} P_{0n}(s,t) + \frac{(s - s_0)(t - t_0)}{(s_n - s_0)(t_n - t_0)} P_{nn}(s,t)$$

(2.23)

satisfies

$$P(s_k, t_l) = P_{kl} \qquad 0 \le k, l \le n.$$

To apply dynamic programming to (2.23), we must first construct all the bilinear interpolants, then all the biquadratic interpolants, then all the bicubic interpolants, and so on, until after n stages we arrive at the bidegree n surface $P(s,t)$. This algorithm has the shape of a square pyramid, so we shall call it the pyramid algorithm (see Figure 2.25).

Both Neville's algorithm and the pyramid algorithm are $O(n^3)$, but Neville's algorithm is more efficient. When $n = m$, Neville's algorithm uses $n + 1$ triangles in the t direction to compute the points $P_0(t), \ldots, P_n(t)$, and then one additional triangle in the s direction to interpolate these $n + 1$ points. Each triangle has $n(n+1)/2$ nodes, and there are $n + 2$ triangles so

$$\text{number of nodes in Neville's algorithm} = \frac{n(n+1)(n+2)}{2}.$$

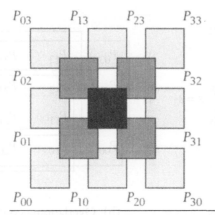

P_{03} P_{13} P_{23} P_{33}

P_{02} P_{32}

P_{01} P_{31}

P_{00} P_{10} P_{20} P_{30}

Figure 2.25 A schematic diagram of the pyramid algorithm for a bicubic patch, viewed from above. Each rectangular panel represents the computation of a point at its center by bilinear interpolation of the points at its corners; the darker the rectangle the higher the degree of the interpolant. Thus the light gray rectangles represent bilinear patches built directly from the control points, the darker gray rectangles represent biquadratic patches built from overlapping 3×3 arrays of control points, and the black rectangle represents the bicubic interpolant. The interior control points are obscured by the panels.

On the other hand, the pyramid algorithm has one node at the top of the pyramid, four at the next level, then nine at the next level, and so on. Thus

$$\text{number of nodes in pyramid algorithm } = \sum_{k=1}^{n} k^2 = \frac{n(n+1)(2n+1)}{6}.$$

Since $(2n+1)/3 < n+2$, there are actually fewer nodes in the pyramid algorithm than in Neville's algorithm. However, each node in the pyramid algorithm represents a bivariate bilinear interpolation, while each node in Neville's algorithm represents only a univariate linear interpolation. Counting just multiplies and divides, we find that each linear interpolation costs two multiplies and one divide, while each bilinear interpolation costs eight multiplies—four for the four arrows entering a node and four more to compute the labels on each arrow—and one divide (since the labels on all the arrows entering a node have the same denominator, we can perform this divide just once at each node; however if the control points lie in three dimensions, the real cost is actually three divides instead of one). Thus

cost per node in Neville's algorithm = 3

cost per node in pyramid algorithm = 9.

Multiplying the cost per node by the total number of nodes, we find

$$\text{total cost of Neville's algorithm} = \frac{3n(n+1)(n+2)}{2}$$

$$\text{total cost of pyramid algorithm} = \frac{3n(n+1)(2n+1)}{2}.$$

Since $n+2 < 2n+1$, Neville's algorithm is somewhat more efficient than the pyramid algorithm. It may also be easier to program, since it uses only univariate interpolation and this code is often already in place to generate curves.

Neville's algorithm has another advantage over the pyramid algorithm that is even more substantial. Typically surfaces are rendered by generating points on the surface along isoparameter lines—that is, along lines of constant s or t. If we fix $t = t^*$ and vary only s, then we can reuse the computation of the points $P_0(t^*),...,P_m(t^*)$. Thus along isoparameter lines, Neville's algorithm for tensor product surfaces reduces to the univariate version of Neville's algorithm, which is only $O(n^2)$. No such reduction occurs for the pyramid algorithm along isoparameter lines. Nevertheless, it is worth taking the time to understand the structure of the pyramid algorithm because in the next section we shall develop a similar algorithm for triangular Lagrange patches where the univariate version of Neville's algorithm is not readily available.

Finally, note that there is an even faster evaluation algorithm for tensor product Lagrange interpolation because the $O(n^2)$ algorithm for the univariate Lagrange basis functions based on the univariate down recurrences leads to a simple $O(n^2)$ algorithm for tensor product Lagrange interpolation (see Exercise 6).

Exercises

1. Prove that the boundary curves of an interpolating tensor product patch are the Lagrange polynomials that interpolate the boundary control points.

2. Complete the analysis of the pyramid algorithm by showing how to implement it when the degree in s is different from the degree in t.

3. Consider an interpolating tensor product patch of bidegree (m,n), where $m < n$.

   a. Show that to compute a single point on the surface it is faster to apply Neville's algorithm first in the s direction and then in the t direction.

   b. Show that to compute many points along the surface it may be faster to apply Neville's algorithm first in the t direction and then in the s direction.

   c. Explain this apparent anomaly.

4. Implement both Neville's algorithm and the pyramid algorithm for tensor product surfaces. Which algorithm do you prefer? Why? Experiment with tensor product surfaces of different degrees.

   a. How does altering the arrangement of the control points affect the shape of the surface?

   b. How does changing the values of the nodes affect the shape of the surface?

5. In Section 2.5 we discussed the up and down recurrences for the univariate Lagrange basis functions. What are the up and down recurrences in the case of tensor products for

   a. Neville's algorithm?

   b. the pyramid algorithm?

6. Use the down recurrence for the univariate Lagrange basis functions to develop an $O(n^2)$ evaluation algorithm for tensor product Lagrange interpolation.

7. Prove that

$$\sum_{k=0}^{m} \sum_{l=0}^{n} L_k^m(s \mid s_0,...,s_m) L_l^n(t \mid t_0,...,t_n) \equiv 1.$$

8. Give necessary and sufficient conditions on the control points for the interpolating tensor product surface to collapse to

   a. a single point

   b. a line

   c. a plane

   Justify your answer.

9. Complete the proof that the surface $P(s,t)$ defined by (2.23) satisfies $P(s_k,t_l) = P_{kl}$ for $0 \le k,l \le n$.

10. The surface generated by Neville's algorithm and the surface generated by the pyramid algorithm both interpolate the control points at the nodes—that is, they both satisfy $P(s_k,t_l) = P_{kl}$. But how do we know that these surfaces are actually identical at every point? Prove that the surface generated by the pyramid algorithm is identical to the surface generated by Neville's algorithm by using (2.23) and induction to show that the coefficient of P_{kl} in the pyramid algorithm is

$$L_k^m(s \mid s_0,...,s_m) L_l^n(t \mid t_0,...,t_n).$$

(See also Exercise 1 of Section 2.13.)

2.12 Triangular Lagrange Patches

To define rectangular interpolating surface patches, we need to choose our parameters to lie on a rectangular grid. Similarly, to construct triangular interpolating surface patches, we must select our parameters to lie on a triangular grid. We can build a triangular grid of size n by selecting three sets of lines $(R_0,...,R_n;\ S_0,...,S_n;\ $ and $T_0,...,T_n)$ in the parameter plane, and insisting that the three lines R_i,S_j,T_k intersect at a common point whenever $i+j+k = n$. Let

$$R_i \cap S_j \cap T_k = Q_{ijk} \qquad i+j+k = n$$

$$= \phi \qquad i+j+k < n\ ;$$

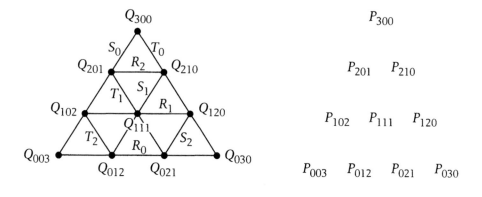

(a) Domain—triangular grid   (b) Range—triangular array of points

Figure 2.26 Data for a cubic triangular interpolant: (a) represents the nodes in the domain and (b) represents the control points in the range. The nodes lie on a triangular grid, but the control points may be in arbitrary positions. The surface $\Delta(s,t)$ must interpolate the control points P_{ijk} at the nodes Q_{ijk}—that is, $\Delta(Q_{ijk}) = P_{ijk}$. Compare to Figure 2.19 for a tensor product surface.

then the points Q_{ijk} are said to form a *triangular grid* (see Figure 2.26). Notice that a triangular grid of size n consists of one point at the apex of the triangle, two points at the next level, three points at the next level, and so on until the final level with $n + 1$ points. Thus the number of points in a triangular array of size n is

$$\sum_{k=1}^{n+1} k = (n+1)(n+2)/2 = \binom{n+2}{2},$$

which is exactly the same as the number of coefficients in a bivariate polynomial of degree n (see Exercise 1 of Section 2.10).

Neville's algorithm for triangular surface patches is a dynamic programming procedure for generating a bivariate polynomial of degree n that interpolates a triangular array of control points P_{ijk} at parameter values Q_{ijk} lying on a triangular grid of size n.

To develop Neville's algorithm, we shall use barycentric coordinates in the parameter plane. (To review barycentric coordinates, see Section 1.2.3.) Given three control points P_1, P_2, P_3 and three noncollinear parameter values $Q_1 = (s_1, t_1)$, $Q_2 = (s_2, t_2)$, $Q_3 = (s_3, t_3)$, we can apply bivariate linear interpolation to interpolate the control points at the specified parameter values. Simply set

$$P(s,t) = \beta_1(s,t)P_1 + \beta_2(s,t)P_2 + \beta_3(s,t)P_3, \tag{2.24}$$

where $\beta_1(s,t), \beta_2(s,t), \beta_3(s,t)$ are the barycentric coordinates with respect to the points Q_1, Q_2, Q_3 of the point Q in the parameter plane with rectangular coordinates (s,t). Equation (2.24) is the analogue in the parameter plane of Equation (2.5) along the parameter line. By the standard properties of barycentric coordinates (Theorem 1.1, Properties 4 and 5), the function $P(s,t)$ is linear in s and t, and

$$P(s_k, t_k) = P_k \qquad k = 1, 2, 3.$$

Now just like in Neville's algorithm for curves, the trick for constructing interpolating surfaces is to build higher-order interpolants by performing linear interpolation on lower-order interpolants. Suppose we can construct degree $n - 1$ polynomials that interpolate triangular data on a triangular grid of size $n - 1$. Let $\Delta_{n00}, \Delta_{0n0}, \Delta_{00n}$ be the triangular interpolants for the overlapping triangular grids of size $n - 1$ with vertices at $Q_{n00}, Q_{0n0}, Q_{00n}$ and corresponding overlapping arrays of control points with corners at $P_{n00}, P_{0n0}, P_{00n}$. That is, each surface has the same index as the unique corner point interpolated by the surface (see Figure 2.27).

Then by construction

$$\Delta_{n00}(Q_{ijk}) = P_{ijk} \qquad i \neq 0$$
$$\Delta_{0n0}(Q_{ijk}) = P_{ijk} \qquad j \neq 0$$
$$\Delta_{00n}(Q_{ijk}) = P_{ijk} \qquad k \neq 0 \ .$$

To build the degree n triangular interpolant Δ on the triangular grid of size n, set

$$\Delta(s,t) = \beta_{n00}(s,t)\Delta_{n00}(s,t) + \beta_{0n0}(s,t)\Delta_{0n0}(s,t) + \beta_{00n}(s,t)\Delta_{00n}(s,t) \qquad (2.25)$$

where $\beta_{n00}, \beta_{0n0}, \beta_{00n}$ are the barycentric coordinates relative to the points Q_{n00}, Q_{0n0}, Q_{00n}.

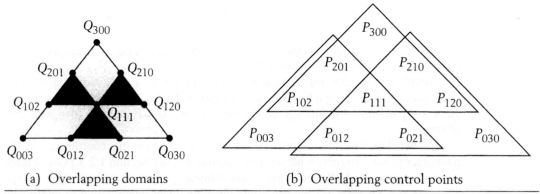

(a) Overlapping domains   (b) Overlapping control points

Figure 2.27 Overlapping data for three quadratic triangular interpolants: (a) represents the three overlapping domains. Each domain contains a triangular grid of size two. The interior node Q_{111} belongs to all three domains; nodes along the edges (dark triangles) belong to two overlapping domains; and nodes at the corners belong to a single domain. (b) represents the three overlapping arrays of control points corresponding to these three domains. The interior control point P_{111} belongs to all three arrays; control points along the edges belong to two arrays; and control points at the corners belong to a single array. Compare to Figure 2.24 for tensor product surfaces.

It is easy to check that $\Delta(s,t)$ interpolates the triangular array of control points $\{P_{ijk}\}$ on the triangular grid of parameter values $\{Q_{ijk}\}$ when $i+j+k=n$. Certainly

$$\Delta(Q_{ijk}) = P_{ijk} \qquad \text{if } i,j,k \neq 0$$

because, by assumption,

$$\Delta_{\lambda\mu\nu}(Q_{ijk}) = P_{ijk} \qquad \text{if } i,j,k \neq 0,$$

and the barycentric coordinates sum to one. It remains to verify interpolation along the boundaries of the grid. Let's check along the boundary defined by $i=0$. If $i=0$ and $j,k \neq 0$, then Q_{0jk} lies on the line joining Q_{0n0}, Q_{00n}, so

$$\beta_{n00}(P_{ijk}) = 0$$

and by construction

$$\Delta_{0n0}(Q_{0jk}) = \Delta_{00n}(Q_{0jk}) = P_{0jk}.$$

Again since the barycentric coordinates sum to one,

$$\Delta(Q_{0jk}) = P_{0jk}.$$

Finally, if $i=j=0$, then

$$\beta_{\lambda\mu\nu}(Q_{00n}) = 0 \qquad \nu = 0$$
$$= 1 \qquad \nu = n$$

so again

$$\Delta(Q_{00n}) = \Delta_{00n}(Q_{00n}) = P_{00n}.$$

Similar arguments apply along the other two boundaries of the triangular grid.

The three boundaries of the triangular patch are the images of the lines R_0, S_0, T_0. Along each of these lines, $\Delta(s,t)$ is a degree n univariate polynomial that interpolates the corresponding boundary control points. Thus each boundary of an interpolating triangular patch is the Lagrange polynomial curve that interpolates the corresponding boundary control points.

Equation (2.25) can be converted into a dynamic programming algorithm for $\Delta(s,t)$ in the usual way by first computing the linear interpolants over triangular grids of size one, then computing the quadratic interpolants over triangular grids of size two, then the cubic interpolants over triangular grids of size three, and so on until finally the degree n interpolant over the triangular grid of size n is generated. This procedure constructs a triangular pyramid of points. Figure 2.28 illustrates this algorithm schematically for cubic interpolants.

Neville's pyramid algorithm guarantees the existence of triangular Lagrange basis functions. That is, there are bivariate polynomials $L_{ijk}(s,t)$ of degree n depending on the grid $\{Q_{ijk}\}$ such that

$$\Delta(s,t) = \sum_{ijk} L_{ijk}(s,t) P_{ijk}.$$

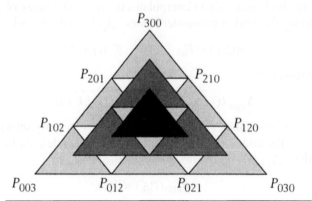

Figure 2.28 A schematic diagram of Neville's pyramid algorithm for cubic surface interpolation. Each triangular panel represents the computation of a point at its center by linear interpolation of the points at its corners. Thus the light gray triangles represent linear patches built directly from the control points, the darker gray triangles represent quadratic patches built from overlapping arrays of control points, and the black triangle represents the cubic interpolant. Notice that interior control points are obscured by the panels, and down-pointing (white) triangles are ignored. Compare to Figure 2.25 for tensor product surfaces.

As in the univariate setting, the basis function $L_{ijk}(s,t)$ is given by the sum over all paths from the point P_{ijk} at the base of the pyramid to the point $\Delta(s,t)$ at the apex. Thus the basis function $L_{ijk}(s,t)$ can be computed using the up recurrence by setting $P_{ijk} = 1$ and all the other control points to zero, or from the down recurrence by placing a one at the apex of the pyramid, reversing all the arrows, and collecting the functions $L_{ijk}(s,t)$ at the base of the pyramid.

As in the case of curves, it is also possible to derive simple explicit formulas for the basis functions $L_{ijk}(s,t)$. In the univariate setting, we observed that, due to the parallel property, all paths from any fixed point at the base to the apex of the recurrence are identical up to constant multiples, and we used this observation to derive explicit expressions for the Lagrange basis functions. Similar observations apply to Neville's pyramid algorithm for interpolating triangular surfaces, although the parallel property may be a bit more difficult to visualize. Therefore, here we shall take a more direct approach to deriving explicit formulas for the Lagrange basis functions.

In Neville's pyramid algorithm all the arrows are labeled with barycentric coordinates, so the basis functions are products of barycentric coordinates. But recall from Section 1.2.3 that barycentric coordinates for a triangle can be constructed from the equations of the lines joining the vertices of the triangle. To construct a triangular grid, we start by selecting three sets of lines $R_0, \ldots, R_n$; $S_0, \ldots, S_n$; and $T_0, \ldots, T_n$ in the parameter plane, and the points $\{Q_{ijk}\}$ in the grid lie on the intersections of these lines (see Figure 2.26). Thus we can use the equations of these lines, properly normalized, to represent all our barycentric coordinates. Define $\Lambda_{ijk}(s,t)$ to consist of the product of n lines that pass through all the points in the triangular grid except for Q_{ijk} by setting

$$\Lambda_{ijk}(s,t) = R_0(s,t) \cdots R_{i-1}(s,t) S_0(s,t) \cdots S_{j-1}(s,t) T_0(s,t) \cdots T_{k-1}(s,t)$$

(see Figure 2.29). We claim that

$$L_{ijk}(s,t) = \frac{\Lambda_{ijk}(s,t)}{\Lambda_{ijk}(s_{ijk}, t_{ijk})},$$

where (s_{ijk}, t_{ijk}) are the rectangular coordinates of Q_{ijk}.

Let's check that this really works. First observe that $\Lambda_{ijk}(s,t)$ is a polynomial of degree n, since $i + j + k = n$. Thus $L_{ijk}(s,t)$ has the correct degree. Next notice that $\Lambda_{ijk}(s_{ijk}, t_{ijk}) \neq 0$ since $Q_{ijk} = R_i \cap S_j \cap T_k$, so Q_{ijk} does not lie on any of the lines in the product for Λ_{ijk}. Finally, we have the cardinal conditions

$$
\begin{aligned}
L_{ijk}(Q_{\alpha\beta\gamma}) &= 0 && (i,j,k) \neq (\alpha,\beta,\gamma) \\
&= 1 && (i,j,k) = (\alpha,\beta,\gamma).
\end{aligned}
\tag{2.26}
$$

The first equality follows because if $(i,j,k) \neq (\alpha,\beta,\gamma)$, then since $\alpha + \beta + \gamma = n = i + j + k$, either $\alpha < i$ or $\beta < j$ or $\gamma < k$. Hence $\Lambda_{ijk}(Q_{\alpha\beta\gamma}) = 0$. The second equality follows by our normalization, since

$$L_{ijk}(Q_{ijk}) = \frac{\Lambda_{ijk}(Q_{ijk})}{\Lambda_{ijk}(Q_{ijk})} = 1.$$

Therefore,

$$\Delta(s,t) = \sum_{ijk} L_{ijk}(s,t) P_{ijk}$$

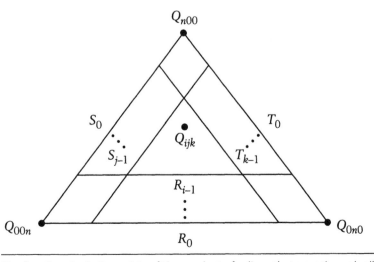

Figure 2.29 The function $\Lambda_{ijk}(s,t)$ consists of the product of n lines that pass through all the points in the triangular grid except for Q_{ijk}.

is indeed an interpolant because by (2.26)

$$\Delta(Q_{\alpha\beta\gamma}) = P_{\alpha\beta\gamma}.$$

In Exercise 9 you will show that this interpolant is exactly the same as the interpolant generated by the pyramid algorithm. Figure 2.30 illustrates the control points and the corresponding basis functions for the cubic triangular interpolant.

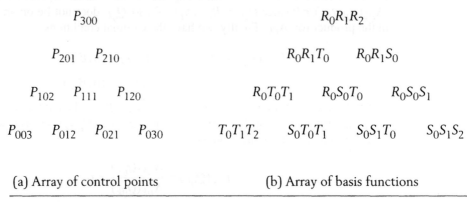

(a) Array of control points   (b) Array of basis functions

Figure 2.30 Arrays of (a) the control points and (b) (unnormalized) Lagrange basis functions for triangular cubic Lagrange surface patches. Here R_i, S_j, T_k are linear functions representing the lines that define the triangular grid in parameter space (see Figure 2.26).

Exercises

1. Use the explicit formula for the triangular Lagrange basis functions to develop a ladder evaluation algorithm for triangular Lagrange surface patches.

2. Suppose that the control points for a triangular surface are taken from the triangular array generated by Neville's algorithm for an interpolating curve. What will the surface look like?

3. Suppose that the control points for a triangular surface are taken from one of the triangular faces of the pyramid generated by Neville's algorithm for a tensor product surface. What will the surface look like?

4. Give necessary and sufficient conditions on the control points for the triangular Lagrange surface patch to collapse to

    a. a single point

    b. a line

    c. a plane

    Justify your answer.

5. Implement Neville's algorithm for triangular surface patches. Experiment with triangular surfaces of different degrees.

   a. How does altering the arrangement of the control points affect the shape of the surface?

   b. How does changing the triangular grid affect the shape of the surface?

6. a. What is the down recurrence for Neville's pyramid algorithm for triangular surface patches?

   b. Explain why this down recurrence is faster for evaluation than the up recurrence.

7. Consider an array $\{P_{ij}\}$ of $\binom{n+2}{2}$ control points in 3-space and another array $\{Q_{ij}\}$ of $\binom{n+2}{2}$ nodes in the parameter plane. Suppose that the nodes are generated by the intersection of $n+2$ lines $L_1,\dots,L_{n+2}$ in the parameter plane (see Figure 2.31).

   a. Prove that there exists a bivariate polynomial of degree n that interpolates the given array of control points at the specified parameter values.

   b. What are the Lagrange basis functions for this interpolant?

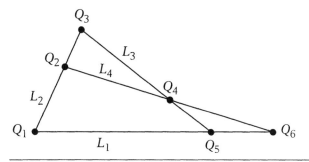

Figure 2.31 Six points generated by the intersection of four lines.

8. Let P_{ijk} be a triangular array of $\binom{n+2}{2}$ control points. Suppose we wish to interpolate these points with a surface at the parameter values (s_i,t_j). For each j, define $P_j(s)$ to be the curve that interpolates the control points $P_{0,j,n-j},\dots,P_{n-j,j,0}$ at the nodes $s_0,\dots,s_{n-j}$. Now define a point on the surface $P(s,t)$ to be the value of the curve $P(t)$ that interpolates the points $P_0(s),\dots,P_n(s)$ at the nodes $t_0,\dots,t_n$ (see Figure 2.32).

   a. Show that the surface $P(s,t)$ interpolates the points P_{ijk} at the nodes (s_i,t_j).

   b. Describe the curves $s = s_0$ and $t = t_0$.

   c. Show that this surface has a singularity at (s_0,t_n).

d. Suppose that $s_i = i$ and $t_j = j$ for all i, j. Then the points $(s_i, t_j) = (i, j)$ lie on a triangular grid.

   i.   Describe the image of the curve $s + t = n$ on the surface $P(s,t)$. In particular, which control points are interpolated by this curve?

   ii.   Show that this surface is not the same as the triangular Lagrange interpolant for the same nodes and control points.

   iii.   Describe a dynamic programming algorithm for generating this surface. Draw the diagram for $n = 3$.

   iv.   Show how to express this surface as a tensor product surface. In particular, describe the nodes and the control points.

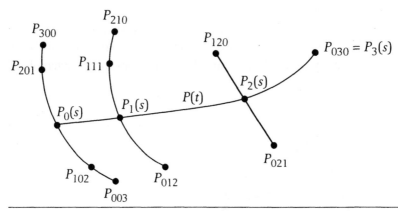

Figure 2.32 The construction of a three-sided surface that interpolates a triangular array of points.

9. Prove that the triangular surface generated by Neville's pyramid algorithm is identical to the triangular surface generated from the explicit basis functions $\{L_{ijk}(s,t)\}$ by using (2.25) and induction to show that the coefficient of P_{ijk} in the pyramid algorithm is $L_{ijk}(s,t)$.

10. Consider a collection of nodes Q_{ijk}, $i + j + k = n$, in the parameter domain such that any three nodes $Q_{ijk}, Q_{pqr}, Q_{uvw}$ are affinely independent provided that neither (i, p, u) nor (j, q, v) nor (k, r, w) are identical.

a. Given a triangular collection of control points P_{ijk}, $i + j + k = n$, show that the same tetrahedral algorithm that defines the Lagrange interpolant over a triangular grid generates a well-defined surface $Q(s,t)$ relative to the nodes $\{Q_{ijk}\}$ even if the nodes $\{Q_{ijk}\}$ do not form a triangular grid.

b. Explain why the surface $Q(s,t)$ need not interpolate the points $\{P_{ijk}\}$ at the nodes $\{Q_{ijk}\}$.

11. *L*-patches are surfaces defined in the following manner. Consider three sets of lines—$R_0,...,R_{n-1}$; $S_0,...,S_{n-1}$; and $T_0,...,T_{n-1}$—such that $R_\alpha \cap S_\beta \cap T_\gamma = \phi$, $\alpha + \beta + \gamma < n \cdot$

a. Show that if $\alpha + \beta + \gamma < n$, then there exist constants $r_{\alpha\beta\gamma}, s_{\alpha\beta\gamma}, t_{\alpha\beta\gamma}$ such that

$$r_{\alpha\beta\gamma}R_\alpha + s_{\alpha\beta\gamma}S_\beta + t_{\alpha\beta\gamma}T_\gamma = 1.$$

Given a triangular collection of control points P_{ijk}, $i + j + k = n$, the *L-patch* $L(s,t)$ is defined recursively by the tetrahedral algorithm

$$P_{ijk}^0 = P_{ijk}$$

$$P_{\alpha\beta\gamma}^d = r_{\alpha\beta\gamma}R_\alpha P_{\alpha+1,\beta\gamma}^{d-1} + s_{\alpha\beta\gamma}S_\beta P_{\alpha,\beta+1\gamma}^{d-1} + t_{\alpha\beta\gamma}T_\gamma P_{\alpha,\beta,\gamma+1}^{d-1}$$

$$L(s,t) = P_{000}^n \qquad\qquad \alpha + \beta + \gamma = n - d$$

Show that

b. There are constants c_{ijk} such that

$$L_{ijk}(s,t) = c_{ijk}R_0(s,t)\cdots R_{i-1}(s,t)S_0(s,t)\cdots S_{j-1}(s,t)T_0(s,t)\cdots T_{k-1}(s,t)$$

$$i + j + k = n,$$

$$L(s,t) = \sum_{i+j+k=n} L_{ijk}(s,t)P_{ijk} \cdot$$

c. If $R_i \cap S_j \cap T_k = Q_{ijk}$, $i + j + k = n$, then $L(s,t)$ is the triangular Lagrange interpolant for the control points $\{P_{ijk}\}$ with nodes $\{Q_{ijk}\}$.

12. A basis $\{L_{ijk}^n(s,t)\}$, $i + j + k = n$, is called a *lineal basis* if there are three sets of linear functions— $R_0,...,R_{n-1}$; $S_0,...,S_{n-1}$; and $T_0,...,T_{n-1}$—such that

$$L_{ijk}^n(s,t) = c_{ijk}R_0(s,t)\cdots R_{i-1}(s,t)S_0(s,t)\cdots S_{j-1}(s,t)T_0(s,t)\cdots T_{k-1}(s,t)$$

$$i + j + k = n \cdot$$

Show that the following bases are lineal bases:

a. $M_{ijk}^n(s,t) = \binom{n}{ijk}s^i t^j$ (monomial basis)

b. $B_{ijk}^n(s,t) = \binom{n}{ijk}s^i t^j(1 - s - t)^k$ (Bernstein basis)

2.13 Uniqueness of the Bivariate Lagrange Interpolant

In the preceding section we provided two distinct constructions for the triangular Lagrange interpolant: a dynamic programming algorithm based on a bivariate version of Neville's algorithm and an explicit formula using bivariate Lagrange basis functions. In Exercise 9 of the preceding section you showed that these two constructions generate identical surfaces. But perhaps there are yet other constructions that lead to different interpolating surfaces of the same degree. Here we shall show that just like in the univariate setting, the bivariate interpolant of fixed degree is unique. We shall concentrate on triangular patches; the rectangular tensor product case is similar so we leave it as an exercise (see Exercise 1). We begin with some simple observations about bivariate polynomials.

LEMMA 2.11   Let $l(s,t) = as + bt + c$, $a \neq 0$, be a linear function and let $f(s,t)$ be a bivariate polynomial of degree n. Then there is a bivariate polynomial $g(s,t)$ of degree $n - 1$ and a univariate polynomial $h(t)$ of degree n such that $f(s,t) = l(s,t)g(s,t) + h(t)$.

Proof   Treating $f(s,t)$ and $l(s,t)$ as polynomials in s with coefficients in t, we can write

$$f(s,t) = f_0(t)s^n + \cdots + f_{n-1}(t)s + f_n(t)$$
$$l(s,t) = l_0(t)s + l_1(t) \; ,$$

where $f_k(t)$ and $l_k(t)$ are polynomials of degree k in t. Since $l_0(t)$ is a non-zero constant, the standard long division algorithm for polynomials in s yields

$$f(s,t) = l(s,t)g(s,t) + h(t),$$

where $g(s,t)$ is a polynomial of degree $n - 1$ and $h(t)$ is a polynomial of degree n.

PROPOSITION 2.12   Let $l(s,t)$ be a linear function and let $f(s,t)$ be a polynomial of degree n. If the two curves $l(s,t) = 0$ and $f(s,t) = 0$ have $n + 1$ points in common, then there is a polynomial $g(s,t)$ of degree $n - 1$ such that $f(s,t) = l(s,t)g(s,t)$.

Proof   By Lemma 2.11, if $l(s,t) = as + bt + c$, $a \neq 0$, then

$$f(s,t) = l(s,t)g(s,t) + h(t)$$

where $g(s,t)$ is a polynomial of degree $n - 1$ and $h(t)$ is a polynomial of degree n. Suppose that $l(s,t) = 0$ and $f(s,t) = 0$ have $n + 1$ points (s_1,t_1), ..., (s_{n+1}, t_{n+1}) in common. Then

$$h(t_k) = 0, \; k = 1, \ldots, n + 1.$$

Thus $h(t)$ is a polynomial of degree n with $n + 1$ distinct roots. Hence by Corollary 2.4 $h(t)$ is identically zero, so $f(s,t) = l(s,t)g(s,t)$. On the other hand, if $a = 0$, then the result is still valid because we can just reverse the roles of s and t.

Proposition 2.12 is actually a special case of Bezout's Theorem, which is a fundamental result in algebraic geometry. Bezout's Theorem asserts that if $f(s,t)$ is a polynomial of degree m and $g(s,t)$ is a polynomial of degree n and the curves $f(s,t) = 0$ and $g(s,t) = 0$ agree at more than mn points, then $f(s,t)$ and $g(s,t)$ have a nonconstant common factor. Here, however, we need only use our special case of Bezout's Theorem to establish the uniqueness of the triangular Lagrange interpolant.

THEOREM 2.13

The degree n Lagrange interpolant on a triangular grid of size n is unique.

Proof   Consider a triangular grid of size n defined by three sets of lines $R_0,...,R_n$; $S_0,...,S_n$; and $T_0,...,T_n$. Suppose that $F(s,t)$ and $G(s,t)$ are two degree n polynomials that interpolate the same data on this grid, and let $H(s,t) = F(s,t) - G(s,t)$. We shall prove that $H(s,t)$ is identically zero by showing that $T_0(s,t),...,T_n(s,t)$ are all factors of $H(s,t)$. To proceed, observe that by construction, $H(s,t)$ vanishes at all the points on the grid; hence the degree n curve $H(s,t) = 0$ and the line $T_0(s,t) = 0$ agree at $n + 1$ points. Therefore, by Proposition 2.12, there is a polynomial $H_1(s,t)$ of degree $n - 1$ such that $H(s,t) = T_0(s,t)H_1(s,t)$. Now suppose that we have already shown that $H(s,t) = T_0(s,t)\cdots T_p(s,t)H_{p+1}(s,t)$, where $H_{p+1}(s,t)$ is a polynomial of degree $n - p - 1$. Then since $H(s,t)$ vanishes at all points on the grid and the lines $T_0,...,T_p$ do not pass through any of the grid points on the line T_{p+1}, $H_{p+1}(s,t) = 0$ and $T_{p+1}(s,t) = 0$ agree at $n - p$ points. Hence $H_{p+1}(s,t)$ is divisible by $T_{p+1}(s,t)$. Therefore it follows by induction on p that $H(s,t) = H_{n+1}(s,t)T_0(s,t)\cdots T_n(s,t)$, where H_{n+1} is a constant. But this is impossible unless $H_{n+1} = 0$, since, by construction, $H(s,t)$ is a polynomial of degree n, so $H(s,t)$ cannot factor into $n + 1$ linear factors. Hence $H(s,t) \equiv 0$. We conclude that $F(s,t) = G(s,t)$, and therefore that the triangular interpolant is unique.

It follows immediately from Theorem 2.13 that the bivariate version of Neville's algorithm and the explicit formula based on the bivariate Lagrange basis functions generate the same interpolating surface.

COROLLARY 2.14

Let $\{Q_{ijk}\}$ be the grid points on a triangular grid of size n. If $P(s,t)$ is a polynomial of degree n, then $P(s,t) = \sum_{ijk} L_{ijk}(s,t)P(Q_{ijk})$.

Proof   The polynomials $P(s,t)$ and $\sum_{ijk} L_{ijk}(s,t)P(Q_{ijk})$ both interpolate the values $\{P(Q_{ijk})\}$ at the nodes $\{Q_{ijk}\}$. Hence by the uniqueness of the polynomial interpolant $P(s,t) = \sum_{ijk} L_{ijk}(s,t)P(Q_{ijk})$.

Exercises

1. Let $f(s,t)$ be a bivariate polynomial of bidegree (m,n), and let a be a fixed constant.

   a. Show that there is a bivariate polynomial $g(s,t)$ of bidegree $(m-1,n)$ and a univariate polynomial $h(t)$ of degree n such that $f(s,t) = (s-a)g(s,t) + h(t)$.

   b. Show that if $f(a,t) = 0$ has $n+1$ roots, then there is a polynomial $g(s,t)$ of bidegree $(m-1,n)$ such that $f(s,t) = (s-a)g(s,t)$.

   c. Show that the tensor product Lagrange interpolant on a rectangular grid of size $(m+1) \times (n+1)$ is unique.

   d. Conclude that Neville's algorithm based on successive univariate interpolation in s and t and the pyramid algorithm based on bilinear interpolation generate the same interpolating surface.

2. a. Prove that the functions $\{L_k^m(s \mid s_0,...,s_m)L_l^n(s \mid t_0,...,t_n)\}$ form a basis for the bivariate polynomials of bidegree (m,n) by using (2.12) to show that these functions form a maximal linearly independent collection of polynomials of bidegree (m,n).

   b. Using part (a), prove that the tensor product Lagrange interpolant on a rectangular grid of size $(m+1) \times (n+1)$ is unique.

3. a. Without appealing to Corollary 2.14, prove that the functions $\{L_{ijk}(s,t)\}$ form a basis for the bivariate polynomials of degree n by using (2.26) to show that $\{L_{ijk}(s,t)\}$ is a maximal linearly independent collection of polynomials of degree n.

   b. Using part (a), prove Theorem 2.13.

4. Prove that $\sum_{ijk} L_{ijk}(s,t) \equiv 1$.

5. Show that the nodes of a triangular grid of size n cannot lie on a polynomial curve $f(s,t) = 0$ of degree n.

6. Let $\{Q_{ijk}\}$ be a triangular grid of size n. What coefficients should we put at the base of Neville's algorithm to generate the polynomials $s, t, s^p t^q$, $0 \le p + q \le n$?

7. Suppose that $\{Q_{ijk}\}$ is a triangular grid of size n. Let $P(s,t)$ be a polynomial of degree n, and let $\Delta(s,t)$ be the polynomial generated by Neville's algorithm that interpolates the control points $P(Q_{ijk})$. Prove that $\Delta(s,t) = P(s,t)$.

8. Without appealing to Proposition 2.12, give an elementary proof that if a line $L(s,t) \equiv as + bt + c = 0$ intersects a degree n polynomial curve $P(s,t) = 0$ in more than n points, then every point on the line lies on the polynomial curve.

9. We say that two surfaces $P(s,t)$ and $Q(s,t)$ are equivalent if $P(s,t) = Q(s,t)$ for all (s,t).

   a. Show that every $(m+1) \times (n+1)$ rectangular grid is embedded in a triangular grid of size $m+n$.

b. Show that the triangular grid $T_n = \{(i,j) \mid i,j \geq 0, \; i+j \leq n\}$ of size n is embedded in an $(n+1) \times (n+1)$ rectangular grid.

c. Show that every tensor product Lagrange interpolant $P(s,t)$ of bidegree (m,n) is equivalent to a unique triangular Lagrange interpolant $\Delta(s,t)$ of degree $m+n$.

d. Show that every triangular Lagrange interpolant $\Delta(s,t)$ of degree n on the grid T_n is equivalent to a unique tensor product Lagrange interpolant $P(s,t)$ of bidegree (n,n).

10. Suppose that $\{F_{ij}(s,t)\}$ is a collection of blending functions and that $\{(s_{ij}, t_{ij})\}$ is a collection of nodes for some fixed finite set of indices $\{(i,j) \in I\}$.

a. Show that for each collection of points $\{P_{ij}\}$, there is a function

$$F(s,t) = \textstyle\sum_{(i,j) \in I} c_{ij} F_{ij}(s,t) \text{ such that } F(s_{ij}, t_{ij}) = P_{ij} \text{ for all } (i,j) \in I,$$

if and only if $\det \left| F_{pq}(s_{kl}, t_{kl}) \right| \neq 0$.

b. Show that if $\det \left| F_{pq}(s_{kl}, t_{kl}) \right| \neq 0$, then the interpolant

$$F(s,t) = \textstyle\sum_{(i,j) \in I} c_{ij} F_{ij}(s,t)$$

that satisfies $F(s_{ij}, t_{ij}) = P_{ij}$ for all $(i,j) \in I$ is unique.

c. Deduce that if an interpolant

$$F(s,t) = \textstyle\sum_{(i,j) \in I} c_{ij} F_{ij}(s,t)$$

exists for every collection of points $\{P_{ij}\}$, $(i,j) \in I$, then the interpolant must be unique.

d. Conclude that

   i. The degree n Lagrange interpolant on a triangular grid of size n is unique.

   ii. The bidegree (m,n) Lagrange interpolant on a rectangular grid of size $(m+1) \times (n+1)$ is unique.

2.14 Rational Lagrange Surfaces

Many common surfaces such as the sphere and the torus cannot be represented exactly by polynomial parametrizations. As with nonpolynomial curves, we could try to approximate these surfaces with polynomials by interpolating lots of points along the surface. Unfortunately, often we would need to use polynomials of quite high degree to generate a good approximation.

The sphere, however, like the circle, has a rational parametrization; in fact, the sphere has several different rational parametrizations. Below we give a rational quadratic and a rational biquadratic parametrization for the unit sphere.

■   *Quadratic parametrization of the sphere*

$$x = \frac{2s}{1+s^2+t^2} \qquad y = \frac{2t}{1+s^2+t^2} \qquad z = \frac{1-s^2-t^2}{1+s^2+t^2} \qquad (2.27)$$

■   *Biquadratic parametrization of the sphere*

$$x = \frac{2s(1-t^2)}{(1+s^2)(1+t^2)} \qquad y = \frac{2t(1+s^2)}{(1+s^2)(1+t^2)} \qquad z = \frac{(1-s^2)(1-t^2)}{(1+s^2)(1+t^2)}. \qquad (2.28)$$

For both of these parametrizations, you can easily check that $x^2 + y^2 + z^2 = 1$, so both parametrizations do indeed represent the same unit sphere.

To represent a rational surface using Lagrange interpolation, we proceed just as in the case of a rational curve: we lift the surface from a rational parametrization in affine space to a polynomial parametrization in Grassmann space by treating the denominator as mass. For example, for the sphere we can consider the four-dimensional surfaces

$$x = 2s \qquad y = 2t \qquad z = 1-s^2-t^2 \qquad w = 1+s^2+t^2$$

$$x = 2s(1-t^2) \qquad y = 2t(1+s^2) \qquad z = (1-s^2)(1-t^2) \qquad w = (1+s^2)(1+t^2)\ .$$

To find the mass-points that are the control points of a polynomial surface in Grassmann space, we first select a grid—a triangular grid for parametrizations of total degree n or a rectangular grid for parametrizations of bidegree (m,n)—and then evaluate the polynomial parametrization at the points of the grid. Dividing by the mass yields the control points along the original rational surface.

We can also use the Lagrange blending functions to write explicit formulas for a rational surface. If $(m_{ijk}P_{ijk}, m_{ijk})$ or $(m_{jk}P_{jk}, m_{jk})$ represent the control points of the corresponding polynomial surface in Grassmann space, then projecting from Grassmann space to affine space, we obtain

■   *Rational triangular Lagrange parametrization*

$$R(s,t) = \frac{\sum_{ijk} m_{ijk} P_{ijk} L_{ijk}(s,t)}{\sum_{ijk} m_{ijk} L_{ijk}(s,t)} \qquad (2.29)$$

■   *Rational tensor product Lagrange parametrization*

$$R(s,t) = \frac{\sum\limits_{k=0}^{m}\sum\limits_{l=0}^{n} m_{kl} P_{kl} L_k^m(s \mid s_0,...,s_m) L_l^n(t \mid t_0,...,t_n)}{\sum\limits_{k=0}^{m}\sum\limits_{l=0}^{n} m_{kl} L_k^m(s \mid s_0,...,s_m) L_l^n(t \mid t_0,...,t_n)}\ . \qquad (2.30)$$

Rational Lagrange surfaces defined by (2.29) and (2.30) interpolate their control points P_{ijk} or P_{jk}, since the corresponding polynomial surfaces in Grassmann space interpolate the mass-points $(m_{ijk}P_{ijk}, m_{ijk})$ or $(m_{jk}P_{jk}, m_{jk})$. Thus the masses serve as shape parameters: they do not affect interpolation, but they do affect the shape of the surface. But just as in the case of rational Lagrange curves, modest changes to a single mass can produce drastic changes in the shape of a surface. Thus care must be taken when modifying the masses of a rational Lagrange surface.

In a rational tensor product Lagrange surface some masses m_{kl} may be set to zero. When $m_{kl} = 0$, the mass-point $(m_{kl}P_{kl}, m_{kl})$ is replaced by a vector $(v_{kl}, 0)$ in Grassmann space. Typically when $m_{kl} = 0$, the rational Lagrange surface $R(s,t)$ has a *singularity* at (s_k, t_l) because when $m_{kl} = 0$ all the Lagrange basis functions in the denominator vanish at (s_k, t_l). If, however, both $m_{kl} = 0$ and $v_{kl} = 0$, the numerator also vanishes and the singularity is replaced by a *base point*, a parameter pair (s_k, t_l) where both the numerator and the denominator vanish. Although, unlike curves, there may be no common factor in the numerator and denominator, base points lower the implicit degree of the parametric surface. Nevertheless, the resulting rational surface still interpolates the control points P_{ij}, $(i,j) \neq (k,l)$. Analogous results hold for triangular surfaces.

If the denominator of a rational Lagrange surface in (2.29) or (2.30) is ever zero and the numerator is nonzero, then we cannot project the surface continuously into affine space. Rather we must project the surface into projective space. Notice, therefore, that for a rational triangular Lagrange surface, the parameter space is an affine space, the control points reside in a Grassmann space, and the surface itself lies in a projective space.

Exercises

1. What is the effect on a rational tensor product Lagrange surface if $m_{kl} = 0$ for all $l = 0,\dots,n$?

2. Experiment with altering the masses in a rational Lagrange surface.

    a. What are the local and global effects of altering a single mass?

    b. What is the effect of a negative mass?

    c. What happens if all the masses are changed simultaneously?

3. a. Find the control points and the masses for the sphere given by the quadratic parametrization (2.27) relative to the triangular grid $T_2 = \{(i,j) \mid i, j \geq 0, \; i + j \leq 2\}$.

    b. Use the results of part (a) together with Neville's pyramid algorithm to render the sphere.

4. a. Find the control points and the masses for the sphere given by the bi-quadratic parametrization (2.28) relative to the rectangular grid $R_{2,2} = \{(i,j) \mid 0 \leq i, j \leq 2\}$.

    b. Use the results of part (a) together with Neville's algorithm to render the sphere.

5. The torus with inner radius $d - a$ and outer radius $d + a$ is the locus of points that satisfy the degree 4 algebraic equation

$$\left(x^2 + y^2 + z^2 - d^2 - a^2\right)^2 + 4d^2z^2 - 4a^2d^2 = 0.$$

a. Verify that

$$x = \frac{d(1 + s^2)(1 - t^2) + a(1 - s^2)(1 - t^2)}{(1 + s^2)(1 + t^2)}$$

$$y = \frac{2d(1 + s^2)t + 2a(1 - s^2)t}{(1 + s^2)(1 + t^2)}$$

$$z = \frac{2as(1 + t^2)}{(1 + s^2)(1 + t^2)}$$

is a rational biquadratic parametrization of the torus.

b. Find the control points and the masses for the torus given by the biquadratic parametrization in part (a) relative to the rectangular grid $R_{2,2} = \{(i, j) \mid 0 \le i, j \le 2\}$.

c. Use the results of part (b) together with Neville's algorithm to render the torus with $d = 5$ and $a = 2$.

6. Let $x = f(s)$, $z = g(s)$ be a curve in the xz-plane.

a. Verify that the surface of revolution generated by rotating this curve around the z-axis can be represented by the parametric equations

$$x = \frac{(1 - t^2)f(s)}{1 + t^2}, \qquad y = \frac{2tf(s)}{1 + t^2}, \qquad z = \frac{(1 - t^2)g(s)}{1 + t^2}.$$

b. Conclude that if the original curve $x = f(s)$, $z = g(s)$ is a rational curve of degree m, then the corresponding surface of revolution is a rational surface of bidegree $(m, 2)$.

c. Use the result of part (a) to generate rational parametrizations for the right circular cylinder and right circular cone by rotating a line about the z-axis.

   i. Find the control points and the masses for the cylinder and cone given by these parametrizations relative to the rectangular grid $R_{1,2} = \{(i, j) \mid 0 \le i \le 1, 0 \le j \le 2\}$.

   ii. Use the results of part (i) together with Neville's algorithm to render the right circular cylinder and right circular cone.

d. Use the result of part (a) to generate rational parametrizations for the sphere and the torus by rotating a circle about the z-axis.

   i. Find the control points and the masses for the sphere and the torus given by these parametrizations relative to the rectangular grid $R_{2,2} = \{(i, j) \mid 0 \le i, j \le 2\}$.

    ii. Use the results of part (i) together with Neville's algorithm to render the sphere and the torus.

7. a. Show that

$$x = \frac{2s(1-t^2)}{(1-s^2)(1+t^2)} \qquad y = \frac{4st}{(1-s^2)(1+t^2)} \qquad z = \frac{(1+s^2)(1+t^2)}{(1-s^2)(1+t^2)}$$

is a parametrization for the hyperboloid $z^2 - x^2 - y^2 = 1$.

b. Find the control points and the masses for this hyperboloid relative to the rectangular grid $R_{2,2} = \{(i,j) \mid 0 \le i, j \le 2\}$.

c. Use the results of part (b) together with Neville's algorithm to render this hyperboloid.

8. Let $R(s,t)$ be a rational Lagrange surface over a grid G (triangular or rectangular) with control points $(m_g P_g, m_g)$, $g \in G$. Let m_h increase and hold m_g fixed for $g \ne h$.

a. Show that $\lim_{m_h \to \infty} R(g) = P_g$.

b. $\lim_{m_h \to \infty} R(s,t) = P_h$ for all $(s,t) \notin G$.

c. Conclude that the limit surface is a disconnected collection of points.

d. What does the limit surface look like if several masses are allowed to increase simultaneously?

9. Given a grid G (triangular or rectangular) and masses $\{m_g\}$, $g \in G$, define

$$R_g(s,t) = \frac{m_g L_g(s,t)}{\sum_{h \in G} m_h L_h(s,t)}, \quad g \in G.$$

Show that these functions behave like rational Lagrange basis functions. In particular,

a. $\sum_{g \in G} R_g(s,t) \equiv 1$.

b. $R_g(h) = 0 \qquad h \ne g$
$ = 1 \qquad h = g.$

c. $R(s,t) = \sum_{g \in G} R_g(s,t) P_g \Rightarrow R(h) = P_h$ for all $h \in G$.

2.15 Ruled, Lofted, and Boolean Sum Surfaces

So far we have performed only discrete interpolation; that is, we have developed curve and surface techniques to interpolate finite collections of control points. But if we replace the control points by curves, then essentially the same techniques can be applied to accomplish *transfinite interpolation*—that is, interpolation of an infinite collection of points on a finite collection of curves.

Suppose, for example, that we are given two curves $U_0(s)$ and $U_1(s)$ and we require a surface to pass through these curves. We can perform linear interpolation on the curves to generate the surface

$$R(s,t) = (1-t)U_0(s) + tU_1(s).$$

This expression is the same linear interpolation formula we first used to interpolate two points, only now the two points have been replaced by two curves. If we fix the value of $s = s^*$, then $R(s^*,t)$ is the line connecting the two points $U_0(s^*)$ and $U_1(s^*)$. Thus a line passes through each point on this surface. For this reason $R(s,t)$ is called a *ruled surface* (see Figure 2.33).

When $U_1(s)$ is a translate of $U_0(s)$—that is, when $U_1(s) = U_0(s) + v$—then

$$R(s,t) = U_0(s) + tv$$

is called a *cylinder* over $U_0(s)$. When $U_1(s)$ collapses to a single point V, then

$$R(s,t) = (1-t)U_0(s) + tV$$

is called the *cone* over $U_0(s)$ with vertex V (see Figure 2.34).

Lofting generalizes ruled surfaces by applying Lagrange interpolation to an arbitrary finite number of curves. Given a collection of curves $U_0(s),...,U_n(s)$, we can construct a surface that interpolates all these curves by setting

$$L_U(s,t) = \sum_{k=0}^{n} L_k^n(t \mid t_0,...,t_n) U_k(s).$$

The surface $L_U(s,t)$ is called the *lofted surface* generated by the *rail curves* $U_0(s),...,U_n(s)$ (see Figure 2.35). It follows immediately from the cardinal conditions (2.12) that

$$L_U(s,t_j) = U_j(s).$$

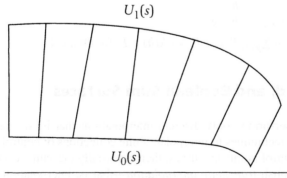

Figure 2.33 A ruled surface interpolating the curves $U_0(s)$ and $U_1(s)$.

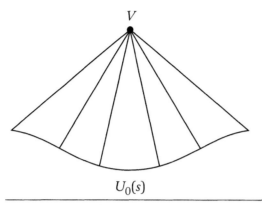

Figure 2.34 The cone over $U_0(s)$ with vertex V.

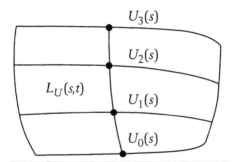

Figure 2.35 A lofted surface $L_U(s,t)$ interpolating the curves $U_0(s),...,U_3(s)$.

Provided we have some procedure for computing points along the rails $U_0(s),...,U_n(s)$, we can compute points on this lofted surface using any of our evaluation algorithms for interpolating curves; we simply replace the control points $P_0,...,P_n$ by the rails $U_0(s),...,U_n(s)$ and apply the method. Thus Neville's algorithm, the ladder algorithm, or the down recurrence can all be used to calculate points along $L_U(s,t)$.

Designers often prefer to specify a surface by a mesh of curves rather than a sequence of curves because a mesh gives them finer control over the shape of their surface. A mesh of curves is the image of a grid of lines in the parameter plane. Thus a mesh can be specified by two sets of curves $U_0(s),...,U_m(s)$ and $V_0(t),...,V_n(t)$, where each U-curve intersects each V-curve at

$$P_{jk} = U_k(s_j) = V_j(t_k).$$

A surface is then required to interpolate all the curves in the mesh or equivalently to fill in the spaces between the grid lines in the parameter plane (see Figure 2.36).

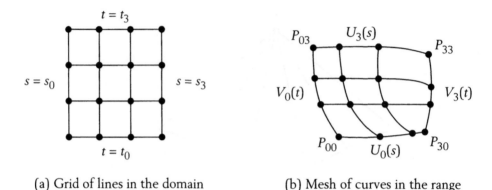

(a) Grid of lines in the domain   (b) Mesh of curves in the range

Figure 2.36 A Boolean sum surface interpolates a mesh of curves in the range that is the image of a grid of lines in the parameter domain.

How do we construct such a surface? We already have a surface that interpolates all the U-curves—the lofted surface $L_U(s,t)$. Similarly, the lofted surface $L_V(s,t)$ interpolates all the V-curves. One idea we might try is to add these surfaces together to generate a surface that interpolates the entire mesh. But this approach cannot be quite right for two reasons. First, surfaces are collections of points in an affine space, and you cannot add points in an affine space. Second, the intersection points P_{jk} would be counted twice, once for the U-curves and once for the V-curves. Thus we need somehow to subtract out one copy of the intersection points P_{jk}. The solution to both problems is to add the lofted surfaces, but then to subtract out the tensor product surface generated by the intersection points. This approach yields the *Boolean sum surface* defined by setting

$$B(s,t) = L_U(s,t) + L_V(s,t) - T_P(s,t), \tag{2.31}$$

where $L_U(s,t)$ is the lofted surface for the U-curves in the s direction, $L_V(s,t)$ is the lofted surface for the V-curves in the t direction, and $T_P(s,t)$ is the interpolating tensor product surface for the control points P_{jk}.

Let's check that the Boolean sum surface actually does interpolate all the curves in the mesh. We can expand (2.31) by substituting in the definitions of the lofted and tensor product surfaces to obtain

$$B(s,t) = \sum_{k=0}^{m} L_k^m(t \mid t_0,...,t_m) U_k(s) + \sum_{j=0}^{n} L_j^n(s \mid s_0,...,s_n) V_j(t)$$

$$- \sum_{k=0}^{m} \sum_{j=0}^{n} L_j^n(s \mid s_0,...,s_n) L_k^m(t \mid t_0,...,t_m) P_{jk} \ .$$

Now let's evaluate along $s = s_i$. By the cardinal conditions (2.12)

$$B(s_i,t) = \sum_{k=0}^{m} L_k^m(t \mid t_0,...,t_m)U_k(s_i) + V_i(t) - \sum_{k=0}^{m} L_k^m(t \mid t_0,...,t_m)P_{ik} \ .$$

But $U_k(s_i) = P_{ik}$, so the first and last sums cancel. Therefore,

$$B(s_i,t) = V_i(t).$$

Similarly, evaluating along $t = t_i$, we obtain

$$B(s,t_i) = U_i(s),$$

so the Boolean sum surface $B(s,t)$ does indeed interpolate the entire mesh of curves.

Exercises

1. Complete the proof that the Boolean sum surface interpolates a mesh of curves by showing that it interpolates the U-curves as well as the V-curves.

2. Show that a lofted surface is equivalent to a tensor product surface if all the rails are polynomial curves.

3. Show that a Boolean sum surface is equivalent to a tensor product surface if all the curves in the mesh are polynomial curves.

4. Using the parametrization of the circle

$$x(t) = \frac{2t}{1+t^2} \qquad y(t) = \frac{1-t^2}{1+t^2}$$

   a. Generate a rational parametrization for a right circular cylinder as a ruled surface.

   b. Generate a rational parametrization for a right circular cone as a ruled surface.

5. Consider three curves U_1,U_2,U_3 defined over the edges of a triangle with vertices Q_1,Q_2,Q_3 such that $U_i(Q_k) = U_j(Q_k) = P_k$, $i \neq j \neq k$—see Figure 2.37. Let β_1,β_2,β_3 be the barycentric coordinates on $\Delta Q_1 Q_2 Q_3$, and let

$$T_P(\beta_1,\beta_2,\beta_3) = \beta_1 P_1 + \beta_2 P_2 + \beta_3 P_3$$

be the plane specified by the three points P_1,P_2,P_3. Define three cones

$$C_{U_i}(\beta_1,\beta_2,\beta_3) = (1 - \beta_i)U_i\left(\frac{\beta_j Q_j + \beta_k Q_k}{1 - \beta_i}\right) + \beta_i P_i, \quad i \neq j \neq k$$

and the surface

$$B(\beta_1,\beta_2,\beta_3) = C_{U_1}(\beta_1,\beta_2,\beta_3) + C_{U_2}(\beta_1,\beta_2,\beta_3) + C_{U_3}(\beta_1,\beta_2,\beta_3)$$
$$- 2T_P(\beta_1,\beta_2,\beta_3).$$

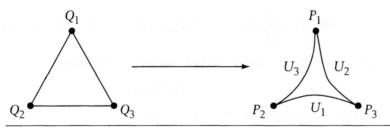

Figure 2.37 Three curves U_1, U_2, U_3 defined over the edges of triangle $\Delta Q_1 Q_2 Q_3$.

Show that

a. $C_{U_i}(\beta_1, \beta_2, \beta_3) = U_i$, when $\beta_i = 0$, $i = 1,2,3$.

b. $B(\beta_1, \beta_2, \beta_3) = U_i$, when $\beta_i = 0$, $i = 1,2,3$.

Thus the surface $B(\beta_1, \beta_2, \beta_3)$ is an analogue for triangles of the Boolean sum surface for rectangles.

6. Suppose that $D_0^n(s),...,D_n^n(s)$ and $E_0^m(t),...,E_m^m(t)$ are collections of functions, not necessarily polynomials, that satisfy the cardinal conditions at $s_0,...,s_n$ and $t_0,...,t_m$, that is,

$$\begin{aligned} D_k^n(s_i) &= 0 & i \neq k & & E_k^m(t_j) &= 0 & j \neq k \\ &= 1 & i = k & \text{and} & &= 1 & j = k \end{aligned}.$$

Consider a mesh of curves $U_0(s),...,U_m(s)$ and $V_0(t),...,V_n(t)$, where each U-curve intersects each V-curve at

$$P_{jk} = U_k(s_j) = V_j(t_k).$$

Define

$$DE_P(s,t) = \sum_{i=0}^{n} \sum_{j=0}^{m} D_i^n(s) E_j^m(t) P_{ij}$$

$$E_U(s,t) = \sum_{k=0}^{m} E_k^m(t) U_k(s) \qquad D_V(s,t) = \sum_{k=0}^{n} D_k^n(s) V_k(t)$$

$$DE_{UV}(s,t) = E_U(s,t) + D_V(s,t) - DE_P(s,t)$$

Show that

a. $DE_P(s_i, t_j) = P_{ij}$

b. $E_U(s, t_j) = U_j(s)$ and $D_V(s_i, t) = V_i(t)$

c. $DE_{UV}(s_i, t) = V_i(t)$ and $DE_{UV}(s, t_j) = U_j(s)$

2.16 **Summary**

In this chapter you have encountered most of the central ideas of this discipline: existence and uniqueness theorems; dynamic programming procedures; pyramid algorithms; up and down recurrences; basis functions; blends of overlapping data; rational schemes; tensor product, triangular, lofted, and Boolean sum surfaces; along with the use of barycentric coordinates to represent points in the domain of triangular surface patches. These themes will recur in various guises throughout this book. If you have understood everything in this chapter, the rest will be easy!

One core tenet of approximation theory and numerical analysis is that all polynomial bases are not equal. To solve problems in interpolation and approximation, *we must use the basis most appropriate to the problem at hand.* In this chapter we have seen that the Lagrange basis, and not the standard monomial basis, is most suited both for point interpolation and for polynomial multiplication. We continue with this theme in the next chapter, where we shall study Hermite interpolation—interpolation of both point and derivative data—by invoking special Hermite basis functions.

CHAPTER 3

Hermite Interpolation and the Extended Neville Algorithm

Lagrange polynomials interpolate positions; Hermite polynomials interpolate positions and directions—points and vectors, function values and derivatives. Hermite interpolation is important for several reasons. Frequently in computational science and engineering we have information about tangents, curvatures, or other higher-order derivatives at various locations, and we need to generate curves and surfaces that fit this data. In geometric design, interpolating derivative data gives us more control over the shape of the curve or surface. Moreover, often we want to connect two or more curves or surfaces; to join them smoothly, we require the ability to interpolate derivatives across common boundaries.

As we did with Lagrange interpolation, we will begin with curve schemes and then extend our techniques to surfaces. Many of the methods developed for Lagrange interpolation, including Neville's algorithm, extend readily to Hermite interpolation.

3.1 Cubic Hermite Interpolation

Two points determine a line, but a point and a direction vector also determine a line. Suppose we want the equation of the line $P(t)$ passing through the affine point P_0 in the direction v_0. Then we can write

$$P_{00}(t) = P_0 + tv_0. \tag{3.1}$$

Notice that $P_{00}(0) = P_0$ and $P_{00}'(0) = v_0$, so $P_{00}(t)$ does indeed interpolate position and derivative data at $t = 0$. We denote the line that interpolates the points P_0 at $t = 0$ and P_1 at $t = 1$ by $P_{01}(t)$; similarly we shall denote the line that interpolates the point P_0 and the direction vector v_0 at the parameter $t = 0$ (or any other parameter $t = t_0$) by $P_{00}(t)$. The double-zero subscript in $P_{00}(t)$ indicates that two pieces of information, position and direction vector—function value and derivative—are interpolated at $t = 0$.

Let's try a slightly harder problem. Suppose we have a pair of points and tangent vectors (P_0, v_0) and (P_1, v_1) that we wish to interpolate with a smooth curve at the parameters $t = 0$ and $t = 1$. How shall we proceed?

We have a way to interpolate (P_0, v_0) at $t = 0$; we can use the straight line $P_{00}(t)$. Similarly, we can interpolate (P_1, v_1) at $t = 1$ with the straight line

$$P_{11}(t) = P_1 + (t-1)v_1.$$

Somehow we need to blend these lines together to form a smooth interpolating curve. From Chapter 2 we know that with a quadratic curve, we can interpolate three data points. Here, however, we have four pieces of data—(P_0, v_0) and (P_1, v_1)—so it is unlikely that we could succeed with just a quadratic curve. Perhaps we should first attack a simpler problem that does have a quadratic solution, interpolating only three pieces of data. Let's try then to find a smooth curve that interpolates the data (P_0, v_0) at $t = 0$ and the point P_1 at $t = 1$.

The line $P_{00}(t)$ interpolates the data (P_0, v_0) at $t = 0$, and the line $P_{01}(t)$ interpolates the points P_0, P_1 at $t = 0, 1$. In Lagrange interpolation the trick for building the quadratic interpolant $P_{012}(t)$ is to perform linear interpolation on the linear interpolants $P_{01}(t)$ and $P_{12}(t)$. Let's try the same tactic here.

Applying linear interpolation to the two curves $P_{00}(t)$ and $P_{01}(t)$ generates the curve

$$P_{001}(t) = (1-t)P_{00}(t) + tP_{01}(t) \tag{3.2}$$

(see Figures 3.1 and 3.2). By substitution it is easy to verify that $P_{001}(t)$ interpolates P_0 at $t = 0$ and P_1 at $t = 1$, since by (3.2)

$$P_{001}(0) = P_{00}(0) = P_0$$
$$P_{001}(1) = P_{01}(1) = P_1 \ .$$

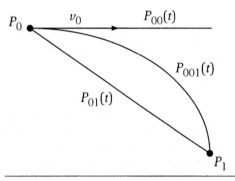

Figure 3.1 The two lines $P_{00}(t)$ and $P_{01}(t)$, and the quadratic interpolant $P_{001}(t)$ generated by linear interpolation on the two lines.

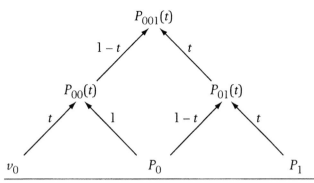

Figure 3.2 A graphical representation of Equation (3.2).

To verify that the derivative of $P_{001}(t)$ also interpolates the vector v_0 at $t = 0$, observe that $P_{00}(t)$ and $P_{01}(t)$ both interpolate the point P_0 at $t = 0$. Therefore,

$$P'_{001}(0) = P'_{00}(0) - P_{00}(0) + P_{01}(0) = v_0.$$

Thus our trick of performing linear interpolation on the linear interpolants actually works. Similarly, if we set

$$P_{011}(t) = (1-t)P_{01}(t) + tP_{11}(t),$$

then it is easy to verify that $P_{011}(0) = P_0$, $P_{011}(1) = P_1$, $P'_{011}(0) = v_1$.

By construction both $P_{001}(t)$ and $P_{011}(t)$ are quadratic curves. If Neville's algorithm really works for Hermite interpolation, then we should be able to form the cubic interpolant for the data (P_0, v_0) and (P_1, v_1) by linear interpolation on the quadratic interpolants $P_{001}(t)$ and $P_{011}(t)$.

Let's try this ploy one more time. Set

$$P_{0011}(t) = (1-t)P_{001}(t) + tP_{011}(t).$$

Using the properties already established for $P_{001}(t)$ and $P_{011}(t)$, we find that

$$P_{0011}(0) = P_{001}(0) = P_0$$
$$P_{0011}(1) = P_{011}(1) = P_1$$
$$P'_{0011}(0) = P'_{001}(0) - P_{001}(0) + P_{011}(0) = v_0$$
$$P'_{0011}(1) = P'_{011}(1) - P_{001}(1) + P_{011}(1) = v_1$$

so $P_{0011}(t)$ really is the desired cubic Hermite interpolant. The repeated subscripts on $P_{0011}(t)$ indicate that this function interpolates both the position and the tangent vector at $t = 0, 1$. We diagram our algorithm for building $P_{0011}(t)$ from the data (P_0, v_0), (P_1, v_1) in our usual triangular fashion in Figure 3.3.

We can also express the cubic Hermite interpolant explicitly in the form

$$P_{0011}(t) = H_0(t)P_0 + H_1(t)P_1 + h_0(t)v_0 + h_1(t)v_1.$$

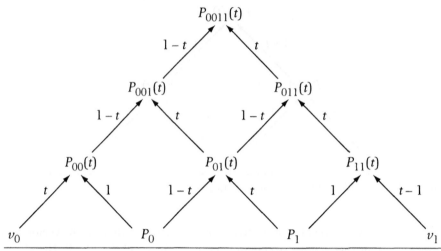

Figure 3.3 Neville's algorithm for cubic Hermite interpolation at $t = 0,1$. Notice that the labels entering a node sum to one only if both arrows emerge from a point or a curve. Labels on arrows emerging from vectors do not need to be normalized, since for vectors we are not required to take affine combinations.

As in Lagrange interpolation, the cubic Hermite basis functions $H_j(t), h_j(t), j = 0,1$, are the sums over all paths from their coefficient at the base of the Neville triangle to $P_{0011}(t)$ at the apex. Thus we can use either the up recurrence—replacing P_j or v_j by 1 and setting all the remaining data to 0—or the down recurrence—placing a 1 at the apex, reversing all the arrows, and collecting the basis functions at the base of the triangle—to find explicit expressions for these Hermite basis functions. Performing these calculations yields

$$H_0(t) = (1-t)^2(1+2t) \qquad h_0(t) = t(1-t)^2$$
$$H_1(t) = t^2(3-2t) \qquad h_1(t) = t^2(t-1) \ .$$

The formulas for the functions $h_0(t)$ and $h_1(t)$ are easy to derive: simply take the products of the labels along the left and right lateral edges of the triangle. The formulas for the functions $H_0(t)$ and $H_1(t)$ are only slightly more difficult to deduce. Since every path from P_0 to the apex of the triangle contains two factors of $(1-t)$, we can factor out $(1-t)^2$ and sum over the remaining paths. By inspection, this yields $H_0(t) = (1-t)^2(1+2t)$. Similarly, every path from P_1 to the apex of the triangle contains two factors of t, so here we can factor out t^2 and sum over the remaining paths. Again by inspection, this yields $H_1(t) = t^2(3-2t)$. Now it is easy to verify that $H_0(t) + H_1(t) \equiv 1$. This constraint is necessary, since these basis functions multiply points; there is no similar constraint on $h_0(t)$ and $h_1(t)$, since these functions multiply vectors.

The Lagrange basis functions are themselves Lagrange interpolants because they satisfy the cardinal conditions (2.12). Similarly, the cubic Hermite basis functions are individually cubic Hermite interpolants because by the up recurrence they satisfy

$$
\begin{array}{llll}
H_0(0) = 1 & H_0(1) = 0 & H_0'(0) = 0 & H_0'(1) = 0 \\
H_1(0) = 0 & H_1(1) = 1 & H_1'(0) = 0 & H_1'(1) = 0 \\
h_0(0) = 0 & h_0(1) = 0 & h_0'(0) = 1 & h_0'(1) = 0 \\
h_1(0) = 0 & h_1(1) = 0 & h_1'(0) = 0 & h_1'(1) = 1
\end{array}
$$

One important application of cubic Hermite interpolation is to generate piecewise cubic curves that join together smoothly. Given point and tangent vector data $(P_0, v_0), ..., (P_n, v_n)$, let $P_j(t)$ be the cubic Hermite interpolant generated by the data $(P_j, v_j), (P_{j+1}, v_{j+1})$. Then the piecewise cubic curve

$$
P(t) = P_j(t - j) \qquad j \le t \le j + 1
$$

has a continuous derivative at every point and interpolates all the data. This construction is one of the most common interpolation techniques in computer graphics and computer-aided design (see Figure 3.4).

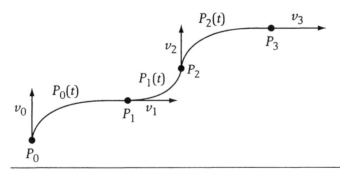

Figure 3.4 A smooth piecewise cubic curve that interpolates points and tangent vectors.

Exercises

1. Using Figure 3.2 for quadratic Hermite interpolation, compute the quadratic Hermite basis functions for the nodes 0,1.

2. Implement cubic Hermite interpolation and use this technique to construct smooth piecewise cubic curves.

3. Draw the diagram of Neville's algorithm for cubic Hermite interpolation at arbitrary nodes. What are the corresponding cubic Hermite basis functions?

4. a. Show that

$$H_0(1-t) = H_1(t) \qquad\qquad h_0(1-t) = -h_1(t)$$
$$H_1(1-t) = H_0(t) \qquad\qquad h_1(1-t) = -h_0(t).$$

  b. Let $P(t)$ be the cubic Hermite curve that interpolates the data (P_0, v_0), (P_1, v_1) and let $Q(t)$ be the cubic Hermite curve that interpolates the data $(P_1, -v_1)$, $(P_0, -v_0)$ at $t = 0, 1$. Conclude from part (a) that $Q(t) = P(1-t)$.

3.2 Neville's Algorithm for General Hermite Interpolation

Now that we understand cubic Hermite interpolation, let's consider a much more general Hermite interpolation problem. Given a collection of points and vectors

$$(P_0, v_{01}, ..., v_{0,\mu_0-1}), ..., (P_n, v_{n1}, ..., v_{n,\mu_n-1}),$$

we want to construct a polynomial curve $P_{\mu_0 \cdots \mu_n}(t)$ that interpolates all this data at the parameter values $t_0, ..., t_n$. By interpolating all the data we mean that

$$P_{\mu_0 \cdots \mu_n}(t_j) = P_j$$
$$P^{(k)}_{\mu_0 \cdots \mu_n}(t_j) = v_{jk} \qquad\qquad 1 \le k \le \mu_j - 1 .$$

Thus the vector v_{jk} represents the kth derivative of the interpolant at the point P_j. In the notation $P_{\mu_0 \cdots \mu_n}(t)$ the integer n tells us there are $n + 1$ points to interpolate, and the integer μ_j tells us we must interpolate μ_j pieces of data—$P_j, v_{j1}, ..., v_{j,\mu_j-1}$—at the parameter t_j. We expect that

$$\text{degree }\{P_{\mu_0 \cdots \mu_n}(t)\} = \sum_{k=0}^{n} \mu_k - 1 ,$$

since there are a total of $\sum_k \mu_k$ pieces of data to interpolate. Note that the subscript μ_j is really shorthand for repeating the subscript t_j a total of μ_j times—that is,

$$P_{\mu_0 \cdots \mu_n}(t) \equiv P_{\underbrace{t_0, ..., t_0}_{\mu_0}, ..., \underbrace{t_n, ..., t_n}_{\mu_n}}(t). \qquad\qquad (3.3)$$

Thus, when the nodes are at $t = 0, 1$ and both nodes have multiplicity two, we write $P_{0011}(t)$, not $P_{22}(t)$.

To take a more generic example, given the data

$$(P_0, v_{01}, v_{02}, v_{03}), (P_1, v_{11}), (P_2, v_{21}, v_{22})$$

at the integer nodes $t_0 = 0, t_1 = 1, t_2 = 2$, we seek a polynomial $P_{000011222}(t)$ of degree 8 such that

$$P_{000011222}(t_0) = P_0 \qquad\qquad P_{000011222}(t_1) = P_1 \qquad\qquad P_{000011222}(t_2) = P_2$$

$$P'_{000011222}(t_0) = v_{01} \qquad\qquad P'_{000011222}(t_1) = v_{11} \qquad\qquad P'_{000011222}(t_2) = v_{21}$$

$$P''_{000011222}(t_0) = v_{02} \qquad\qquad\qquad\qquad\qquad\qquad\quad P''_{000011222}(t_2) = v_{22}$$

$$P'''_{000011222}(t_0) = v_{03}$$

As in cubic Hermite interpolation, we are going to proceed by extending Neville's algorithm to general Hermite interpolation. First, however, we need to review the Taylor expansion because the Taylor polynomial will now appear as the base case in Neville's algorithm.

When $n = 0$, Hermite interpolation requires a curve to fit the data $P, v_1, ..., v_{\mu-1}$ at the parameter t_0. The Taylor expansion at $t = t_0$ of a polynomial $P(t)$ of degree $\mu - 1$ is

$$P(t) = P(t_0) + P'(t_0)(t - t_0) + \cdots + \frac{P^{(\mu-1)}(t_0)}{(\mu-1)!}(t - t_0)^{\mu-1}$$

(see Section 2.4). By matching constraints to coefficients we see that to construct a polynomial curve $P_\mu(t)$ that interpolates the data $P, v_1, ..., v_{\mu-1}$ at the parameter t_0, we must set

$$P_\mu(t) = P + v_1(t - t_0) + \cdots + \frac{v_{\mu-1}}{(\mu-1)!}(t - t_0)^{\mu-1}.$$

We shall see shortly that this polynomial is the base case of a recurrence for $P_{\mu_0 \cdots \mu_n}(t)$.

Now to develop Neville's algorithm for Hermite interpolation, we proceed almost exactly as we did for Lagrange interpolation. We need to be aware, however, of two subtleties. In the notation $P_{\mu_0 \cdots \mu_n}(t)$ if some $\mu_j = 0$, then we simply ignore the parameter t_j because there is no data to fit at t_j. For example, if $\mu_0 = 0$, then $P_{\mu_0 \cdots \mu_n}(t) = P_{\mu_1 \cdots \mu_n}(t)$. Also, as we have already mentioned, the base cases for the Hermite version of Neville's algorithm are the Taylor expansions $P_{\mu_j}(t)$, not the constants P_j. After we derive Neville's algorithm, we shall describe a method for efficiently calculating all the Taylor polynomials that appear in Neville's algorithm from the original data.

THEOREM 3.1

Given data $(P_0, v_{01}, ..., v_{0,\mu_0-1}), ..., (P_n, v_{n1}, ..., v_{n,\mu_n-1})$ and distinct parameters $t_0, ..., t_n$, there exists a unique polynomial curve $P_{\mu_0 \cdots \mu_n}(t)$ of degree $\sum_k \mu_k - 1$ that interpolates the given data at the specified parameters. That is,

$$P_{\mu_0 \cdots \mu_n}(t_j) = P_j$$

$$P^{(k)}_{\mu_0 \cdots \mu_n}(t_j) = v_{jk} \qquad\qquad 1 \le k \le \mu_j - 1 .$$

Proof We shall leave uniqueness as an exercise (see Exercise 3); here we will concentrate on proving existence. If $\mu_k = 0$ for all $k \neq j$, then we can apply the Taylor expansion at $t = t_j$ to construct the polynomial interpolant $P_{\mu_j}(t)$. Otherwise the proof proceeds by induction on $\sum_k \mu_k$. Suppose then that there are at least two indices i, j such that $\mu_i, \mu_j \neq 0$. Reindexing if necessary, we can assume, without loss of generality, that these indices are μ_0 and μ_n. Moreover, we can also assume, without loss of generality, that $\mu_k > 0$ for all k; if not, simply remove the indices where $\mu_k = 0$ and reindex. Now by the inductive hypothesis, there are polynomial curves $P_{\mu_0 \cdots \mu_n - 1}(t)$ and $P_{\mu_0 - 1 \cdots \mu_n}(t)$ of degree $\sum_k \mu_k - 2$ that, respectively, interpolate the data

$$(P_0, v_{01}, \ldots, v_{0, \mu_0 - 1}), \ldots, (P_n, v_{n1}, \ldots, v_{n, \mu_n - 2})$$

$$(P_0, v_{01}, \ldots, v_{0, \mu_0 - 2}), \ldots, (P_n, v_{n1}, \ldots, v_{n, \mu_n - 1})$$

at the parameters $t_0, \ldots, t_n$. Define

$$P_{\mu_0 \cdots \mu_n}(t) = \frac{t_n - t}{t_n - t_0} P_{\mu_0 \cdots \mu_n - 1}(t) + \frac{t - t_0}{t_n - t_0} P_{\mu_0 - 1 \cdots \mu_n}(t). \tag{3.4}$$

Then by the inductive hypothesis

$$P_{\mu_0 \cdots \mu_n - 1}(t_k) = P_{\mu_0 - 1 \cdots \mu_n}(t_k) = P_k \qquad\qquad k = 1, \ldots, n - 1$$

so by (3.4)

$$P_{\mu_0 \cdots \mu_n}(t_k) = P_k \qquad\qquad k = 1, \ldots, n - 1.$$

Moreover, again by the inductive hypothesis and (3.4)

$$P_{\mu_0 \cdots \mu_n}(t_0) = P_{\mu_0 \cdots \mu_n - 1}(t_0) = P_0$$

$$P_{\mu_0 \cdots \mu_n}(t_n) = P_{\mu_0 - 1 \cdots \mu_n}(t_n) = P_n$$

so $P_{\mu_0 \cdots \mu_n}(t)$ certainly interpolates all the data points at the specified parameter values. It remains to check the derivatives. By Leibniz's rule, differentiating (3.4) k times yields

$$P_{\mu_0 \cdots \mu_n}^{(k)}(t) = \frac{t_n - t}{t_n - t_0} P_{\mu_0 \cdots \mu_n - 1}^{(k)}(t) - \frac{k}{t_n - t_0} P_{\mu_0 \cdots \mu_n - 1}^{(k-1)}(t)$$

$$+ \frac{t - t_0}{t_n - t_0} P_{\mu_0 - 1 \cdots \mu_n}^{(k)}(t) + \frac{k}{t_n - t_0} P_{\mu_0 - 1 \cdots \mu_n}^{(k-1)}(t)$$

(see Exercise 2). Evaluating the right-hand side at $t = j, j \neq 0, n$, and applying the inductive hypothesis, we find that both kth derivatives are v_{jk} and both $(k-1)$st derivatives are $v_{j,k-1}$. Since the coefficients of the $(k-1)$st derivatives are negatives of each other and the coefficients of the kth derivatives sum to one, we conclude that

$$P_{\mu_0\cdots\mu_n}^{(k)}(t_j) = v_{jk} \qquad 1 \le k \le \mu_j - 1 \qquad j \ne 0, n.$$

Similarly, evaluating at $t = t_0, t_n$, we obtain

$$P_{\mu_0\cdots\mu_n}^{(k)}(t_0) = P_{\mu_0\cdots\mu_n-1}^{(k)}(t_0) = v_{0k} \qquad 1 \le k \le \mu_0 - 1$$

$$P_{\mu_0\cdots\mu_n}^{(k)}(t_n) = P_{\mu_0-1\cdots\mu_n}^{(k)}(t_n) = v_{nk} \qquad 1 \le k \le \mu_n - 1.$$

Thus $P_{\mu_0\cdots\mu_n}(t)$ does indeed interpolate all the data at the specified para-meter values. Finally, by the inductive hypothesis, $P_{\mu_0\cdots\mu_n-1}(t)$ and $P_{\mu_0-1\cdots\mu_n}(t)$ are polynomials of degree $\sum_k \mu_k - 2$, so it follows from (3.4) that $P_{\mu_0\cdots\mu_n}(t)$ is a polynomial of degree $\sum_k \mu_k - 1$.

Starting with the Taylor polynomials $P_{\mu_j}(t)$, $j = 0,\ldots,n$, we can apply dynamic programming to build the Hermite interpolant $P_{\mu_0\cdots\mu_n}(t)$ in the usual way, con-structing higher-order interpolants from lower-order ones using (3.4). If $\mu_j = 1$, then the Taylor interpolant $P_{\mu_j}(t) = P_j$. We illustrate Neville's algorithm for $P_{0112}(t)$ in Figure 3.5.

Recall from (3.3) that in our notation the subscript μ_j in $P_{\mu_0\cdots\mu_n}(t)$ is shorthand for repeating the node t_j as a subscript a total of μ_j times. When the nodes and mul-tiplicities are known, instead of using μ_j as a subscript, we simply repeat t_j a total of μ_j times. Thus $P_{0112}(t)$ means the Hermite interpolant for the nodes $t = 0,1,2$, where $t = 0$ has multiplicity one, $t = 1$ has multiplicity two, and $t = 2$ has multiplicity one.

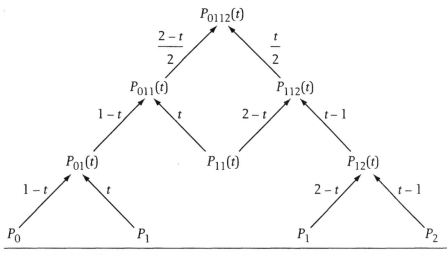

Figure 3.5 Neville's algorithm for the cubic Hermite interpolant $P_{0112}(t)$. Here the nodes are at the inte-gers: $t_0 = 0, t_1 = 1, t_2 = 2$. Notice that the Taylor polynomial $P_{11}(t)$ is one of the leaf nodes— that is, one of the base cases of the algorithm.

Since the dynamic programming algorithm bottoms out at the Taylor polynomials, we may need to compute several Taylor polynomials of different degrees for the same node t_j. If the node t_j has multiplicity μ_j, then the Taylor polynomials

$$P_{\mu_j}(t), P_{\mu_j-1}(t), \dots, P_j$$

will all appear in Neville's triangle. It is wasteful to calculate all these polynomials independently, since

$$P_{\mu_j}(t) = P_{\mu_j-1}(t) + \frac{(t-t_j)^{\mu_j-1}}{(\mu_j-1)!} v_{j,\mu_j-1}. \tag{3.5}$$

Using (3.5), we can apply dynamic programming to bootstrap ourselves up from the original data, calculating only one new term for each Taylor polynomial (see Figure 3.6).

Neville's algorithm for Hermite interpolation now has the same general structure as Neville's algorithm for Lagrange interpolation. Start with $P_{\mu_0\cdots\mu_n}(t)$ at the apex. Strip off the last index and place $P_{\mu_0\cdots\mu_n-1}(t)$ below it to the left; strip off the first index and place $P_{\mu_0-1\cdots\mu_n}(t)$ below it to the right. Since the index t_n was removed to the left, label the left arrow with $t_n - t$; since the index t_0 was removed to the right, label the right arrow with $t - t_0$. Now proceed recursively stripping off labels from $P_{\mu_0\cdots\mu_n-1}(t)$ and $P_{\mu_0-1\cdots\mu_n}(t)$ and labeling the arrows accordingly. Remember to join $P_{\mu_0-1\cdots\mu_n-1}(t)$ to both $P_{\mu_0\cdots\mu_n-1}(t)$ and $P_{\mu_0-1\cdots\mu_n}(t)$ to generate a dynamic programming algorithm instead of a recursive procedure. When you arrive

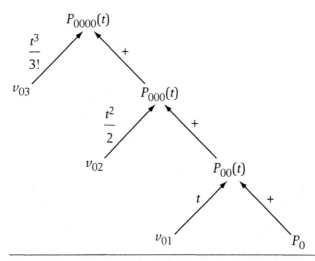

Figure 3.6 Applying dynamic programming to compute higher-order Taylor polynomials efficiently from lower-order Taylor polynomials at the node $t = 0$. Only one new term is computed for each successive Taylor polynomial. Here we illustrate the cubic case.

at $P_{\mu_k}(t)$ you are at a Taylor polynomial, so you should apply the dynamic programming algorithm for computing Taylor polynomials depicted in Figure 3.6.

We illustrate this version of Neville's algorithm for $P_{000112}(t)$ in Figure 3.7; you should also look at Figure 3.3, where we have already adopted this strategy for $P_{0011}(t)$. All the leaf nodes in Figure 3.7 now contain point or vector data rather than Taylor polynomials. We shall see shortly that this convention makes it easy to compute the Hermite basis functions from the up or down recurrence. Notice too that the parallel property (see Section 2.3) of Neville's algorithm remains valid for Hermite interpolation.

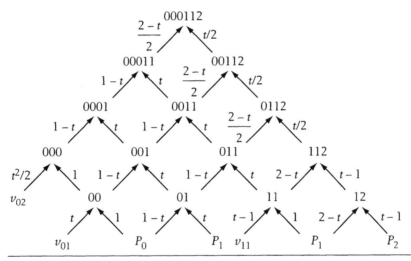

Figure 3.7 Neville's algorithm for $P_{000112}(t)$. The interpolants are represented by their subscripts, and the Taylor polynomials are computed efficiently from lower-order Taylor polynomials. Here we have explicitly normalized the labels along the arrows to sum to one. The labels on arrows emerging from nodes that contain vectors do not need to be normalized. Notice that due to the presence of the Taylor polynomial $P_{11}(t)$, the point P_1 now appears at two different leaf nodes.

Exercises

1. Draw Neville's triangle for $P_{01112}(t)$.

2. Prove by induction on k that if $L(t)$ is linear in t, then

$$\{L(t)P(t)\}^{(k)} = L(t)P^{(k)}(t) + kL'(t)P^{(k-1)}(t).$$

3. Prove that the Hermite interpolant $P_{\mu_0\dots\mu_n}(t)$ is unique—that is, there is no other polynomial of the same degree that interpolates the same data at the same nodes. (Hint: Use Exercise 2 in Section 2.4.)

4. Implement Neville's algorithm for Hermite interpolation. Experiment with how changing the nodes or the data affects the shape of the curve.

5. Let $P(t)$ be the Hermite interpolating polynomial for the nodes $t_0,...,t_n$ and the data

$$(P_0, v_{01},...,v_{0,\mu_0-1}),...,(P_n, v_{n1},...,v_{n,\mu_n-1}).$$

Form a new Hermite interpolating curve $Q(t)$ by replacing each node t_k by the node $\tau_k = t_k + b$, $k = 0,...,n$, for some fixed constant b.

a. Show that changing all the nodes in this way has no affect on the shape of the interpolating curve. In particular, using Neville's algorithm, show that $Q(t + b) = P(t)$.

b. Form a new Hermite interpolating curve $R(t)$ by replacing each node t_k by the node $\tau_k = at_k + b$ for some fixed constants $a > 0$ and b. Show that $R(t)$ is not a reparametrized version of $P(t)$. What goes wrong in Neville's algorithm?

c. Compare these results for Hermite interpolation to similar results for Lagrange interpolation in Section 2.2, Exercise 4.

3.3   **The Hermite Basis Functions**

In Hermite interpolation, we are given a collection of points and vectors

$$(P_0, v_{01},...,v_{0,\mu_0-1}),...,(P_n, v_{n1},...,v_{n,\mu_n-1}),$$

and we construct a polynomial curve $P_{\mu_0\cdots\mu_n}(t)$ of degree $\sum_k \mu_k - 1$ that interpolates this data at the parameter values $t_0,...,t_n$. For general Hermite interpolation, we can always write $P_{\mu_0\cdots\mu_n}(t)$ explicitly by setting

$$P_{\mu_0\cdots\mu_n}(t) = \sum_{j=0}^{n} H_j(t \mid t_0,...,t_n)P_j + \sum_{j=0}^{n} \sum_{k=0}^{\mu_j-1} h_{jk}(t \mid t_0,...,t_n)v_{jk}.$$

As with Lagrange interpolation, the polynomials $H_j(t \mid t_0,...,t_n)$ and $h_{jk}(t \mid t_0,...,t_n)$ can be found by Neville's algorithm from either the up recurrence or the down recurrence. Notice, however, that in the Hermite version of Neville's algorithm, a control point P_j can appear at more than one leaf node (see Figure 3.7). Therefore in calculating $H_j(t \mid t_0,...,t_n)$ using the up recurrence, we must take care to place a one at each of the leaf nodes where P_j appears. Similarly, when applying the down recurrence, we must sum all the values at the leaf nodes where P_j would appear in order to calculate the value of $H_j(t \mid t_0,...,t_n)$.

Although it is difficult to compute simple explicit expressions for the general Hermite basis functions, there are four particularly important special cases of Hermite interpolation where fairly elementary explicit formulas are available:

i.   Lagrange interpolation—$\mu_j = 1$, $j = 0,...,n$

ii. Taylor interpolation—$n = 0$

iii. Hermite interpolation, one derivative at each point—$\mu_j = 2$, $j = 0,...,n$

iv. Hermite interpolation at two points—$n = 1$, $\mu_0 = \mu_1$

We have already studied Lagrange interpolation extensively in Chapter 2, and by now you should be familiar as well with the Taylor polynomial. The last two cases of Hermite interpolation are straightforward generalizations of the cubic Hermite interpolant discussed in Section 3.1. Here we focus our attention on case iii, which we shall apply in Section 3.5 to construct tensor product, lofted, and Boolean sum Hermite surfaces. You will analyze case iv in Exercise 2.

The problem in case iii is to interpolate the data $(P_0, v_0),...,(P_n, v_n)$ at the parameter values $t_0,...,t_n$. Since there are $2n + 2$ pieces of data, the interpolant

$$P_{t_0 t_0 \cdots t_n t_n}(t) = \sum_{k=0}^{n} H_k(t \mid t_0, t_0, ..., t_n, t_n) P_k + \sum_{k=0}^{n} h_k(t \mid t_0, t_0, ..., t_n, t_n) v_k \qquad (3.6)$$

is of degree $2n + 1$. To find the Hermite basis functions $H_k(t \mid t_0, t_0, ..., t_n, t_n)$ and $h_k(t \mid t_0, t_0, ..., t_n, t_n)$, we must compute all paths from P_k or v_k at the leaf nodes to the interpolant at the apex of the Neville triangle. Recall, however, that in general Hermite interpolation a point P_k may appear at more than one leaf of the triangle (see Figure 3.7).

To get from the apex $P_{t_0 t_0 \cdots t_n t_n}(t)$ of the triangle to any leaf node containing P_k, we must strip off each index $j \neq k$ exactly twice (see Figure 3.8). Each time we remove the index j, we introduce a factor of $t - t_j$ along an arrow. Thus any path

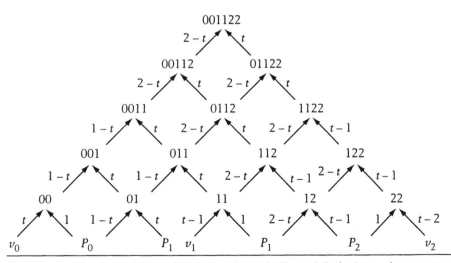

Figure 3.8 Neville's algorithm for the interpolant $P_{001122}(t)$. As in Figure 3.7, the interpolants are represented by their subscripts, and higher-order Taylor polynomials are computed efficiently from lower-order Taylor polynomials. Unlike Figure 3.7, the functions along the arrows have not been normalized.

from the apex to a leaf containing P_k must contain two factors of $t - t_j$ for each $j \neq k$. Therefore,

$$H_k(t \mid t_0,t_0,...,t_n,t_n) = \lambda_k(t) \prod_{j \neq k}(t - t_j)^2 \,,$$

where $\lambda_k(t)$ must be linear in t, since degree$\{H_k(t \mid t_0,t_0,...,t_n,t_n)\} = 2n+1$. Recalling the explicit formula (2.11) for the Lagrange basis functions, we can rewrite this equation as

$$H_k(t \mid t_0,t_0,...,t_n,t_n) = \lambda_k(t)\{L_k^n(t \mid t_0,...,t_n)\}^2 \,,$$

where we have absorbed the constant in the denominator of the Lagrange basis function into the function $\lambda_k(t)$. Similarly, we can argue that

$$h_k(t \mid t_0,t_0,...,t_n,t_n) = \omega_k(t)\{L_k^n(t \mid t_0,...,t_n)\}^2 \,,$$

where $\omega_k(t)$ is linear in t. It remains only to find $\lambda_k(t)$ and $\omega_k(t)$.

It is easy to find $\omega_k(t)$ because each vector v_k appears at only one leaf node and the arrow emerging from this node is labeled $t - t_k$. Thus we must have

$$\omega_k(t) = c_k(t - t_k)$$

for some constant c_k. Moreover, since

$$h_k'(t_k \mid t_0,t_0,...,t_n,t_n) = 1,$$

it follows that

$$\omega_k'(t_k)\{L_k^n(t_k \mid t_0,...,t_n)\}^2 + 2\omega_k(t_k)L_k^n(t_k \mid t_0,...,t_n)\frac{dL_k^n}{dt}\mid_{t=t_k} = 1.$$

Therefore, since $\omega_k(t_k) = 0$ and $L_k^n(t_k \mid t_0,...,t_n) = 1$,

$$c_k = \omega_k'(t_k) = 1.$$

To find $\lambda_k(t)$, we proceed in a similar fashion. Since $\lambda_k(t)$ is linear, we can write

$$\lambda_k(t) = d_1(t - t_k) + d_0.$$

But we know that

$$H_k(t_k \mid t_0,t_0,...,t_n,t_n) = L_k^n(t_k \mid t_0,...,t_n) = 1,$$

so certainly

$$d_0 = \lambda_k(t_k) = 1.$$

Similarly since

$$H_k'(t_k \mid t_0,t_0,...,t_n,t_n) = 0 \,,$$

we must have

$$\lambda_k'(t_k)\{L_k^n(t_k\mid t_0,...,t_n)\}^2 + 2\lambda_k(t_k)L_k^n(t_k\mid t_0,...,t_n)\frac{dL_k^n}{dt}\mid_{t=t_k} = 0.$$

Thus, since $L_k^n(t_k\mid t_0,...,t_n) = 1$ and $\lambda_k(t_k) = 1$,

$$d_1 = \lambda_k'(t_k) = -2\frac{dL_k^n}{dt}\mid_{t=t_k}.$$

From the preceding analysis we conclude that

$$H_k(t\mid t_0,t_0,...,t_n,t_n) = \left\{1 - 2\frac{dL_k^n}{dt}\mid_{t=t_k} (t-t_k)\right\}\{L_k^n(t\mid t_0,...,t_n)\}^2$$

$$h_k(t\mid t_0,t_0,...,t_n,t_n) = (t-t_k)\{L_k^n(t\mid t_0,...,t_n)\}^2 \ .$$

(3.7)

Notice how these formulas generalize the formulas for the cubic Hermite basis functions derived in Section 3.1.

Exercises

1. Prove the identity

$$(x-t)^{2n+1} = \sum_{k=0}^{n} H_k(t\mid t_0,t_0,...,t_n,t_n)(x-t_k)^{2n+1}$$

$$+ (2n+1)\sum_{k=0}^{n} h_k(t\mid t_0,t_0,...,t_n,t_n)(x-t_k)^{2n} \ .$$

(Compare to Section 2.5, Exercise 1.)

2. Consider the special case of two-point Hermite interpolation, where we interpolate the data $(P_0, v_{01},..., v_{0,\mu-1})$, $(P_1, v_{11},..., v_{1,\mu-1})$ at the parameter values $t = t_0, t_1$.

   a. Show that this interpolant can be written as

   $$P_{\mu\mu}(t) = H_0(t\mid t_0,...,t_1)P_0 + H_1(t\mid t_0,...,t_1)P_1$$

   $$+ \sum_{k=1}^{\mu-1} h_{0k}(t\mid t_0,...,t_1)v_{0k} + \sum_{k=1}^{\mu-1} h_{1k}(t\mid t_0,...,t_1)v_{1k} \ .$$

   b. Explain why each control point and each vector appear at only a single node in Neville's triangle.

   c. Find explicit formulas for the basis functions $H_0(t\mid t_0,...,t_1)$ and $H_1(t\mid t_0,...,t_1)$.

   d. Using the formulas from part (c), verify that
   $$H_0(t\mid t_0,...,t_1) + H_1(t\mid t_0,...,t_1) \equiv 1.$$

e. Find explicit formulas for the basis functions $h_{0k}(t \mid t_0,...,t_1)$ and $h_{1k}(t \mid t_0,...,t_1)$.

3. Suppose we are given the data $(P_0,v_{01},...,v_{0,\mu_0-1}),...,(P_n,v_{n1},...,v_{n,\mu_n-1})$ to interpolate at the parameters $t_0,...,t_n$. Then the general Hermite interpolant can be written as

$$P_{\mu_0\cdots\mu_n}(t) = \sum_{j=0}^{n} H_j(t \mid t_0,...,t_n)P_j + \sum_{j=0}^{n} \sum_{k=1}^{\mu_j-1} h_{jk}(t \mid t_0,...,t_n)v_{jk}.$$

Use Neville's triangle to show that

a. $H_j(t \mid t_0,...,t_n) = \lambda_j(t) \prod_{i \neq j}(t - t_i)^{\mu_i}$

b. $h_{jk}(t \mid t_0,...,t_n) = \omega_{jk}(t)(t - t_j)^k \prod_{i \neq j}(t - t_i)^{\mu_i}$,

where $\lambda_j(t)$ is a polynomial of degree at most $\mu_j - 1$, and $\omega_{jk}(t)$ is a polynomial of degree at most $\mu_j - k - 1$.

4. Prove that $\sum_{k=0}^{n} H_k(t \mid t_0,t_0,...,t_n,t_n) \equiv 1$.

5. Let $P(t) = a_{2n+1}t^{2n+1} + \cdots + a_1 t + a_0$. Then $P(t)$ interpolates the data (P_0,v_0), ...,(P_n,v_n) at the parameters $t_0,...,t_n$ if and only if

$$a_{2n+1}t_0^{2n+1} + \cdots + a_1 t_0 + a_0 = P_0$$
$$(2n+1)a_{2n+1}t_0^{2n} + \cdots + 2a_2 t_0 + a_1 = v_0$$
$$\vdots$$
$$a_{2n+1}t_n^{2n+1} + \cdots + a_1 t_n + a_0 = P_n$$
$$(2n+1)a_{2n+1}t_n^{2n} + \cdots + 2a_2 t_n + a_1 = v_n \ .$$

Prove that this system of linear equations in the unknowns $a_0,...,a_{2n+1}$ has a unique solution by showing that the determinant of the coefficients

$$\begin{vmatrix} t_0^{2n+1} & \cdots & t_0^2 & t_0 & 1 \\ (2n+1)t_0^{2n} & \cdots & 2t_0 & 1 & 0 \\ \vdots & \vdots & \vdots & \vdots & \vdots \\ t_n^{2n+1} & \cdots & t_n^2 & t_n & 1 \\ (2n+1)t_n^{2n} & \cdots & 2t_n & 1 & 0 \end{vmatrix} \neq 0 \ .$$

Conclude that $P_{t_0 t_0 \cdots t_n t_n}(t)$ exists and is unique. (Compare to Section 2.4, Exercise 5.)

6. Develop an $O(n)$ ladder evaluation algorithm to compute

$$P_{t_0 t_0 \cdots t_n t_n}(t) = \sum_{k=0}^{n} H_k(t \mid t_0, t_0, \ldots, t_n, t_n) P_k + \sum_{k=0}^{n} h_k(t \mid t_0, t_0, \ldots, t_n, t_n) v_k.$$

7. Consider the special case of two-point Hermite interpolation, where we interpolate the data $(P_0, v_{01}, \ldots, v_{0,\mu-1})$, $(P_1, v_{11}, \ldots, v_{1,\mu-1})$ at the parameter values $t = 0,1$. Show that

$$H_0(t \mid \underbrace{0, \ldots, 0}_{\mu}, \underbrace{1, \ldots, 1}_{\mu}) = \frac{\int_t^1 x^{\mu-1}(1-x)^{\mu-1} dx}{\int_0^1 x^{\mu-1}(1-x)^{\mu-1} dx}$$

$$H_1(t \mid \underbrace{0, \ldots, 0}_{\mu}, \underbrace{1, \ldots, 1}_{\mu}) = \frac{\int_0^t x^{\mu-1}(1-x)^{\mu-1} dx}{\int_0^1 x^{\mu-1}(1-x)^{\mu-1} dx}.$$

3.4 Rational Hermite Curves

When we studied Lagrange interpolation, we observed that many common curves are not polynomial curves, but rather are rational curves. Therefore, we resorted to rational Lagrange interpolation to represent these curves in Lagrange form. Now that we are investigating Hermite interpolation, we would like to develop a similar approach to represent rational curves in Hermite form.

To simplify the discussion, let's first consider cubic Hermite interpolation. Suppose that we have a rational cubic curve $R(t) = P(t)/Q(t)$ that we wish to represent in rational Hermite form. As usual we lift this curve from a rational curve $R(t) = P(t)/Q(t)$ in affine space to a polynomial curve $S(t) = (P(t), Q(t))$ in Grassmann space by treating the denominator $Q(t)$ as mass. To find the cubic Hermite representation of $S(t)$ in Grassmann space, we compute

$$(m_0 P_0, m_0) = (P(0), Q(0)) \qquad (\lambda_0 v_0, \lambda_0) = (P'(0), Q'(0))$$
$$(m_1 P_1, m_1) = (P(1), Q(1)) \qquad (\lambda_1 v_1, \lambda_1) = (P'(1), Q'(1)) .$$

Then in Grassmann space, we have the cubic Hermite representation

$$S(t) = (m_0 P_0, m_0) H_0(t) + (m_1 P_1, m_1) H_1(t) + (\lambda_0 v_0, \lambda_0) h_0(t) + (\lambda_1 v_1, \lambda_1) h_1(t),$$

and by construction

$$S(0) = (m_0 P_0, m_0) \qquad S'(0) = (\lambda_0 v_0, \lambda_0)$$
$$S(1) = (m_1 P_1, m_1) \qquad S'(1) = (\lambda_1 v_1, \lambda_1) .$$

Now let's project this curve back into affine space. Then we get the rational representation

$$R(t) = \frac{m_0 P_0 H_0(t) + m_1 P_1 H_1(t) + \lambda_0 v_0 h_0(t) + \lambda_1 v_1 h_1(t)}{m_0 H_0(t) + m_1 H_1(t) + \lambda_0 h_0(t) + \lambda_1 h_1(t)}. \tag{3.8}$$

Is this formula really a Hermite representation for the original rational cubic curve $R(t)$? It is easy to check that indeed $R(0) = P_0$ and $R(1) = P_1$. What about derivatives? We hope to get $R'(0) = v_0$ and $R'(1) = v_1$, but, in fact, after a bit of algebra, we find that

$$R'(0) = \frac{\lambda_0 v_0 - \lambda_0 P_0}{m_0} \qquad \text{and} \qquad R'(1) = \frac{\lambda_1 v_1 - \lambda_1 P_1}{m_1}.$$

Not the answers we expected. What did we do wrong?

In Grassmann space we found that $S'(0) = (\lambda_0 v_0, \lambda_0)$ and $S'(1) = (\lambda_1 v_1, \lambda_1)$, and these derivatives do indeed project to the values v_0 and v_1 in affine space. The problem is that differentiation and projection do not commute (see Section 1.1.6, Exercise 2)—the derivative of the quotient is not equal to the quotient of the derivatives—so this approach to rational Hermite interpolation cannot hope to succeed.

Worse yet, the right-hand sides of $R'(0)$ and $R'(1)$ are not well-defined expressions in affine space. In fact, looking back, we see that the expression for $R(t)$ on the right-hand side of (3.8) is also not a well-defined expression in affine space, since the coefficients of P_0 and P_1 do not sum to one. So this approach is doomed to failure from the very beginning. We need a fresh start.

In cubic Hermite interpolation, the coefficients of $H_0(t)$ and $H_1(t)$ are points; the coefficients of $h_0(t)$ and $h_1(t)$ are vectors. To maintain these constraints in Grassmann space, we must write

$$S(t) = (m_0 P_0, m_0) H_0(t) + (m_1 P_1, m_1) H_1(t) + (\lambda_0 v_0, 0) h_0(t) + (\lambda_1 v_1, 0) h_1(t).$$

Projecting this formula into affine space yields

$$R(t) = \frac{m_0 P_0 H_0(t) + m_1 P_1 H_1(t) + \lambda_0 v_0 h_0(t) + \lambda_1 v_1 h_1(t)}{m_0 H_0(t) + m_1 H_1(t)}. \tag{3.9}$$

Let's see if this works.

Certainly the expression for $R(t)$ is now well defined, since the coefficients of P_0 and P_1 sum to one. Also it is easy to verify that $R(0) = P_0$ and $R(1) = P_1$, so we still interpolate the point data. What about derivatives? To simplify our computation, let

$$P(t) = m_0 P_0 H_0(t) + m_1 P_1 H_1(t) + \lambda_0 v_0 h_0(t) + \lambda_1 v_1 h_1(t)$$
$$w(t) = m_0 H_0(t) + m_1 H_1(t).$$

Then $R(t) = P(t)/w(t)$ so by the quotient rule

$$R'(t) = \frac{w(t)P'(t) - w'(t)P(t)}{w^2(t)} = \frac{P'(t) - w'(t)R(t)}{w(t)}. \tag{3.10}$$

Using the properties of the cubic Hermite basis functions provided in Section 3.1, it is simple to check that

$$w(0) = m_0, \ w'(0) = 0, \ P'(0) = \lambda_0 v_0,$$
$$w(1) = m_1, \ w'(1) = 0, \ P'(1) = \lambda_1 v_1.$$

Substituting $t = 0,1$ into (3.10), we get

$$R'(0) = \lambda_0 v_0 / m_0$$
$$R'(1) = \lambda_1 v_1 / m_1 \ .$$

Thus if we want $R'(0) = v_0$ and $R'(1) = v_1$, we need to choose $\lambda_0 = m_0$ and $\lambda_1 = m_1$.

These observations lead us to define a rational cubic Hermite curve by setting

$$R(t) = \frac{m_0 P_0 H_0(t) + m_1 P_1 H_1(t) + m_0 v_0 h_0(t) + m_1 v_1 h_1(t)}{m_0 H_0(t) + m_1 H_1(t)} \ . \tag{3.11}$$

With this definition, the function $R(t)$ interpolates the data (P_0, v_0) and (P_1, v_1) at the parameters $t = 0$ and $t = 1$ independent of the choice of the masses m_0 and m_1.

For general Hermite interpolation, we have basis functions

$$H_j(t \mid t_0, \ldots, t_n), \ j = 0, \ldots, n \text{ and } h_{jk}(t \mid t_0, \ldots, t_n), \ k = 0, \ldots, \mu_{j-1},$$

so we define the general rational Hermite curve by setting

$$R(t) = \frac{\sum_{j=0}^{n} H_j(t \mid t_0, \ldots, t_n) m_j P_j + \sum_{j=0}^{n} \sum_{k=0}^{\mu_j - 1} h_{jk}(t \mid t_0, \ldots, t_n) m_j v_{jk}}{\sum_{j=0}^{n} m_j H_j(t \mid t_0, \ldots, t_n)} \ . \tag{3.12}$$

THEOREM 3.2

Let $R(t)$ be defined as by Equation (3.12). Then

$$R(t_i) = P_i \qquad\qquad 0 \le i \le n$$
$$R^{(k)}(t_i) = v_{ik} \qquad\qquad 1 \le k \le \mu_i - 1 \ .$$

Proof To simplify the computation, let

$$P(t) = \sum_{j=0}^{n} H_j(t \mid t_0, \ldots, t_n) m_j P_j + \sum_{j=0}^{n} \sum_{k=0}^{\mu_j - 1} h_{jk}(t \mid t_0, \ldots, t_n) m_j v_{jk}$$

$$w(t) = \sum_{j=0}^{n} m_j H_j(t \mid t_0, \ldots, t_n) \ .$$

Then $R(t) = P(t) / w(t)$. Now by the defining properties of the Hermite basis functions, $P(t_i) = m_i P_i$ and $w(t_i) = m_i$, so certainly $R(t_i) = P_i$, $0 \le i \le n$. Moreover, for $1 \le k \le \mu_i - 1$,

$$w^{(k)}(t_i) = \sum_{j=0}^{n} m_j H_j^{(k)}(t_i \mid t_0,...,t_n) = 0$$

$$P^{(k)}(t_i) = \sum_{j=0}^{n} H_j^{(k)}(t_i \mid t_0,...,t_n)m_j P_j + \sum_{j=0}^{n}\sum_{l=0}^{\mu_j-1} h_{jl}^{(k)}(t_i \mid t_0,...,t_n)m_j v_{jl} = m_i v_{ik} \ .$$

In particular, since $w^{(k)}(t_i) = 0$, $1 \le k \le \mu_i - 1$, it follows by induction (see Exercise 12) that

$$\left(w(t)R(t)\right)^{(k)}(t_i) = w(t_i)R^{(k)}(t_i) \quad 1 \le k \le \mu_i - 1.$$

But $w(t)R(t) = P(t)$, so

$$w(t_i)R^{(k)}(t_i) = P^{(k)}(t_i) \quad 1 \le k \le \mu_i - 1,$$

or equivalently

$$R^{(k)}(t_i) = \frac{P^{(k)}(t_i)}{w(t_i)} \quad 1 \le k \le \mu_i - 1.$$

Now $P^{(k)}(t_i) = m_i v_{ik}$ and $w(t_i) = m_i$, so $R^{(k)}(t_i) = v_{ik}$, $1 \le k \le \mu_i - 1$.

Thus, as in rational Lagrange interpolation, the masses in rational Hermite interpolation serve as shape parameters. That is, the masses do not affect interpolation at the nodes, but the masses do alter the shape of the interpolating curve. Again as with rational Lagrange interpolation, these shape parameters can be hard to control, so we must handle them with care (see Exercises 1 and 6). Nevertheless, for two-point Hermite interpolation, the effect of the masses is fairly well behaved. In Figure 3.9 we illustrate the effect of the masses on the rational quadratic Hermite representation of the quarter circle. The effect of the masses on rational cubic Hermite curves as well as the effect of the masses on other examples of two-point rational Hermite interpolation is investigated in Exercises 8 and 9.

To provide a concrete example, let's now represent the quarter circle in rational quadratic Hermite form. Recall from Section 2.7 that the unit circle has the rational quadratic parametrization

$$R(t) = \left(\frac{2t}{1+t^2}, \frac{1-t^2}{1+t^2} \right).$$

From Section 3.1, Exercise 1, the quadratic Hermite basis functions for the nodes 0,1 are

$$H_0(t) = 1 - t^2 \qquad H_1(t) = t^2 \qquad h_0(t) = t(1-t).$$

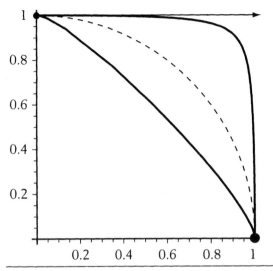

Figure 3.9 The quarter circle (dashed) as a rational quadratic Hermite curve with masses $m_0 = 1, m_1 = 2$ at the control points $P_0 = (0,1), (P_1 = (1,0)$ with control vector $v_0 = (2,0)$. (The control vector is not drawn to scale in order not to dwarf the rest of the figure.) The outer curve represents the case where $m_0/m_1 = 20$ and the inner curve where $m_0/m_1 = 1/40$.

Now we must find the points P_0, P_1, the masses m_0, m_1 and the vector v_0 so that

$$R(t) = \frac{m_0 P_0 H_0(t) + m_1 P_1 H_1(t) + m_0 v_0 h_0(t)}{m_0 H_0(t) + m_1 H_1(t)}.$$

By Theorem 3.2 we need to choose

$$P_0 = R(0) = (0,1) \qquad P_1 = R(1) = (1,0) \qquad v_0 = R'(0) = (2,0) \ .$$

Moreover, $m_0 H_0(t) + m_1 H_1(t) = 1 + t^2$, so

$$m_0 = 1 \qquad m_1 = 2.$$

What happens if we alter the masses? Dividing the numerator and denominator by m_0 or m_1, we can rewrite $R(t)$ as

$$R(t) = \frac{P_0 H_0(t) + \dfrac{m_1}{m_0} P_1 H_1(t) + v_0 h_0(t)}{H_0(t) + \dfrac{m_1}{m_0} H_1(t)} = \frac{\dfrac{m_0}{m_1} P_0 H_0(t) + P_1 H_1(t) + \dfrac{m_0}{m_1} v_0 h_0(t)}{\dfrac{m_0}{m_1} H_0(t) + H_1(t)}.$$

Thus we find that

$$m_0 \to 0 \Rightarrow R(t) \to P_1 \text{ and } R(0) = P_0,$$

so the curve collapses to two points;

$$m_0 \to \infty \Rightarrow R(t) \to P_0 + \frac{h_0(t)}{H_0(t)} v_0,$$

which is the line through P_0 in the direction v_0, and

$$R(1) = P_1,$$

so the curve splits into a point and a line.

Notice that only the ratios m_1/m_0 and m_0/m_1 matter, and not the particular values of the individual masses. Thus we get similar, but reciprocal, behavior as m_1 approaches zero or infinity. We illustrate this behavior in Figure 3.9.

Despite all the similarities between Lagrange and Hermite interpolation, there is one very important difference between the rational Lagrange and rational Hermite representations. In rational Lagrange interpolation all the basis functions appear in the denominator, so we can represent any rational function using the rational Lagrange representation. But in rational Hermite interpolation, only the basis functions $H_j(t \mid t_0,...,t_n)$ $j = 0,...,n$ appear in the denominator. Since the functions $H_j(t \mid t_0,...,t_n)$ by themselves do not form a polynomial basis, it is not possible to represent arbitrary denominators of degree $\sum_k \mu_k - 1$ with a rational Hermite representation. Thus there are many rational curves that have no rational Hermite form. For example, the circle has no rational cubic Hermite representation (see Exercise 3).

Nevertheless, we can still apply Neville's algorithm for Hermite interpolation to compute values along a rational Hermite curve. As usual, we perform the computation in Grassmann space and then divide by the mass to get values along the curve in affine space. The only restriction is that the vectors representing the derivatives in Grassmann space must really be vectors in affine space, not arbitrary vectors in Grassmann space—that is, their mass coordinate must be set to zero—to ensure that the vector components of the Hermite function appearing in the denominator of the rational function are all zero.

If the mass of a Hermite curve in Grassmann space is ever zero, then the projection of the curve into affine space is not continuous. As with rational Lagrange interpolation, we can avoid these discontinuities by projecting the curve instead into projective space.

Exercises

1. Implement Neville's algorithm for rational Hermite interpolation, and experiment with altering the masses in a rational Hermite curve.

   a. What are the local and global effects of altering a single mass?

   b. What happens when one of the masses is set to zero?

   c. What is the effect of a negative mass?

   d. What happens if all the masses are changed simultaneously?

2. Consider the rational quadratic Hermite representation of the quarter circle given by the masses and control points in the text.

   a. Plot the point with $t = .99$ for larger and larger values of m_0.

   b. Plot the point with $t = .99$ for larger and larger values of m_1.

   c. What do you observe?

   d. Explain what is happening.

3. Show that it is not possible to represent the circle parametrized by

$$R(t) = \left(\frac{2t}{1+t^2}, \frac{1-t^2}{1+t^2} \right)$$

   in rational cubic Hermite form.

4. Find the rational quadratic Hermite representation for the ellipse parametrized by

$$R(t) = \left(\frac{2at}{1+t^2}, \frac{b(1+t^2)}{1+t^2} \right)$$

   with respect to the nodes $t = 0,1$, and use Neville's algorithm to draw this segment of the ellipse for $a = 2, b = 5$.

5. Find the rational quadratic Hermite representation for the hyperbola parametrized by

$$R(t) = \left(\frac{a(1+t^2)}{1-t^2}, \frac{2bt}{1-t^2} \right)$$

   with respect to the nodes $t = -0.5, 0.5$; $t = -4, -2$; $t = -1, 1$; and use Neville's algorithm to draw different parts of this hyperbola for $a = 2, b = 5$.

6. Let $R(t)$ be a rational Hermite curve with nodes $t_0, \ldots, t_n$, control points $(m_0 P_0, m_0), \ldots, (m_n P_n, m_n)$, and control vectors $\{v_{ik}\}$. Let m_j increase and hold m_i fixed for $i \neq j$. Show that

   a. $\lim_{m_j \to \infty} R(t_i) = P_i, \ i = 0, \ldots, n$.

   b. $\lim_{m_j \to \infty} R(t)$ lies in the space spanned by P_j and $\{v_{jl}\}$ for all $t \neq t_i, \ i = 0, \ldots, n$.

   c. $\lim_{m_j \to \infty} R^{(k)}(t_i) = v_{ik}, \ 1 \leq k \leq \mu_i - 1, \ i = 0, \ldots, n$.

   d. $\lim_{m_j \to \infty} R^{(k)}(t)$ lies in the space spanned by $\{v_{jl}\}$ for all $t \neq t_i, \ i = 0, \ldots, n$.

   (Hint: See Section 2.7, Exercises 8 and 9.)

7. Let $R(t)$ be a rational Hermite curve with nodes $t_0, \ldots, t_n$, control points $(m_0 P_0, m_0), \ldots, (m_n P_n, m_n)$, and control vectors $\{v_{ik}\}$. What does the limit curve look like if

    a. two masses are allowed to increase simultaneously while the other masses are held fixed?

    b. three or more masses are allowed to increase simultaneously while the other masses are held fixed?

8. Experiment with rational cubic Hermite interpolation at the nodes $0,1$.

    a. What happens to the shape of the curve as you increase m_0 and leave m_1 fixed?

    b. Show that $\lim_{m_0 \to \infty} R(t)$ lies on the line through P_0 in the direction v_0 for all $t \neq 1$.

    c. Explain how it is possible for $\lim_{m_0 \to \infty} R(t)$ to lie on the line through P_0 in the direction v_0 for all $t \neq 1$, even though the curve still interpolates the data (P_1, v_1) at $t = 1$.

9. Generalize the results in Exercise 8 to arbitrary two-point rational Hermite interpolation.

10. Implement Neville's algorithm for rational cubic Hermite interpolation at the nodes $t = 0,1$ and explore the different geometric effects of changing

    a. control points

    b. control vectors

    c. masses

11. Experiment with rational quintic Hermite interpolation, where one point and one derivative is interpolated at each of the three nodes $t_0 < t_1 < t_2$.

    a. What happens to the shape of the curve as you increase m_1 and leave m_0 and m_2 fixed?

    b. What happens to the shape of the curve as you increase m_0 and leave m_1 and m_2 fixed?

    c. Show that in both cases $\lim_{m_j \to \infty} R(t)$ lies on the line through P_j in the direction v_j for all $t \neq t_k$, $k \neq j$.

    d. Explain in each case how it is possible for $\lim_{m_j \to \infty} R(t)$ to lie on the line through P_j in the direction v_j for all $t \neq t_k$, $k \neq j$, even though the curve still interpolates the data (P_k, v_k) at the nodes $t = t_k$, $k \neq j$.

    e. Does your explanation account for the difference in the behavior of the curve that you observe when you increase the masses m_0 and m_1?

12. Let

$$w(t) = \sum_{j=0}^{n} m_j H_j(t \mid t_0, \ldots, t_n)$$

Prove by induction on k that for any function $R(t)$

$$\left(w(t) R(t) \right)^{(k)}(t_i) = w(t_i) R^{(k)}(t_i) = m_i R^{(k)}(t_i) \quad 0 \leq k \leq \mu_i - 1.$$

13. Given a collection of nodes $t_0,\dots,t_n$ and masses $m_0,\dots,m_n$, define

$$R_j(t) = \frac{m_j H_j(t \mid t_0,\dots,t_n)}{\displaystyle\sum_{i=0}^{n} m_i H_i(t \mid t_0,\dots,t_n)}, \qquad k = 0,\dots,n$$

$$r_{jk}(t) = \frac{m_j h_{jk}(t \mid t_0,\dots,t_n)}{\displaystyle\sum_{i=0}^{n} m_i H_i(t \mid t_0,\dots,t_n)}, \qquad 1 \le k \le \mu_j - 1 \ .$$

Show that these functions behave like rational Hermite basis functions. In particular,

a. $\displaystyle\sum_{j=0}^{n} R_j(t) \equiv 1$

b. $R_j(t_i) = 0 \qquad i \ne j$

$\qquad\quad = 1 \qquad i = j$

c. $r_{jk}^{(p)}(t_i) = 0 \qquad i \ne j \qquad\qquad 0 \le p \le \mu_i - 1$

$\qquad\qquad\ \ = 0 \qquad i = j, \ p \ne k \qquad 0 \le k \le \mu_j - 1$

$\qquad\qquad\ \ = 1 \qquad i = j, \ p = k \qquad 1 \le k \le \mu_j - 1$

d. $\displaystyle R(t) = \sum_{j=0}^{n} R_j(t \mid t_0,\dots,t_n)P_j + \sum_{j=0}^{n}\sum_{k=0}^{\mu_j-1} r_{jk}(t \mid t_0,\dots,t_n)v_{jk}$

interpolates the data $(P_0, v_{01},\dots,v_{0,\mu_0-1}),\dots,(P_n, v_{n1},\dots,v_{n,\mu_n-1})$ at the nodes $t_0,\dots,t_n$.

(Compare to Section 2.7, Exercise 10.)

3.5 Hermite Surfaces

Tensor product, lofted, and Boolean sum surfaces can all be generalized from Lagrange to Hermite interpolation. Here we shall briefly examine each of these surface schemes, beginning with the tensor product construction. Triangular and rational Hermite surfaces can also be developed, but since these Hermite surfaces are less common in practice they are relegated to the exercises.

3.5.1 Tensor Product Hermite Surfaces

In tensor product Hermite interpolation, we start with a rectangular grid of parameter values (s_i, t_j) in the domain and a rectangular array of control points $\{P_{ij}\}$ in the range $0 \le i \le m, \ 0 \le j \le n$. Associated with each control point P_{ij} is a set of vectors

$\{v_{ij}^{kl}\}$, $0 \le k \le \mu - 1$, $0 \le l \le v - 1$, $(k,l) \ne (0,0)$, that represent mixed partial derivatives of the surface at P_{ij}. For the construction given below to work, we require the same amount of derivative data to be associated with each control point. The general problem is to construct a surface that interpolates this data at the grid points; that is, to build a surface $P(s,t)$ such that

$$P(s_i, t_j) = P_{ij}$$

$$P^{(k,l)}(s_i, t_j) = v_{ij}^{kl} \qquad\qquad 0 \le k \le \mu - 1, \;\; 0 \le l \le v - 1, \; (k,l) \ne (0,0) \;.$$

This generic problem is difficult to visualize and not very important in practice, so we shall not attempt to study it in such generality here (see Exercise 2). Rather we shall concentrate our attention on a simple, but important, special case that illustrates the general procedure. We will then apply this special case to construct lofted and Boolean sum Hermite surfaces.

Suppose then that at each point P_{ij} we have only three vectors (u_{ij}, v_{ij}, t_{ij}), and we seek a surface that interpolates this data (see Figure 3.10). That is, we seek a surface $P(s,t)$ such that

$$P(s_i, t_j) = P_{ij}, \;\; P^{(1,0)}(s_i, t_j) = u_{ij}, \;\; P^{(0,1)}(s_i, t_j) = v_{ij}, \;\; P^{(1,1)}(s_i, t_j) = t_{ij} \;.$$

To solve this problem, we can apply the Hermite basis functions

$$H_i(s \mid s_0, s_0, ..., s_m, s_m), \;\; h_i(s \mid s_0, s_0, ..., s_m, s_m)$$

$$H_j(t \mid t_0, t_0, ..., t_n, t_n), \;\; h_j(t \mid t_0, t_0, ..., t_n, t_n)$$

developed in Section 3.3 for interpolating one derivative at each point.

In the Hermite tensor product construction we form the bivariate Hermite basis functions by taking the product of each univariate Hermite basis function in s with each univariate Hermite basis function in t. Thus the tensor product Hermite surface is defined by setting

$$P(s,t) = \sum_{i=0}^{m} \sum_{j=0}^{n} H_i(s \mid s_0, s_0, ..., s_m, s_m) H_j(t \mid t_0, t_0, ..., t_n, t_n) P_{ij}$$

$$+ \sum_{i=0}^{m} \sum_{j=0}^{n} h_i(s \mid s_0, s_0, ..., s_m, s_m) H_j(t \mid t_0, t_0, ..., t_n, t_n) u_{ij}$$

$$\hspace{10cm} (3.13)$$

$$+ \sum_{i=0}^{m} \sum_{j=0}^{n} H_i(s \mid s_0, s_0, ..., s_m, s_m) h_j(t \mid t_0, t_0, ..., t_n, t_n) v_{ij}$$

$$+ \sum_{i=0}^{m} \sum_{j=0}^{n} h_i(s \mid s_0, s_0, ..., s_m, s_m) h_j(t \mid t_0, t_0, ..., t_n, t_n) t_{ij} \;.$$

It follows easily from the properties of the Hermite basis functions that the surface $P(s,t)$ does indeed interpolate the given data at the specified parameter values (see Exercise 1).

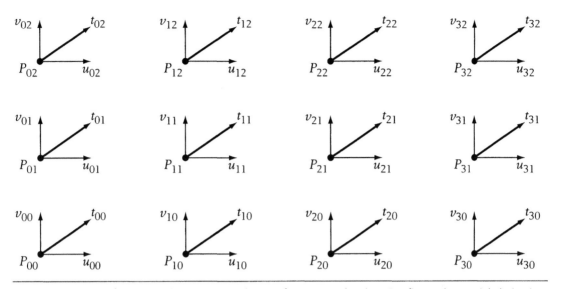

Figure 3.10 Data for a Hermite tensor product surface: control points P_{ij}; first-order partial derivatives u_{ij}, v_{ij}; and twists t_{ij}.

The vectors t_{ij} are called *twists*. The twists represent mixed partials of the surface. If the data $(P_{ij}, u_{ij}, v_{ij}, t_{ij})$ is taken off a surface we are trying to approximate with a polynomial, then the twists can be computed directly from the surface equation. If, however, we are simply designing a free-form surface using Hermite interpolation, then often it is unclear what values these twists should take. Nevertheless, simply setting them to zero—omitting the last summation in the definition of the tensor product surface—is not a good strategy, since zeros can cause flat spots to appear on the surface (see Exercise 3(b)). The problem of how to set the twists in general is a difficult one; we shall not attempt to deal with it here.

We can rewrite (3.13) in the following manner:

$$P(s,t) = \sum_{i=0}^{m} H_i(s \mid s_0, s_0, \dots, s_m, s_m) \sum_{j=0}^{n} \{H_j(t \mid t_0, t_0, \dots, t_n, t_n) P_{ij} + h_j(t \mid t_0, t_0, \dots, t_n, t_n) v_{ij}\}$$

$$+ \sum_{i=0}^{m} h_i(s \mid s_0, s_0, \dots, s_m, s_m) \sum_{j=0}^{n} \{H_j(t \mid t_0, t_0, \dots, t_n, t_n) u_{ij} + h_j(t \mid t_0, t_0, \dots, t_n, t_n) t_{ij}\}.$$

For $i = 0, \dots, m$, let

$$P_i(t) = \sum_{j=0}^{n} \{H_j(t \mid t_0, t_0, \dots, t_n, t_n) P_{ij} + h_j(t \mid t_0, t_0, \dots, t_n, t_n) v_{ij}\}$$

$$v_i(t) = \sum_{j=0}^{n} \{H_j(t \mid t_0, t_0, \dots, t_n, t_n) u_{ij} + h_j(t \mid t_0, t_0, \dots, t_n, t_n) t_{ij}\} .$$

Then

$$P(s,t) = \sum_{i=0}^{m} H_i(s \mid s_0, s_0, ..., s_m, s_m) P_i(t) + \sum_{i=0}^{m} h_i(s \mid s_0, s_0, ..., s_m, s_m) v_i(t) \ .$$

If we fix the value of $t = t^*$, then $P(s,t^*)$ is simply the univariate Hermite polynomial that interpolates the Hermite data $\{(P_0(t^*), v_0(t^*)), ..., (P_m(t^*), v_m(t^*))\}$ at the parameter values $s_0, ..., s_m$. Similarly, each curve $P_i(t)$ interpolates the Hermite data $\{(P_{i0}, v_{i0}), ..., (P_{in}, v_{in})\}$ at the nodes $t_0, ..., t_n$, and each vector field $v_i(t)$ interpolates the Hermite data $\{(u_{i0}, t_{i0}), ..., (u_{in}, t_{in})\}$ at the nodes $t_0, ..., t_n$ (see Figure 3.11). Notice that if we restrict to the domain $s_0 \le s \le s_m$ and $t_0 \le t \le t_n$, then we get a four-sided surface patch. Moreover, it is easy to see that the boundary curves of this rectangular patch are the Hermite polynomial curves that interpolate the Hermite data along the boundaries.

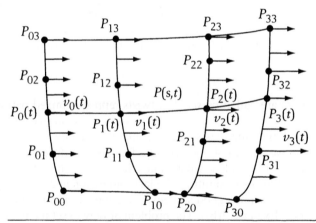

Figure 3.11 Tensor product Hermite interpolation. The curve $P_i(t)$ interpolates the Hermite data $\{(P_{i0}, v_{i0}),$ $(P_{i1}, v_{i1}), (P_{i2}, v_{i2}), (P_{i3}, v_{i3})\}$, and the vector field $v_i(t)$ interpolates the Hermite data $\{(u_{i0}, t_{i0}),$ $(u_{i1}, t_{i1}), (u_{i2}, t_{i2}), (u_{i3}, t_{i3})\}$ at the nodes t_0, t_1, t_2, t_3. The surface $P(s,t)$ interpolates the Hermite data $\{(P_0(t), v_0(t)), (P_1(t), v_1(t)), (P_2(t), v_2(t)), (P_3(t), v_3(t))\}$ at the parameters s_0, s_1, s_2, s_3. The boundary curves are interpolating curves for the boundary data.

Exercises

1. Consider the tensor product Hermite surface $P(s,t)$ defined in (3.13).

   a. Show that this surface $P(s,t)$ interpolates the data $(P_{ij}, u_{ij}, v_{ij}, t_{ij})$ at the parameter values (s_i, t_j).

   b. Explain how to use Neville's algorithm for univariate Hermite interpolation to evaluate points on this tensor product Hermite surface.

   c. What data lies at the leaves of the graph constructed in part (b)?

2. Use the general tensor product construction to define a surface that interpolates an array of control points $\{P_{ij}\}$ and a set of vectors $\{v_{ij}^{kl}\}$, $0 \le k \le \mu - 1$, $0 \le l \le \nu - 1$, that represent the mixed partial derivatives of the surface at P_{ij}. That is, construct a surface $P(s,t)$ such that

$$P(s_i, t_j) = P_{ij}$$
$$P^{(k,l)}(s_i, t_j) = v_{ij}^{kl} \qquad\qquad (k,l) \ne (0,0) \ .$$

3. Implement the tensor product Hermite surface for the data $(P_{ij}, u_{ij}, v_{ij}, t_{ij})$.

   a. Experiment with how changing the data affects the shape of the surface.

   b. What is the effect on the shape of the surface when all the twists t_{ij} are set to zero?

4. Define a tensor product rational Hermite surface $R(s,t)$ by setting

$$P(s,t) = \sum_{i=0}^{m} \sum_{j=0}^{n} H_i(s \mid s_0, s_0, ..., s_m, s_m) H_j(t \mid t_0, t_0, ..., t_n, t_n) m_{ij} P_{ij}$$

$$+ \sum_{i=0}^{m} \sum_{j=0}^{n} h_i(s \mid s_0, s_0, ..., s_m, s_m) H_j(t \mid t_0, t_0, ..., t_n, t_n) m_{ij} u_{ij}$$

$$+ \sum_{i=0}^{m} \sum_{j=0}^{n} H_i(s \mid s_0, s_0, ..., s_m, s_m) h_j(t \mid t_0, t_0, ..., t_n, t_n) m_{ij} v_{ij}$$

$$+ \sum_{i=0}^{m} \sum_{j=0}^{n} h_i(s \mid s_0, s_0, ..., s_m, s_m) h_j(t \mid t_0, t_0, ..., t_n, t_n) m_{ij} t_{ij}$$

$$w(s,t) = \sum_{i=0}^{m} \sum_{j=0}^{n} H_i(s \mid s_0, s_0, ..., s_m, s_m) H_j(t \mid t_0, t_0, ..., t_n, t_n) m_{ij}$$

$$R(s,t) = \frac{P(s,t)}{w(s,t)} \cdot$$

Show that

$$R(s_i, t_j) = P_{ij}, \quad R^{(1,0)}(s_i, t_j) = u_{ij}, \quad R^{(0,1)}(s_i, t_j) = v_{ij}, \quad R^{(1,1)}(s_i, t_j) = t_{ij} \ .$$

5. Let $\{P_{ij}\}$ be a rectangular array of control points, and let $\{v_{ij}^{kl}\}$ be a set of vectors that represent the mixed partial derivatives of a surface at P_{ij} up to order μ_{ij}, $i,j = 0,...,n$. We would like to construct a surface that interpolates this data at the parameters (s_i, t_j). That is, we seek a surface $P_\mu(s,t)$ such that

$$P_\mu(s_i, t_j) = P_{ij} \qquad\qquad i,j = 0,...,n$$
$$P_\mu^{(k,l)}(s_i, t_j) = v_{ij}^{kl} \qquad\qquad 0 \le k+l \le \mu_{ij} \qquad (k,l) \ne (0,0) \ .$$

Let $P_\mu^{i,j}(s,t)$ denote a surface that interpolates the same data, except that along the lines $s = s_i$ and $t = t_j$ we replace μ_{il} by $\mu_{il} - 1$ and μ_{hj} by $\mu_{hj} - 1$, $i,j = 0,n$.

a. Show that $P_\mu(s,t)$ satisfies the recurrence

$$P_\mu(s,t) = \frac{(s - s_n)(t - t_n)}{(s_n - s_0)(t_n - t_0)} P_\mu^{0,0}(s,t) + \frac{(s - s_0)(t_n - t)}{(s_n - s_0)(t_n - t_0)} P_\mu^{n,0}(s,t)$$

$$+ \frac{(s_n - s)(t - t_0)}{(s_n - s_0)(t_n - t_0)} P_\mu^{0,n}(s,t) + \frac{(s - s_0)(t - t_0)}{(s_n - s_0)(t_n - t_0)} P_\mu^{n,n}(s,t) \ .$$

b. What are the base cases for the recurrence in part (a)?

c. Use the recurrence in part (a) to develop a rectangular pyramid algorithm for the interpolant $P_\mu(s,t)$.

6. Let $\{P_{ijk}\}$, $i + j + k = n$, be a triangular array of control points, and let $\{v_{ijk}^{pq}\}$ be a set of vectors that represent the mixed partial derivatives of a surface at P_{ijk} up to order μ_{ijk}. We would like to construct a surface that interpolates this data at the points of a triangular grid $\{Q_{ijk}\}$. That is, we seek a surface $P_\mu(s,t)$ such that

$$P_\mu(Q_{ijk}) = P_{ijk} \qquad\qquad i + j + k = n$$

$$P_\mu^{(p,q)}(Q_{ijk}) = v_{ijk}^{pq} \qquad 0 \le p + q \le \mu_{ijk} \qquad (p,q) \ne (0,0) \ .$$

Let $P_\mu^1(s,t)$ denote a surface that interpolates the same data, except that at the points Q_{0jk} we replace μ_{0jk} by $\mu_{0jk} - 1$. Let the surfaces $P_\mu^2(s,t)$ and $P_\mu^3(s,t)$ be similarly defined. Finally, let $\beta_1(s,t), \beta_2(s,t), \beta_3(s,t)$, denote the barycentric coordinate functions of $\Delta Q_{n00} Q_{0n0} Q_{00n}$.

a. Show that $P_\mu(s,t)$ satisfies the recurrence

$$P_\mu(s,t) = \beta_1(s,t) P_\mu^1(s,t) + \beta_2(s,t) P_\mu^2(s,t) + \beta_3(s,t) P_\mu^3(s,t).$$

b. What are the base cases for the recurrence in part (a)?

c. Use the recurrence in part (a) to develop a triangular pyramid algorithm for the interpolant $P_\mu(s,t)$.

3.5.2   Lofted Hermite Surfaces

In the basic Hermite lofting problem, we are given a sequence of curves $U_0(s),\ldots,U_n(s)$ and a sequence of vector fields $u_0(s),\ldots,u_n(s)$ representing cross-boundary derivatives along these curves (see Figure 3.12). We seek a surface $H_U(s,t)$ to interpolate this data. That is, we want to construct a surface $H_U(s,t)$ such that for $k = 0,\ldots,n$

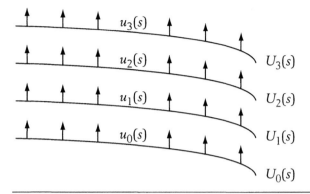

Figure 3.12 Data for a lofted Hermite surface: curves $U_k(s)$ and vector fields $u_k(s)$, $k = 0,...,3$.

$$H_U(s,t_k) = U_k(s)$$

$$H_U^{(0,1)}(s,t_k) = u_k(s) .$$

Again we can apply the Hermite basis functions $H_k(t \mid t_0,t_0,...,t_m,t_m)$, $h_k(t \mid t_0,t_0,...,t_m,t_m)$ developed in Section 3.3 to construct this surface by setting

$$H_U(s,t) = \sum_{k=0}^{m} H_k(t \mid t_0,t_0,...,t_m,t_m)U_k(s) + \sum_{k=0}^{m} h_k(t \mid t_0,t_0,...,t_m,t_m)u_k(s). \quad (3.14)$$

Here we have simply replaced the points P_k with the curves $U_k(s)$ and the vectors v_k with the vector fields $u_k(s)$ in (3.6). Hence it follows immediately from the properties of the Hermite basis functions that this lofted surface has the desired interpolation properties (see Exercise 1).

One nice feature of this lofting procedure is that it allows us to piece surfaces together with smooth partial derivatives across common boundaries. Two lofted surfaces with a common boundary curve and a common boundary vector field will, by construction, join continuously, and their partial derivative will also be continuous across the common boundary. More significantly, suppose we have two disjoint surface patches $P_1(s,t)$ and $P_2(s,t)$ that we wish to connect smoothly by joining them with a third surface $Q(s,t)$. Then we can construct $Q(s,t)$ as a lofted surface with one boundary curve given by $P_1(1,t)$ with the corresponding vector field $\partial P_1/\partial s$, and the other boundary curve given by $P_2(0,t)$ with the corresponding vector field $\partial P_2/\partial s$. By the interpolation properties of lofted surfaces, $Q(s,t)$ and $P_j(s,t)$, $j = 1,2$, will join smoothly across their common boundary (see Figure 3.13).

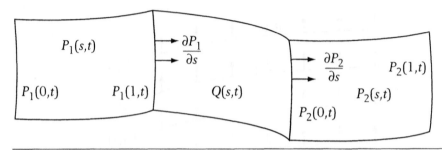

Figure 3.13   Joining the two surfaces $P_1(s,t)$ and $P_2(s,t)$ with a third surface $Q(s,t)$ so that derivatives are continuous across the common boundaries.

Exercises

1. Show that the lofted Hermite surface defined in (3.14) interpolates the curves $U_k(s)$ and the vector fields $u_k(s)$, $k = 0,...,n$.

2. Given a collection of curves $U_k(s)$ and vector fields $u_k(s)$, $k = 0,...,n$, use the lofted surface construction to define a surface $P(s,t)$ that is piecewise cubic in t and interpolates the given curves and vector fields.

3. Define a lofted rational Hermite surface $R_U(s,t)$ by setting

$$R_U(s,t) = \frac{\sum\limits_{k=0}^{m} H_k(t \mid t_0,t_0,...,t_m,t_m) m_k U_k(s) + \sum\limits_{k=0}^{m} h_k(t \mid t_0,t_0,...,t_m,t_m) m_k u_k(s)}{\sum\limits_{k=0}^{m} m_k H_k(t \mid t_0,t_0,...,t_m,t_m)}.$$

Show that

a. $R_U(s,t_k) = U_k(s)$

b. $R_U^{(0,1)}(s,t_k) = u_k(s)$

3.5.3   Boolean Sum Hermite Surfaces

We can also construct Boolean sum Hermite surfaces. Here we are given a mesh of curves $U_0(s),...,U_m(s)$ and $V_0(t),...,V_n(t)$ together with a collection of vector fields $u_0(s),...,u_m(s)$ and $v_0(t),...,v_n(t)$ representing cross-boundary derivatives along these curves. We are also provided with an array of control points P_{ij} and an array of twists t_{ij}. The problem is to interpolate all of this data: curves, cross-boundary derivatives, control points, and twists along a grid with prespecified nodes (see Figure 3.14).

Even in the Boolean sum Lagrange construction, we could not solve the interpolation problem without some compatibility conditions (see Section 2.15). For Lagrange interpolation, we need to assume that

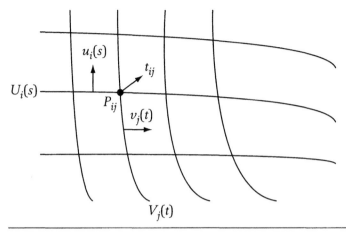

Figure 3.14 Data for a Boolean sum Hermite surface: a mesh of curves $\{U_i(s), V_j(t)\}$, a band of vector fields $\{u_i(s), v_j(t)\}$, an array of control points $\{P_{ij}\}$, and an array of twists $\{t_{ij}\}$.

$$P_{ij} = U_j(s_i) = V_i(t_j);$$

that is, we require that the mesh of curves actually intersect. Here we need compatibility conditions as well between the cross-boundary derivatives. Indeed we require the following compatibility conditions:

1. $\quad P_{ij} = U_j(s_i) = V_i(t_j)$ $\hspace{5cm}$ (3.15a)

2. $\quad U_j'(s_i) = v_i(t_j)$ $\hspace{5.5cm}$ (3.15b)

3. $\quad V_i'(t_j) = u_j(s_i)$ $\hspace{5.5cm}$ (3.15c)

4. $\quad t_{ij} = u_j'(s_i) = v_i'(t_j)$ $\hspace{4.8cm}$ (3.15d)

The first condition requires that the mesh of curves intersect at the nodes. The second and third conditions assert that the cross-boundary derivatives agree with the curve tangents at the grid points, and the last condition says that the mixed partials agree at the nodes.

Now the Boolean sum Hermite surface is defined in a manner analogous to the Boolean sum Lagrange surface by setting

$$B(s,t) = H_U(s,t) + H_V(s,t) - T_P(s,t),$$

where the Hermite tensor product surface $T_P(s,t)$ is defined by (3.13) with respect to the points P_{ij}, the derivative vectors $u_{ij} = U_j'(s_i)$ and $v_{ij} = V_i'(t_j)$, and the twists t_{ij}. Expanding this formula by substituting in the definitions of the lofted and tensor product surfaces, we find that

$$B(s,t) = \sum_{j=0}^{m} H_j(t \mid t_0,t_0,...,t_m,t_m)U_j(s) + \sum_{j=0}^{m} h_j(t \mid t_0,t_0,...,t_m,t_m)u_j(s)$$

$$+ \sum_{i=0}^{n} H_i(s \mid s_0,s_0,...,s_n,s_n)V_i(t) + \sum_{j=0}^{n} h_i(s \mid s_0,s_0,...,s_n,s_n)v_i(t)$$

$$- \sum_{i=0}^{n}\sum_{j=0}^{m} H_i(s \mid s_0,s_0,...,s_n,s_n)H_j(t \mid t_0,t_0,...,t_m,t_m)P_{ij}$$

$$- \sum_{i=0}^{n}\sum_{j=0}^{m} h_i(s \mid s_0,s_0,...,s_n,s_n)H_j(t \mid t_0,t_0,...,t_m,t_m)U'_j(s_i)$$ (3.16)

$$- \sum_{i=0}^{n}\sum_{j=0}^{m} H_i(s \mid s_0,s_0,...,s_n,s_n)h_j(t \mid t_0,t_0,...,t_m,t_m)V'_i(t_j)$$

$$- \sum_{i=0}^{n}\sum_{j=0}^{m} h_i(s \mid s_0,s_0,...,s_n,s_n)h_j(t \mid t_0,t_0,...,t_m,t_m)t_{ij} .$$

By applying the properties of the Hermite basis functions together with the compatibility conditions, you can check that this Boolean sum surface indeed has the desired interpolation properties (see Exercise 1).

Using this Boolean sum construction, we can fill a four-sided hole so that the surface patches join with smooth cross-boundary derivatives, provided that the data from the bounding surfaces is compatible at the four corner points. We illustrate this construction in Figure 3.15.

Exercises

1. Show that when the compatibility conditions (3.15a–d) are satisfied, the Boolean sum Hermite surface defined in (3.16) interpolates all of the data—curves, cross-boundary derivatives, control points, and twists—along the parameter lines $s = s_i$ and $t = t_j$.

2. Let $\beta_1, \beta_2, \beta_3$ be the barycentric coordinate functions for the triangle $\Delta Q_1 Q_2 Q_3$. Define three functions $r_i(\beta_1, \beta_2, \beta_3)$, $i = 1,2,3$, by setting

$$r_i(\beta_1,\beta_2,\beta_3) = \frac{\beta_j^2 \beta_k^2}{\beta_1^2\beta_2^2 + \beta_2^2\beta_3^2 + \beta_1^2\beta_3^2} \qquad i \neq j \neq k.$$

   Show that

   a. $r_i(\beta_1,\beta_2,\beta_3) = 0$ if $\beta_j = 0$ or $\beta_k = 0$, $i \neq j \neq k$

   b. $r_i(\beta_1,\beta_2,\beta_3) = 1$ if $\beta_i = 0$

   c. $\dfrac{\partial r_i}{\partial \beta_j} = 0$, if $\beta_p = 0$, $p = 1,2,3$

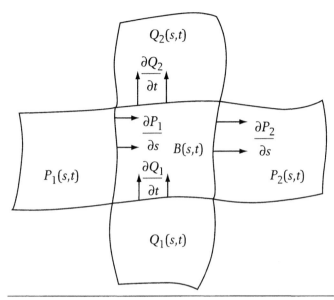

$Q_2(s,t)$

$\dfrac{\partial Q_2}{\partial t}$

$\dfrac{\partial P_1}{\partial s}$ $B(s,t)$ $\dfrac{\partial P_2}{\partial s}$

$P_1(s,t)$ $\dfrac{\partial Q_1}{\partial t}$ $P_2(s,t)$

$Q_1(s,t)$

Figure 3.15 Filling the four-sided hole surrounded by the surface patches $P_1(s,t)$, $P_2(s,t)$, $Q_1(s,t)$, $Q_2(s,t)$ with a Boolean sum Hermite surface $B(s,t)$ so that derivatives are continuous across common boundaries. This construction only succeeds when the data from the patches surrounding the hole satisfy the compatibility conditions (3.15a–d) at the four corners.

d. $\displaystyle\sum_{i=1}^{3} r_i(\beta_1,\beta_2,\beta_3) \equiv 1$

3. Consider three curves U_1, U_2, U_3 and three vector fields v_1, v_2, v_3 defined over the edges of a triangle with vertices Q_1, Q_2, Q_3 satisfying the compatibility conditions

$$U_i(Q_j) = U_k(Q_j) = P_j, \text{ and } v_i(Q_j) = v_k(Q_j) = w_j, \ i \neq j \neq k.$$

Let $\beta_1, \beta_2, \beta_3$ be the barycentric coordinate functions of $\Delta Q_1 Q_2 Q_3$, and let $H_0(t), H_1(t), h_0(t)$ be the quadratic Hermite basis functions for the nodes $0, 0, 1$ (see Section 3.1, Exercise 1). Construct three quadratic cones

$$C_{U_i}(\beta_1,\beta_2,\beta_3) = H_0(\beta_i)U_i\left(\frac{\beta_j Q_j + \beta_k Q_k}{1 - \beta_i}\right) + H_1(\beta_i)P_i$$
$$+ h_0(\beta_i)v_i\left(\frac{\beta_j Q_j + \beta_k Q_k}{1 - \beta_i}\right), \ i \neq j \neq k$$

and the surface

$$B(\beta_1,\beta_2,\beta_3) = \sum_{i=1}^{3} r_i(\beta_1,\beta_2,\beta_3) C_{U_i}(\beta_1,\beta_2,\beta_3),$$

where the functions $r_i(\beta_1, \beta_2, \beta_3)$, $i = 1, 2, 3$, are defined in Exercise 2. Using the properties of the quadratic Hermite basis functions and the functions $r_i(\beta_1, \beta_2, \beta_3)$, $i = 1, 2, 3$, developed in Exercise 2, show that

a. $C_{U_i}(\beta_1, \beta_2, \beta_3) = U_i$, when $\beta_i = 0$, $i = 1, 2, 3$

b. $\dfrac{\partial C_{U_i}}{\partial \beta_i} = v_i$, when $\beta_i = 0$, $i = 1, 2, 3$

c. $B(\beta_1, \beta_2, \beta_3) = U_i$, when $\beta_i = 0$, $i = 1, 2, 3$

d. $\dfrac{\partial B}{\partial \beta_i} = v_i$ along $\beta_i = 0$, $i = 1, 2, 3$

e. $B(\beta_1, \beta_2, \beta_3)$ is well defined at $\beta_i = 1$, $i = 1, 2, 3$

Thus the surface $B(\beta_1, \beta_2, \beta_3)$ interpolates the same data over triangles that the Boolean sum Hermite surface interpolates over rectangles.

4. Develop an analogue of the construction in Exercise 3, where the three cones are replaced by three surfaces interpolating pairs of edges and corresponding vector fields, and the quadratic Hermite basis functions are replaced by cubic Hermite basis functions.

3.6 Summary

In this chapter we have extended the ideas and techniques from Chapter 2 on Lagrange interpolation of control points to Hermite interpolation of control points and derivatives. Most of the result on Lagrange interpolation including existence and uniqueness theorems, Neville's algorithm, dynamic programming procedures, up and down recurrences, basis functions, rational schemes, and tensor product, lofted, and Boolean sum surfaces extend readily to the Hermite setting. If you understood Chapter 2 well, this chapter will have been mostly a review with some modest extensions.

We mentioned at the end of Chapter 2 that to solve problems in interpolation and approximation, *we must use the basis most appropriate to the problem at hand.* While the Lagrange and Hermite bases are improvements over the standard monomial basis for performing Lagrange and Hermite interpolation, they are not as efficient computationally as the monomial scheme. In the next chapter we introduce the Newton basis, a basis that is quite suitable for performing interpolation and as efficient computationally as the monomial basis.

CHAPTER 4

Newton Interpolation and Difference Triangles

We are going to revisit polynomial interpolation one more time. So far we have encountered several important polynomial bases, including

- *Monomial basis*: $1, t, ..., t^n$
- *Taylor basis*: $1, (t - t_0), ..., (t - t_0)^n$
- *Lagrange basis*: $L_0^n(t \mid t_0, ..., t_n), ..., L_n^n(t \mid t_0, ..., t_n)$
- *Hermite basis*: $H_0(t \mid t_0, t_0, ..., t_n, t_n), ..., H_n(t \mid t_0, t_0, ..., t_n, t_n),$

$$h_0(t \mid t_0, t_0, ..., t_n, t_n), ..., h_n(t \mid t_0, t_0, ..., t_n, t_n).$$

Here we plan to study yet another polynomial basis: the *Newton basis*.

Each of these bases has some good features and some bad features. Univariate polynomials written in terms of the monomial or Taylor bases can be evaluated quickly using Horner's method (see Section 2.8, Exercise 3), but computing the monomial or Taylor coefficients from Lagrange or Hermite data requires inverting a matrix, a slow and numerically unstable procedure. On the other hand, given Lagrange or Hermite data, we do not need to perform any computation to find the Lagrange or Hermite coefficients, since these coefficients are precisely the data we want to interpolate. This fact is one of the reasons we introduced these two bases in the first place. But Neville's evaluation algorithm for polynomials of degree n written in terms of the Lagrange or Hermite bases is $O(n^2)$, whereas Horner's method for polynomials written in terms of the monomial or Taylor bases is $O(n)$. Thus polynomial evaluation is relatively slow for polynomials expressed in terms of the Lagrange or Hermite bases. Even the ladder algorithm for Lagrange polynomials (Section 2.6, Figure 2.12), which is $O(n)$, is slower than Horner's method because the ladder algorithm requires $3n$ multiplications compared to only n multiplications for Horner's approach.

The Newton basis combines the best of both worlds. Given Lagrange or Hermite data, the coefficients of the interpolating polynomial relative to the Newton basis are easy to find using a simple recursive procedure. Moreover, we shall see that Horner's method for polynomial evaluation extends readily to the Newton basis. Thus the Newton basis combines some of the best features of the Lagrange and monomial bases: easy-to-compute coefficients and a fast evaluation algorithm.

4.1   The Newton Basis

To construct the Newton basis, we begin by fixing a set of nodes $t_0 \le t_1 \le \cdots \le t_n$. The Newton basis for these nodes is then defined by

$$N_0(t) = 1$$
$$N_1(t) = t - t_0$$
$$\vdots$$
$$N_n(t) = (t - t_0) \cdots (t - t_{n-1}) \ .$$

The nodes $t_0, \ldots, t_n$ need not be distinct. When all the nodes are identical, the Newton basis reduces to the Taylor basis at $t = t_0$. Thus the Taylor basis is a special case of the Newton basis.

The Newton basis has several rather obvious but important properties:

1. $N_k(t) = (t - t_0) \cdots (t - t_{k-1})$ is a polynomial of exact degree k.

2. $N_k^{(p)}(t_j) = 0, 0 \le j < k, 0 \le p < \mu_{jk}$,
    where μ_{jk} is the number of times t_j appears in $t_0, \ldots, t_{k-1}$.

3. $N_0(t), \ldots, N_n(t)$ form a basis for the polynomials of degree n.

4. $\displaystyle\sum_{k=0}^{n} c_k N_k(t)$ has an $O(n)$ Horner evaluation algorithm.

The first two properties are immediate from the definition of the Newton basis, and the third property follows easily from the first (see Exercises 2 and 3). The fourth property is what currently interests us here.

We illustrate the $O(n)$ Horner evaluation algorithm for cubic polynomials written in terms of a Newton basis in Figure 4.1. Notice that the labels along edges entering each node do not sum to one. This phenomenon is related to the fact that the Newton basis functions themselves do not sum to one. Thus, except for the coefficient of $N_0(t)$, the coefficients of the Newton basis functions are vectors, not points, so we need not take affine combinations of these coefficients for our results to make sense in affine space.

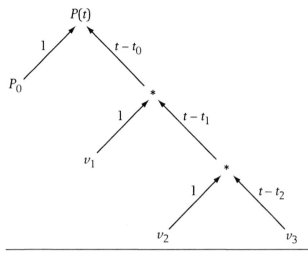

Figure 4.1 Horner's $O(n)$ evaluation algorithm for cubic polynomials $(P(t) = P_0 N_0(t) + v_1 N_1(t) + v_2 N_2(t) + v_3 N_3(t)$ written in Newton form. The labels along the edges are not normalized, since the Newton coefficients are vectors, not points. Multiplication is performed only along the lateral right edge. Edges labeled with a 1 represent additions, not multiplications. Compare to Figure 2.18, which is Horner's method for the monomial basis.

Exercises

1. What is the Newton basis when $t_k = 0$, $k = 0,...,n$?
2. Prove that

$$N_k^{(p)}(t_j) = 0, 0 \leq j < k, 0 \leq p < \mu_{jk},$$

   where μ_{jk} is the number of times t_j appears in $t_0,...,t_{k-1}$.
3. Prove that the Newton polynomials form a basis for the polynomials of degree n.
4. Diagram Horner's method for polynomials written in terms of the Taylor basis.

4.2 **Divided Differences**

Although we can quickly evaluate polynomials expressed in Newton form, it still remains to compute the Newton coefficients of the polynomial interpolant in an efficient manner. Let's begin then by trying a few simple calculations.

Given an arbitrary curve $F(t)$, suppose we want to interpolate the points $F(t_0),...,F(t_n)$ with a degree n polynomial curve $P_{0...n}(t)$. We know how to compute

$P_{0 \cdots n}(t)$ using Neville's algorithm, but now, for fast evaluation, we want to find the Newton coefficients of $P_{0 \cdots n}(t)$. If $n = 0$, then

$$P_0(t) = F(t_0) = F(t_0)N_0(t),$$

so this case is easy. If $n = 1$, then we have

$$P_{01}(t) = c_0 N_0(t) + c_1 N_1(t). \tag{4.1}$$

Since $N_1(t_0) = 0$, it follows once again that

$$c_0 = P_{01}(t_0) = F(t_0).$$

To find c_1, substitute t_1 into (4.1) to obtain

$$F(t_1) = P_{01}(t_1) = c_0 N_0(t_1) + c_1 N_1(t_1) = F(t_0) + c_1(t_1 - t_0).$$

Solving for c_1, we find that

$$c_1 = \frac{F(t_1) - F(t_0)}{t_1 - t_0}. \tag{4.2}$$

What if $t_1 = t_0$? Then instead of $P_{01}(t)$, we must consider the interpolating polynomial $P_{00}(t)$. But $P_{00}(t)$ is the Taylor polynomial

$$P_{00}(t) = F(t_0) + F'(t_0)(t - t_0).$$

Thus we can simply read off the Newton coefficients $c_0 = F(t_0)$ and $c_1 = F'(t_0)$. Notice that this result is consistent with our previous formulas, since

$$F'(t_0) \;=\; \lim_{t_1 \to t_0} \frac{F(t_1) - F(t_0)}{t_1 - t_0}.$$

We could go on to consider higher-order Newton coefficients, but although these computations are straightforward, they would not be very enlightening. Instead we shall soon take a less direct, but more revealing, approach. One thing you should notice now, however, is that

$$P_{0 \cdots n}(t) = P_{0 \cdots n-1}(t) + c_n N_n(t)$$

because $N_n(t_k) = 0$, $k = 0,\ldots,n-1$. Therefore, once we know the Newton coefficients for the nodes $t_0,\ldots,t_{n-1}$, we need not recalculate them for the nodes $t_0,\ldots,t_n$; all we need to do is to calculate the last coefficient c_n. This observation remains valid even if some nodes t_j have multiplicities $\mu_j > 1$ because

$$N_n^{(p)}(t_j) = 0, \;\; 0 \le p < \mu_j.$$

The Newton coefficient in (4.2) is a ratio of two differences. The general nth-order Newton coefficients are called *divided differences*. We begin with a recursive definition of the divided difference, and then argue from Neville's algorithm that these divided differences do indeed represent the Newton coefficients of the polynomial interpolant.

DEFINITION   *Divided Differences I*

$$F[t_0] = F(t_0)$$

$$F[t_0,t_1] = \frac{F(t_1) - F(t_0)}{t_1 - t_0} \qquad\qquad t_1 \neq t_0$$

$$= F'(t_0) \qquad\qquad t_1 = t_0$$

$$\vdots \qquad\qquad\qquad\qquad \vdots$$

$$F[t_0,...,t_n] = \frac{F[t_1,...,t_n] - F[t_0,...,t_{n-1}]}{t_n - t_0} \qquad t_n \neq t_0$$

$$= \frac{F^{(n)}(t_0)}{n!} \qquad\qquad t_n = t_0.$$

In this notation we do not assume that the nodes $t_0,...,t_n$ are distinct, and we repeat a node inside the bracket of $F[t_0,...,t_n]$ as often as its multiplicity.

When the nodes are distinct, this definition of divided difference looks a lot like a discrete version of differentiation. We shall have more to say about the connection between divided differences and derivatives in Section 4.3. We illustrate the computation of the divided difference for four distinct nodes in Figure 4.2. Notice the familiar triangular structure of this computation. In fact, if we look just at the indices

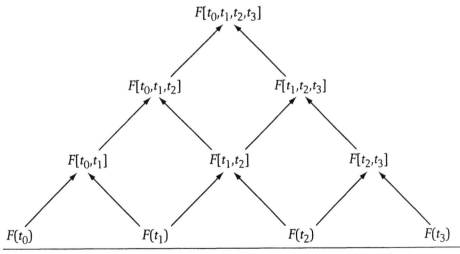

Figure 4.2 The triangular computation of the divided difference. Arrows entering the node $F[t_j,...,t_k]$ are labeled $\pm 1/(t_k - t_j)$. These labels have been suppressed here to avoid cluttering the diagram. Notice that the indices in the nodes are identical to the indices in the nodes for Neville's algorithm; compare this diagram to Figure 2.5.

in the nodes and ignore the labels on the arrows, the structure of this algorithm is identical to the structure of Neville's algorithm. This connection between these two algorithms is not a coincidence; we shall investigate the link between the divided difference recurrence and Neville's algorithm shortly.

Although this definition of divided difference is easy to understand and simple to compute, it does not, at first glance, seem to have much to do with polynomial interpolation. There are, however, two subtle clues that this construction might actually generate the Newton coefficients of the polynomial interpolant. First, we have seen that $F[t_0]$ and $F[t_0,t_1]$ are the coefficients of $N_0(t)$ and $N_1(t)$, so the recursion starts out in the right way. Second, the structure of the divided difference computation exactly mirrors the structure of Neville's algorithm.

There is another definition of the divided difference that, although somewhat more abstract, is much more closely tied to interpolation. It also turns out that this more abstract definition is often much easier to apply in deriving additional mathematical properties of the divided difference. We shall now give this alternative definition and then prove that our two definitions are equivalent.

DEFINITION   *Divided Differences II*

Let $F(t)$ be an arbitrary curve, and let $t_0 \leq t_1 \leq \cdots \leq t_n$ be a set of $n + 1$ nodes, not necessarily distinct. Denote by μ_k the multiplicity of t_k in the sequence $t_0 \leq t_1 \leq \cdots \leq t_n$. Let $P_{0 \cdots n}(t)$ be the unique degree n polynomial that interpolates the data

$$F(t_0),\ldots,F^{(\mu_0-1)}(t_0),\ldots,F(t_n),\ldots,F^{(\mu_n-1)}(t_n).$$

Then

$$F[t_0,\ldots,t_n] = \text{coefficient of } t^n \text{ in the monomial representation of the interpolant } P_{0 \cdots n}(t). \qquad (4.3)$$

In this notation, we repeat the node t_k a total of μ_k times inside the brackets of $F[t_0,\ldots,t_n]$.

For example, suppose that

$$F(t) = 3t^4 - 5t^3 + 2t^2 - 2t + 3,$$

and let $t_0 = 0$, $t_1 = 0$, $t_2 = 1$, $t_3 = 1$. Then in the cubic Hermite basis (see Section 3.1)

$$P_{0011}(t) = F(0)(1-t)^2(1+2t) + F(1)t^2(3-2t) + F'(0)t(1-t)^2 + F'(1)t^2(t-1).$$

Substituting $F(0) = 3$, $F'(0) = -2$, $F(1) = 1$, $F'(1) = -1$ and expanding in the monomial basis yields

$$P_{0011}(t) = t^3 - t^2 - 2t + 3 \ .$$

Therefore, $F[0,0,1,1] = $ coefficient of $t^3 = 1$.

THEOREM
4.1

Let $F[t_0,...,t_n]$ be the divided difference as defined by (4.3). Then $F[t_0,...,t_n]$ satisfies the recurrence

$$F[t_0,...,t_n] = \frac{F[t_1,...,t_n] - F[t_0,...,t_{n-1}]}{t_n - t_0} \qquad t_n \neq t_0$$

$$= \frac{F^{(n)}(t_0)}{n!} \qquad t_n = t_0 .$$

Thus Definition I and Definition II for the divided difference are equivalent.

Proof If $t_n \neq t_0$, then by Neville's algorithm (3.4)

$$P_{0 \cdots n}(t) = \frac{t - t_0}{t_n - t_0} P_{1 \cdots n}(t) + \frac{t_n - t}{t_n - t_0} P_{0 \cdots n-1}(t).$$

Comparing the coefficients of t^n on both sides of this equation yields

$$F[t_0,...,t_n] = \frac{F[t_1,...,t_n] - F[t_0,...,t_{n-1}]}{t_n - t_0} \qquad t_n \neq t_0.$$

If, on the other hand, $t_n = t_0$, then $P_{0 \cdots n}(t) = P_{\underbrace{0 \cdots 0}_{n+1}}(t)$ is the Taylor polynomial of degree n at $t = t_0$. Thus

$$P_{0 \cdots n}(t) = \sum_{k=0}^{n} \frac{F^{(k)}(t_0)}{k!}(t - t_0)^k.$$

Comparing the coefficients of t^n on both sides of this equation yields

$$F[\underbrace{t_0,...,t_0}_{n+1}] = \frac{F^{(n)}(t_0)}{n!}.$$

Although we now see a direct connection between divided differences, recursion, and interpolation, we have yet to link the divided difference to the Newton basis. Our next result states that divided differences are indeed the Newton coefficients of the polynomial interpolant.

COROLLARY
4.2

Let $F(t)$ be an arbitrary curve, and let $t_0 \leq t_1 \leq \cdots \leq t_n$ be a set of $n+1$ nodes, not necessarily distinct. Denote by μ_k the multiplicity of t_k in the sequence $t_0 \leq t_1 \leq \cdots \leq t_n$. Let $P_{0\cdots n}(t)$ be the unique degree n polynomial that interpolates the data

$$F(t_0),\ldots,F^{(\mu_0-1)}(t_0),\ldots,F(t_n),\ldots,F^{(\mu_n-1)}(t_n)$$

at the parameter values $t_0,\ldots,t_n$. Then

$$P_{0\cdots n}(t) = \sum_{k=0}^{n} F[t_0,\ldots,t_k]N_k(t).$$

That is, the Newton coefficients of the polynomial interpolant are divided differences.

Proof This result follows by induction on n. If $n = 0$, then by definition $P_0(t) = F(t_0)$, so the result is certainly true. Now suppose the result is valid for all natural numbers less than n. Since the polynomials $N_0(t),\ldots,N_n(t)$ form a basis,

$$P_{0\cdots n}(t) = \sum_{k=0}^{n-1} c_k N_k(t) + c_n N_n(t).$$

Moreover, since

$$N_n^{(p_k)}(t_k) = 0, \ \ p_k = 0,\ldots,\mu_k - 1, \ \ k = 0,\ldots,n-1, \ \ p_n = 0,\ldots,\mu_n - 2,$$

it follows that

$$P_{0\cdots n-1}(t) = \sum_{k=0}^{n-1} c_k N_k(t)$$

because both sides are polynomials of degree $n-1$ that interpolate the data

$$F(t_0),\ldots,F^{(\mu_0-1)}(t_0),\ldots,F(t_{n-1}),\ldots,F^{(\mu_{n-1}-1)}(t_{n-1}),F(t_n),\ldots,F^{(\mu_n-2)}(t_n) \ .$$

Hence by the inductive hypothesis

$$c_k = F[t_0,\ldots,t_k] \ , \quad k = 0,\ldots,n-1.$$

Finally, observe that since degree$\{N_k(t)\} = k$, c_n is the coefficient of t^n in the monomial basis of the polynomial interpolant $P_{0\cdots n}(t)$; hence by (4.3), $c_n = F[t_0,\ldots,t_n]$.

Theorem 4.1 and Corollary 4.2 provide us with a fast way to compute the polynomial interpolant: first use the divided difference recurrence to find the Newton coefficients of the interpolant; then apply Horner's method to evaluate the interpolant in the Newton basis. The divided difference recurrence is $O(n^2)$, but we need to apply this recurrence only once to find the Newton coefficients. We can then apply Horner's method, which is only $O(n)$, to evaluate the interpolant at as many parameter values as we like.

For fixed parameters $t_0,...,t_n$ we can think of the divided difference as an operator that assigns to each function $F(t)$ the constant $F[t_0,...,t_n]$. The divided difference is a linear operator, since it is easy to show either by induction from Definition I or more directly from Definition II that

$$(F+G)[t_0,...,t_n] = F[t_0,...,t_n] + G[t_0,...,t_n]$$
$$(cF)[t_0,...,t_n] = c(F[t_0,...,t_n]) \ .$$

A linear operator that vanishes on all but one of a fixed set of basis functions and yields the value one on a single basis function is called a *dual functional*. For example, polynomial evaluation is the linear operator that provides the dual functionals with respect to the Lagrange basis $\{L_k^n(t \mid t_0,...,t_n)\}$ because by (2.12)

$$L_k^n(t_j \mid t_0,...,t_n) = 0 \qquad j \neq k$$
$$= 1 \qquad j = k.$$

Similarly, divided difference is the linear operator that provides the dual functionals for the Newton basis. Indeed, if we set $F(t) = N_k(t)$ in Corollary 4.2, then it follows from the uniqueness of the polynomial interpolant that $P_{0...n}(t) = N_k(t)$ (see Section 2.4, Exercise 3), so

$$N_j[t_0,....,t_k] = 0 \qquad j \neq k$$
$$= 1 \qquad j = k \ .$$

Dual functionals are convenient because if we know the dual functionals for a particular basis, then we can compute the coefficients of an arbitrary element with respect to this basis (see Exercise 11). For example, if $P(t)$ is a polynomial of degree n, then by Theorem 2.7 and Corollary 4.2

$$P(t) = \sum_{k=0}^{n} P(t_k) L_k^n(t \mid t_0,...,t_n)$$

$$P(t) = \sum_{k=0}^{n} P[t_0,...,t_k] N_k(t) \ .$$

Dual functionals are important tools in interpolation and approximation. We shall return to this theme again in Chapter 6, where we discuss blossoming, which provides the dual functionals for the Bernstein and B-spline bases.

Exercises

1. Draw the diagram of the divided difference recurrence for $F[0,1,1,2]$. Compare this diagram to Figure 3.5.

2. Draw the diagram of the divided difference recurrence for $F[0,0,0,1,1,2]$ and $F[0,0,1,1,2,2]$. Compare your diagrams to Figures 3.7 and 3.8. Why do these figures differ from your diagram?

3. Use the mean value theorem to prove that if $F(t)$ is a differentiable function, then

   a. $F[t_0, t_1] = F'(c)$ for some constant c such that $t_0 \leq c \leq t_1$.

   b. $F[t_0, \ldots, t_n] = \dfrac{F^{(n)}(c)}{n!}$ for some constant c such that $t_0 \leq c \leq t_n$.

4. Prove that $(F \circ G)[t_0, t_1] = F[G(t_0), G(t_1)]G[t_0, t_1]$.

5. Prove that if all the nodes are distinct, then

$$F[t_0, \ldots, t_n] = \sum_{k=0}^{n} \frac{F(t_k)}{\prod_{j \neq k}(t_k - t_j)} \ .$$

(Hint: Consider Lagrange interpolation.)

6. Let $G(t) = F(t + b)$. Prove that $G[t_0, \ldots, t_n] = F[t_0 + b, \ldots, t_n + b]$.

7. Let $P(t)$ be the Newton interpolating polynomial for $F(t)$ relative to the nodes $t_0, \ldots, t_n$. Form a new Newton interpolating polynomial $Q(t)$ by replacing each node t_k by the node $\tau_k = t_k + b$, $k = 0, \ldots, n$, for some fixed constant b.

   a. Show that changing all the nodes in this way has no effect on the shape of the Newton interpolating curve. In particular, show that $Q(t + b) = P(t)$.

   b. Form a new Newton interpolating polynomial $R(t)$ by replacing each node t_k by the node $\tau_k = at_k + b$ for some fixed constants $a > 0$ and b. Show that $R(t)$ is not, in general, a reparametrized version of $P(t)$.

   c. Compare these results for Newton interpolation to similar results for Hermite interpolation in Section 3.2, Exercise 5, and Lagrange interpolation in Section 2.2, Exercise 4. Why do these results for Newton interpolation resemble the corresponding results for Hermite interpolation rather than the corresponding results for Lagrange interpolation?

8. Prove that if all the nodes are distinct, then

$$F[t_0, \ldots, t_n] = \frac{\begin{vmatrix} F(t_0) & t_0^{n-1} & \cdots & t_0 & 1 \\ \vdots & \vdots & \vdots & \vdots & \vdots \\ F(t_n) & t_n^{n-1} & \cdots & t_n & 1 \end{vmatrix}}{\begin{vmatrix} t_0^n & t_0^{n-1} & \cdots & t_0 & 1 \\ \vdots & \vdots & \vdots & \vdots & \vdots \\ t_n^n & t_n^{n-1} & \cdots & t_n & 1 \end{vmatrix}} \ .$$

(Hint: Consider interpolation using the monomial basis.)

9. Let $t_0 < \cdots < t_n$ be distinct nodes with multiplicities $\mu_0, \ldots, \mu_n$. Generalize the result in Exercise 8 by replacing the row

$$R(t_j) = \{F(t_j) \quad t_j^{n-1} \quad \cdots \quad t_j \quad 1\}$$

by the rows $R(t_j),\ldots,R^{(\mu_j-1)}(t_j)$, and the row

$$r(t_j) = \{t_j^n \quad t_j^{n-1} \quad \cdots \quad t_j \quad 1\}$$

by the rows $r(t_j),\ldots,r^{(\mu_j-1)}(t_j)$ for $j = 0,\ldots,n$.

10. What are the dual functionals for the following bases?

   a. Taylor basis

   b. Hermite basis

11. Let $D_0^n(t),\ldots,D_n^n(t)$ be a basis for the polynomials of degree n. Suppose that $\lambda_0^n,\ldots,\lambda_n^n$ are linear functionals that assign a real number to each polynomial $P(t)$ of degree n. Show that the following two properties are equivalent:

   i. $\lambda_j^n(D_k^n) = 0 \qquad j \neq k$

   $\qquad\qquad\;\; = 1 \qquad j = k.$

   ii. $P(t) = \displaystyle\sum_{j=0}^{n} \lambda_j^n(P)D_j^n(t) \cdot$

   (Compare to Section 2.5, Exercise 4.)

4.3 Properties of Divided Differences

Divided differences have many interesting properties. When the nodes are distinct, the divided difference is a discrete version of the derivative, and when the nodes are identical, the divided difference is a derivative. Thus divided difference shares many of the familiar properties of differentiation. But divided differences are also the coefficients of a polynomial interpolant; thus they also possess properties related to interpolation. We collect a dozen of the most important properties of divided differences and list them in the next theorem. Additional intriguing formulas can be found in the exercises. For easy reference many of these divided difference identities, and others as well, are listed together at the end of this chapter.

THEOREM
4.3

Properties of the Divided Difference

Let $F(t)$ be an arbitrary curve, and let $t_0 \leq t_1 \leq \cdots \leq t_n$ be a set of $n + 1$ nodes, not necessarily distinct. Denote by μ_k the multiplicity of t_k in the sequence $t_0 \leq t_1 \leq \cdots \leq t_n$. Let $P_{0\cdots n}(t)$ be the unique degree n polynomial that interpolates the data

$$F(t_0),\ldots,F^{(\mu_0-1)}(t_0),\ldots,F(t_n),\ldots,F^{(\mu_n-1)}(t_n)$$

at the parameter values $t_0,\ldots,t_n$. Then the divided difference $F[t_0,\ldots,t_n]$, where each distinct node t_j is repeated μ_j times, satisfies the following properties:

1. *Recursion*

$$F[t_0,\ldots,t_n] = \frac{F[t_1,\ldots,t_n] - F[t_0,\ldots,t_{n-1}]}{t_n - t_0} \qquad t_n \neq t_0$$

$$= \frac{F^{(n)}(t_0)}{n!} \qquad t_n = t_0 \; .$$

2. *Symmetry*

$F[t_0,\ldots,t_n] = F[t_{\sigma(0)},\ldots,t_{\sigma(n)}]$, where σ is any permutation of $\{0,\ldots,n\}$.

3. *Recursion Revisited*

$$F[t_0,\ldots,t_n] = \frac{F[t_0,\ldots,t_{i-1},t_{i+1},\ldots,t_n] - F[t_0,\ldots,t_{j-1},t_{j+1},\ldots,t_n]}{t_j - t_i} \qquad t_j \neq t_i \; .$$

4. *Linearity*

$$(F + G)[t_0,\ldots,t_n] = F[t_0,\ldots,t_n] + G[t_0,\ldots,t_n]$$
$$(cF)[t_0,\ldots,t_n] = c(F[t_0,\ldots,t_n]) \; .$$

5. *Cancellation*

$$F[t_0,\ldots,t_n] = \{(t - t_{n+1})F(t)\}[t_0,\ldots,t_n,t_{n+1}].$$

6. *Leibniz's Rule*

$$(FG)[t_0,\ldots,t_n] = \sum_{k=0}^{n} F[t_0,\ldots,t_k]G[t_k,\ldots,t_n] \; .$$

7. *Highest-Order Coefficient of the Polynomial Interpolant*

$F[t_0,\ldots,t_n] = $ coefficient of t^n in the monomial representation of the interpolant $P_{0\cdots n}(t)$.

8. *Highest-Order Coefficient of the Newton Interpolant*

$F[t_0,\ldots,t_n] = $ coefficient of $N_n(t)$ in the Newton representation of the interpolant $P_{0\cdots n}(t)$.

9. *Newton Coefficients of Polynomial Interpolant*

$$P_{0\cdots n}(t) = \sum_{k=0}^{n} F[t_0,\ldots,t_k]N_k(t).$$

10. *Dual Functionals for the Newton Basis*

$$N_j[t_0,\ldots,t_k] = 0 \quad j \neq k$$
$$= 1 \quad j = k \ .$$

11. *Equality Conditions*

$$F^{(p)}(t_j) = G^{(p)}(t_j) \quad 0 \leq p \leq \mu_j - 1, \quad j = 0,\ldots,n$$
$$\Rightarrow \ F[t_0,\ldots,t_n] = G[t_0,\ldots,t_n].$$

12. *Value on Low-Order Polynomials*
    a. If $F(t)$ is a polynomial of degree $n-1$, then $F[t_0,\ldots,t_n] = 0$.
    b. If $F(t)$ is a polynomial of degree n, then $F[t_0,\ldots,t_n]$ is the coeffi-
        cient of t^n in the monomial representation for $F(t)$. Thus, in this
        case, $F[t_0,\ldots,t_n]$ is a constant independent of $t_0,\ldots,t_n$.

Proof We are already familiar with some of these properties, and most of the remainder are fairly simple to derive. We shall take Property 7 as the definition of the
divided difference and deduce the other properties as a consequence.

1. Follows from 7 by Theorem 4.1.

2. Also follows immediately from 7 because the interpolant $P_{0\cdots n}(t)$ is
    independent of the order of the nodes $t_0,\ldots,t_n$.

3. Follows immediately from Properties 1 and 2.

4. Follows easily by induction from Property 1 or more directly from
    Property 7 by the linearity of the polynomial interpolant.

5. Follows from Property 7 because if $P_{0\cdots n}(t)$ is the polynomial interpo-
    lant for $F(t)$, then $(t - t_{n+1})P_{0\cdots n}(t)$ is the polynomial interpolant for
    $(t - t_{n+1})F(t)$.

6. This result is the analogue of Leibniz's rule for the nth derivative of the
    product of two functions. We shall prove this result separately below in
    Proposition 4.4.

7. This property is the definition of the divided difference.

8. Follows directly from Corollary 4.2.

9. This result is Corollary 4.2.

10. Follows easily from Property 9 with $F(t) = N_j(t)$.

11. Again this result follows easily by induction from Property 1 or more
    directly from Property 7 by the uniqueness of the polynomial interpolant.

12. Part (a) is a consequence of Property 7 because if $F(t)$ is a polynomial of degree $n - 1$, then $F(t)$ is the polynomial interpolant to the data generated by $F(t)$. But since $F(t)$ is a polynomial of degree $n - 1$, the coefficient of t^n in the monomial representation of $F(t)$ is zero, so $F[t_0,...,t_n] = 0$. Part (b) is also a consequence of Property 7 because again $F(t)$ is the polynomial interpolant to the data generated by $F(t)$. Thus by Property 7, $F[t_0,...,t_n]$ is the coefficient of t^n in the monomial representation of $F(t)$. It follows that $F[t_0,...,t_n]$ is independent of the nodes $t_0,...,t_n$.

Properties 1–3 and 7–11 are directly related to interpolation. For example, we have seen that the structure of the divided difference recurrence in Property 1 is identical to the structure of Neville's algorithm for polynomial interpolation. Properties 4, 6, and 12 are reminiscent of similar properties for differentiation. Here Property 6 reminds us of the product rule for higher-order derivatives (see Exercise 1), and Property 12 recalls the fact that the nth derivative of a degree $n - 1$ polynomial is zero while the nth derivative of a degree n polynomial is a constant. It remains then only to prove Leibniz's rule.

PROPOSITION
4.4

Leibniz's Rule

$$(FG)[t_0,...,t_n] = \sum_{k=0}^{n} F[t_0,...,t_k]G[t_k,...,t_n]$$

Proof This result can be proved directly from the recurrence by a hard induction. Here we shall adopt a much simpler proof due to E. T. Y. Lee. To shorten our notation, let $P(t) = P_{0\cdots n}(t)$ be the polynomial interpolant for $F(t)$. Then by Property 9 of Theorem 4.3,

$$P(t) = \sum_{k=0}^{n} F[t_0,...,t_k](t - t_0)\cdots(t - t_{k-1}).$$ (4.4)

Now since FG and PG agree at the nodes, it follows from Property 11 of Theorem 4.3 that

$$(FG)[t_0,...,t_n] = (PG)[t_0,...,t_n].$$

Therefore, by Equation (4.4) and Properties 4 and 5 of Theorem 4.3,

$$(FG)[t_0,...,t_n] = \sum_{k=0}^{n} F[t_0,...,t_k]\{(t - t_0)\cdots(t - t_{k-1})G(t)\}[t_0,...,t_n]$$

$$= \sum_{k=0}^{n} F[t_0,...,t_k]G[t_k,...,t_n].$$

Exercises

1. Using Leibniz's rule for divided differences, prove Leibniz's rule for differentiation:

$$(FG)^{(n)} = \sum_{k=0}^{n} \binom{n}{k} F^{(k)} G^{(n-k)}.$$

2. Show that Property 5 (cancellation) in Theorem 4.3 is a special case of Leibniz's rule.

3. Prove that

a. $\dfrac{\partial F[t_0,\ldots,t_k,\ldots,t_n]}{\partial t_k} = \mu_k F[t_0,\ldots,t_k,t_k,\ldots,t_n],$

   where μ_k is the multiplicity of t_k

b. $F'[t_0,\ldots,t_n] = \displaystyle\sum_{k=0}^{n} F[t_0,\ldots,t_k,t_k,\ldots,t_n]$

c. $F'[t_0,\ldots,t_n] = \displaystyle\sum_{k=0}^{n} \dfrac{\partial F[t_0,\ldots,t_k,\ldots,t_n]}{\partial t_k}$ if all the nodes are distinct

4. Prove that the divided difference of a polynomial is a polynomial. That is, prove that if $P(t)$ is a polynomial in t, then $P[t_0,\ldots,t_n]$ is a polynomial in the variables $t_0,\ldots,t_n$.

5. Prove that $\{t^n\}[v_1,\ldots,v_{n-k}] = \sum_{i_1 \le i_2 \le \ldots \le i_{k+1}} v_{i_1} v_{i_2} \cdots v_{i_{k+1}},\ 0 \le k \le n-1.$

6. Prove that $\left\{\dfrac{1}{x-t}\right\}[t_0,\ldots,t_n] = \dfrac{1}{(x-t_0)\cdots(x-t_n)}.$

   Here x is treated as a constant and the divided difference is taken with respect to t.

7. Generalize the result in Exercise 6 by showing that if $P(t)$ is a polynomial with degree$(P) \le n$, then

$$\left\{\dfrac{P(t)}{x-t}\right\}[t_0,\ldots,t_n] = \dfrac{P(x)}{(x-t_0)\cdots(x-t_n)}.$$

   (Hint: Observe that $r - t = (r - x) + (x - t)$ and apply Theorem 4.3, Property 12a.)

8. Prove that $\left\{\dfrac{1}{(x-t)^2}\right\}[t_0,\ldots,t_n] = \displaystyle\sum_{k=0}^{n} \dfrac{1}{(x-t_0)\cdots(x-t_k)^2\cdots(x-t_n)}.$

   (Hint: See Exercise 3(b) and Exercise 6.)

9. Use the recurrence for the divided difference to derive the Hermite-Genocchi formula:

$$G[x_0,...,x_n] = \int_{\Delta^n} G^{(n)}\big(x_0 + v_1(x_1 - x_0) + \cdots + v_n(x_n - x_0)\big)dv_1 \cdots dv_n,$$

where $\Delta^n = \{(v_1,...,v_n) \mid v_j \geq 0 \text{ and } \sum_j v_j \leq 1\}$.

10. Use the recurrence for the divided difference to prove that

$$F[\underbrace{t,...,t}_{n},t+h] = \frac{F[t+h] - \sum\limits_{k=1}^{n} h^{k-1}F[\underbrace{t,...,t}_{k}]}{h^n}.$$

4.4   **An Axiomatic Approach to Divided Difference**

The divided difference has a bewildering array of remarkable properties. In Theorem 4.3 we list a dozen such properties, and additional formulas and identities are provided in the exercises at the end of Sections 4.2 and 4.3. It seems natural to ask, Which of these properties are most important? That is, which properties are primary and which are derived? Of course, the two definitions—the recursion formula and the highest-order coefficient of the polynomial interpolant—are fundamental. But there is another powerful mathematical paradigm for selecting basic properties: axiomatic systems. Below we provide an axiomatic approach to divided differences.

We have another, ulterior motive for introducing these axioms here. We observed at the end of Section 4.2 that the divided difference provides the dual functionals for the Newton basis. In Chapter 6, we are going to encounter another important linear operator called the *blossom,* which furnishes the dual functionals for the Bernstein and B-spline bases. The blossom of a polynomial is typically introduced by a set of elementary axioms. Here we provide a simple axiomatic characterization of the divided difference to prepare the way for blossoming and to emphasize the close connection between the blossom and the divided difference.

AXIOMS

Axioms for the Divided Difference

1. *Symmetry*

$F[t_0,...,t_n] = F[t_{\sigma(0)},...,t_{\sigma(n)}]$, where σ is any permutation of $\{0,...,n\}$

2. *Linearity*

$$(F + G)[t_0,...,t_n] = F[t_0,...,t_n] + G[t_0,...,t_n]$$
$$(cF)[t_0,...,t_n] = c(F[t_0,...,t_n])$$

3. *Cancellation*

$$F[t_0,...,t_n] = \{(t - t_{n+1})F(t)\}[t_0,...,t_n,t_{n+1}]$$

4. *Differentiation*

$$F[\underbrace{t_0,...,t_0}_{n+1}] = \frac{F^{(n)}(t_0)}{n!}$$

The last axiom is a diagonal property that specifies how the divided difference behaves when all the nodes are the same. We shall see in Theorem 4.5 that the first three axioms completely characterize the divided difference when some of the nodes are distinct. Thus we can take these four axioms as the primary properties of the divided difference; all the other formulas and identities can be derived from these four axioms.

THEOREM
4.5

The divided difference is the unique operator satisfying the four axioms of symmetry, linearity, cancellation, and differentiation.

Proof By Theorem 4.3, the divided difference operator certainly satisfies these four axioms. To prove that the divided difference is the only operator that satisfies these axioms, we shall derive the divided difference recurrence from these axioms. This derivation is straightforward, since by linearity, symmetry, and the cancellation axiom:

$$\begin{aligned}(t_n - t_0)F[t_0,...,t_n] &= \{(t - t_0) - (t - t_n)F(t)\}[t_0,...,t_n] \\ &= \{(t - t_0)F(t)\}[t_0,...,t_n] - \{(t - t_n)F(t)\}[t_0,...,t_n] \\ &= F[t_1,...,t_n] - F[t_0,...,t_{n-1}] \ .\end{aligned}$$

Dividing both sides by $t_n - t_0$ yields the recurrence.

Notice that in the proof of Theorem 4.5 the cancellation axiom does most of the vital work. Often the cancellation axiom is easier to apply than the divided difference recurrence; see, for example, the proof of Leibniz's rule in Proposition 4.4. To further exhibit the power of these axioms, especially the cancellation axiom, we prove the following result, which can be used to derive a generalization of the de Boor recurrence as well as Boehm's knot insertion formula for B-spline curves (see Section 7.7.4, Exercises 7 and 8).

PROPOSITION
4.6

Let $\tau = \sum_k \lambda_k t_k$, where $\sum_k \lambda_k = 1$. Then

$$F[t_0,\ldots,t_n] = \sum_{k=0}^{n} \lambda_k F[t_0,\ldots,t_{k-1},\tau,t_{k+1},\ldots,t_n].$$

Proof If $t_0 = t_1 = \cdots = t_n$, then $\tau = t_k$, $k = 0,\ldots,n$, and both sides reduce to $F[t_0,\ldots,t_0]$. Otherwise using only cancellation, linearity, and symmetry, we have

$$F[t_0,\ldots,t_n] = \{(t-\tau)F(t)\}[t_0,\ldots,t_n,\tau]$$

$$= \left\{ \sum_{k=0}^{n} \lambda_k(t-t_k)F(t) \right\}[t_0,\ldots,t_n,\tau]$$

$$= \sum_{k=0}^{n} \{\lambda_k(t-t_k)F(t)\}[t_0,\ldots,t_k,\ldots,t_n,\tau]$$

$$= \sum_{k=0}^{n} \lambda_k F[t_0,\ldots,t_{k-1},\tau,t_{k+1},\ldots,t_n] \ .$$

Exercises

1. In this exercise you will need the following two results from complex analysis. Let C be a simple closed curve, and let t be any point in the complex plane that lies inside of C. Suppose that $F(z)$ is analytic inside C. Then the following results are classical:

THEOREMS

Cauchy's Integral Formula

$$F(t) = \frac{1}{2\pi i} \int_C \frac{F(z)\,dz}{z-t}$$

Cauchy's Integral Formula for Derivatives

$$\frac{F^{(n-1)}(t)}{(n-1)!} = \frac{1}{2\pi i} \int_C \frac{F(z)\,dz}{(z-t)^n}$$

In this exercise you are going to generalize these results from complex analysis to develop the following complex contour integration formula for the divided difference.

THEOREM   *Complex Contour Integration Formula for the Divided Difference*

$$F[t_0,\ldots,t_n] = \frac{1}{2\pi i}\int_C \frac{F(z)\,dz}{(z-t_0)\cdots(z-t_n)} \tag{4.5}$$

where C is any simple closed curve containing $t_0,\ldots,t_n$ and F is analytic inside C.

---

    a. Use the axioms for the divided difference to prove (4.5). What is the diagonal property?

    b. Use the divided difference recurrence to prove (4.5). What is the base case?

2. Use Equation (4.5) to prove that

    a. $\dfrac{\partial F[t_0,\ldots,t_k,\ldots,t_n]}{\partial t_k} = \mu_k F[t_0,\ldots,t_k,t_k,\ldots,t_n]$, where μ_k is the multiplicity of t_k

    b. $F'[t_0,\ldots,t_n] = \displaystyle\sum_{k=0}^{n} F[t_0,\ldots,t_k,t_k,\ldots,t_n]$

3. Prove that the linearity axiom for divided difference can be replaced by the following affinity axiom:

$$\{(x-(1-\alpha)u-\alpha v)F(x)\}[t_0,\ldots,t_n]$$
$$= (1-\alpha)\{(x-u)F(x)\}[t_0,\ldots,t_n] + \alpha\{(x-v)F(x)\}[t_0,\ldots,t_n] \ .$$

4. Prove that the affinity axiom for divided difference in Exercise 3 can be replaced by the identity in Proposition 4.6.

5. Prove that the differentiation axiom can be replaced by the following pair of axioms:

    i. *Evaluation* $F[t_0] = F(t_0)$

    ii. *Continuity* $\lim_{t_i \to \tau} F[t_0,\ldots,t_i,\ldots,t_n] = F[t_0,\ldots,\tau,\ldots,t_n]$.

    (Hint: Use the result of Section 4.3, Exercise 10.)

4.5 Forward Differencing

The fastest way to evaluate a polynomial at a single point is to apply Horner's method. But remarkably, if we wish to evaluate a polynomial at a great many points, then there are faster ways to compute these values. The fastest method is a technique called *fast forward differencing*, which is closely related to the divided difference and the Newton basis. This method computes points along a polynomial curve at equally spaced parameter values. After an initial start-up step, fast forward differencing requires only n additions and zero multiplications to evaluate each new point on a degree n polynomial curve.

The recursive definition of the forward difference $\Delta^n F(t_0,\dots,t_n)$ is similar to the recursive definition of the divided difference $F[t_0,\dots,t_n]$, but without the bothersome denominator:

$$\Delta^0 F(t_0) = F(t_0)$$
$$\Delta F(t_0,t_1) = F(t_1) - F(t_0)$$
$$\vdots \qquad\qquad \vdots$$
$$\Delta^n F(t_0,\dots,t_n) = \Delta^{n-1} F(t_1,\dots,t_n) - \Delta^{n-1} F(t_0,\dots,t_{n-1}) \ .$$

(4.6)

We illustrate the recursive computation of $\Delta^3 F(t_0,\dots,t_3)$ in Figure 4.3. Taking sums or differences in this way is closely related to Pascal's triangle (see Chapter 5). Not surprisingly, then, if we write out the first few differences explicitly, binomial coefficients begin to appear:

$$\Delta^0 F(t_0) = F(t_0)$$
$$\Delta F(t_0,t_1) = F(t_1) - F(t_0)$$
$$\Delta^2 F(t_0,t_1,t_2) = F(t_2) - 2F(t_1) + F(t_0)$$
$$\Delta^3 F(t_0,t_1,t_2,t_3) = F(t_3) - 3F(t_2) + 3F(t_1) - F(t_0) \ .$$

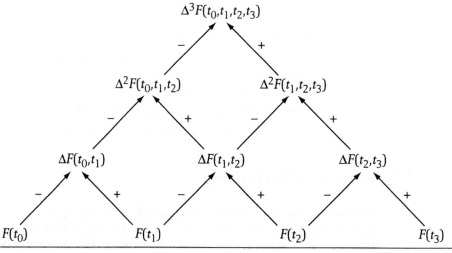

Figure 4.3 The triangular computation of the forward difference. Notice once again that the indices in the nodes are identical to the indices in the nodes for Neville's algorithm; compare this diagram to Figures 2.5 and 4.2.

Proceeding by induction (see Exercise 1), we can express the nth forward difference explicitly in terms of the function values $F(t_0),\ldots,F(t_n)$ by the formula

$$\Delta^n F(t_0,\ldots,t_n) = \sum_{k=0}^{n}(-1)^{n-k}\tbinom{n}{k}F(t_k). \tag{4.7}$$

What is most important to us here is how the forward difference behaves when the values $t_0,\ldots,t_n$ are evenly spaced along the parameter line. In this case it turns out that there is a simple relationship between the forward difference and the divided difference.

PROPOSITION 4.7

Suppose that the values $t_0,\ldots,t_n$ are evenly spaced along the parameter line—that is, $t_k = t_0 + k\Delta t,\ k = 0,\ldots,n$. Then

$$\Delta^n F(t_0,\ldots,t_n) = n!(\Delta t)^n F[t_0,\ldots,t_n]. \tag{4.8}$$

Proof We proceed by induction on n. Clearly the result is valid for $n = 0,1$, since

$$\Delta^0 F(t_0) = F(t_0) = F[t_0]$$

$$\Delta F(t_0,t_1) = F(t_1) - F(t_0) = \left\{\frac{F(t_1)-F(t_0)}{t_1-t_0}\right\}(t_1 - t_0) = F[t_0,t_1]\Delta t \ .$$

Suppose the result is valid for all natural numbers less than n. Then by (4.6) and the inductive hypothesis

$$\begin{aligned}
\Delta^n F(t_0,\ldots,t_n) &= \Delta^{n-1}F(t_1,\ldots,t_n) - \Delta^{n-1}F(t_0,\ldots,t_{n-1}) \\
&= (n-1)!(\Delta t)^{n-1}F[t_1,\ldots,t_n] - (n-1)!(\Delta t)^{n-1}F[t_0,\ldots,t_{n-1}] \\
&= (n-1)!(\Delta t)^{n-1}\left\{\frac{F[t_1,\ldots,t_n]-F[t_0,\ldots,t_{n-1}]}{t_n-t_0}\right\}(t_n - t_0) \\
&= (n-1)!(\Delta t)^{n-1}F[t_0,\ldots,t_n](n\Delta t) \\
&= n!(\Delta t)^n F[t_0,\ldots,t_n] \ .
\end{aligned}$$

Now recall that if $F(t)$ is a degree n polynomial, then $F[t_0,\ldots,t_n]$ is a constant independent of the parameters $t_0,\ldots,t_n$ (Theorem 4.3, Property 12b). Consequently, by Proposition 4.7, when the parameter values $t_0,\ldots,t_n$ are evenly spaced along the parameter line—that is, when Δt is fixed—then $\Delta^n F(t_0,\ldots,t_n)$ is also a constant independent of $t_0,\ldots,t_n$. We shall now apply this observation to develop a fast evaluation algorithm for polynomial curves.

Consider a degree n polynomial $F(t)$ and a sequence $t_0,\ldots,t_p$ of equally spaced parameter values. Suppose that somehow, by whatever means, we have already calculated the values

$$\Delta^n F(t_0,\ldots,t_n), \Delta^{n-1}F(t_1,\ldots,t_n),\ldots,\Delta F(t_{n-1},t_n),F(t_n).$$

Then we can calculate the new values

$$\Delta^n F(t_1,\ldots,t_{n+1}), \Delta^{n-1}F(t_2,\ldots,t_{n+1}),\ldots,\Delta F(t_n,t_{n+1}),F(t_{n+1})$$

using only addition in the following manner:

$$\Delta^n F(t_1,\ldots,t_{n+1}) = \Delta^n F(t_0,\ldots,t_n)$$
$$\Delta^{n-1}F(t_2,\ldots,t_{n+1}) = \Delta^{n-1}F(t_1,\ldots,t_n) + \Delta^n F(t_1,\ldots,t_{n+1})$$
$$\vdots$$
$$\Delta F(t_n,t_{n+1}) = \Delta F(t_{n-1},t_n) + \Delta^2 F(t_{n-1},t_n,t_{n+1})$$
$$F(t_{n+1}) = F(t_n) + \Delta F(t_n,t_{n+1}) \ .$$

The final equation gives us the value of $F(t_{n+1})$ (see Figure 4.4). Iterating this procedure, we can calculate additional values $F(t_{n+2}),F(t_{n+3}),\ldots,F(t_p)$ of the polynomial $F(t)$ still using only addition.

To obtain the initial sequence $\Delta^n F(t_0,\ldots,t_n), \Delta^{n-1}F(t_1,\ldots,t_n),\ldots,F(t_n)$, we simply compute the difference triangle for $\Delta^n F(t_0,\ldots,t_n)$ (Figure 4.3). In order to perform this computation, we must calculate the initial values $F(t_0),\ldots,F(t_n)$ at the base

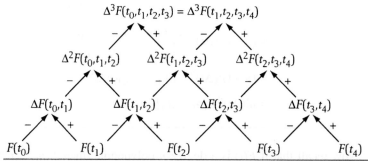

Figure 4.4 Two overlapping difference triangles for a cubic polynomial curve. The values at the apexes of the two triangles are identical by Proposition 4.7. Each value along the far right lateral edge can be computed by adding the values above it and to the left. Thus moving up the first triangle and then down the right lateral edge of the second triangle, we eventually calculate $F(t_4)$ from $F(t_0),\ldots,F(t_3)$ using only addition and subtraction. Iterating the second part of this procedure for more and more points is the fast forward differencing algorithm for polynomial evaluation. The triangle on the left represents the start-up step (see also Figure 4.3), and the right lateral edge represents the fast forward differencing action (see also Figure 4.5).

of the triangle, and for these calculations we can use Horner's method. Therefore, the start-up for the algorithm requires $n(n+1)$ multiplications—n multiplications for each of the $n + 1$ values $F(t_k)$, $k = 0,\ldots,n$—but once the algorithm gets going no further multiplications are required to compute additional points. Thus from $F(t_0),\ldots,F(t_n)$, we can calculate arbitrarily many points along the curve using only addition. This evaluation algorithm is called *fast forward differencing*. While forward differencing is very fast, care must be taken when using this evaluation technique because fast forward differencing is numerically unstable. We illustrate this fast forward differencing algorithm for cubic polynomials in Figures 4.4 and 4.5.

There is an interesting way to think about fast forward differencing that involves change of basis algorithms. Since by Corollary 4.2 the divided differences represent the coefficients of the polynomial interpolant relative to the Newton basis, it follows by Proposition 4.7 that when the parameter values are evenly spaced, the forward differences represent the coefficients of the polynomial interpolant relative to the rescaled Newton basis

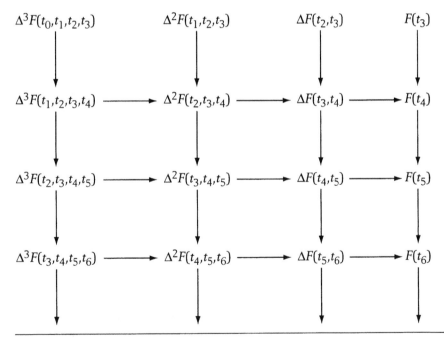

Figure 4.5 The fast forward differencing algorithm for a cubic polynomial curve $F(t)$. The top row is computed from a difference triangle (Figure 4.3). Each value in subsequent rows is computed by adding the values at the nodes from the arrows that point into the node (the two values directly above it and to the left). The values in the leftmost column are all identical, and the values in the rightmost column are points on the curve at equally spaced parameter values.

$$N_0(t) = 1$$

$$N_1(t) = \frac{t - t_0}{\Delta t}$$

$$\vdots$$

$$N_n(t) = \frac{(t - t_0)\cdots(t - t_{n-1})}{n!(\Delta t)^n} \quad .$$

Thus what fast forward differencing really does is to convert from the coefficients of one rescaled Newton basis to another closely related rescaled Newton basis.

Forward differencing has another widespread application: IQ tests. In many mathematical IQ tests, the student is given a short sequence of numbers and asked to find the next value. Assuming the sequence is generated from a polynomial, we can find the next number using differencing. We illustrate with an example.

Consider the sequence: 4, 13, 28, 49, 76. To find the next value, take differences. This generates the sequences

$$\Delta^2 F: \qquad 6 \quad 6 \quad 6$$

$$\Delta F: \qquad 9 \quad 15 \quad 21 \quad 27$$

$$F: \quad 4 \quad 13 \quad 28 \quad 49 \quad 76 \quad .$$

It is trivial to find the next value of $\Delta^2 F$: this value must be 6 because evidently $\Delta^2 F$ is the constant sequence. To find the next value of ΔF, just add the last value of the second difference to the last value of the first difference, giving $27 + 6 = 33$. Finally, to find the next value of the original sequence, add the newly calculated value of the first difference to the last value of the original sequence; this yields $76 + 33 = 109$. Here is this computation, where the values added to the old sequences are underlined:

$$\Delta^2 F: \qquad 6 \quad 6 \quad 6 \quad \underline{6}$$

$$\Delta F: \qquad 9 \quad 15 \quad 21 \quad 27 \quad \underline{33}$$

$$F: \quad 4 \quad 13 \quad 28 \quad 49 \quad 76 \quad \underline{109} \quad .$$

Notice that the procedure we just used to find the next value of the sequence F is identical to the fast forward differencing algorithm we presented for computing new values of a polynomial F at evenly spaced parameter values.

The validity of this process is based on the simple assumption that the original sequence represents the values of a polynomial F at the integers 1,2,.... Of course, even if this assumption is true, we may need to take more than two differences before we arrive at a constant sequence. How many differences we must take depends on the degree of the polynomial from which the initial sequence is generated; a degree n polynomial will generate a constant sequence after n differences. The polynomial in our problem is of degree 2, since the second difference is a constant. Moreover, we can easily retrieve an expression for this polynomial in Newton form because the differences represent the rescaled Newton coefficients. Since such a sequence represents values at the integers, $\Delta t = 1$, so

$$F(t) = \frac{6(t-1)(t-2)}{2} + 9(t-1) + 4 .$$

The reader should check that this polynomial does indeed generate the original sequence for $t = 1,\ldots,5$.

If the original sequence does not represent the values of a polynomial of degree n at the integers, then, of course, this technique for generating new values will fail. For example, given the sequence $1\ \ 2\ \ 4\ \ 8\ \ 16\ \ 32\ \ldots$, we find that

$$\Delta^2 F: \quad 1\ \ 2\ \ 4\ \ 8\ \ldots$$
$$\Delta F: \quad 1\ \ 2\ \ 4\ \ 8\ \ 16\ \ldots$$
$$F: \quad 1\ \ 2\ \ 4\ \ 8\ \ 16\ \ 32\ \ldots .$$

For an exponential sequence, we never reach a constant difference, and other tricks must be employed to generate the next value.

Exercises

1. Prove that $\Delta^n F(t_0,\ldots,t_n) = \sum_{k=0}^{n}(-1)^{n-k}\binom{n}{k}F(t_k)$.

2. Prove that

   $$\Delta^j\{\Delta^k F(t_0,\ldots,t_{j+k})\} = \Delta^{j+k}F(t_0,\ldots,t_{j+k}) = \Delta^k\{\Delta^j F(t_0,\ldots,t_{j+k})\} .$$

3. Prove that forward difference is a linear operator. That is, prove that

   a. $\Delta^n(F+G)(t_0,\ldots,t_n) = \Delta^n F(t_0,\ldots,t_n) + \Delta^n G(t_0,\ldots,t_n)$

   b. $\Delta^n(cF)(t_0,\ldots,t_n) = c\Delta^n F(t_0,\ldots,t_n)$

4. If the parameter values are evenly spaced, then the differencing triangle can be used to convert between the coefficients of which two bases?

5. Consider the infinite Fibonacci sequence: $1\ \ 1\ \ 2\ \ 3\ \ 5\ \ 8\ \ 13\ \ldots$ defined by the recurrence

   $$a_1 = a_0 = 1$$
   $$a_{n+1} = a_n + a_{n-1} .$$

   Prove that there is no polynomial $F(t)$ such that $a_n = F(n)$ for all n.

4.6 **Summary**

The principal focus of this chapter is the divided difference, which provides the dual functionals for the Newton basis. The Newton basis allows us to use Horner's method for fast polynomial evaluation, and the divided difference generates the coefficients for the polynomial interpolant relative to the Newton basis. We developed three approaches to the divided difference:

- Computational: a recurrence based on difference quotients
- Theoretical: the highest-order coefficient of the polynomial interpolant
- Axiomatic: a system of four properties (symmetry, linearity, cancellation, and differentiation) that completely characterize the divided difference

We used these different approaches to derive properties, formulas, and identities for the divided difference. We also considered fast forward differencing, a technique for fast polynomial evaluation at equally spaced parameter values.

This chapter concludes our study of interpolation. Although interpolation is a classical topic in approximation theory and numerical analysis, computer graphics and computer-aided design often deal with approximation as well as with interpolation, so it is to approximation schemes that we shall turn our attention in subsequent chapters. Many of the topics encountered in interpolation, including dynamic programming procedures, up and down recurrence, basis functions, dual functionals, divided differences, rational schemes, and tensor product, triangular, lofted, and Boolean sum surfaces will reappear in approximation theory. A good grounding in the principles of polynomial interpolation will serve you well when you go on to the study of polynomial approximation.

4.6.1 **Identities for the Divided Difference**

It is difficult to remember all the interesting identities for the divided difference that we have encountered in the text and in the exercises. For quick recall, we have collected the most important of these formulas here in one place. A few of these identities will not be proved till later in the book, when we encounter blossoming and B-splines, but we list them here anyway for the sake of completeness.

1. *Highest-Order Coefficient of the Polynomial Interpolant*

    $F[t_0,\ldots,t_n]$ = coefficient of t^n in the monomial representation of the polynomial interpolant $P_{0\ldots n}(t)$

2. *Recursion*

$$F[t_0,\ldots,t_n] = \frac{F[t_1,\ldots,t_n] - F[t_0,\ldots,t_{n-1}]}{t_n - t_0} \qquad t_n \neq t_0$$

3. *Symmetry*

$F[t_0,\ldots,t_n] = F[t_{\sigma(0)},\ldots,t_{\sigma(n)}]$, where σ is any permutation of $\{0,\cdots,n\}$

4. *Linearity*

$$(F+G)[t_0,\ldots,t_n] = F[t_0,\ldots,t_n] + G[t_0,\ldots,t_n]$$

$$(cF)[t_0,\ldots,t_n] = c(F[t_0,\ldots,t_n])$$

5. *Cancellation*

$$F[t_0,\ldots,t_n] = \{(t - t_{n+1})F(t)\}[t_0,\ldots,t_n,t_{n+1}]$$

6. *Differentiation*

$$F[\underbrace{t,\ldots,t}_{n+1}] = \frac{F^{(n)}(t)}{n!}$$

7. *Leibniz's Rule*

$$(FG)[t_0,\ldots,t_n] = \sum_{k=0}^{n} F[t_0,\ldots,t_k]G[t_k,\ldots,t_n]$$

8. *Equality Conditions*

$$F^{(p)}(t_j) = G^{(p)}(t_j) \quad 0 \le p \le \mu_j - 1, \ j = 0,\ldots,n$$

$$\Rightarrow \ F[t_0,\ldots,t_n] = G[t_0,\ldots,t_n]$$

9. *Value on Low-Order Polynomials*
    a. If $F(t)$ is a polynomial of degree $n-1$, then $F[t_0,\ldots,t_n] = 0$.
    b. If $F(t)$ is a polynomial of degree n, then $F[t_0,\ldots,t_n]$ is the coefficient of t^n in the monomial representation for $F(t)$. Thus, in this case, $F[t_0,\ldots,t_n]$ is a constant independent of $t_0,\ldots,t_n$.

10. *Affine Combinations*

$$F[t_0,\ldots,t_n] = \sum_{k=0}^{n} \lambda_k F[t_0,\ldots,t_{k-1},\tau,t_{k+1},\ldots,t_n]$$

$$\tau = \sum_k \lambda_k t_k \text{ and } \sum_k \lambda_k = 1$$

11. *Values on Monomials*

$$\{t^n\}[v_1,\ldots,v_{n-k}] = \sum_{i_1 \le i_2 \le \ldots \le i_{k+1}} v_{i_1} v_{i_2} \cdots v_{i_{k+1}}$$

12. *Value on $(x - t)^{-1}$*

$$\left\{\frac{1}{x-t}\right\}[t_0,\ldots,t_n] = \frac{1}{(x-t_0)\cdots(x-t_n)}$$

13. *Newton Coefficients of Polynomial Interpolant*

$$P_{0\cdots n}(t) = \sum_{k=0}^{n} F[t_0,\ldots,t_k]N_k(t)$$

14. *Dual Functionals for Newton Basis*

$$N_j[t_0,\ldots,t_k] = 0 \quad j \neq k$$
$$= 1 \quad j = k$$

15. *Determinant Formula*

$$F[t_0,\ldots,t_n] = \frac{\begin{vmatrix} F(t_0) & t_0^{n-1} & \cdots & t_0 & 1 \\ \vdots & \vdots & \vdots & \vdots & \vdots \\ F(t_n) & t_n^{n-1} & \cdots & t_n & 1 \end{vmatrix}}{\begin{vmatrix} t_0^n & t_0^{n-1} & \cdots & t_0 & 1 \\ \vdots & \vdots & \vdots & \vdots & \vdots \\ t_n^n & t_n^{n-1} & \cdots & t_n & 1 \end{vmatrix}} \qquad t_0,\ldots,t_n \text{ distinct}$$

16. *Lagrange Coefficients*

$$F[t_0,\ldots,t_n] = \sum_{k=0}^{n} \frac{F(t_k)}{\prod_{j \neq k}(t_k - t_j)} \qquad t_0,\ldots,t_n \text{ distinct}$$

17. *Partial Derivatives with Respect to the Nodes*

$$\frac{\partial F[t_0,\ldots,t_k,\ldots,t_n]}{\partial t_k} = \mu_k F[t_0,\ldots,t_k,t_k,\ldots,t_n],$$

where μ_k is the multiplicity of t_k

18. *Antidifferentiation*

$$F'[t_0,\ldots,t_n] = \sum_{k=0}^{n} F[t_0,\ldots,t_k,t_k,\ldots,t_n]$$

19. *Relation to Blossoming*

$$P[v_1,\ldots,v_k] = \frac{(n-k+1)!}{n!} \sum p^{(k-1)}(v_{j_1},\ldots,v_{j_{n-k+1}})$$

20. *B-splines as Divided Differences*

$$N_{k,n}(t) = \{(t_{k+n+1} - t_k)(x-t)_+^n\}[t_k,\ldots,t_{k+n+1}]$$

21. *B-spline Integration*

$$F[t_k,\ldots,t_{k+n+1}] = \int_{Support} \left\{ \frac{N_{k,n}(t)}{Support} \right\} \left\{ \frac{F^{(n+1)}(t)}{n!} \right\} dt$$

22. *Complex Contour Integration*

$$F[t_0,\ldots,t_n] = \frac{1}{2\pi i} \int_C \frac{F(z)dz}{(z-t_0)\cdots(z-t_n)}$$

- *C* is any simple closed curve containing $\{t_0,\ldots,t_n\}$
- *F* is analytic inside *C*

23. *Hermite-Genocchi Formula*

$$F[t_0,\ldots,t_n] = \int_{\Delta^n} F^{(n)}\big(t_0 + v_1(t_1 - t_0) + \cdots + v_n(t_n - t_0)\big)dv_1\cdots dv_n$$

where $\Delta^n = \{(v_i,\ldots,v_n) \mid V_j \geq 0 \text{ and } \sum_j v_j \leq 1\}$

Part II

Approximation

CHAPTER 5

Bezier Approximation and Pascal's Triangle

In previous chapters, we used interpolation to specify shape. But interpolation is not always a good way to describe the contour of a curve or surface. To accurately reproduce complicated shapes, we may need to interpolate lots of data. Polynomial interpolation for many points is impractical because the degree of the interpolant can get extremely high, leading to slow and numerically unstable computations. Also polynomial interpolants may oscillate unnecessarily and fail to reproduce the desired shapes (see Figure 5.1). Thus, even if we were to specify more and more points, there is no guarantee that the polynomial interpolants would converge to the curves or surfaces we wish to represent.

Spline interpolation—that is, interpolation by piecewise polynomial functions— is better computationally because splines allow us to keep the degree low. But interpolating splines may still oscillate unnecessarily and fail to reproduce the desired shapes. Our approach from here on will be to abandon interpolation altogether and to take a very different approach to describing the shape of a curve or surface.

Given a relatively small collection of points in affine space, we are going to investigate methods for generating polynomial and rational curves and surfaces that

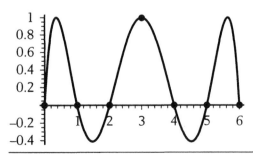

Figure 5.1 Lagrange interpolation. Notice the oscillations in the interpolating polynomial curve, even though there is no oscillation in the original data points.

approximate the shape described by these points. We shall not insist that our curves and surfaces go through these points, but we shall insist that these curves and surfaces capture in some mathematically precise way the shape defined by these points. As usual we begin with schemes for curves and later extend our techniques to surfaces.

5.1 De Casteljau's Algorithm

Let's return for a moment to where we began our investigation of polynomial curves and surfaces: Lagrange interpolation and Neville's algorithm. Recall that Neville's algorithm (Figure 5.2) is a dynamic programming procedure for computing points along a polynomial interpolant. We are going to start our investigation of approximation schemes by using the same basic triangular structure but simplifying the computations along the edges.

The simplest thing—you might almost say the only thing—we know how to do is linear interpolation. All our interpolation procedures, and especially Neville's algorithm, are based on this simple idea or some variant thereof. What makes Neville's algorithm the least bit complicated is that we perform a different linear interpolation at each node of the diagram. To take the same triangular structure and make the evaluation algorithm as easy as possible, we will perform the same linear interpolation at each node. This idea generates the algorithm represented in Figure 5.3.

The algorithm represented in Figure 5.3 is called *de Casteljau's evaluation algorithm,* and the curves that emerge at the apex of this diagram are called *Bezier curves.* Intermediate nodes marked $\diamond$ and $\blacklozenge$ also represent Bezier curves, but of lower degree. Thus the de Casteljau algorithm is a dynamic programming algorithm

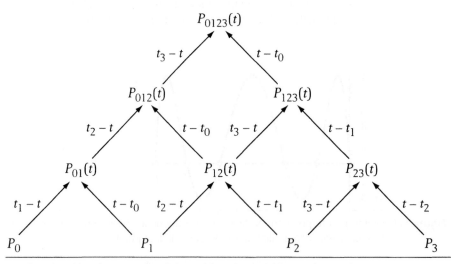

Figure 5.2 Neville's algorithm (unnormalized) for cubic polynomial interpolation.

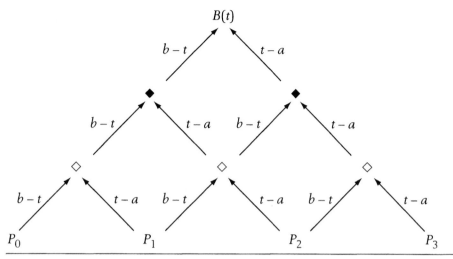

Figure 5.3 The de Casteljau algorithm for a cubic Bezier curve $B(t)$ in the interval $[a,b]$. The label on every edge must be normalized by dividing by $b - a$, so that the labels along arrows entering each node sum to one.

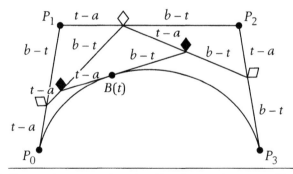

Figure 5.4 Geometric construction algorithm for a point on a cubic Bezier curve based on a geometric interpretation of the de Casteljau evaluation algorithm. At the parameter t, each line segment in the trellis is split in the ratio $(t - a)/(b - t)$.

for computing points on a Bezier curve. Typically, for reasons that will become clear in the next section, Bezier curves are restricted to the interval $[a,b]$. Usually, for simplicity, we take $a = 0$ and $b = 1$, but there are cases, as we shall see later on, where it is useful to allow a and b to be arbitrary as long as $b > a$. Notice that when $a = 0$ and $b = 1$, no normalization is required.

The de Casteljau algorithm has the following elegant geometric interpretation. Since each node represents a linear interpolation, each node symbolizes a point on the line segment joining the two points whose arrows point into the node. Drawing all these line segments generates the trellis in Figure 5.4.

We are going to study the geometric characteristics of curves generated by de Casteljau's algorithm. We begin with some simple features before going on to derive the basis functions associated with Bezier curves. We shall then use these basis functions to develop the more advanced mathematical properties of this approximation scheme.

Exercise

1. Implement the de Casteljau algorithm for Bezier curves. Experiment with how moving control points affects the shape of the curve.

5.2   Elementary Properties of Bezier Curves

Bezier curves have the following elementary properties:

1. Polynomial parametrization
2. Affine invariance
3. Convex hull
4. Symmetry
5. Interpolation of end points

Below we briefly discuss and derive each of these properties in turn, and we explain as well why these features are important for geometric design.

Most of these properties can be proved by direct observation or by easy inductive arguments using the de Casteljau algorithm. To set up these inductive arguments, let $B[P_0,...,P_n](t)$ denote the Bezier curve over the interval $[a,b]$ with affine control points $P_0,...,P_n$. Then the last stage of the de Casteljau algorithm can be written as

$$B[P_0,...,P_n](t) = \frac{b-t}{b-a}B[P_0,...,P_{n-1}](t) + \frac{t-a}{b-a}B[P_1,...,P_n](t). \qquad (5.1)$$

We are now ready to proceed with our derivations:

1. *Polynomial Parametrization*

   In the de Casteljau algorithm, the only operations we perform involving the functions along the edges are addition and multiplication (see Figure 5.3). Since the functions along the edges are linear polynomials, it follows that a Bezier curve with $n + 1$ control points is a polynomial curve of degree n because there are n levels from the control points at the base to the curve at the apex of the triangle. (This result also follows by an easy induction from (5.1).) Since Bezier curves are polynomial curves, all the tools we know for polynomials apply.

2. *Affine Invariance*

A curve is said to be *affinely invariant* if it consists of a collection of points in affine space. We can prove by induction on the number of control points that all Bezier curves are affinely invariant. Clearly a Bezier curve with only two control points is affinely invariant, since this Bezier curve is just the line segment joining the two affine control points. Suppose that all Bezier curves with n control points are affinely invariant. Since by the inductive hypothesis $B[P_0,...,P_{n-1}](t)$ and $B[P_1,...,P_n](t)$ are affinely invariant, it follows by (5.1) that $B[P_0,...,P_n](t)$ is also affinely invariant, since it is formed by taking an affine combination of affinely invariant curves. Thus Bezier curves make sense in affine space.

Affine invariance is a crucial feature for any curve scheme because it asserts that the curve is independent of the choice of the coordinate system. This property is essential for a good approximation scheme, since in a typical geometric model many different coordinate systems are available. Affine invariance guarantees that the curve will be the same no matter which coordinate system is invoked.

3. *Convex Hull Property*

A set S of points in affine space is said to be *convex* if, whenever P and Q are points in S, the entire line segment from P to Q lies in S (see Figure 5.5). The intersection S of a collection of convex sets $\{S_i\}$ is a convex set because if P and Q are points in S, they must also be points in each of the sets S_i. Since, by assumption, the sets S_i are convex, the entire line segment from P to Q lies in each set S_i. Hence the entire line segment from P to Q lies in the intersection S, so S too is convex.

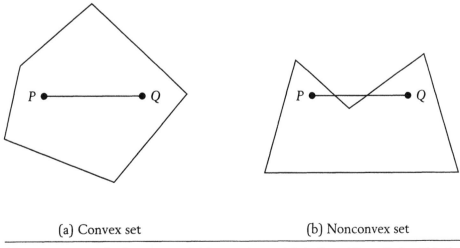

(a) Convex set   (b) Nonconvex set

Figure 5.5 (a) In a convex set, the line segment joining any two points in the set lies entirely within the set. (b) In a nonconvex set, part of the line segment joining two points in the set may lie outside the set.

The *convex hull* of a collection of points in affine space is the intersection of all the convex sets containing the points. Since the intersection of convex sets is a convex set, the convex hull is the smallest convex set containing the points. For two points, the convex hull is the line segment joining the points. For three noncollinear points, the convex hull is the triangle whose vertices are the three points. The convex hull of a finite collection of points in the plane can be found mechanically by placing a nail at each point, stretching a rubber band so that its interior contains all the nails, and then releasing the rubber band. When the rubber band comes to rest on the nails, the interior of the rubber band is the convex hull of the points.

Since the convex hull of two points is the line segment joining the two points,

$$ConvexHull\{P_0,P_1\} = \{c_0P_0 + c_1P_1 \mid c_0 + c_1 = 1 \text{ and } c_0,c_1 \geq 0\}.$$

More generally it follows by a simple inductive argument (see Exercise 3) that

$$ConvexHull\{P_0,...,P_n\} = \{ \sum_{k=0}^{n} c_kP_k \mid \sum_{k=0}^{n} c_k = 1 \text{ and } c_k \geq 0\}.$$

Bezier curves always lie in the convex hull of their control points. That is,

$$B[P_0,...,P_n](t) \subseteq ConvexHull\{P_0,...,P_n\}.$$

Again we can prove this assertion by a simple inductive argument. First recall that, by convention, we always restrict the Bezier curve $B[P_0,...,P_n](t)$ in (5.1) to the parameter interval $a \leq t \leq b$. With this restriction, the convex hull property is certainly true for a Bezier curve with only two control points since, by construction, this curve is the line segment joining the two control points. More generally suppose that this result is valid for Bezier curves with n control points. By (5.1), $B[P_0,...,P_n](t)$ lies on the line segment joining the points $B[P_0,...,P_{n-1}](t)$ and $B[P_1,...,P_n](t)$, and by the inductive hypothesis $B[P_0,...,P_{n-1}](t)$ and $B[P_1,...,P_n](t)$ both lie in the convex hull of the points $P_0,...,P_n$. But if two points lie in a convex set, the entire line segment joining them also lies in the set; thus the entire Bezier curve $B[P_0,...,P_n](t)$, $a \leq t \leq b$, must lie in $ConvexHull\{P_0,...,P_n\}$.

The convex hull property is important because it constrains Bezier curves to lie in the proximity of their control points. This property is a vital feature for an approximation scheme. Designers not only require curves that approximate the shape defined by their control points, they also demand curves that lie in the same region of space as their control points. To be useful in design, the curves must be visible to the designer. The convex hull property guarantees that if all the control points are visible on the graphics terminal, then the entire curve is visible as well. The restriction $a \leq t \leq b$ on the parameter t is there precisely to guarantee the convex hull property.

4. *Symmetry*

Replacing t by $a+b-t$ reverses the order of the parameter domain. As the parameter t varies from a to b, the curve $B[P_0,...,P_n](a+b-t)$ traverses the same points as $B[P_0,...,P_n](t)$ but in the direction from b to a rather than from a to b. Thus $B[P_0,...,P_n](a+b-t)$ is essentially the same curve as $B[P_0,...,P_n](t)$ but with opposite orientation. Similarly, reversing the order of the control points of a Bezier curve generates the same Bezier curve but with opposite orientation. Analytically this means that

$$B[P_n,...,P_0](t) = B[P_0,...,P_n](a+b-t) \qquad a \le t \le b. \qquad (5.2)$$

This symmetry property is what professional designers and most naive users would naturally expect of a simple approximation scheme, so it is gratifying to see that it holds for all Bezier curves.

To prove (5.2), simply replace t by $a+b-t$ in the de Casteljau diagram and observe that the new diagram is the mirror image of the de Casteljau diagram for $B[P_n,...,P_0](t)$ (see Figure 5.6).

5. *Interpolation of End Points*

Unlike Lagrange polynomials, Bezier curves generally do not interpolate all their control points. But Bezier curves always interpolate their first and last control points. In fact,

$$B[P_0,...,P_n](a) = P_0 \text{ and } B[P_0,...,P_n](b) = P_n.$$

The first result follows easily from setting $t = a$ in de Casteljau's algorithm and observing that all the labels on left-pointing arrows become zero while

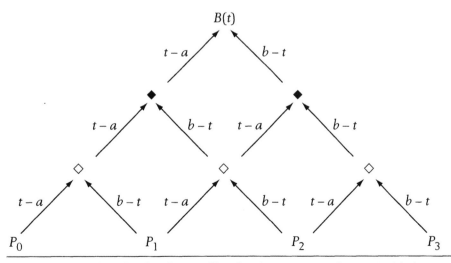

Figure 5.6 The de Casteljau algorithm for $B[P_0,...,P_n](a+b-t)$. Compare to Figure 5.3 with the control points in reverse order.

all the labels on right-pointing arrows become one. If $k \neq 0$, then any path from P_k to the apex of the triangle must traverse at least one left-pointing arrow, so there is no contribution from P_k to the value of the curve at $t = a$. On the other hand, when $t = a$ all the labels on the single path from P_0 to the apex of the triangle are one. Hence $B[P_0,...,P_n](a) = P_0$. A similar argument for $t = b$ shows that $B[P_0,...,P_n](b) = P_n$. Again an easy inductive argument based on (5.1) yields the same results.

Interpolating end points is important because we often want to connect two curves. To assure that two Bezier curves join at their end points, all we need to do is to make sure that the first control point of the second curve is the same as the last control point of the first curve. This device ensures continuity. Later on, in Section 5.6.2, we shall develop techniques for guaranteeing higher-order smoothness between adjacent Bezier curves.

Exercises

1. Let $P_0,...,P_n$ be a collection of points. Prove that for every vector v

$$B[P_0 + v,...,P_n + v](t) = B[P_0,...,P_n](t) + v.$$

That is, translating the control points by a vector v translates every point on the Bezier curve by the same vector v.

2. Show that every affine space is convex.

3. Let $P_0,...,P_n$ be a collection of points in an affine space. Prove that

$$ConvexHull\{P_0,...,P_n\} = \{ \sum_{k=0}^{n} c_k P_k \mid \sum_{k=0}^{n} c_k = 1 \text{ and } c_k \geq 0\}.$$

4. Prove that $B[P_0,...,P_n](b) = P_n$.

5. Give an example to show that a degree n Lagrange interpolating polynomial with nodes $t_0 < t_1 < \cdots < t_n$ does not necessarily satisfy the convex hull property on the interval $[t_0,t_n]$.

5.3   The Bernstein Basis Functions and Pascal's Triangle

The de Casteljau algorithm is a dynamic programming algorithm for computing points on a Bezier curve from points on Bezier curves of lower degree. Here we shall also develop an explicit formula for evaluating points on a Bezier curve.

We begin by observing that there must exist polynomials $B_0^n(t),...,B_n^n(t)$ of degree n such that the Bezier curve $B(t) = B[P_0,...,P_n](t)$ is given by

$$B(t) = \sum_{k=0}^{n} B_k^n(t) P_k. \tag{5.3}$$

This result is just a restatement of the polynomial property we proved in the previous section. The functions $B_0^n(t),...,B_n^n(t)$ are called the *Bernstein basis functions* of degree n.

There are many ways to compute the Bernstein basis functions. In the de Casteljau algorithm if we set

$$P_j = 0 \quad j \neq k$$
$$= 1 \quad j = k,$$

then by (5.3) $B(t) = B_k^n(t)$. This algorithm is called the *up recurrence* for $B_k^n(t)$. Observe that by the up recurrence, the Bernstein basis function $B_k^n(t)$ is simply the sum of all paths from P_k at the base to $B(t)$ at the apex of the de Casteljau triangle. Recall that in Chapter 2 we had a similar up recurrence for the Lagrange basis functions.

We can apply this insight about paths to compute all the Bernstein basis functions up to degree n simultaneously. Paths from a node to the apex of the triangle are identical to paths from the apex to the node. Thus if we place a one at the apex and reverse all the arrows, then the Bernstein basis functions emerge at the nodes of the triangle. In particular, the Bernstein basis functions of degree n emerge at the base of a de Casteljau triangle with n levels (see Figure 5.7). This algorithm is called the *down recurrence* for the Bernstein basis functions and is similar to the down recurrence for the Lagrange basis functions we encountered in Section 2.6. But there is one very important difference between the down recurrence for the Bernstein basis functions and the down recurrence for the Lagrange basis functions. In the down recurrence for the Lagrange basis functions, the intermediate nodes do not contain

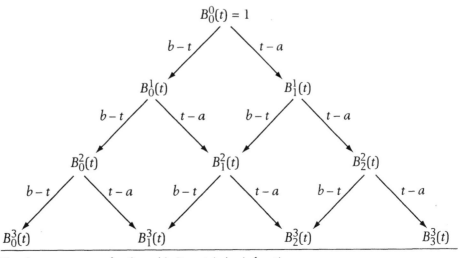

Figure 5.7 The down recurrence for the cubic Bernstein basis functions.

Lagrange basis functions of lower degree; in the down recurrence for the Bernstein basis functions the intermediate nodes are precisely the Bernstein basis functions of lower degree.

Since the intermediate nodes in the down recurrence contain the Bernstein basis functions of lower degree, the down recurrence yields the following standard recursion formula for the Bernstein basis functions:

$$B_0^0(t) = 1$$

$$B_k^n(t) = \frac{b-t}{b-a} B_k^{n-1}(t) + \frac{t-a}{b-a} B_{k-1}^{n-1}(t).$$

(5.4)

This recursion formula is just a restatement of the down recurrence. We shall see later that many results about Bernstein basis functions follow easily by induction from this recursion formula.

Finally, as promised, we can also use path arguments to find explicit expressions for the Bernstein basis functions $B_k^n(t)$, $k = 0,...,n$. Observe that all paths between the apex and $B_k^n(t)$ are identical. Indeed to get to $B_k^n(t)$ from the apex of the triangle, we must make exactly k right turns and $n-k$ left turns (see Figure 5.7). Now every left-pointing arrow carries the label $(b-t)/(b-a)$, and every right-pointing arrow carries the label $(t-a)/(b-a)$. Thus we discover that

$$B_k^n(t) = P(n,k) \frac{(t-a)^k (b-t)^{n-k}}{(b-a)^n},$$

where $P(n,k)$ denotes the number of paths from the apex of the triangle to the kth position on the nth level.

We can find $P(n,k)$ from Pascal's triangle. From the structure of the de Casteljau triangle we observe that the only way to arrive at the kth position on the nth level is to arrive first at either the $(k-1)$st or kth position on the $(n-1)$st level. This observation yields the recurrence

$$P(0,0) = 1$$

$$P(n,k) = P(n-1,k-1) + P(n-1,k).$$

But this is exactly the recurrence in Pascal's triangle (Figure 5.8(a)):

$$\binom{0}{0} = 1$$

$$\binom{n}{k} = \binom{n-1}{k-1} + \binom{n-1}{k}.$$

Thus the elements $\binom{n}{k}$ in Pascal's triangle compute the number of paths $P(n,k)$ from the apex to the node at which the entry $\binom{n}{k}$ appears. Therefore,

$$P(n,k) = \binom{n}{k} = \frac{n!}{k!(n-k)!}.$$

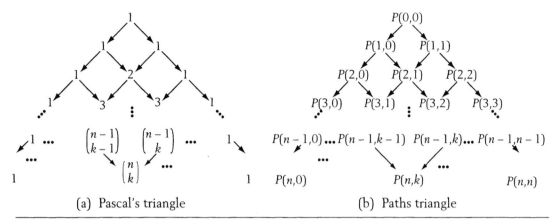

(a) Pascal's triangle   (b) Paths triangle

Figure 5.8 (a) Pascal's triangle and (b) the paths triangle represent the same recurrence. Therefore, the values in Pascal's triangle represent the number of paths from the apex to the node in which the value appears.

It follows immediately that

$$B_k^n(t) = \binom{n}{k} \frac{(t-a)^k (b-t)^{n-k}}{(b-a)^n}.$$ (5.5)

Another way to see that (5.5) is correct is to observe that in traversing the paths from the apex to the node containing $B_k^n(t)$, you must select exactly k right turns out of n possible choices. Thus there are exactly n choose k paths from the apex to $B_k^n(t)$, which again accounts for the coefficient $\binom{n}{k}$ in the explicit expression for $B_k^n(t)$.

The explicit formula for the Bernstein basis functions given by (5.5) should look familiar, since it comes up in many other areas of mathematics. For example, by the binomial theorem

$$(x+y)^n = \sum_{k=0}^{n} \binom{n}{k} x^k y^{n-k}.$$

Setting $x = \dfrac{t-a}{b-a}$ and $y = \dfrac{b-t}{b-a}$ yields

$$1^n = \left\{ \frac{t-a}{b-a} + \frac{b-t}{b-a} \right\}^n = \sum_{k=0}^{n} \binom{n}{k} \frac{(t-a)^k (b-t)^{n-k}}{(b-a)^n}.$$

Thus the binomial theorem gives us a quick proof that the Bernstein basis functions sum to one, and hence that Bezier curves are affinely invariant. (For an alternative, recursive proof, see Exercise 1.)

Another place the Bernstein basis functions appear in mathematics is in probability theory. If $a = 0$ and $b = 1$, then

$$B_k^n(t) = \binom{n}{k} t^k (1-t)^{n-k}, \quad k = 0,...,n,$$

is the familiar binomial distribution. Discrete distributions are important in approximation theory and computer-aided design because the convex hull property requires that the blending functions must be positive and sum to one (see Section 5.2, Exercise 3). Thus discrete distributions are natural candidates for blending functions, so it is no accident that one of the most common distributions has been chosen to represent one of the most common approximation schemes. We shall have occasion to take advantage of this connection to probability theory in Section 5.5.4, where we discuss Bezier subdivision, and again in Chapter 6, when we introduce blossoming via random walks.

Exercises

1. Use the down recurrence to prove that

$$\sum_{k=0}^{n} B_k^n(t) \equiv 1$$

by showing that the functions on each level of the recurrence sum to one. Conclude that Bezier curves are affinely invariant.

2. Prove that $B_k^n(t) \geq 0$ for $a \leq t \leq b$, and use this result together with the results of Exercise 1 and Section 5.2, Exercise 3, to conclude that Bezier curves lie in the convex hull of their control points.

3. Prove that $B_k^n(a + b - t) = B_{n-k}^n(t)$, and use this identity to conclude that Bezier curves have the symmetry property described in Section 5.2.

4. Prove that $B_k^n(a) = \delta_{0k}$ and $B_k^n(b) = \delta_{nk}$, and use this identity to conclude that Bezier curves interpolate their first and last control points.

5. Let $c_0,...,c_n$ be a collection of arbitrary constants. By reversing the arrows in Pascal's triangle, show how to compute

$$\sum_{k=0}^{n} \binom{n}{k} c_k$$

without any multiplication.

6. Let $\{c_{km}\}$, $k = 0,...,m$, $m = 0,...,n$, be a collection of arbitrary constants. Show how to use Pascal's triangle to compute

$$\sum_{m=0}^{n} \sum_{k=0}^{m} \binom{m}{k} c_{km}$$

without any multiplications and with only $O(n^2)$ additions.

7. Let $\{c_{km}\}$, $k = 0,...,m$, $m = 0,...,n$, be a collection of arbitrary constants. Show how to use de Casteljau's algorithm to compute

$$\sum_{m=0}^{n} \sum_{k=0}^{m} c_{km} B_k^n(t)$$

with a minimal amount of multiplication.

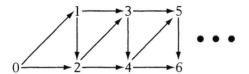

Figure 5.9 A graph with the structure of Clenshaw's algorithm for the evaluation of orthogonal polynomials.

8. Consider Figure 5.9, which arises in Clenshaw's algorithm for the evaluation of orthogonal polynomials.

    a. Find a recursive formula for the number of paths $p(k)$ from the node containing 0 to the node containing an arbitrary integer k in Figure 5.9.

    b. Let $c_0, ..., c_n$ be a collection of arbitrary constants. Show how to compute

$$\sum_{k=0}^{n} p(k) c_k$$

without any multiplication.

9. Prove by induction from the recurrence (5.4) that

$$B_k^n(t) = \binom{n}{k} \frac{(t-a)^k (b-t)^{n-k}}{(b-a)^n}.$$

10. Prove that $\displaystyle\sum_{k=0}^{n} (-1)^k B_k^n(t) = \left\{ \frac{(b+a-2t)}{(b-a)} \right\}^n.$

11. Define the Bernstein basis functions of negative degree by setting

$$B_k^{-n}(t) = \binom{-n}{k} t^k (1-t)^{-(n+k)} \qquad k = 0, 1, ...$$

$$\binom{-n}{k} = \frac{(-n)(-n-1)\cdots(-n-k+1)}{k!} = (-1)^k \binom{n+k-1}{k}.$$

    a. Using the binomial theorem for negative integer exponents, prove that

$$\sum_{k=0}^{\infty} B_k^{-n}(t) \equiv 1.$$

    b. For what values of t does this series converge?

12. Prove that $\displaystyle\left(\frac{x-t}{b-a} \right)^n = \sum_{k=0}^{n} (-1)^k \binom{n}{k}^{-1} B_{n-k}^n(x) B_k^n(t).$

(Hint: Use the identity $(b-a)(x-t) = (x-a)(b-t) - (b-x)(t-a)$.)

5.4   More Properties of Bernstein/Bezier Curves

There is a natural correlation between geometric properties of Bezier curves and algebraic properties of Bernstein basis functions. Thus one way to study the geometry of Bezier curves is to investigate the algebra of Bernstein basis functions. This we now proceed to do (see also Section 5.3, Exercises 1–4). To simplify our notation from here on out, we shall let $a = 0$ and $b = 1$, although the proofs do not change much for arbitrary values of a and b.

We begin with a simple trick that is worth remembering because we shall use it several times. Observe that

$$\frac{B_k^n(t)}{(1-t)^n} = \frac{\binom{n}{k} t^k (1-t)^{n-k}}{(1-t)^n} = \binom{n}{k} \frac{t^k}{(1-t)^k}.$$

Making the substitution $u = t/(1-t)$, we obtain

$$\frac{B_k^n(t)}{(1-t)^n} = \binom{n}{k} u^k \equiv M_k^n(u). \tag{5.6}$$

Thus the Bernstein basis functions are readily transformed into the monomial basis functions. We shall use this simple device to show that many of the properties of the monomial basis carry over to the Bernstein basis. Conversely, many properties of the Bernstein basis are inherited by this scaled monomial basis, simply by replacing each factor of $1 - t$ by the 1.

Exercise

1. Let $M_k^n(t) = \binom{n}{k} t^k$, $k = 0,\dots,n$.

    a. Show that these functions satisfy the recurrence
    $$M_k^n(t) = M_k^{n-1}(t) + t M_{k-1}^{n-1}(t).$$

    b. Describe the analogue of the de Casteljau algorithm for polynomials $P(t) = \sum_k c_k M_k^n(t)$.

5.4.1   Linear Independence and Nondegeneracy

Let's now prove that the Bernstein basis functions of degree n do indeed form a basis for the polynomials of degree n. To show that they are linearly independent, suppose that there are constants c_k, $k = 0,\dots,n$, such that

$$\sum_{k=0}^{n} c_k B_k^n(t) \equiv 0;$$

we must show that $c_k = 0$, $k = 0,\dots,n$. Dividing this equation by $(1-t)^n$ and applying (5.6), we obtain

$$\sum_{k=0}^{n} c_k M_k^n(u) \equiv 0.$$

Since the monomials $M_0^n(u),...,M_n^n(u)$ are linearly independent, it follows that $c_k = 0$, $k = 0,...,n$. Thus the polynomials $B_0^n(t),...,B_n^n(t)$ are indeed linearly independent. Therefore, these polynomials must form a basis for the polynomials of degree n, since the space of polynomials of degree n has dimension $n + 1$ and there are $n + 1$ Bernstein basis functions of degree n.

A curve scheme is said to be *nondegenerate* if the curve never collapses to a single point unless all the control points are located at that point. Bezier curves are nondegenerate because the Bernstein basis functions are linearly independent. Indeed suppose that some Bezier curve $B(t)$ collapses to a single point Q. Then

$$B(t) = \sum_{k=0}^{n} B_k^n(t) P_k \equiv Q.$$

Since the Bernstein polynomials sum to one, it follows that

$$\sum_{k=0}^{n} B_k^n(t)(P_k - Q) \equiv 0.$$

Dotting both sides with any vector v yields

$$\sum_{k=0}^{n} B_k^n(t)\{(P_k - Q) \bullet v\} \equiv 0.$$

Since the functions $B_0^n(t),...,B_n^n(t)$ are linearly independent, we can conclude that

$$(P_k - Q) \bullet v = 0 \qquad k = 0,...,n$$

for every vector v. But this can happen only if $P_k = Q$ for $k = 0,...,n$, which establishes that Bezier curves are indeed nondegenerate.

5.4.2   Horner's Evaluation Algorithm for Bezier Curves

The de Casteljau evaluation algorithm for Bezier curves is $O(n^2)$. But Horner's evaluation algorithm for polynomials written in terms of the monomial basis is $O(n)$. We would like to have an $O(n)$ algorithm to evaluate polynomials written in Bernstein/Bezier form. We can achieve this goal by applying (5.6) to transform the Bernstein basis into the monomial basis.

Let $B(t) = \sum_{k=0}^{n} B_k^n(t) P_k$ be a Bezier curve. Using (5.6), we find that

$$\frac{B(t)}{(1-t)^n} = \frac{\sum_{k=0}^{n} B_k^n(t) P_k}{(1-t)^n} = \sum_{k=0}^{n} Q_k u^k,$$

where $u = t/(1-t)$ and $Q_k = \binom{n}{k}P_k$, $k = 0,...,n$. Now we can apply Horner's method to evaluate

$$\frac{B(t)}{(1-t)^n} = \sum_{k=0}^{n} Q_k u^k,$$

and this computation costs only $O(n)$ multiplications. Notice that Q_k, $k = 0,...,n$, needs to be computed only once per curve, and setting $u = t/(1-t)$ costs just one division. Finally, $(1-t)^n$ can be calculated in $O(\lg n)$ time and multiplying $B(t)/(1-t)^n$ by $(1-t)^n$ costs only one more multiplication. Putting this all together, we can compute $B(t)$ using only $O(n)$ multiplications.

There is one slight difficulty with this algorithm. If t is close to 1, then $u = t/(1-t)$ gets very large and the computation is numerically unstable. To correct this problem, use the identities

$$\frac{B_k^n(t)}{(1-t)^n} = \binom{n}{k}u^k \qquad u = \frac{t}{1-t} \qquad 0 \le t \le 0.5$$

$$\frac{B_k^n(t)}{t^n} = \binom{n}{k}v^k \qquad v = \frac{1-t}{t} \qquad 0.5 \le t \le 1.$$

Now proceed as before applying Horner's method to compute

$$\frac{B(t)}{(1-t)^n} = \sum_{k=0}^{n} Q_k u^k, \qquad 0 \le t \le 0.5$$

$$\frac{B(t)}{t^n} = \sum_{k=0}^{n} Q_{n-k} v^k, \qquad 0.5 \le t \le 1.$$

This procedure is numerically stable and yields an $O(n)$ evaluation algorithm for Bezier curves.

Exercises

1. Implement both de Casteljau's algorithm and Horner's method for evaluating points on a Bezier curve.

2. Derive an $O(n)$ ladder evaluation algorithm for Bezier curves (see Section 2.6). Which approach is faster, the ladder algorithm or Horner's method?

5.4.3 Unimodality

A function is said to be *unimodal* if it has only one local maximum. The Bernstein basis functions $B_k^n(t)$ are unimodal in the parameters k and t. That is, if we fix k, then each polynomial $B_k^n(t)$ is unimodal in t over the interval $[0,1]$. Furthermore, if we fix $t \in [0,1]$ and let $B(k,t) = B_k^n(t)$, then $B(k,t)$ takes on the discrete values $B_0^n(t),...,B_n^n(t)$ and these values are unimodal in k. Here we shall explore each of these forms of unimodality for the Bernstein basis functions and examine their consequences for Bezier curves.

The Bernstein basis functions $B_0^n(t) = (1-t)^n$ and $B_n^n(t) = t^n$ are monotonic on the interval $[0,1]$ because the function $(1-t)^n$ decreases monotonically from 1 to 0 while the function t^n increases monotonically from 0 to 1. Hence the Bernstein basis functions $B_0^n(t)$ and $B_n^n(t)$ are certainly unimodal in t on the interval $[0,1]$.

To prove that each basis function $B_k^n(t)$, $k = 1,...,n-1$, is also unimodal in t, we need to differentiate $B_k^n(t)$ to find its local maxima. This computation is straightforward, since we have a simple explicit formula for $B_k^n(t)$. Recall that by (5.5)

$$B_k^n(t) = \binom{n}{k}t^k(1-t)^{n-k}.$$

Differentiating this formula yields

$$\frac{dB_k^n(t)}{dt} = k\binom{n}{k}t^{k-1}(1-t)^{n-k} - (n-k)\binom{n}{k}t^k(1-t)^{n-k-1}$$

$$= \binom{n}{k}t^{k-1}(1-t)^{n-k-1}\{k(1-t)-(n-k)t\}.$$

Since $k(1-t)-(n-k)t = k-nt$, we find that $t = k/n$ is the only solution of the equation

$$\frac{dB_k^n(t)}{dt} = 0$$

for $0 < t < 1$. Since the Bernstein basis functions $B_k^n(t)$, $k = 1,...,n-1$, are positive in the open interval $(0,1)$ and zero at the end points of the interval, it follows that each Bernstein basis function $B_k^n(t)$ has a single maximum in $[0,1]$; hence the Bernstein basis functions are unimodal in t (see Figure 5.10).

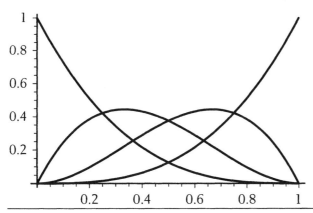

Figure 5.10 The four Bernstein basis functions of degree 3. Each basis function is unimodal in t on the interval $[0,1]$.

This unimodality of the basis functions localizes the effect of the control points on the shape of a Bezier curve. As t increases from 0 to k/n, the value of the basis function $B_k^n(t)$ increases and the control point P_k has more and more influence on the position of the curve. This influence peaks at $t = k/n$ and recedes thereafter. Thus even though, in general, a Bezier curve does not interpolate its control points, we usually associate the control point P_k with the parameter value k/n (see also Section 5.5.1, Exercise 2).

The Bernstein basis functions are also unimodal in the discrete parameter k. This means that if we fix a parameter $t^* \in [0,1]$, then there is an index k depending on t^* such that

$$B_0^n(t^*) \leq \cdots \leq B_{k-1}^n(t^*) \leq B_k^n(t^*) \geq B_{k+1}^n(t^*) \geq \cdots \geq B_n^n(t^*).$$

This result is easy to prove by induction on n. Certainly the result is true for $n = 0,1$. Suppose that it is true up to degree $n - 1$. Recall that by (5.4)

$$B_j^n(t) = (1 - t)B_j^{n-1}(t) + tB_{j-1}^{n-1}(t).$$

Thus $B_j^n(t^*)$ is a convex combination of $B_j^{n-1}(t^*)$ and $B_{j-1}^{n-1}(t^*)$. Hence $B_j^n(t^*)$ lies between $B_j^{n-1}(t^*)$ and $B_{j-1}^{n-1}(t^*)$. Together with the inductive hypothesis, this observation is enough to conclude that the sequence $B_0^n(t^*),\dots,B_n^n(t^*)$ is unimodal, since we now have

$$B_0^n(t^*) \leq B_0^{n-1}(t^*) \leq \cdots \leq B_{k-1}^{n-1}(t^*) \leq B_k^n(t^*) \leq B_k^{n-1}(t^*)$$

$$\geq B_{k+1}^n(t^*) \geq B_{k+1}^{n-1}(t^*) \geq \cdots \geq B_{n-1}^{n-1}(t^*) \geq B_n^n(t^*) \ .$$

We illustrate these inequalities in Figure 5.11.

This unimodality property tells us something important about how the control points influence a Bezier curve at a fixed parameter value t^*. For each parameter t^* there is some basis function $B_k^n(t)$ such that

$$B_k^n(t^*) \geq B_j^n(t^*)$$

for all j. (See Exercise 1 for how to determine k from t^*.) Thus at t^* the control point P_k has the most influence on the curve. Moreover, by unimodality, the influence of the other control points recedes as their index recedes from k because the values of the basis functions $B_j^n(t^*)$ get smaller as j recedes from k. Thus if we want to change the position of the curve at t^* by moving the control points as little as possible, we should first move the control point P_k, then the adjacent control points P_{k-1} and P_{k+1}, continuing in this manner until we have adjusted the curve to the desired location.

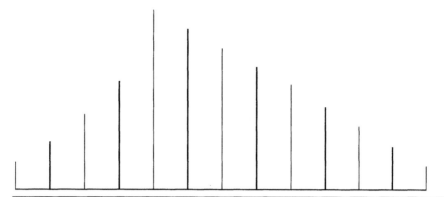

Figure 5.11 The Bernstein basis functions of degree $n - 1$ are indicated schematically by thin vertical lines, and the Bernstein basis functions of degree n by thick vertical lines. The height of a line represents the value of the corresponding basis function. Because each basis function of degree n is a convex combination of two adjacent basis functions of degree $n - 1$, the height of any thick line must lie between the heights of the two surrounding thin lines. Thus the unimodality of the Bernstein basis functions of degree $n - 1$ implies the unimodality of the Bernstein basis functions of degree n.

Exercises

1. Prove that $B_k^n(t) \geq B_j^n(t)$ for all $j \Leftrightarrow k/(n+1) \leq t \leq (k+1)/(n+1)$.

2. Prove that the cubic Ball basis functions

$$b_0(t) = (1-t)^2 \qquad b_1(t) = 2t(1-t)^2 \qquad b_2(t) = 2t^2(1-t) \qquad b_3(t) = t^2$$

   are unimodal in both t and k for $0 \leq t \leq 1$.

3. Prove that the Taylor basis functions $\{t^k / k!\}$ are unimodal in both t and k for $0 \leq t < \infty$.

4. Define the Poisson basis functions $\{b_k(t)\}$ by setting

$$b_k(t) = \frac{e^{-t} t^k}{k!}, \quad k = 0,1,\dots.$$

   Prove that the Poisson basis functions $\{b_k(t)\}$ are unimodal in both t and k for $0 \leq t < \infty$.

5. Consider again the Bernstein functions of negative degree from Section 5.3, Exercise 11.

   a. Graph a few of the Bernstein basis functions of negative degree for $t \leq 0$.

   b. Prove that the Bernstein basis functions of negative degree are unimodal in both t and k for $t \leq 0$.

6. Let $L_k^n(t)$, $k = 0,...,n$, denote the Lagrange basis functions of degree n for the nodes $t_k = k/n$, $k = 0,...,n$. Show by example that these functions are not unimodal over the interval $[0,1]$ in either k or t for $n \geq 2$.

5.4.4   Descartes' Law of Signs and the Variation Diminishing Property

One of the drawbacks of Lagrange interpolation is that the interpolating polynomial may oscillate too much and fail to capture the shape defined by its control points (see Figure 5.1). We have abandoned Lagrange interpolation in favor of Bezier approximation because, unlike the Lagrange interpolant, a Bezier curve never oscillates more than its control polygon. We shall now provide a precise mathematical definition of this oscillation property and prove that it holds for all Bezier curves.

We can measure the oscillations of a continuous curve in the plane with respect to a straight line by counting the number of intersections of the curve with the line (Figure 5.12).

Of course a curve may intersect some lines more often than others, so we shall want to measure the oscillations of a curve with respect to every line. We say that a curve C_1 oscillates no more than a curve C_2 if for every line L

number of intersections of C_1 and $L \leq$ number of intersections of C_2 and L.

Let $P(t)$ denote the polygon defined by the control points $P_0,...,P_n$. Then $P(t)$ can be parametrized as a continuous piecewise linear function over the interval $[0,1]$ by setting

$$P(t) = (k + 1 - nt)P_k + (nt - k)P_{k+1} \qquad 0 \leq \frac{k}{n} \leq t \leq \frac{k+1}{n} \leq 1.$$

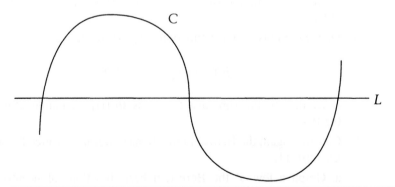

Figure 5.12 The curve C oscillates up, then down, then up again. These oscillations can be detected by observing that the curve C intersects the line L three times: once on the way up, then again on the way down, and yet again one more time on the way up.

We say that a curve scheme $D(t) = \sum_k D_k(t)P_k$, $0 \le t \le 1$, has the *variation diminishing property* if the curve $D(t)$ oscillates no more than the control polygon $P(t)$ for every choice of control points $P_0, ..., P_n$ (see Figure 5.13). Thus the curve scheme is variation diminishing if for every control polygon $P(t)$ and every line L

number of intersections of $D(t)$ and $L \le$ number of intersections of $P(t)$ and L.

We are now going to show that Bezier curves have this variation diminishing property.

It is easy to measure the number of intersections of a control polygon with a straight line. Let L be the line defined by the point Q and the normal vector v. Then a point P lies on the line L if $v \cdot (P - Q) = 0$. Moreover, as we can see from Figure 5.14, two points R and S lie on opposite sides of L if and only if

$$sign\{v \cdot (R - Q)\} = -sign\{v \cdot (S - Q)\}.$$

If the line L intersects the control polygon $P(t)$ along the edge joining P_k and P_{k+1}, then P_k and P_{k+1} must lie on opposite sides of the line L. Hence for every sign change in the sequence $\{v \cdot (P_0 - Q), ..., v \cdot (P_n - Q)\}$, there must be an intersection between the line and the control polygon. Therefore,

number of sign alternations of $\{v \cdot (P_0 - Q), ..., v \cdot (P_n - Q)\}$
$\le$ number of intersections of $P(t)$ and L.

We have inequality here rather than equality because the line may intersect the control polygon at a vertex P_k or along an edge $P_k P_{k+1}$.

To simplify the rest of this discussion, we shall adopt the following notation:

$Zeros_{(a,b)}(B(t))$ = the number of roots of the function $B(t)$ in the interval (a,b)

$SA(c_0, ..., c_n)$ = the number of sign alternations of the sequence $(c_0, ..., c_n)$.

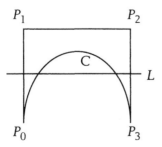

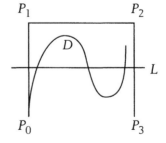

(a) Variation diminishing   (b) Not variation diminishing

Figure 5.13 (a) The curve C is variation diminishing with respect to its control polygon, since C intersects any line L no more often than its control polygon intersects L. (b) On the other hand, the curve D is not variation diminishing with respect to its control polygon, since D intersects the line L three times while its control polygon intersects L only twice.

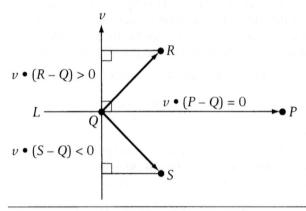

Figure 5.14 Two points R and S lie on opposite sides of the line $v \bullet (P-Q) = 0$ if and only if $sign\{v \bullet (R-Q)\} = -sign\{v \bullet R - Q\} = -sign\{v \bullet (S-Q)\}$.

Here roots are counted with multiplicity, and zeros are ignored in sign alternations.

To prove that the Bezier curve

$$B(t) = \sum_k B_k^n(t) P_k$$

oscillates no more than its control polygon $P(t)$, we shall show that

$$\text{number of intersections of } B(t) \text{ and } L \le SA\{v \bullet (P_0 - Q),...,v \bullet (P_n - Q)\}.$$

It will then follow immediately from our previous discussion that

$$\text{number of intersections of } B(t) \text{ and } L \le \text{number of intersections of } P(t) \text{ and } L.$$

But

$$\text{number of intersections of } B(t) \text{ and } L = Zeros_{(0,1)}\{(B(t) - Q) \bullet v\}$$

$$= Zeros_{(0,1)}\left\{\sum_k B_k^n(t)(P_k - Q) \bullet v\right\}.$$

Thus to prove the variation diminishing property for Bezier curves, we need to relate the number of zeros of a polynomial to the number of sign alternations of its Bernstein coefficients. That is, we need to show that

$$Zeros_{(0,1)}\left\{\sum_k B_k^n(t)(P_k - Q) \bullet v\right\} \le SA\{v \bullet (P_0 - Q),...,v \bullet (P_n - Q)\}.$$

To establish this result, we introduce *Descartes' Law of Signs.*

A sequence of functions $D_0(t),...,D_n(t)$ is said to satisfy Descartes' Law of Signs in the interval (a,b) if for every sequence of constants $c_0,...,c_n$

$$Zeros_{(a,b)}\left\{\sum_{k=0}^{n} c_k D_k(t)\right\} \le SA(c_0,...,c_n).$$

For example, the monomial basis $1, t, ..., t^n$ is known to satisfy Descartes' Law of Signs in the interval $(0, \infty)$. Therefore, the polynomial $7t^3 + 3t^2 - 3t + 11$ can have at most two positive roots. Similarly, the Bernstein basis $B_0^n(t), ..., B_n^n(t)$ is known to satisfy Descartes' Law of Signs in the interval $(0,1)$. Therefore, the polynomial $7(1 - t)^3 + 3t(1 - t)^2 - 3t^2(1 - t) + 11t^3$ can have at most two roots in the interval $(0,1)$.

We are going to prove that the monomial basis $1, t, ..., t^n$ satisfies Descartes' Law of Signs in the interval $(0, \infty)$ and that the Bernstein basis $B_0^n(t), ..., B_n^n(t)$ satisfies Descartes' Law of Signs in the interval $(0,1)$. We shall then apply Descartes' Law of Signs for the Bernstein basis functions to conclude that Bezier curves are variation diminishing. We begin with the following lemma.

LEMMA 5.1

Let $f(t) = \sum_{j=0}^{n} c_j t^j$ and $g(t) = \sum_{k=0}^{n+1} b_k t^k$. If $g(t) = (t - r)f(t)$ and $r > 0$, then

$$SA(b_{n+1}, ..., b_0) > SA(c_n, ..., c_0).$$

Proof If $g(t) = (t - r)f(t)$, then

    i. $b_{n+1} = c_n$

    ii. $b_k = c_{k-1} - rc_k$

    iii. $b_0 = -rc_0$

Consider the two sequences $c_n, ..., c_0$ and $b_{n+1}, ..., b_0$. By (i) the sequences start out with the same sign. Moreover, from (ii) and the assumption that $r > 0$, it follows that

    iv. $sign(c_k) \neq sign(c_{k-1}) \implies sign(b_k) = sign(c_{k-1})$.

Therefore, between any two indices where the c's change sign exactly once, the b's must change sign at least once. That is, the two sequences look like

    c's: +,...,+,−,...,−,+,...,+,−,...

    b's: +,...,−, ... ,+, ... , −,...

Now let c_k, c_{k-1}, $k \geq 1$, be the last sign change in the c's. Then

$$SA(b_{n+1}, ..., b_k) \geq SA(c_n, ..., c_{k-1}) = SA(c_n, ..., c_0).$$

But since $r > 0$, it follows by (iii) and (iv) that

$$sign(b_k) = sign(c_{k-1}) = \cdots = sign(c_0) \neq sign(b_0).$$

Hence there is at least one sign change between b_k and b_0. Therefore,

$$SA(b_{n+1}, ..., b_0) \geq SA(b_n, ..., b_{k+1}) + 1 \geq SA(c_n, ..., c_0) + 1.$$

PROPOSITION
5.2

Descartes' Law of Signs for the Monomial Basis

$$Zeros_{(0,\infty)}\left\{\sum_{k=0}^{n}a_k t^k\right\} \le SA\{a_n,...,a_0\}.$$

Proof

Let $g(t) = \sum_{k=0}^{n}a_k t^k$ and let $r_1,...,r_m$ be the positive roots of $g(t)$. Then

$$g(t) = (t - r_m)\cdots(t - r_1)g_0(t).$$

Define the polynomials $g_1(t),...,g_m(t)$ by setting

$$g_k(t) = (t - r_k)\cdots(t - r_1)g_0(t) \qquad k = 1,...,m.$$

Then by construction

$$g_k(t) = (t - r_k)g_{k-1}(t)$$
$$g_m(t) = g(t) \ .$$

We are going to show by induction on k that

$$\text{sign alternations}\{\text{coefficients } g_k(t)\} \ge k.$$

By Lemma 5.1

sign alternations {coefficients $g_1(t)$} > sign alternations {coefficients $g_0(t)$}.

Therefore,

$$\text{sign alternations \{coefficients } g_1(t)\} \ge 1.$$

Moreover, again by Lemma 5.1

$$\text{sign alternations \{coefficients } g_k(t)\}$$
$$> \text{sign alternations \{coefficients } g_{k-1}(t)\}.$$

Therefore, it follows by induction on k that

$$\text{sign alternations \{coefficients } g_k(t)\} \ge k.$$

Hence

$$\text{sign alternations \{coefficients } g(t)\}$$
$$= \text{sign alternations \{coefficients } g_m(t)\} \ge m.$$

But m is the number of positive roots of $g(t)$. Therefore, we have proved that

$$SA\{a_n,...,a_0\} \ge Zeros_{(0,\infty)}\left\{\sum_{k=0}^{n}a_k t^k\right\}.$$

COROLLARY
5.3

Descartes' Law of Signs for the Bernstein Basis

$$Zeros_{(0,1)}\left\{ \sum_{k=0}^{n} a_k B_k^n(t) \right\} \leq SA\{a_n,...,a_0\}.$$

Proof Once again we shall use Equation (5.6) to convert from the Bernstein basis to the monomial basis. Let $u = t/(1-t)$. Then by Equation (5.6) and Proposition 5.2

$$Zeros_{(0,1)}\left\{ \sum_{k=0}^{n} a_k B_k^n(t) \right\} = Zeros_{(0,1)}\left\{ \frac{\sum_{k=0}^{n} a_k B_k^n(t)}{(1-t)^n} \right\}$$

$$= Zeros_{(0,\infty)}\left\{ \sum_{k=0}^{n} \binom{n}{k} a_k u^k \right\} \leq SA\{a_n,...,a_0\}.$$

COROLLARY
5.4

Bezier curves are variation diminishing.

Proof Consider a Bezier $B(t) = \sum_k B_k^n(t) P_k$ with control polygon $P(t)$. Let L be the line defined by the point Q and the normal vector v. If L intersects the control polygon $P(t)$ along the edge between P_k and P_{k+1}, then P_k and P_{k+1} must lie on opposite sides of the line L. Hence

$$SA\{v \bullet (P_0 - Q),...,v \bullet (P_n - Q)\} \leq \text{number of intersections of } P(t) \text{ and } L.$$

On the other hand,

number of intersections of $B(t)$ and $L = Zeros_{(0,1)}\left\{ \sum_k B_k^n(t)(P_k - Q) \bullet v \right\}.$

Therefore, by Corollary 5.3

number of intersections of $B(t)$ and $L \leq SA\{v \bullet (P_0 - Q),...,v \bullet (P_n - Q)\}$

$\leq$ number of intersections of $P(t)$ and L.

This proof of the variation diminishing property based on Descartes' Law of Signs is highly algebraic. We shall give two geometric proofs of the variation diminishing property for Bezier curves in Section 5.5 when we study subdivision and degree elevation.

Finally, notice that the definition and analysis of the variation diminishing property for Bezier curves was carried out here only for planar Bezier curves. But much the same analysis applies to Bezier curves in three dimensions. The only change is that the lines with respect to which oscillations are measured must be replaced by planes. The remaining details are unchanged (see Exercise 6).

Exercises

1. Consider the Bernstein basis functions

$$B_k^n(t) = \binom{n}{k} \frac{(t-a)^k (b-t)^{n-k}}{(b-a)^n} \qquad k = 0,...,n.$$

   a. Prove that these basis functions satisfy Descartes' Law of Signs in the interval (a,b).

   b. Conclude that Bezier curves defined over the interval $[a,b]$ are variation diminishing.

2. Prove that the cubic Ball basis functions

$$b_0(t) = (1-t)^2 \qquad b_1(t) = 2t(1-t)^2 \qquad b_2(t) = 2t^2(1-t) \qquad b_3(t) = t^2$$

   satisfy Descartes' Law of Signs in the interval $(0,1)$.

3. Show that the Taylor basis $1,(t-t_0),...,(t-t_0)^n/n!$ satisfies Descartes' Law of Signs in the interval (t_0,∞).

4. Show that the Newton basis for the nodes $t_0 < t_1 < \cdots < t_n$ satisfies Descartes' Law of Signs in the interval (t_0,∞).

5. Show by example that the Lagrange basis for the nodes $t_0 < t_1 < \cdots < t_n$ does not satisfy Descartes' Law of Signs in the interval (t_0,t_n).

6. Replace lines by planes in the definition of the variation diminishing property. That is, say that a curve scheme $D(t) = \sum_k D_k(t)P_k$ in three dimensions is variation diminishing if for every control polygon $P(t)$ and every plane L

   number of intersections of $D(t)$ and L
   $\leq$ number of intersections of $P(t)$ and L.

   Prove that the variation diminishing property holds for Bezier curves in 3-space.

5.5   Change of Basis Procedures and Principles of Duality

Consider two polynomial bases—a basis for polynomials of degree n:

$$B(t) = \big(B_0(t),...,B_n(t)\big)$$

and a basis for polynomials of degree $m \geq n$:

$$D(t) = \big(D_0(t),...,D_m(t)\big)$$

Let $P(t) = \sum_k B_k(t)P_k$ be a polynomial curve. Since $m \geq n$, there must be coefficients $\{Q_j\}$ such that $P(t) = \sum_j D_j(t)Q_j$. The general problem we shall address here is how to find the D-coefficients $\{Q_j\}$ given the B-coefficients $\{P_k\}$. We shall see in subsequent sections that the solution to this generic problem has substantial applications in the theory of Bezier curves, including conversion between Bezier and monomial form (Section 5.5.1), degree elevation (Section 5.5.3), and subdivision (Section 5.5.4).

To simplify our notation, let $P = (P_0, ..., P_n)$ and $Q = (Q_0, ..., Q_m)$. Then

$$P(t) = D(t) \bullet Q^T = B(t) \bullet P^T,$$

where the superscript T denotes transpose and $\bullet$ denotes matrix multiplication. Since $D(t)$ is a polynomial basis and $m \geq n$, there must be a matrix $M = (M_{jk})$ such that $M: D(t) \rightarrow B(t)$; that is,

$$B_k(t) = \sum_{j=0}^{m} M_{jk} D_j(t) \quad k = 0, ..., n$$

$$B(t) = D(t) \bullet M \ .$$

Therefore,

$$D(t) \bullet Q^T = B(t) \bullet P^T = (D(t) \bullet M) \bullet P^T = D(t) \bullet (M \bullet P^T).$$

Since the polynomials $D(t)$ are linearly independent, we conclude that

$$Q^T = M \bullet P^T$$

or equivalently taking the transpose of both sides:

$$Q = P \bullet M^T$$

$$Q_j = \sum_{k=0}^{n} M_{kj}^T P_k = \sum_{k=0}^{n} M_{jk} P_k \quad j = 0, ..., m.$$

Thus $M^T: P \rightarrow Q$. Therefore, we have established that

$$M: D(t) \rightarrow B(t) \Leftrightarrow M^T: P \rightarrow Q.$$

These observations can be summarized by the following rule.

RULE

First Principle of Duality

$$M: D-basis \ \rightarrow \ B-basis \ \Leftrightarrow \ M^T: B-coefficients \ \rightarrow \ D-coefficients$$

$$B_k(t) = \sum_{j=0}^{m} M_{jk} D_j(t) \quad k = 0, ..., n \ \Leftrightarrow \ Q_j = \sum_{k=0}^{n} M_{jk} P_k \quad j = 0, ..., m$$

Hence, the same change of basis matrix M used for representing the basis $B(t)$ in terms of the basis $D(t)$ can be used to convert from the B control points P to the D control points Q. In Section 5.5.1 we shall employ this strategy to convert between the Bernstein and monomial bases and in Section 5.5.3 we shall use it as well to construct a degree elevation algorithm for Bezier curves.

This principle of duality has an interesting interpretation in terms of triangular computations. We shall see shortly that many change of basis algorithms can be represented by diagrams such as the one in Figure 5.15(a). The input P is placed at the base of the triangle, and the output Q emerges along the left lateral edge. Let M denote the matrix that represents this transformation from P to Q. Then there is a closely related diagram that represents the transformation matrix M^T: just reverse the orientation of all the arrows in the diagram that represents the transformation M, leaving the labels along the arrows unchanged, and place the input at the nodes along the left lateral edge of the triangle. Now the output emerges at the base of the triangle, and the matrix representing this new transformation is M^T (see Figure 5.15(b)). These observations can be condensed into the following rule.

RULE   *Second Principle of Duality*

Reversing the arrows in a triangle that represents a transformation M yields a triangle that represents the transformation M^T.

Notice that in Figure 5.15(b) values already exist at the nodes R_0, R_1, R_2 where we are computing new values. When this occurs, we perform the usual computation and then add the value residing at the node as in Figure 5.16.

Why does this work? Since M represents the transformation from P to Q,

$$Q_j = \sum_{k=0}^{n} M_{jk} P_k.$$

Therefore, it follows from the diagram that

$$M_{jk} = \text{the sum of the products along all paths from } P_k \text{ to } Q_j.$$

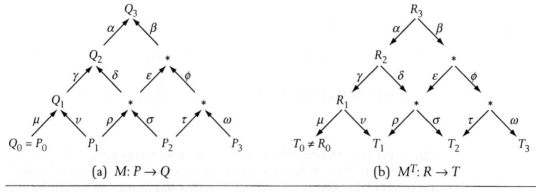

(a) $M: P \rightarrow Q$ (b) $M^T: R \rightarrow T$

Figure 5.15   Schematic depiction of the Second Principle of Duality. If (a) represents the transformation M, then (b) represents the transformation M^T.

Figure 5.16 The value computed at the node labeled T is $\alpha R + \beta S + T$.

Now let N denote the transformation from R to T depicted in Figure 5.15(b). Then

$$T_k = \sum_{k=0}^{n} N_{kj} R_j \, ,$$

and from the diagram we have

$$N_{kj} = \text{the sum of the products along all paths from } R_j \text{ to } T_k.$$

But R_j and Q_j occupy the same node in these diagrams, and so too do P_k and T_k. Therefore, the paths between R_j and T_k are identical to the paths between P_k and Q_j. Thus

$$N_{kj} = M_{jk}$$
$$N = M^T \, .$$

Combining our first two principles of duality produces the following rule.

RULE   *Third Principle of Duality*

Reversing the arrows in a triangle that represents a transformation from a D-basis to a B-basis yields the transformation from the B-coefficients to the D-coefficients and vice versa.

For example, recall from Section 4.5 that the forward differences of a polynomial represent the coefficients of the polynomial with respect to a normalized Newton basis. Thus Figure 4.3 is a transformation algorithm from the Lagrange coefficients (evaluation at the nodes) to the normalized Newton coefficients (differences). By the Third Principle of Duality, reversing the arrows and placing the normalized Newton basis functions along the left lateral edge of the triangle generates the transformation from the normalized Newton basis to the Lagrange basis with the same nodes (see Figure 5.17 and Exercise 2).

In the next few sections we shall apply only the First Principle of Duality. We will have occasion to apply the other two duality principles in Chapter 7 when we study change of basis algorithms for B-splines.

Exercises

1. Suppose that $\sum_j D_j(t) = \sum_k B_k(t) = 1$ and $B(t) = D(t) \cdot M$.

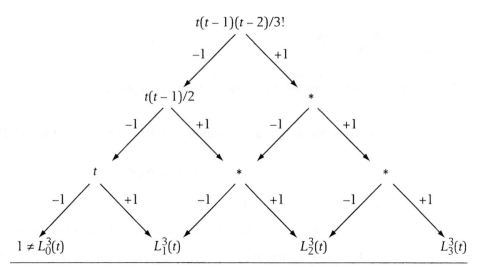

Figure 5.17 Computation of the cubic Lagrange basis functions with nodes 0,1,2,3 from the normalized cubic Newton basis functions with the same nodes by the Third Principle of Duality. The Newton basis functions are placed along the left lateral edge, the arrows of the difference triangle are reversed, and the Lagrange basis functions emerge along the base of the triangle.

    a. Prove that $\sum_k M_{jk} = 1$ for every j. That is, prove that when the curve schemes are affine invariant, every row of the change of basis matrix sums to one.

    b. When the rows of a matrix are nonnegative and sum to one, then the matrix is called a *Markov chain*. Markov chains play a prominent role in probability theory. Prove that if the change of basis matrix M is a Markov chain, then the D control points lie inside the convex hull of the B control points.

2. Verify by direct calculation that in Figure 5.17 the cubic Lagrange basis functions with nodes at 0,1,2,3 emerge at the base of the diagram.

3. Let M denote the matrix that represents the transformation where the input is placed at the base of the triangle, and the output emerges along the right lateral edge. What diagram represents the transformation M^T?

4. Show that

    a. If M represents the transformation in Figure 5.15(a), then M is lower triangular.

    b. If N represents the transformation in Figure 5.15(b), then N is upper triangular.

5.5.1   **Conversion between Bezier and Monomial Form**

To convert between the Bezier and monomial forms of a curve, we must represent the Bernstein basis functions in terms of the monomial basis and represent the monomial basis functions in terms of the Bernstein basis. Then we can apply the techniques from Section 5.5. Here we shall derive both types of identities.

To represent the Bernstein basis in terms of the monomial basis, start with the explicit formula (5.5):

$$B_k^n(t) = \binom{n}{k}t^k(1-t)^{n-k}.$$

Expanding the term $(1-t)^{n-k}$ by the binomial theorem yields

$$B_k^n(t) = \sum_{j=0}^{n-k}(-1)^j\binom{n}{k}\binom{n-k}{j}t^{j+k}.$$

To represent $B_k^n(t)$ in terms of powers of t^j rather than powers of t^{j+k}, reindex by replacing j by $j-k$ to obtain

$$B_k^n(t) = \sum_{j=k}^{n}(-1)^{j-k}\binom{n}{k}\binom{n-k}{j-k}t^j.$$

Next observe that

$$\binom{n}{k}\binom{n-k}{j-k} = \binom{j}{k}\binom{n}{j})$$

so

$$B_k^n(t) = \sum_{j=k}^{n}(-1)^{j-k}\binom{j}{k}\binom{n}{j}t^j. \tag{5.7}$$

Because the factor $\binom{n}{j}$ appears repeatedly in these conversion formulas, it is more convenient to consider the scaled monomial basis $M_j^n(t) = \binom{n}{j}t^j$ rather than the standard monomial basis $t^j, j = 0,...,n$. Now by the First Principle of Duality from Section 5.5,

$$\sum_{j=0}^{n}Q_jM_j^n(t) = \sum_{k=0}^{n}B_k^n(t)P_k \Leftrightarrow Q_j = \sum_{k=0}^{j}(-1)^{j-k}\binom{j}{k}P_k. \tag{5.8}$$

Thus (5.8) converts polynomials from Bezier to monomial form.

Conversely, to convert from monomial to Bezier form, we must represent the monomial basis in terms of the Bernstein basis. To do so, observe that by the binomial theorem

$$((1-t)+tz)^n = \sum_{j=0}^{n}\binom{n}{j}t^jz^j(1-t)^{n-j} = \sum_{j=0}^{n}B_j^n(t)z^j.$$

(The expression $\sum_{j=0}^{n} B_j^n(t)z^j$ is called the *generating function* for the Bernstein basis functions.)

Differentiating both sides of this equation with respect to z a total of k times, we obtain

$$\frac{n!}{(n-k)!}t^k((1-t)+tz)^{n-k} = \sum_{j=k}^{n} \frac{j!}{(j-k)!}B_j^n(t)z^{j-k}.$$

Dividing by $k!$ and setting $z = 1$, we arrive at the identity

$$M_k^n(t) = \sum_{j=k}^{n} \binom{j}{k}B_j^n(t). \tag{5.9}$$

Hence again from the First Principle of Duality,

$$\sum_{j=0}^{n} B_j^n(t)P_j = \sum_{k=0}^{n} Q_k M_k^n(t) \Leftrightarrow P_j = \sum_{k=0}^{j} \binom{j}{k}Q_k. \tag{5.10}$$

Thus the $n + 1$ rows of the change of basis matrix are the first $n + 1$ rows of Pascal's triangle.

Formulas (5.8) and (5.10) are inverses. It turns out that these two formulas both have interesting interpretations in terms of triangular computations.

Formula (5.8) is the expression for forward differencing (compare to Equation (4.7)). Thus if we place the Bezier control points at the base of a triangle and run the forward differencing computation (subtraction), the monomial coefficients emerge along the left edge of the triangle. This dynamic programming algorithm is illustrated for cubic curves in Figure 5.18.

The coefficients $\binom{j}{k}$ in (5.10) are the entries in Pascal's triangle. Thus if we place the monomial coefficients at the base of Pascal's triangle and reverse all the arrows (addition), then the Bezier coefficients emerge along the left edge of the triangle (see Section 5.3, Exercise 5). This dynamic programming algorithm is illustrated for cubic curves in Figure 5.19.

Exercises

1. Suppose that the control points of a Bezier curve are evenly spaced along a straight line. That is, suppose that the control points are given by $P_k = P_0 + (k/n)v$, $k = 0,...,n$. Show that the Bezier curve $P(t) = \sum_k B_k^n(t)P_k$ is a straight line with a linear parametrization.

2. Let $B(t)$ be a polynomial with Bezier coefficients P_k, $k = 0,...,n$.
   a. Show that the graph of $B(t)$—that is, the curve $(t,B(t))$—has control points $(k/n,P_k)$, $k = 0,...,n$.

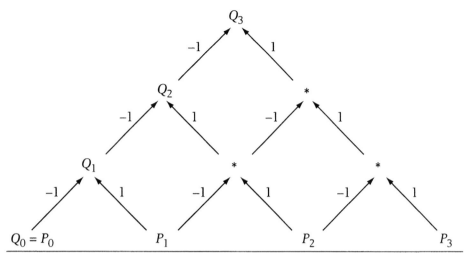

Figure 5.18 Differencing can be used to convert from Bezier to monomial form. Place the Bezier coefficients $\{P_k\}$ at the base of the triangle and take differences. Then the coefficients $\{Q_j\}$ with respect to the monomial basis $M_j^n(t) = \binom{n}{j}t^j$ emerge along the left edge of the triangle. Here the algorithm is illustrated for $n = 3$. Compare to Figure 4.3 for forward differencing.

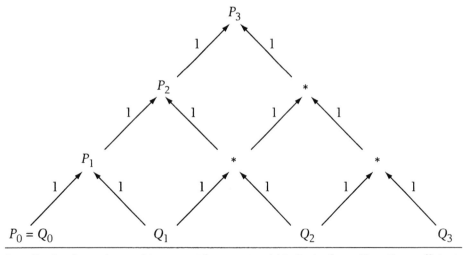

Figure 5.19 Pascal's triangle can be used to convert from monomial to Bezier form. Place the coefficients $\{Q_k\}$ with respect to the monomial basis $M_k^n(t) = \binom{n}{k}t^k$ at the base of the triangle and add up. Then the Bezier coefficients $\{P_j\}$ emerge along the left edge of the triangle. Here the algorithm is illustrated for $n = 3$. This algorithm is the inverse of the algorithm illustrated in Figure 5.18.

(Note that this is another reason that the control point P_k is usually associated with the parameter value k/n.)

    b. Where are the control points for the graph of the basis function $B_k^n(t)$?

3. Consider the generating function $G(z) = \sum_k B_k^n(t)e^{kz}$. Prove that

    a. $\displaystyle\sum_{k=0}^{n} B_k^n(t)e^{kz} = ((1-t)+te^z)^n$

    b. $\displaystyle\sum_{k=0}^{n} k^j B_k^n(t) = \frac{n!}{(n-j)!}t^j + \text{lower-order terms}$

4. Suppose that the control points of a Bezier curve of degree n are evenly spaced points in parameter space on a polynomial curve of degree $m \le n$. That is, suppose that $P_k = P(k/n)$ where $P(t)$ is a polynomial curve of degree m. Using the result of Exercise 3(b), show that the Bezier curve $P(t) = \sum_k B_k^n(t)P_k$ is a polynomial curve of degree m.

5. Develop change of basis algorithms to convert between the Taylor basis

$$\binom{n}{j}(t-a)^j, \quad j = 0,\dots,n,$$

and the Bernstein basis

$$B_k^n(t) = \binom{n}{k}\frac{(t-a)^k(b-t)^{n-k}}{(b-a)^n}, \quad k = 0,\dots,n.$$

6. Explain why the same algorithm that converts curves from Bezier to monomial form also converts polynomials from Lagrange to Newton form when the nodes are at the integers and the Newton basis function $N_k(t)$ is normalized by dividing by $k!$. (Hint: Compare Figures 4.3 and 5.18.)

7. Consider the functions $\{B_j^{-n}(t)\}$ defined in Section 5.3, Exercise 11. Prove that

$$\binom{-n}{k}t^k = \sum_{j=k}^{\infty}\binom{j}{k}B_j^{-n}(t).$$

8. Using the Third Principle of Duality from Section 5.5, show how to use Figure 5.18 to compute the Bernstein basis functions $\{B_k^n(t)\}$ from the monomial basis functions $\{M_k^n(t)\}$.

5.5.2 The Weierstrass Approximation Theorem

Every continuous curve on a closed interval can be approximated to within any desired tolerance by some polynomial curve. This statement is the *Weierstrass Approximation Theorem*. The Weierstrass Theorem justifies our preoccupation with polynomial curves because it asserts that to within any arbitrary tolerance all continuous curves are essentially polynomials. The approximating polynomials in the Weierstrass Theorem need not be interpolating; indeed, a sequence of polynomials that interpolate more and more points on a fixed curve does not necessarily converge to the curve that is interpolated.

For analytic functions we can resort to Taylor polynomials to approximate the curve to any desired tolerance, but for arbitrary continuous functions we need to invoke other techniques. Here we are going to use Bezier approximation to prove the Weierstrass Theorem. Our main technical tool will be the change of basis formula (5.9) that represents the monomial basis in terms of the Bernstein basis. We begin with a simple identity that comes up in the main body of the proof.

LEMMA
5.5

$$\sum_{k=0}^{n}(k-nt)^2 B_k^n(t) = nt(1-t)$$

Proof   Begin by recalling (5.9):

    i. $M_j^n(t) = \sum_{k=j}^{n}\binom{k}{j}B_k^n(t).$

When $j = 0,1,2$, this formula specializes to

    ii. $\sum_{k=0}^{n}B_k^n(t) = 1$

    iii. $\sum_{k=0}^{n}kB_k^n(t) = nt$

    iv. $\sum_{k=0}^{n}k(k-1)B_k^n(t) = n(n-1)t^2.$

Adding (iii) and (iv), we obtain

    v. $\sum_{k=0}^{n}k^2 B_k^n(t) = n(n-1)t^2 + nt.$

Therefore, by (ii), (iii), and (v),

$$\sum_{k=0}^{n}(k-nt)^2 B_k^n(t) = \sum_{k=0}^{n}(k^2 - 2knt + n^2t^2)B_k^n(t)$$

$$= \sum_{k=0}^{n}k^2 B_k^n(t) - 2nt\sum_{k=0}^{n}kB_k^n(t) + n^2t^2\sum_{k=0}^{n}B_k^n(t)$$

$$= n(n-1)t^2 + nt - 2n^2t^2 + n^2t^2$$

$$= nt(1-t) \ .$$

Let $F(t)$ be a continuous curve on the interval $[0,1]$. Define the *nth Bernstein approximation* to $F(t)$ by setting

$$B_n[F](t) = \sum_{k=0}^{n} F(k/n) B_k^n(t). \tag{5.11}$$

The graph of the polynomial $B_n[F](t)$ is the Bezier curve with control points at equally spaced parameter values along the continuous curve $F(t)$. In general, the functions $B_n[F](t)$ interpolate only the values $F(0), F(1)$; nevertheless, we shall now show that as the degree n increases, these approximating polynomials $B_n[F](t)$ actually converge uniformly to the continuous function $F(t)$.

THEOREM
5.6

$\lim_{n \to \infty} B_n[F](t) = F(t) \{\text{uniform convergence}\}$

Proof Consider the difference between $F(t)$ and $B_n[F](t)$. Since the Bernstein basis functions sum to 1,

$$F(t) - B_n[F](t) = \sum_{k=0}^{n} \{F(t) - F(k/n)\} B_k^n(t). \tag{5.12}$$

Thus to analyze the difference between $F(t)$ and $B_n[F](t)$, we need to analyze the difference between $F(t)$ and $F(k/n)$. To begin, notice that since $F(t)$ is continuous on the closed interval $[0,1]$, $F(t)$ is bounded on this interval. Therefore, there is a constant $M > 0$ such that $|F(t)| \le M$, so for all $0 \le x, t \le 1$

$$|F(x) - F(t)| \le 2M.$$

Now choose any $\varepsilon > 0$. Again since $F(t)$ is continuous on the closed interval $[0,1]$, there is some $\delta > 0$ such that for all $0 \le x, t \le 1$

$$|x - t| < \delta \quad \Rightarrow \quad |F(x) - F(t)| < \varepsilon.$$

Let us fix the value of t. Then for each $k = 0,...,n$, either $|k/n - t| < \delta$ or $|k/n - t| \ge \delta$. Therefore, we can split the sum on the right-hand side of (5.12) into two sums: the first containing those terms where $|k/n - t| < \delta$ and the second containing those terms where $|k/n - t| \ge \delta$. Hence by (5.12),

$|(F(t) - B_n[F](t)|$

$= \left| \sum_{|k/n-t|<\delta} \{F(t) - F(k/n)\} B_k^n(t) + \sum_{|k/n-t|\ge\delta} \{F(t) - F(k/n)\} B_k^n(t) \right|$

$\le \sum_{|k/n-t|<\delta} |F(t) - F(k/n)| B_k^n(t) + \sum_{|k/n-t|\ge\delta} |F(t) - F(k/n)| B_k^n(t)$

$\le \sum_{|k/n-t|<\delta} \varepsilon B_k^n(t) + \sum_{|k/n-t|\ge\delta} 2M B_k^n(t)$

$\le \varepsilon \sum_{|k/n-t|<\delta} B_k^n(t) + 2M \sum_{|k/n-t|\ge\delta} B_k^n(t)$

$\le \varepsilon + 2M \sum_{|k/n-t|\ge\delta} B_k^n(t) \ . \tag{5.13}$

It remains only to bound the sum $\sum_{|k/n-t|\geq\delta} B_k^n(t)$. But if $|k/n-t|\geq\delta$, then

$$(k-nt)^2 \geq n^2\delta^2,$$

so

$$\frac{(k-nt)^2}{n^2\delta^2} \geq 1.$$

Therefore,

$$\sum_{|k/n-t|\geq\delta} B_k^n(t) \leq \sum_{|k/n-t|\geq\delta} \frac{(k-nt)^2}{n^2\delta^2} B_k^n(t) \leq \frac{1}{n^2\delta^2}\sum_k (k-nt)^2 B_k^n(t).$$

Hence by Lemma 5.5

$$\sum_{|k/n-t|\geq\delta} B_k^n(t) \leq \frac{nt(1-t)}{n^2\delta^2}.$$

But the product $t(1-t)$ takes its maximum value at $t=1/2$, so $t(1-t)\leq 1/4$. Therefore, for all t

$$\sum_{|k/n-t|\geq\delta} B_k^n(t) \leq \frac{1}{4n\delta^2}.$$

Substituting this bound back into (5.13), we find that

$$\left|F(t)-B_n[F](t)\right| = \varepsilon + \frac{M}{2n\delta^2}.$$

Now choose $n > \dfrac{M}{2\delta^2\varepsilon}$; then

$$\left|F(t)-B_n[F](t)\right| \leq 2\varepsilon,$$

so for n sufficiently large $B_n[F](t)$ is arbitrarily close to $F(t)$.

COROLLARY
5.7

Weierstrass Approximation Theorem

Let $F(t)$ be a continuous function on the interval $[0,1]$. Then $F(t)$ can be approximated to within any desired tolerance by some polynomial curve.

Exercises

1. Prove that if $F(t)$ is a continuous function on the interval $[a,b]$, then $F(t)$ can be approximated by a polynomial to any desired tolerance.

2. Prove that B_n is a linear operator—that is,

$$B_n[aF+bG](t) = aB_n[F](t) + bB_n[G](t).$$

3. Prove that if $F(t)$ is a polynomial of degree m, then the Bernstein approximation $B_n[F](t)$ is always a polynomial of degree m for $n \geq m$. (Hint: See Section 5.5.1, Exercise 4.)

4. Plot the Taylor and Bernstein polynomial approximations to the functions $\cos(\pi t)$, $\sin(\pi t)$ for different degrees n. Use these results to generate polynomial approximations to the semicircle $(\cos(\pi t), \sin(\pi t))$ for $0 \leq t \leq 1$. Which approximation scheme converges faster over the interval $[0,1]$?

5. Show that

a. $\dfrac{dB_k^n(t)}{dt} = n\left(B_{k-1}^{n-1}(t) - B_k^{n-1}(t)\right)$ $\qquad k = 0,...,n$

b. $\lim_{n \to \infty}\left\{B_n[F](t)\right\}' = F'(t)$.

(Hint: Apply part (a) and the mean value theorem together with Theorem 5.6.)

5.5.3 Degree Elevation for Bezier Curves

Degree elevation is another change of basis algorithm that is important in the study of Bezier curves. Because the Bernstein polynomials form a basis, every degree n Bezier curve is also a Bezier curve of degree $n + 1$. Since higher-degree curves have more control points, we can degree elevate a curve to attain additional control over the shape of the curve. Given the control points $\{P_0,...,P_n\}$ for a Bezier curve of degree n, we would like an algorithm to find the control points $\{Q_0,...,Q_{n+1}\}$ for the same curve represented now as a Bezier curve of degree $n + 1$.

In Section 5.5 we showed that to find this change of basis algorithm, we need to represent the Bernstein basis functions of degree n in terms of the Bernstein basis functions of degree $n + 1$. There is a simple trick for doing this. Observe from (5.5) that $tB_k^n(t)$ has the same powers of t and $1 - t$ as $B_{k+1}^{n+1}(t)$, and $(1 - t)B_k^n(t)$ has the same powers of t and $1 - t$ as $B_k^{n+1}(t)$ Thus, up to constant multiples, $tB_k^n(t)$ is the same as $B_{k+1}^{n+1}(t)$ and $(1 - t)B_k^n(t)$ is the same as $B_k^{n+1}(t)$. Adding these results will give us $B_k^n(t)$ as a linear combination of $B_k^{n+1}(t)$ and $B_{k+1}^{n+1}(t)$.

Working through this algebra, we find that

$$(1-t)B_k^n(t) = \frac{n+1-k}{n+1} B_k^{n+1}(t) \tag{5.14}$$

$$tB_k^n(t) = \frac{k+1}{n+1} B_{k+1}^{n+1}(t). \tag{5.15}$$

Adding these equations yields

$$B_k^n(t) = \frac{n+1-k}{n+1} B_k^{n+1}(t) + \frac{k+1}{n+1} B_{k+1}^{n+1}(t). \tag{5.16}$$

Thus the change of basis matrix $M = (M_{jk})$ is given by

$$M_{jk} = \frac{n+1-j}{n+1} \qquad j = k$$

$$= \frac{j}{n+1} \qquad j = k+1$$

$$= 0 \qquad j \neq k, k+1.$$

Hence by the First Principle of Duality in Section 5.5,

$$\sum_j Q_j B_j^{n+1}(t) = \sum_k P_k B_k^n(t) \Leftrightarrow Q_j = \frac{j}{n+1} P_{j-1} + \frac{n+1-j}{n+1} P_j. \qquad (5.17)$$

Equation (5.17) is the degree elevation formula for Bezier curves. It expresses the degree $n + 1$ control points in terms of the degree n control points. Notice that the point Q_j lies on the line joining P_{j-1} and P_j. We illustrate this degree elevation algorithm for cubic curves in Figure 5.20.

Degree elevation is a corner-cutting procedure. Notice in Figure 5.20 that the degree elevated control polygon $Q = (Q_0, Q_1, Q_2, Q_3, Q_4)$ is obtained from the original control polygon $P = (P_0, P_1, P_2, P_3)$ by cutting off the corners at P_1 and P_2. Since every Bezier curve lies in the convex hull of its control points, the degree elevated control polygon is closer to the Bezier curve than the original control polygon. Thus if we continue to degree elevate a Bezier curve, the control polygon gets closer and

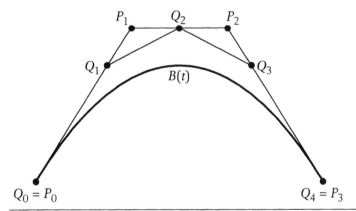

Figure 5.20 Degree elevation algorithm for a cubic Bezier curve. The points $\{P_0, P_1, P_2, P_3\}$ are the control points for the cubic Bezier curve $B(t)$, and the points $\{Q_0, Q_1, Q_2, Q_3, Q_4\}$ represent the same curve $B(t)$ as a quartic Bezier curve.

closer to the original curve. Moreover, in the limit, these control polygons converge uniformly to some continuous curve. We shall now apply the Weierstrass Approximation Theorem to show that this limit curve is, in fact, the original Bezier curve.

THEOREM 5.8

Let $B(t)$ be a Bezier curve of degree n and let $Q_m(t)$ denote the control polygon that represents $B(t)$ as a Bezier curve of degree $m \geq n$. Then $\lim_{m\to\infty}Q_m(t) = B(t)$. That is, the control polygons generated by degree elevation converge uniformly to the original Bezier curve.

Proof By construction $Q_m(t)$ is a piecewise linear curve over the interval $[0,1]$ and $Q_m(k/m)$ is the kth vertex of $Q_m(t)$. Since $Q_m(t)$ is the control polygon that represents $B(t)$ as a Bezier curve of degree m, $Q_m(k/m)$ is the kth control point of $B(t)$ considered as a polynomial curve of degree m. Let $B_m[F](t)$ denote the mth Bernstein approximation to $F(t)$ (see (5.11)). Then

$$B_m[Q_m](t) = \sum_{k=0}^{m}Q_m(k/m)B_k^m(t) = B(t).$$

Let $Q(t) = \lim_{m\to\infty}Q_m(t)$. To show that $Q(t) = B(t)$, we need to examine the difference between $B(t)$ and $Q(t)$. By Theorem 5.6,

$$B(t) - Q(t) = \lim_{m\to\infty} B_m[B - Q](t)$$
$$= \lim_{m\to\infty} \{B_m[B](t) - B_m[Q](t)\}$$
$$= \lim_{m\to\infty} \{B_m[B](t) - B_m[Q_m](t) + B_m[Q_m](t) - B_m[Q](t)\}$$
$$= \lim_{m\to\infty} \{B_m[B](t) - B(t)\} + \lim_{m\to\infty}B_m[Q_m - Q](t).$$

But again by Theorem 5.6,

$$\lim_{m\to\infty}\{B_m[B](t) - B(t)\} = 0,$$

and the convergence is uniform. Moreover, since $Q(t) = \lim_{m\to\infty}Q_m(t)$, given any $\varepsilon > 0$ for m sufficiently large

$$|Q_m(t) - Q(t)| < \varepsilon.$$

Therefore,

$$|B_m[Q_m - Q](t)| \leq \sum_{k=0}^{m}|Q_m(k/m) - Q(k/m)|B_k^m(t) < \varepsilon,$$

so

$$\lim_{m\to\infty}B_m[Q_m - Q](t) = 0.$$

Hence $B(t) - Q(t) = 0$, so $B(t) = \lim_{m\to\infty}Q_m(t)$ and the convergence is uniform on the interval $[0,1]$.

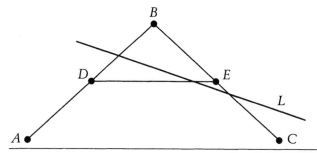

Figure 5.21 Polygon *ADEC* is formed from polygon *ABC* by cutting off the corner at *B*. The line *L* intersects the edge *DE* and therefore must also intersect either edge *DB* or edge *BE* in the polygon *ABC* because if a line intersects one side of a triangle, then it must intersect one of the other two sides.

Corner cutting reduces the oscillation of a polygon. If $Q(t)$ is the piecewise linear curve obtained by cutting corners off of another piecewise linear curve $P(t)$, then any line L intersects $P(t)$ at least as often as it intersects $Q(t)$ because if a line intersects one side of a triangle, then it must intersect one of the other two sides (see Figure 5.21). Thus *corner cutting is a variation diminishing process*. Since degree elevation is a corner-cutting procedure (see Figure 5.20), we can use this observation along with Theorem 5.8 to give a simple geometric proof that Bezier curves satisfy the variation diminishing property.

COROLLARY   *Bezier curves are variation diminishing.*
5.9

Proof   Since corner cutting reduces oscillation, the limit curve of a corner-cutting procedure must be variation diminishing with respect to the original control polygon. But by Theorem 5.8, each Bezier curve is the limit curve of a degree elevation process, which is a corner-cutting procedure. Hence Bezier curves are variation diminishing.

Exercises

1. Prove that

a. $B_j^m(t)B_k^n(t) = \dfrac{\binom{m}{j}\binom{n}{k}}{\binom{m+n}{j+k}} B_{j+k}^{m+n}(t)$

b. $B_k^n(t) = \displaystyle\sum_{j=0}^{m} \dfrac{\binom{m}{j}\binom{n}{k}}{\binom{m+n}{j+k}} B_{j+k}^{m+n}(t)$

2. Let $B(t)$ be a Bezier curve of degree n with control points $\{P_k\}$. Use the results in Exercise 1 to find the control points $\{Q_j\}$ that represent $B(t)$ as a Bezier curve of degree $m + n$.

3. Consider the negative binomial distribution defined in Section 5.3, Exercise 11. Prove that for $n > 1$

$$B_k^{-n}(t) = \frac{n+k-1}{n-1} B_k^{-n+1}(t) - \frac{k+1}{n-1} B_{k+1}^{-n+1}(t).$$

4. Prove that the arc length of a Bezier curve is always less than or equal to the perimeter of its control polygon.

5. Use the degree elevation formula to derive Descartes' Law of Signs for the Bernstein basis. (Hint: First use degree elevation to prove an analogue of Lemma 5.1 for the Bernstein basis.)

6. Show that

$$M_k^n(t) = \left(\frac{k+1}{n+1}\right)\frac{M_{k+1}^{n+1}(t)}{t} = \left(\frac{n+1-k}{n+1}\right)M_k^{n+1}(t).$$

7. Let $t_0 < t_1 < \cdots < t_{2n+1}$ and define

$$\psi_k^n(t) = (-1)^{n-k}(t_{k+1}-t)\cdots(t_{k+n}-t) \qquad\qquad k = 0,...,n$$

$$\psi_k^{n+1}(t) = (-1)^{n+1-k}(t_k-t)\cdots(t_{k+n}-t) \qquad\qquad k = 0,...,n+1.$$

   a. Show that

$$(t_k-t)\psi_k^n(t) = -\psi_k^{n+1}(t)$$

$$(t_{k+n+1}-t)\psi_k^n(t) = \psi_{k+1}^{n+1}(t).$$

   b. Use the degree elevation formulas in part (a) to show that the functions

$$\psi_k^n(t), \ k = 0,...,n,$$

satisfy Descartes' Law of Signs in the interval (t_n, t_{n+1}).

8. Suppose that $\{P_0,...,P_n\}$ are the control points for a degree n Bezier curve, and that $\{Q_0,...,Q_{n+1}\}$ are the control points for the degree-elevated version of the same curve. Let $s_{max}(P)$ be the largest side of the control polygon generated by the control points $\{P_0,...,P_n\}$, and let $s_{max}(Q)$ be the largest side of the control polygon generated by the control points $\{Q_0,...,Q_{n+1}\}$.

   a. Prove that $s_{max}(Q) \le \dfrac{n}{n+1} s_{max}(P)$.

   b. Conclude that as we continue to degree elevate a Bezier curve, the lengths of the sides of the control polygons approach zero.

5.5.4   Subdivision

The control points of a Bezier curve describe the curve with respect to the parameter interval $[0,1]$. Sometimes, however, we are interested only in the part of the curve in some subinterval $[a,b]$. For example, when rendering a Bezier curve we may need to clip the curve to a window. Since any segment of a Bezier curve is a polynomial curve, we should be able to represent such a segment as a complete Bezier curve with a new set of control points. Splitting a Bezier curve into smaller pieces is also useful as a divide and conquer strategy for intersection algorithms (see below). The process of splitting a Bezier curve into two or more Bezier curves that represent the exact same curve is called *subdivision*.

We begin with an important special case. Consider a Bezier curve $B(t) = \sum_k B_k^n(t)P_k$, where $t \in [0,1]$. The curve $B(rt) = \sum_k B_k^n(rt)P_k$, where $t \in [0,1]$, represents the segment of $B(t)$ for which $t \in [0,r]$ because as t varies from 0 to 1, $B(rt)$ varies from $B(0)$ to $B(r)$. Thus by the results in Section 5.5, to subdivide a Bezier curve at $t = r$, we must represent the basis functions $B_0^n(rt),...,B_n^n(rt)$ in terms of the standard basis $B_0^n(t),...,B_n^n(t)$. A subdivision algorithm, then, is nothing more than a change of basis algorithm from the Bernstein basis $B_0^n(rt),...,B_n^n(rt)$ to the standard Bernstein basis $B_0^n(t),...,B_n^n(t)$.

Using the binomial theorem, we could derive such a change of basis formula by purely formal algebraic methods (see Exercise 1). It happens, however, that this change of basis formula has a simple probabilistic interpretation that provides further insight into its meaning. Since probability also simplifies the derivation of this identity, we shall adopt a probabilistic approach. Later on we shall see that many other identities involving the Bernstein polynomials also have simple probabilistic interpretations that simplify their derivation (see Exercises 2 and 3).

Bezier curves are related to probability theory because the Bernstein basis functions represent the binomial distribution. There are many ways to model this distribution: coin tossings, random walks, or urn models. For our purposes we shall adopt an urn model based on sampling with replacement.

5.5.4.1   Sampling with Replacement

Consider an urn containing w white balls and b black balls. One ball at a time is drawn at random from this urn, its color inspected, and then returned to the urn. Since we are performing sampling with replacement, the probability of choosing a white ball on any trial is $t = w/(w + b)$, and the probability of choosing a black ball on any trial is $1 - t$. Now there are $\binom{n}{k}$ ways of selecting exactly k white balls in n trials, so the probability of choosing *exactly* k white balls in n trials is given by the binomial distribution

$$B_k^n(t) = \binom{n}{k}t^k(1-t)^{n-k}.$$

**LEMMA
5.10**

$$B_k^n(rt) = \sum_{j=k}^{n} B_k^j(r)B_j^n(t) \qquad\qquad (5.18)$$

Proof   Consider two urns: one with red and blue balls and another with white and black balls. Let

   r = probability of selecting a red ball from urn #1

   t = probability of selecting a white ball from urn #2.

Applying the binomial distribution for sampling with replacement, we know that

   $B_k^j(r)$ = probability of choosing exactly k red balls in j trials from urn #1

   $B_j^n(t)$ = probability of choosing exactly j white balls in n trials from urn #2.

Place these two urns into a super urn (see Figure 5.22). A selection from the super urn consists of selecting one ball at random from each regular urn, inspecting their colors, and returning the balls to their respective urns.

The super urn also models sampling with replacement, where

   rt = probability of selecting a red-white combination

   $B_k^n(rt)$ = probability of selecting exactly k red-white combinations in n trials.

To select a red-white combination from the super urn on any trial, we must certainly select a white ball from urn #2 on this trial; if we select a black ball from urn #2, we need not even inspect the color of the ball selected from urn #1. Suppose that we choose exactly j white balls in n trials from urn #2. Then to select exactly k red-white combinations in n trials from the super urn, we must select exactly k red balls from urn #1 during the j trials that we selected a white ball from urn #2. Thus

   $B_k^n(rt)$ = probability of selecting exactly k red-white combinations in n trials

   $= \sum_{j\geq k}$ (probability of selecting exactly j white balls in n trials from urn #2)

   $\times$ (probability of selecting exactly k red balls from urn #1 during the j trials in which white balls were selected from urn #2)

   $= \sum_{j=k}^{n} B_k^j(r)B_j^n(t).$

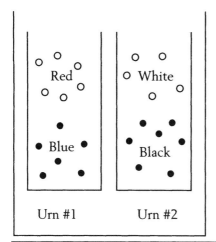

Figure 5.22 The super urn contains two ordinary urns. A selection from the super urn consists of selecting one ball at random from each regular urn, inspecting their colors, and returning the balls to their respective urns.

Equation (5.18) represents the scaled Bernstein basis $B_0^n(rt),...,B_n^n(rt)$ in terms of the standard Bernstein basis $B_0^n(t),...,B_n^n(t)$. Hence by the First Principle of Duality in Section 5.5,

$$\sum_j Q_j B_j^n(t) = \sum_k P_k B_k^n(rt) \Leftrightarrow Q_j = \sum_{k=0}^{j} B_k^j(r) P_k. \qquad (5.19)$$

This formula leads to the following subdivision algorithm for Bezier curves.

5.5.4.2 Subdivision Algorithm

Let $B(t)$ be a Bezier curve with control points $P_0,...,P_n$. To subdivide $B(t)$ at $t = r$, run the de Casteljau algorithm at $t = r$. The points $Q_0,...,Q_n$ that emerge along the left lateral edge of the triangle are the Bezier control points of the segment of the curve from $t = 0$ to $t = r$, and the points $R_0,...,R_n$ that emerge along the right lateral edge of the triangle are the Bezier control points of the segment of the curve from $t = r$ to $t = 1$ (see Figure 5.23).

This construction works because the point Q_j that emerges along the left lateral edge of the triangle at level j is the point at parameter $t = r$ along the Bezier curve with control points $P_0,...,P_j$. These points are given by Equation (5.19), so they represent the Bezier curve $B(t)$ between $t = 0$ to $t = r$. To find the points that represent the Bezier curve $B(t)$ from $t = r$ to $t = 1$, consider the curve $B(1-t)$ from $t = 0$ to $t = 1 - r$. We know by symmetry (Section 5.2, Property 4) that the curve $B(1-t)$ can be represented as a Bezier curve with control points $P_n,...,P_0$. To subdivide this curve from $t = 0$ to $t = 1 - r$, we can use our previous algorithm; that is, simply run the de Casteljau algorithm at $t = 1 - r$ and read off the results from the left lateral

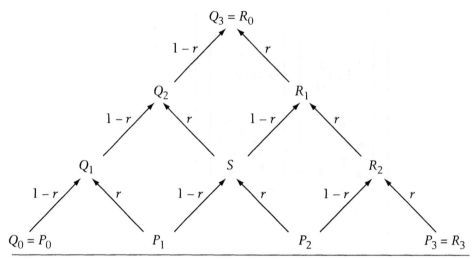

Figure 5.23 The de Casteljau subdivision algorithm for a cubic Bezier curve with control points P_0,P_1,P_2,P_3. The points Q_0,Q_1,Q_2,Q_3 that emerge along the left edge of the triangle are the Bezier control points of the segment of the original curve from $t = 0$ to $t = r$, and the points R_0,R_1,R_2,R_3 that emerge along the right edge of the triangle are the Bezier control points of the segment of the original curve from $t = r$ to $t = 1$.

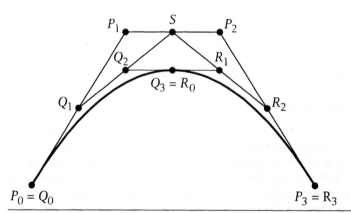

Figure 5.24 Geometric interpretation of the de Casteljau subdivision algorithm for a cubic Bezier curve with control points P_0,P_1,P_2,P_3. The points Q_0,Q_1,Q_2,Q_3 are the Bezier control points of the segment of the original curve from $t = 0$ to $t = r$, and the points R_0,R_1,R_2,R_3 are the Bezier control points of the segment of the original curve from $t = r$ to $t = 1$.

edge of the diagram. But by symmetry, these points are precisely the points on the right lateral edge of the de Casteljau triangle for the original curve evaluated at $t = r$.

To find the control points of the segment of a Bezier curve between $t = r$ and $t = s$, subdivide the original Bezier curve at $t = s$ and then subdivide the first segment of the subdivided curve at $t = r/s$ (see Exercise 9).

The de Casteljau algorithm is a corner-cutting procedure (see Figure 5.24). Therefore, subdivision, like degree elevation, is also a corner-cutting procedure. If we continue recursively subdividing a Bezier curve, the control polygons get closer and closer to the original curve. Moreover, in the limit, these control polygons converge to a continuous curve. We shall now show that this limit curve is, in fact, the original Bezier curve. For convenience, we restrict our attention to recursive subdivision at $r = 1/2$, though the results are much the same for any value of r between 0 and 1 (see too Exercise 4).

THEOREM
5.11

The control polygons generated by recursive subdivision converge to the original Bezier curve.

Proof Suppose that the maximum distance between any two adjacent control points is d. By construction, the points on any level of the de Casteljau algorithm for $t = 1/2$ lie at the midpoints of the edges of the polygons generated by the previous level. Therefore, it follows easily by induction that adjacent points on any level of the de Casteljau diagram for $t = 1/2$ are no further than d apart (see Figure 5.25).

By the same midpoint argument, as we proceed up the diagram adjacent points along the left (right) lateral edge of the triangle can be no further than $d/2$ apart. Hence as we apply recursive subdivision, the distance between the control points of any single control polygon must converge to zero. Since the first and last control points of a Bezier control polygon always lie on the curve, these control polygons must converge to points along the original curve.

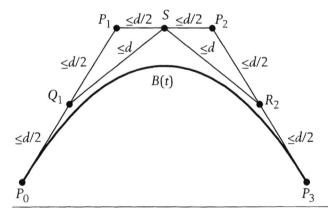

Figure 5.25 One level of the de Casteljau algorithm for a cubic Bezier curve. If adjacent control points P_0, P_1, P_2, P_3 are no further than d units apart, then adjacent points Q_1, S, R_2 on the second level of the de Casteljau algorithm can be no further than d units apart.

The convergence of recursive subdivision gives us yet another proof that Bezier curves are variation diminishing.

COROLLARY
5.12

Bezier curves are variation diminishing.

Proof Since recursive subdivision is a corner-cutting procedure, the limit curve must be variation diminishing with respect to the original control polygon. But by Theorem 5.11, the Bezier curve is the limit curve generated from the original control polygon by recursive subdivision, so Bezier curves are variation diminishing.

The convergence of recursive subdivision is much faster than the convergence of degree elevation. Degree elevation converges at a rate of $O(1/n)$, where n is degree; recursive subdivision converges at a rate of $O(h^2)$, where h is the length of the parameter interval. This rapid convergence of recursive subdivision leads to the following important recursive algorithms for rendering and intersecting Bezier curves.

Rendering Algorithm

1. If the Bezier curve can be approximated to within tolerance by the straight line joining its first and last control points, then draw this straight line.
2. Otherwise subdivide the curve and render the segments recursively.

Intersection Algorithm

1. If the convex hulls of two Bezier curves fail to intersect, then the curves themselves do not intersect.
2. Otherwise if each Bezier curve can be approximated within tolerance by the straight line joining its first and last control points, then intersect these straight lines.
3. Otherwise subdivide the two curves and intersect the segments recursively.

To determine whether a Bezier curve can be approximated to within tolerance by the straight line joining its first and last control points, it is sufficient, by the convex hull property, to test whether all the interior control points lie within tolerance of this straight line. By the Pythagorean theorem (see Figure 5.26), the distance between a point P and a line L determined by a point Q and a unit direction vector v is given by

$$dist^2(P,L) = |P - Q|^2 - ((P - Q) \cdot v)^2.$$

It may happen that a control point P_k is close to the line L determined by the first and last control points P_0, P_n even though the projection of P_k does not lie inside the line segment $P_0 P_n$; that is, P_k may be close to the line L even though it is not close to the line segment $P_0 P_n$. To be sure that this is not the case, we need only check that

$$0 \le (P_k - P_0) \cdot (P_n - P_0) \le |P_n - P_0|^2.$$

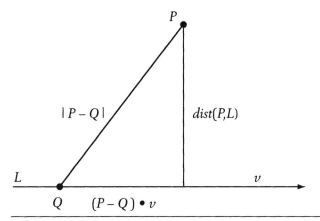

Figure 5.26 A line L determined by a point Q and a unit direction vector v, and a point P not on L. By the Pythagorean theorem $dist^2(P,L) = |P - Q|^2 - ((P - Q) \cdot v)^2$.

Thus it is relatively easy to test whether or not a Bezier curve can be approximated to within some tolerance by a straight line segment. On the other hand, finding and intersecting the convex hulls of two Bezier curves can be quite difficult and time consuming. In practice, the convex hulls in the intersection algorithm are replaced by bounding boxes, which are much easier to compute and intersect than the actual convex hulls. Since the subdivision algorithm converges rapidly, not much time is lost by replacing convex hulls with bounding boxes.

Exercises

1. Use the identity $1 - rt = (1 - r)t + (1 - t)$ together with the explicit formula for the Bernstein basis functions to give a direct derivation of Equation (5.18) without appealing to probability theory.

2. Give a probabilistic interpretation for each of the following identities:

   a. $\sum_{k=0}^{n} B_k^n(t) = 1.$

   b. $B_k^n(t) \geq 0.$

   c. $B_k^n(0) = 0 \qquad k \neq 0$
   $ = 1 \qquad k = 0 .$

   d. $B_k^n(1) = 0 \qquad k \neq n$
   $ = 1 \qquad k = n .$

   e. $B_{n-k}^n(1 - t) = B_k^n(t).$

   f. $B_k^n(t) = (1 - t)B_k^{n-1}(t) + tB_{k-1}^{n-1}(t).$

g. $\sum_{k=0}^{n} k B_k^n(t) = nt \cdot$

3. a. Give a probabilistic proof of the identity

$$B_k^n((1-t)r + t) = \sum_{j=0}^{k} B_{k-j}^{n-j}(r) B_j^n(t) \cdot$$

b. Use the result of part (a) to show that the points that emerge along the right lateral edge of the de Casteljau triangle are the Bezier control points of the segment of the curve from $t = r$ to $t = 1$.

4. Bezier subdivision at $t = 1/2$ generates the following binary tree, whose nodes are control polygons. Denote the original control polygon by P, and let this polygon be the root of the tree. Let P_0—the left child of P—denote the control polygon for the left segment of the Bezier curve (from $t = 0$ to $t = 1/2$), and let P_1—the right child of P—denote the control polygon for the right portion of the curve (from $t = 1/2$ to $t = 1$). Continue to build the binary tree recursively in this fashion. Thus if P_b is a node in the tree, then P_b represents the control polygon for a portion of the curve, and P_{b0}—the left child of P_b—represents the control polygon for the left half of the Bezier segment represented by P_b, while P_{b1}—the right child of P_b—represents the control polygon for the right half of the Bezier segment represented by P_b (see Figure 5.27).

a. Prove that $P_{b_1 \cdots b_n}$ is the control polygon for the curve from $t = b$ to $t = b + 2^{-n}$, where b is the binary fraction represented by $0.b_1 \cdots b_n$.

b. Prove that the sequence of control polygons $P_{b_1}, P_{b_1 b_2}, \ldots, P_{b_1 \cdots b_n}, \ldots$, converges to the point on the Bezier curve at parameter value b = $\lim_{n \to \infty} 0.b_1 \ldots b_n$.

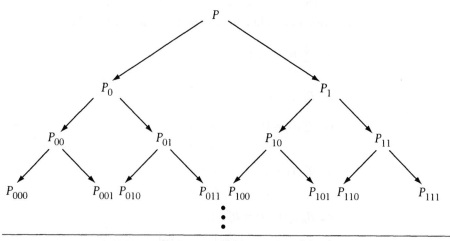

Figure 5.27 The binary tree generated by recursive subdivision of a Bezier curve at $t = 1/2$.

5. Let $P(t)$ be a degree n Bezier curve with control points $P = (P_0,...,P_n)$. Define

$$L(r) = \begin{pmatrix} B_0^0(r) & 0 & 0 & 0 \\ B_0^1(r) & B_0^1(r) & 0 & 0 \\ \vdots & \vdots & \vdots & \vdots \\ B_0^n(r) & B_1^n(r) & \cdots & B_n^n(r) \end{pmatrix} \quad \text{and} \quad M(r) = \begin{pmatrix} B_0^n(r) & B_1^n(r) & \cdots & B_n^n(r) \\ 0 & B_0^{n-1}(r) & \cdots & B_{n-1}^{n-1}(r) \\ \vdots & \vdots & \vdots & \vdots \\ 0 & 0 & 0 & B_0^0(r) \end{pmatrix}.$$

Show that the points generated by the de Casteljau subdivision algorithm for Bezier curves are given by $Q = L(r) * P^T$ and $R = M(r) * P^T$, where the superscript T denotes transpose and $*$ denotes matrix multiplication.

6. Implement the recursive subdivision algorithm for rendering a Bezier curve.

7. Implement the recursive subdivision algorithm for intersecting two Bezier curves.

8. Prove the identity

$$B_k^n((1-t)r + ts) = \sum_{j=0}^{n} \left(\sum_{p+q=k} B_p^{n-j}(r)B_q^j(s) \right) B_j^n(t),$$

and use this identity to derive an explicit formula for the control points that represent the Bezier curve $\sum_k B_k^n(t)P_k$ from $t = r$ to $t = s$.

9. Show that to find the control points of the segment of a Bezier curve between $t = r$ and $t = s$, we can subdivide the original Bezier curve at $t = s$ and then subdivide the first segment of the subdivided curve at $t = r/s$.

10. Consider the negative binomial distribution $\{B_k^{-n}(t)\}$ defined in Section 5.3, Exercise 11. Prove the identity

$$B_k^{-n}(rt) = \sum_{j=k}^{\infty} B_k^j(r)B_j^{-n}(t).$$

11. Consider the Poisson distribution $\{b_k(t)\}$ defined in Section 5.4.3, Exercise 4. Prove the identity

$$b_k(rt) = \sum_{j=k}^{\infty} B_k^j(r)b_j(t).$$

12. The following urn model is due to Polya: consider an urn containing w white balls and b black balls. One ball at a time is drawn at random from the urn and its color inspected. It is then returned to the urn and a constant number c of balls of the same color are added to the urn. Let $t = w/(w + b)$ and let $a = c/(w + b)$. If we hold a constant and allow t to vary, we obtain a discrete probability distribution

$$D_k^n(t) = \text{probability of selecting exactly } k \text{ white balls in } n \text{ trials.}$$

Given a collection of control points $P_0,...,P_n$, we can define Polya curves by setting $D(t) = \sum_k D_k^n(t)P_k$.

a. Find explicit and recursion formulas for $D_k^n(t)$, $k = 0,...,n$.

b. Show that most of the geometric properties of Bezier curves carry over to Polya curves. Describe as many geometric properties of Polya curves as you can.

c. Develop rendering and intersection algorithms for Polya curves.

d. Experiment with the free parameter a. What is the geometric effect of increasing or decreasing the value of a?

13. Generalize Polya's urn model in Exercise 12 by adding a different number of balls of the same color after every trial.

a. Find explicit and recursion formulas for these new distributions.

b. Use these new distributions to define generalized Polya curves, and show that most of the geometric properties of Bezier curves still carry over to these generalized Polya curves.

c. Develop rendering and intersection algorithms for these generalized Polya curves.

d. Experiment with the new free parameters in this generalized urn model. What are the geometric effects of increasing or decreasing these parameters?

e. Show that the Lagrange basis functions can be generated from the generalized Polya urn model by subtracting instead of adding balls of the same color to the urn.

14. Experiment with other urn models or with more general stochastic models to develop new curve schemes. Determine the analytic and geometric properties of these new schemes.

5.6 Differentiation and Integration

Given any number of control points, we can construct a smooth approximation to the shape of the control polygon using a single Bezier curve. But if there are a large number of control points, this Bezier curve will be a polynomial of high degree, leading to slow and numerically unstable computations. Rather than using a single polynomial of high degree, it would be better numerically to construct a sequence of low-degree curves that join together smoothly. Thus we need a way to ensure that two Bezier curves meet smoothly at their join.

Bezier curves interpolate their first and last control points. Thus it is easy to connect two Bezier curves; all we need to do is to make sure that the first control point of the second curve is the same as the last control point of the first curve. This device ensures continuity, but what about smoothness? Smoothness depends on the differential properties of Bezier curves, so it is to these properties that we now turn our attention.

5.6.1   Discrete Convolution and the Bernstein Basis Functions

To simplify our investigation of differentiation, we are going to introduce another technique for studying Bernstein polynomials and Bezier curves: discrete convolution. Let $A(t) = \{A_i(t)\}$, $i = 0,...,m$, and Let $B(t) = \{B_j(t)\}$, $j = 0,...,n$, be two sequences of functions. Define the *discrete convolution* sequence $(A \otimes B)(t) = \{(A \otimes B)_k(t)\}$, $k = 0,...,m+n$, by setting

$$(A \otimes B)_k(t) = \sum_{i+j=k} A_i(t)B_j(t), \qquad k = 0,...,m+n. \tag{5.20}$$

The following two properties of discrete convolution follow easily from (5.20):

i. $A(t) \otimes B(t) = B(t) \otimes A(t)$ \qquad (Commutativity)

ii. $A(t) \otimes (B(t) \otimes C(t)) = (A(t) \otimes B(t)) \otimes C(t)$ \qquad (Associativity)

What does discrete convolution have to do with Bernstein polynomials and Bezier curves? Consider the following example:

$$\{(1-t),t\} \otimes \{(1-t),t\} = \{(1-t)^2, 2t(1-t), t^2\}.$$

Thus if we convolve the degree 1 Bernstein basis functions with themselves, we get the degree 2 Bernstein basis functions. Moreover, it is easy to verify that

$$\{(1-t),t\} \otimes \{(1-t)^2, 2t(1-t), t^2\} = \{(1-t)^3, 3t(1-t)^2, 3t^2(1-t), t^3\};$$

so convolving the degree 1 Bernstein basis functions with the degree 2 Bernstein basis functions yields the degree 3 Bernstein basis functions. More generally, let $B^n(t) = (B_0^n(t),...,B_n^n(t))$ denote the sequence of degree n Bernstein basis functions. Then we have the following results.

PROPOSITION 5.13 $B^{n+1}(t) = B^1(t) \otimes B^n(t)$.

Proof Let $C(t) = B^1(t) \otimes B^n(t)$. Then by construction

$$C_k(t) = (1-t)B_k^n(t) + tB_{k-1}^n(t).$$

But by Equation (5.4)

$$B_k^{n+1}(t) = (1-t)B_k^n(t) + tB_{k-1}^n(t).$$

Hence $C_k(t) = B_k^{n+1}(t)$.

COROLLARY 5.14 $(B_0^n(t),...,B_n^n(t)) = \underbrace{\{(1-t),t\} \otimes \cdots \otimes \{(1-t),t\}}_{n \ factors}.$

Proof This result follows by induction from Proposition 5.13 and the associativity of discrete convolution.

Corollary 5.14 asserts that the Bernstein basis functions of degree n can be constructed by convolving the sequence $\{(1-t),t\}$ with itself n times. As the proof of Proposition 5.13 shows, this result is just a restatement of the two-term recursion formula (5.4) for the Bernstein basis functions, which is itself just a reformulation of the down recurrence for the Bernstein basis functions (Figure 5.7).

Before we can show how to apply convolution to differentiate Bezier curves, we first need to generalize the down recurrence for the Bernstein basis functions to discrete convolutions of arbitrary pairs. Consider the sequences

$$C^{n+1}(t) = \underbrace{\{L_0(t),R_0(t)\} \otimes \cdots \otimes \{L_n(t),R_n(t)\}}_{n+1 \ \ factors}.$$

Then by the definition of discrete convolution

$$C_k^{n+1}(t) = L_n(t)C_k^n(t) + R_n(t)C_{k-1}^n(t),$$

which is a two-term recurrence for the functions

$$C^{n+1}(t) = (C_0^{n+1}(t),...,C_{n+1}^{n+1}(t)).$$

Iterating this recurrence for the sequences $C^n(t),C^{n-1}(t),...,C^1(t)$ generates a down recurrence for the functions $C^{n+1}(t)$. We illustrate this recurrence for $n = 2$ in Figure 5.28.

Since discrete convolution is commutative, the order of the pairs $\{L_k(t),R_k(t)\}$ in the convolution, and hence too in Figure 5.28, does not matter. Thus $\{L_0(t),R_0(t)\}$ could appear on the third level and $\{L_2(t),R_2(t)\}$ on the first level of the diagram, and the functions

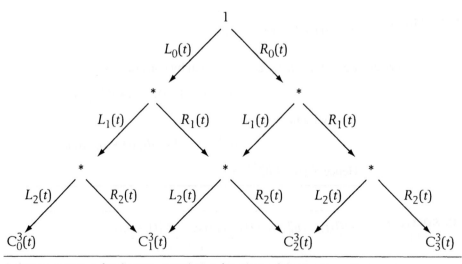

Figure 5.28   A down recurrence for discrete convolution functions of degree 3.

$$C^3(t) = (C_0^3(t), C_1^3(t), C_2^3(t), C_3^3(t))$$

emerging at the base would not change.

As in any down recurrence, the function $C_k^{n+1}(t)$ is simply the sum of all paths from the apex to the kth position at the base of the diagram. Thus to evaluate an expression of the form

$$C(t) = \sum_k C_k^{n+1}(t) P_k,$$

we simply place the values $\{P_k\}$ at the base and reverse all the arrows in the diagram. This generates the evaluation algorithm in Figure 5.29. Notice that here we have not assumed that the functions $L_k(t)$ and $R_k(t)$ necessarily sum to one for all values of k.

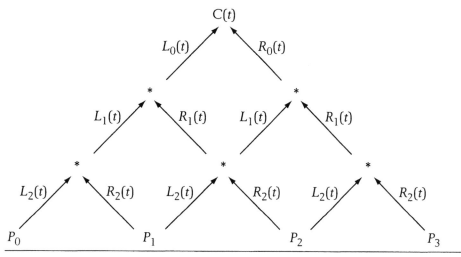

Figure 5.29 Evaluation algorithm for cubic polynomials represented with a convolution basis.

Exercises

1. Let $B_k^n(t) = \binom{n}{k} \dfrac{(t-a)^k (b-t)^{n-k}}{(b-a)^n}$, $k = 0,...,n$,

   denote the Bernstein basis functions over the interval $[a,b]$. Show that

   $$(B_0^n(t),...,B_n^n(t)) = \underbrace{\left\{ \frac{b-t}{b-a}, \frac{t-a}{b-a} \right\} \otimes \cdots \otimes \left\{ \frac{b-t}{b-a}, \frac{t-a}{b-a} \right\}}_{n \ \ factors}.$$

2. Let $B^p(t) = \left(B_0^p(t),...,B_p^p(t) \right)$. Prove that

   a. $B^{m+n}(t) = B^m(t) \otimes B^n(t)$

   b. $B_k^{m+n}(t) = \sum_{i+j=k} B_i^m(t) B_j^n(t)$

3. Let $B^{-n}(t) = \left(B_0^{-n}(t), B_1^{-n}(t), \ldots\right)$ (see Section 5.3, Exercise 11.) Prove that

   a. $B^{-n}(t) = \underbrace{B^{-1}(t) \otimes \cdots \otimes B^{-1}(t)}_{n \text{ factors}}$

   b. $B^n(t) \otimes B^{-n}(t) = \{1, 0, 0, \ldots\}$

   c. $B^{\pm m}(t) \otimes B^{\pm n}(t) = B^{\pm m \pm n}(t)$

4. Let $M^n(t) = \left(M_0^n(t), \ldots, M_n^n(t)\right)$,

   where $M_k^n(t) = \binom{n}{k} t^k$, $k = 0, \ldots, n$. Prove that

   a. $M^n(t) = \underbrace{\{1, t\} \otimes \cdots \otimes \{1, t\}}_{n \text{ factors}} = \underbrace{M^1(t) \otimes \cdots \otimes M^1(t)}_{n \text{ factors}}$

   b. $M^{m+n}(t) = M^m(t) \otimes M^n(t)$

   c. $\binom{m+n}{k} = \sum_{i+j=k} \binom{m}{i}\binom{n}{j}$

5. Let $M^{-n}(t) = \left(M_0^{-n}(t), M_1^{-n}(t), \ldots\right)$, where $M_k^{-n}(t) = \binom{-n}{k} t^k$, $k = 0, 1, \ldots$.
   Prove that

   a. $M^{-n}(t) = \underbrace{M^{-1}(t) \otimes \cdots \otimes M^{-1}(t)}_{n \text{ factors}}$

   b. $M^n(t) \otimes M^{-n}(t) = \{1, 0, \ldots\}$

   c. $M^{\pm m}(t) \otimes M^{\pm n}(t) = M^{\pm m \pm n}(t)$

   d. $\binom{\pm m \pm n}{k} = \sum_{i+j=k} \binom{\pm m}{i}\binom{\pm n}{j}$

6. Let $C^{n+1}(t) = \{L_0(t), R_0(t)\} \otimes \cdots \otimes \{L_n(t), R_n(t)\}$. Show that

   $$C_k^{n+1}(t) = \sum L_{i_0}(t) \cdots L_{i_{n-k}}(t) R_{j_0}(t) \cdots R_{j_{k-1}}(t),$$

   where the sum is taken over all sets $I = \{i_0, \ldots, i_{n-k}\}$ and $J = \{j_0, \ldots, j_{k-1}\}$
   such that $I \cup J = \{0, \ldots, n\}$ and $I \cap J = \phi$.

7. Define the generating function of a sequence $A(t) = \{A_k(t)\}$ by setting

   $$G_A(x) = \sum_{k=0}^n A_k(t) x^k.$$

   a. Prove that $G_{A \otimes B}(x) = G_A(x) G_B(x)$.

   b. Using part (a), prove the following results:

      i. $B^{\pm n}(t) = \left(B_0^{\pm n}(t), B_1^{\pm n}(t), \ldots\right) \Rightarrow G_{B^{\pm n}}(x) = ((1-t) + tx)^{\pm n}$.

ii. $M^{\pm n}(t) = \left(M_0^{\pm n}(t), M_1^{\pm n}(t), \ldots\right) \Rightarrow G_{M^{\pm n}}(x) = (1 + tx)^{\pm n}$.

iii. $C^{n+1}(t) = \{L_0(t), R_0(t)\} \otimes \cdots \otimes \{L_n(t), R_n(t)\}$

$\Rightarrow G_{C^{n+1}}(x) = \left(L_0(t) + R_0(t)x\right) \cdots \left(L_n(t) + R_n(t)x\right)$.

8. Let $b_k(t)$, $k = 0,1\ldots$ denote the Poisson basis functions defined in Section 5.4.3, Exercise 4, and let $b(t) = \left(b_0(t), b_1(t), \ldots\right)$. Show that

   a. $b(t) \otimes b(t) = b(2t)$

   b. $G_b(x) = e^{-t(1-x)}$.

9. Show that discrete convolution distributes through addition. That is, show that

$$A(t) \otimes \left(B(t) + C(t)\right) = A(t) \otimes B(t) + A(t) \otimes C(t)$$

5.6.2 Differentiating Bernstein Polynomials and Bezier Curves

By Corollary 5.14 the Bernstein basis functions of degree $n + 1$ can be constructed by convolving the sequence $\{(1-t), t\}$ with itself n times. Thus to differentiate the Bernstein basis functions, all we need to know is how to differentiate discrete convolutions. But it follows easily from (5.20) that

$$(A(t) \otimes B(t))' = A'(t) \otimes B(t) + A(t) \otimes B'(t)$$

where $A'(t) = \left(A_0'(t), \ldots, A_m'(t)\right)$. Thus the rule for differentiating a discrete convolution is identical to the rule for differentiating an ordinary product. More generally, it follows by induction on n that

$$(A_1(t) \otimes \cdots \otimes A_n(t))' = \sum_{k=1}^{n} A_1(t) \otimes \cdots \otimes A_k'(t) \otimes \cdots \otimes A_n(t). \tag{5.21}$$

Now we can use Corollary 5.14 together with our differentiation formula (5.21) to differentiate all the Bernstein basis functions simultaneously.

**COROLLARY
5.15**

$$\left(\frac{dB_0^n}{dt}, \ldots, \frac{dB_n^n}{dt}\right) = n\{-1, 1\} \otimes \underbrace{\{(1-t), t\} \otimes \cdots \otimes \{(1-t), t\}}_{n-1 \ \ factors}.$$

Proof This result is an immediate consequence of Corollary 5.14, Equation (5.21), and the commutativity of discrete convolution.

By repeated differentiation we can generalize Corollary 5.15 to an arbitrary number of derivatives. Let

$$A^r = \underbrace{A \otimes \cdots \otimes A}_{r \ \ factors}.$$

Then we have the following generalization of Corollary 5.15.

COROLLARY
5.16

$$\left(\frac{d^r B_0^n}{dt^r}, \dots, \frac{d^r B_n^n}{dt^r}\right) = \frac{n!}{(n-r)!}\{-1,1\}^r \otimes \{(1-t,t)\}^{n-r}.$$

Proof This result follows from Corollary 5.15 by induction on r.

We can interpret Corollaries 5.15 and 5.16 in several ways. First since

$$((1-t),t)^{n-1} = \left(B_0^{n-1}(t), \dots, B_{n-1}^{n-1}(t)\right),$$

Corollary 5.15 gives us the explicit formula

$$\frac{dB_k^n(t)}{dt} = n\left(B_{k-1}^{n-1}(t) - B_k^{n-1}(t)\right) \qquad\qquad k = 0, \dots, n. \qquad (5.22)$$

More generally, since

$$((1-t),t)^{n-r} = \left(B_0^{n-r}(t), \dots, B_{n-r}^{n-r}(t)\right)$$

$$(-1,1)^r = \left((-1)^r, \dots, (-1)^{r-j}\binom{r}{j}, \dots, 1\right)$$

(see Exercise 7), it follows from Corollary 5.15 that

$$\frac{d^r B_k^n(t)}{dt^r} = \frac{n!}{(n-r)!}\sum_{j=0}^{r}(-1)^{r-j}\binom{r}{j}B_{k-j}^{n-r}(t).$$

We can also apply Corollaries 5.15 and 5.16 to generate recurrences to evaluate the derivatives of a Bezier curve. Exploiting the interpretation of discrete convolution given in Figure 5.28, we observe that by Corollary 5.15 the derivatives of the Bernstein basis functions can be computed from the down recurrence for the Bernstein basis if we replace $\{(1-t),t\}$ on one level of the algorithm by $\{-1,1\}$ and multiply the result by n. So the same argument we used to generate Figure 5.29—reversing the arrows in this down recurrence—gives us an evaluation algorithm for the derivative of a Bezier curve (see Figure 5.30).

There are several things to notice about Figure 5.30. First, by running only the lowest level of the algorithm, we see that, up to a constant multiple, we can think of the derivative of a cubic Bezier curve with control points $\{P_k\}$ as a quadratic Bezier polynomial with coefficients $\{P_{k+1} - P_k\}$. In general, by the same argument

$$B(t) = \sum_{k=0}^{n} B_k^n(t)P_k \quad\Rightarrow\quad B'(t) = n\sum_{k=0}^{n-1} B_k^{n-1}(t)(P_{k+1} - P_k). \qquad (5.23)$$

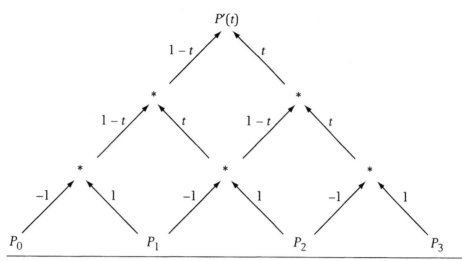

Figure 5.30 The first derivative of a cubic Bezier curve with control points P_0, P_1, P_2, P_3. To get the correct derivative, we must multiply the output $P'(t)$ of this algorithm by $n = 3$.

Thus, up to a constant multiple, the derivative of a Bezier curve with control points $\{P_k\}$ is a Bezier polynomial of one lower degree with coefficients $\{P_{k+1} - P_k\}$.

Because convolution is commutative, we can place $\{-1,1\}$ on any level of the algorithm. If we place $\{-1,1\}$ at the top level and let $B[P_0,...,P_n](t)$ denote the Bezier curve with control points $P_0,...,P_n$, then we find that

$$B'[P_0,...,P_n](t) = n\big(B[P_1,...,P_n](t) - B[P_0,...,P_{n-1}](t)\big);$$

that is, we can compute the derivative of a Bezier curve by subtracting Bezier curves of one lower degree.

We can generate Figure 5.30 from Figure 5.3 (with $a = 0$ and $b = 1$) by differentiating one level of the algorithm. To find the rth derivative of a Bezier curve, we see from Corollary 5.16 that we can simply replace $\{(1-t),t\}$ by $\{-1,1\}$ on r levels of the de Casteljau algorithm; in other words, we simply differentiate r levels of the de Casteljau algorithm. Moreover, because convolution is commutative, it does not matter which r levels we differentiate; any r levels will do the job. If we differentiate the r lowest levels, then, up to a constant multiple, we can express the rth derivative of a degree n Bezier curve with control points $P_0,...,P_n$ as a Bezier polynomial of degree $n - r$. Since

$$\{-1,1\}^r = \{(-1)^{r-j}\tbinom{r}{j}\},$$

the Bezier coefficients are given by

$$v_k = \sum_{j=0}^{r}(-1)^{r-j}\tbinom{r}{j}P_{j+k}.$$

That is,

$$B(t) = \sum_{k=0}^{n} B_k^n(t)P_k \Rightarrow B^{(r)}(t) = \frac{n!}{(n-r)!} \sum_{k=0}^{n-r} B_k^{n-r}(t)\left(\sum_{j=0}^{r}(-1)^{r-j}\binom{r}{j}P_{j+k} \right). \tag{5.24}$$

We began our discussion of differentiation by saying that we wanted to be able to join two Bezier curves together smoothly. To do so, we need to calculate derivatives at their end points. From Equation (5.24), we have

$$B^{(r)}(0) = v_0 = \sum_{j=0}^{r}(-1)^{r-j}\binom{r}{j}P_j \tag{5.25}$$

$$B^{(r)}(1) = v_{n-r} = \sum_{j=0}^{r}(-1)^{r-j}\binom{r}{j}P_{j+n-r}. \tag{5.26}$$

It follows that the rth derivative of a Bezier curve at $t = 0$ depends only on the first $r + 1$ control points $P_0, \ldots, P_r$ and the rth derivative at $t = 1$ depends only on the last $r + 1$ control points $P_{n-r}, \ldots, P_n$.

Suppose then that we are given a Bezier curve $B(t) = \sum_k B_k^n(t)P_k$ and we want to construct another Bezier curve $C(t) = \sum_k B_k^n(t)Q_k$ that meets $B(t)$ and matches its first r derivatives at its end point. Then from Equations (5.25) and (5.26) we get

$r = 0$: $Q_0 = P_n$

$r = 1$: $Q_1 - Q_0 = P_n - P_{n-1} \Rightarrow Q_1 = P_n + (P_n - P_{n-1})$

$r = 2$: $Q_2 - 2Q_1 + Q_0 = P_n - 2P_{n-1} + P_{n-2} \Rightarrow Q_2 = P_{n-2} + 4(P_n - P_{n-1})$

and so on. Each additional derivative allows us to solve for one additional control point. We could go on in this manner solving for one point at a time, but there is a better way that avoids all this tedious computation.

Since $B(t)$ is a polynomial curve, we can extend $B(t)$ past the interval $[0,1]$. Now certainly at $t = 1$ the derivatives of the segment of the curve $B(t)$ for $t \in [1,2]$ must exactly match the derivatives of the curve $B(t)$ for $t \in [0,1]$ because these two curves are the same polynomial. But we can use subdivision to find the Bezier control points of $B(t)$ for the interval $[1,2]$; in fact, by the subdivision algorithm all we need to do is run the de Casteljau algorithm at $t = 2$ and read the control points off the right lateral edge of the triangle (see Figure 5.31). Explicitly these control points are given by

$$Q_k = \sum_{j=0}^{k} B_j^k(2)P_{n-k+j}, \, k = 0, \ldots, n. \tag{5.27}$$

By Equation (5.25) the rth derivative of a Bezier curve at $t = 0$ is uniquely determined by its first $r + 1$ control points. Thus the first r derivatives of a Bezier curve $C(t)$ at $t = 0$ can match the first r derivatives of the Bezier curve $B(t)$ at $t = 1$ if and only if the first $r + 1$ control points of $C(t)$ are given by Equation (5.27).

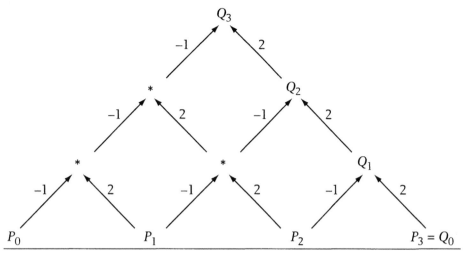

Figure 5.31 Algorithm for finding the first $r + 1$ control points Q_k, $k = 0,...,r$, for a cubic Bezier curve whose first r derivatives at $t = 0$ match the first r derivatives at $t = 1$ of the cubic Bezier curve with control points P_j, $j = 0,...,3$. Notice that because of the labels -1 appearing along the edges, the points Q_k do not lie in the convex hull of the points P_j.

Exercises

1. By differentiating the explicit formula $B_k^n(t) = \binom{n}{k} t^k (1-t)^{n-k}$, show directly that

   a. $\dfrac{dB_k^n(t)}{dt} = n\Big(B_{k-1}^{n-1}(t) - B_k^{n-1}(t) \Big)$.

   b. $\dfrac{d^r B_k^n(t)}{dt^r} = \dfrac{n!}{(n-r)!} \displaystyle\sum_{j=0}^{r} (-1)^{r-j} \binom{r}{j} B_{k-j}^{n-r}(t)$.

   c. Using parts (a) and (b), derive Formulas (5.23) and (5.24).

2. Let $B(t)$ be a Bezier curve with control points $\{P_0,...,P_n\}$, and let $P(t)$ be the control polygon of $B(t)$—that is, $P(t)$ is the piecewise linear function with $P(k/n) = P_k$. Show that

   a. $B^{(r)}(0) = \dfrac{n!}{(n-r)!} \Delta^r P\Big(0, \dfrac{1}{n}, ..., \dfrac{r}{n}\Big)$

   b. $B^{(r)}(1) = \dfrac{n!}{(n-r)!} \Delta^{n-r} P\Big(\dfrac{n-r}{n}, ..., \dfrac{n}{n}\Big)$

   c. $B^{(r)}(t) = \dfrac{n!}{(n-r)!} \displaystyle\sum_{k=0}^{n} B_k^{n-r}(t) \Delta^r P\Big(\dfrac{k}{n}, ..., \dfrac{k+r}{n}\Big)$

3. Let $B^{-n}(t) = (B_0^{-n}(t), B_1^{-n}(t),...)$ (see Section 5.3, Exercise 11). Prove that

$$\frac{dB_k^{-n}(t)}{dt} = n\left(B_k^{-(n+1)}(t) - B_{k-1}^{-(n+1)}(t)\right).$$

4. Let $M^n(t) = \left(M_0^n(t),...,M_n^n(t)\right)$, where $M_k^n(t) = \binom{n}{k}t^k$, $k = 0,...,n$. Show that

a. $\left(\dfrac{dM_0^n}{dt},...,\dfrac{dM_n^n}{dt}\right) = n\{0,1\} \otimes \{1,t\}^{n-1}.$

b. $\left(\dfrac{d^r M_0^n}{dt^r},...,\dfrac{d^r M_n^n}{dt^r}\right) = \dfrac{n!}{(n-r)!}\{0,1\}^r \otimes \{1,t\}^{n-r}.$

c. Develop de Casteljau–like algorithms to compute

$$\left(M_0^n(t),...,M_n^n(t)\right)^{(r)} \text{ and } \left(\Sigma_k M_k^n(t)P_k\right)^{(r)}.$$

(Hint: See Section 5.6.1, Exercise 4.)

5. Let

$$C^{n+1}(t) = \left(C_0(t),...,C_{n+1}(t)\right) = \{L_0(t), R_0(t)\} \otimes \cdots \otimes \{L_n(t), R_n(t)\},$$

where $L_k(t)$ and $R_k(t)$ are linear functions in t for $k = 0,...,n$. Develop de Casteljau–like algorithms to compute $\left(C_0(t),...,C_{n+1}(t)\right)^{(r)}$ and $\left(\Sigma_k C_k(t)P_k\right)^{(r)}$.

6. Let

$$B_k^n(t) = \binom{n}{k}\frac{(t-a)^k(b-t)^{n-k}}{(b-a)^n}, \quad k = 0,...,n,$$

be the Bernstein basis functions over the interval $[a,b]$, and let

$$B^n(t) = \left(B_0^n(t),...,B_n^n(t)\right).$$

a. Prove the identity

$$B^{(r)}(t) = \frac{n!}{(n-r)!}\left\{\frac{-1}{b-a},\frac{1}{b-a}\right\}^r \otimes \left\{\frac{b-t}{b-a},\frac{t-a}{b-a}\right\}^{n-r}.$$

b. Find a de Casteljau–like algorithm for computing the derivatives of a Bezier curve defined over the interval $[a,b]$ instead of over the interval $[0,1]$.

7. Prove by induction on r that $(-1,1)^r = \left((-1)^r,...,(-1)^{r-j}\binom{r}{j},...,1\right).$

8. Let

$$A(t) = \left(A_1(t),...,A_p(t)\right)$$

$$A'(t) = \left(A_1'(t),...,A_p'(t)\right)$$

$$A^n(t) = \underbrace{A(t) \otimes \cdots \otimes A(t)}_{n \ \text{factors}}.$$

Show that

$$\left(A^n(t)\right)' = n\,A^{n-1}(t) \otimes A'(t).$$

9. Use Taylor's Theorem and Equation (5.25) to derive a change of basis formula from Bezier to Taylor form. Compare your result to Equation (5.8).

10. Let $P(t)$ be a Bezier curve with control points $P_0,\ldots,P_n$.

   a. Find a necessary and sufficient condition on the control points $P_0,\ldots,P_n$ so that $P(t)$ degenerates to a polynomial curve of degree $n-1$.

   b. Develop an algorithm to determine whether or not $P(t)$ represents a polynomial curve of degree $n-1$.

   (Hint: A Bezier curve $P(t)$ represents a polynomial curve of degree $n-1$, if and only if the nth derivative of $P(t)$ is zero.)

   c. If $P(t)$ degenerates to a polynomial curve of degree $n-1$, develop an algorithm to find the control points $Q_0,\ldots,Q_{n-1}$ that represent $P(t)$ as a Bezier curve of degree $n-1$ from the control points $P_0,\ldots,P_n$ that represent $P(t)$ as a Bezier curve of degree n.

11. Given point and derivative data $(P_0,v_0),\ldots,(P_n,v_n)$, explain how to place Bezier control points to generate a piecewise cubic Hermite interpolant for this data.

12. The formulas for the unit tangent $U(t)$, the curvature $K(t)$, and the torsion $T(t)$ of a parametric curve $P(t)$ are given by

   ■ $$U(t) = \frac{P'(t)}{|\,P'(t)\,|}$$

   ■ $$K(t) = \frac{|\,P'(t) \times P''(t)\,|}{|\,P'(t)\,|^3}$$

   ■ $$T(t) = \frac{P'(t) \bullet (P''(t) \times P'''(t))}{|\,P'(t) \times P''(t)\,|^2}$$

   a. Compute the unit tangent, curvature, and torsion of a Bezier curve at $t = 0,1$.

   b. Find conditions on the control points of a Bezier curve $C(t)$ so that it matches a given Bezier curve $B(t)$ with continuous unit tangent, curvature, and torsion.

13. Let $P(t)$ be a Bezier curve of degree d.

   a. Show that the Bezier control points of $P'(t)$ generated by first degree elevating and then differentiating $P(t)$ are identical to the Bezier control points of $P'(t)$ generated by first differentiating $P(t)$ and then degree elevating $P'(t)$.

b. Let $P_0^n,\ldots,P_n^n$ denote the degree n Bezier control points of $P(t)$—that is, the control points of $P(t)$ generated by degree elevating the curve $n - d$ times. Using part (a) and Theorem 5.8 on the convergence of degree elevation, show that the slopes of the degree-elevated control polygons converge to the tangents of the original Bezier curve. That is, prove that

$$\text{if } k/n \to t, \text{ then } \lim_{n\to\infty} \frac{P_{k+1}^n - P_k^n}{1/n} = P'(t)$$

5.6.3 Wang's Formula

Recursive subdivision is a powerful tool for rendering and intersecting Bezier curves. Since subdivision at $t = 0.5$ involves only averaging, subdivision itself is very fast. Thus in the rendering and intersection algorithms presented in Section 5.5.4.2 most of the time is spent in testing whether or not each curve segment can be approximated to within some tolerance by a straight line. The purpose of Wang's formula is to avoid all these tests by computing in advance how many levels of subdivision are required to assure that every segment will be approximated to within some prespecified tolerance by the straight line joining its end points. Wang's formula is based on bounds on the second derivative of a Bezier curve. To derive Wang's formula, we begin with a technical result from numerical analysis.

PROPOSITION 5.17

Let $P(t)$ be any twice-differentiable parametric curve on the interval $[a,b]$, and let $L(t) = ((b-t)/(b-a))P(a) + ((t-a)/(b-a))P(b)$ be a parametrization of the straight line through the points $P(a)$ and $P(b)$. Then

$$\max | P(t) - L(t) | \le \frac{(b-a)^2}{8} \max | P''(t) |.$$

Proof To simplify our notation, we introduce the vector-valued function $E(t) = P(t) - L(t)$. By construction,

$$E(a) = E(b) = 0 \Rightarrow E(a) \bullet E(a) = E(b) \bullet E(b) = 0.$$

Thus by Rolle's Theorem applied to the real-valued function $E(t) \bullet E(t)$, there is a parameter $\tau \in [a,b]$ where $E(\tau) \bullet E(\tau)$ is maximal; hence $(E(\tau) \bullet (E(\tau))' = 0$. Therefore, $E(\tau) \bullet E'(\tau) = 0$.

Now by the integral version of Taylor's Theorem with remainder

$$E(t) = E(\tau) + E'(\tau)(t - \tau) + \int_\tau^t (t - x)E''(x)dx.$$

Dotting both sides with $E(\tau)$ and recalling that $E(\tau) \bullet E'(\tau) = 0$, we get

$$E(\tau) \bullet E(t) = E(\tau) \bullet E(\tau) + E(\tau) \bullet \int_\tau^t (t - x)E''(x)dx.$$

Without loss of generality, we can assume that $a \leq \tau \leq (a+b)/2$ (the proof is symmetric if $(a+b)/2 \leq \tau \leq b$). Now substituting $t = a$, recalling that $E(a) = 0$, and noting that $E''(x) = P''(x)$ because $L(x)$ is linear, we obtain

$$-E(\tau) \bullet E(\tau) = E(\tau) \bullet \int_\tau^a (a-x)P''(x)dx.$$

Hence, since $|a \bullet b| \leq |a| \|b|$,

$$|E(\tau)|^2 \leq |E(\tau)| \left| \int_\tau^a (a-x)P''(x)dx \right|.$$

Thus either $E(\tau) = 0$ and there is nothing to prove, or since $\tau \leq (a+b)/2$,

$$|E(\tau)| \leq \left| \int_\tau^a (a-x)P''(x)dx \right|$$

$$\leq \int_\tau^a (a-x)|P''(x)|dx$$

$$\leq \max|P''(x)| \int_\tau^a (a-x)dx$$

$$\leq \max|P''(x)| \frac{(\tau-a)^2}{2}$$

$$\leq \max|P''(x)| \frac{(b-a)^2}{8} .$$

To apply Proposition 5.17 to Bezier curves, we need to bound the second derivative of a Bezier curve.

LEMMA 5.18   Let $B(t)$ be a Bezier curve with control points $P_0,...,P_n$. Then

$$Max\,|\,B''(t)\,| \leq n(n-1)Max_{0 \leq k \leq n-2}|P_{k+2} - 2P_{k+1} + P_k| .$$

Proof By Equation (5.24) and the triangular inequality,

$$|\,B''(t)\,| \leq \left| n(n-1) \sum_{k=0}^{n-2} B_k^{n-2}(t)\{P_{k+2} - 2P_{k+1} + P_k\} \right|$$

$$\leq n(n-1) \sum_{k=0}^{n-2} B_k^{n-2}(t)|P_{k+2} - 2P_{k+1} + P_k|$$

$$\leq n(n-1)Max|P_{k+2} - 2P_{k+1} + P_k| \sum_{k=0}^{n-2} B_k^{n-2}(t)$$

$$\leq n(n-1)Max|P_{k+2} - 2P_{k+1} + P_k| .$$

THEOREM
5.19

Let $B(t)$ be a Bezier curve with control points $P_0,...,P_n$. Given any $\varepsilon > 0$ define

$$m = Max_{0 \leq k \leq n-2}|P_{k+2} - 2P_{k+1} + P_k|$$

$$l \geq Log_4\left(\frac{n(n-1)m}{8\varepsilon}\right). \text{ (Wang's formula)}$$

Let $C(t)$ be any segment of $B(t)$ after at least l levels of subdivision at $t = 1/2$, and let $L(t)$ be the straight line joining the end points of $C(t)$. Then $Dist(C(t),L(t)) \leq \varepsilon$. Thus after l levels of subdivision, the straight lines joining the subdivision points approximate the Bezier curve $B(t)$ to within a tolerance of ε.

Proof   Consider the curve segment $C(t)$ between any two subdivision parameters t_1 and t_2 after l levels of subdivision. By Proposition 5.17 and Lemma 5.18,

$$Dist(C(t),L(t)) \leq Max|C''(t)|\frac{(t_2 - t_1)^2}{8}$$

$$\leq n(n-1)m\frac{\left((0.5)^l\right)^2}{8}$$

$$\leq n(n-1)m\frac{4^{-l}}{8}$$

$$\leq \varepsilon \ .$$

We can apply Theorem 5.19 to speed up substantially the algorithms for rendering and intersecting Bezier curves based on recursive subdivision. The theorem asserts that if we subdivide down to a prespecified level l, then we are guaranteed that the curve segments we generate will be approximated to within ε by the straight line segments joining their end points. Thus by subdividing to l levels, we can avoid all testing. Although the value of l given in Theorem 5.19 is necessarily conservative—some curve segments may require fewer levels of subdivision than others—nevertheless, avoiding all testing substantially speeds up standard algorithms based on recursive subdivision.

Exercises

1. Implement the recursive subdivision algorithm for rendering a Bezier curve given in Section 5.5.4.2 with and without Wang's formula. How much does Wang's formula speed up this algorithm?

2. Implement the recursive subdivision algorithm for intersecting two Bezier curves given in Section 5.5.4.2 with and without Wang's formula. How much does Wang's formula speed up this algorithm?

5.6.4   Integrating Bernstein Polynomials and Bezier Curves

Many important geometric properties of curves such as arc length and total curvature are integral properties. Here we shall use the differentiation formulas derived in Section 5.6.2 to develop integration formulas for Bernstein polynomials and Bezier curves.

We begin by computing the antiderivative of a Bernstein basis function. From Equation (5.22),

$$\frac{dB_{k+1}^{n+1}(t)}{dt} = (n+1)\{B_k^n(t) - B_{k+1}^n(t)\}$$

$$\frac{dB_{k+2}^{n+1}(t)}{dt} = (n+1)\{B_{k+1}^n(t) - B_{k+2}^n(t)\}$$

$$\vdots$$

$$\frac{dB_n^{n+1}(t)}{dt} = (n+1)\{B_{n-1}^n(t) - B_n^n(t)\}$$

$$\frac{dB_{n+1}^{n+1}(t)}{dt} = (n+1)\{B_n^n(t) - B_{n+1}^n(t)\} \ .$$

The terms on the right-hand side form a telescoping series. Summing and observing that $B_{n+1}^n(t)$ is identically zero, we obtain

$$(n+1)B_k^n(t) = \sum_{j=k+1}^{n+1} \frac{dB_j^{n+1}(t)}{dt}.$$

Integrating both sides, dividing by $n + 1$, and dropping the constant of integration yields

$$\int B_k^n(t)dt = \frac{1}{n+1} \sum_{j=k+1}^{n+1} B_j^{n+1}(t).$$

A similar argument starting the summation from $\dfrac{dB_0^{n+1}}{dt}$ shows that

$$\int B_k^n(t)dt = \frac{-1}{n+1} \sum_{j=0}^{k} B_j^{n+1}(t).$$

Definite integrals are also easy to compute. Integrating Equation (5.22) yields

$$(n+1)\int_0^1 \{B_k^n(t) - B_{k+1}^n(t)\}dt = \int_0^1 \left\{\frac{dB_{k+1}^{n+1}(t)}{dt}\right\}dt$$

$$= B_{k+1}^{n+1}(1) - B_{k+1}^{n+1}(0)$$

$$= 0 \ .$$

Therefore,

$$\int_0^1 B_k^n(t)dt = \int_0^1 B_{k+1}^n(t)dt \qquad k = 0,...,n-1.$$

Thus these definite integrals of the Bernstein basis functions are all the same; we can compute them all by computing any one. Since $B_n^n(t) = t^n$, we have

$$\int_0^1 B_n^n(t)dt = \int_0^1 t^n dt = \frac{1}{n+1}.$$

Hence

$$\int_0^1 B_k^n(t)dt = \frac{1}{n+1} \qquad k = 0,...,n. \tag{5.28}$$

Equation (5.28) has some interesting consequences. For example, we can use this formula to prove that the arc length of a Bezier curve is bounded above by the perimeter of its control polygon. For a smooth curve $B(t)$ defined on the interval $[a,b]$,

$$\text{arc length } \{B(t) = \int_a^b | B'(t) | dt.$$

Therefore, for Bezier curves we have the following result.

THEOREM 5.20   Let $B(t)$ be a Bezier curve on $[0,1]$ with control points $P_0,...,P_n$. Then

$$\int_0^1 | B'(t) | dt \le \sum_{k=0}^{n-1} | P_{k+1} - P_k |.$$

Proof By Equation (5.23) if $B(t) = \sum_k B_k^n(t)P_k$, then

$$B'(t) = n \sum_{k=0}^{n-1} B_k^{n-1}(t)(P_{k+1} - P_k).$$

Therefore, by the triangular inequality and Equation (5.28),

$$\int_0^1 | B'(t) | dt \le \int_0^1 \left| n \sum_{k=0}^{n-1} B_k^{n-1}(t)(P_{k+1} - P_k) \right| dt$$

$$\le \int_0^1 n \sum_{k=0}^{n-1} B_k^{n-1}(t) | P_{k+1} - P_k | dt$$

$$= n \sum_{k=0}^{n-1} | P_{k+1} - P_k | \int_0^1 B_k^{n-1}(t)dt$$

$$= \sum_{k=0}^{n-1} | P_{k+1} - P_k |.$$

COROLLARY 5.21

Let $B(t)$ be a Bezier curve on $[0,1]$ with control points $P_0,...,P_n$. Then length of chord $P_0 P_n \leq$ arc length $B(t) \leq$ perimeter of the control polygon.

Proof Since a Bezier curve interpolates its first and last control points, the lower bound follows because a straight line is the shortest distance between two points. The upper bound is just a restatement of Theorem 5.20.

Exercises

1. Let $B(t)$ be a Bezier curve on $[0,1]$ with control points $P_0,...,P_n$. Prove that

$$\int_0^1 B(t)dt = \text{center of mass of } \{P_0,...,P_n\}.$$

2. Prove that $\int B_k^n(t)dt = \dfrac{-1}{n+1}\sum_{j=0}^{k} B_j^{n+1}(t).$

3. Prove that

   a. $\int_0^t B_k^n(\tau)d\tau = \dfrac{1}{n+1}\sum_{j=k+1}^{n+1} B_j^{n+1}(t)$

   b. $\int_t^1 B_k^n(\tau)d\tau = \dfrac{1}{n+1}\sum_{j=0}^{k} B_j^{n+1}(t)$

4. Let $P(t)$ be a Bezier curve with control points $P_0,...,P_n$. Show that

$$\int_0^1 |P''(t)|dt \leq n\sum_{k=0}^{n-2}| P_{k+2} - 2P_{k-1} + P_k |.$$

5.7 Rational Bezier Curves

Although the Weierstrass Approximation Theorem (Section 5.5.2) guarantees that every continuous curve on a closed interval can be approximated to within any desired tolerance by a polynomial curve, these approximating polynomials may have arbitrarily high degree. Moreover, there are some simple curves like circles that we would rather not approximate, but which cannot be represented exactly by polynomial parametrizations. As we have already seen in Chapters 2 and 3 in the context of Lagrange and Hermite interpolation, by resorting to rational functions we can greatly expand the range of curves with exact representations. Here we shall introduce rational Bezier curves in much the same way that we constructed rational Lagrange curves in Chapter 2, as projections of polynomial curves from a higher-dimensional Grassmann space.

Indeed, by definition, a *rational Bezier curve* in affine space is the projection of a polynomial Bezier curve

$$P(t) = \sum_{k=0}^{n} B_k^n(t)(w_k P_k, w_k) \qquad 0 \le t \le 1$$

in Grassmann space. The polynomial curve $P(t)$ projects to the rational curve

$$R(t) = \frac{\sum_{k=0}^{n} B_k^n(t) w_k P_k}{\sum_{k=0}^{n} w_k B_k^n(t)} \qquad 0 \le t \le 1. \tag{5.29}$$

Thus to represent a rational curve $R(t)$ in Bezier form, we associate with each control point P_k a scalar mass or weight w_k. Notice that we have also associated to each point $R(t)$ on a rational Bezier curve the scalar weight $w(t)$ defined by

$$w(t) = \sum_{k=0}^{n} w_k B_k^n(t).$$

Thus a rational Bezier curve is more than just a continuous collection of points in affine space; there is also a scalar field, a mass distribution, associated with each rational Bezier curve.

If all the weights are equal and nonzero, then the rational Bezier curve $R(t)$ reduces to an ordinary Bezier curve. To distinguish these polynomial Bezier curves from rational Bezier curves, we call such polynomial curves *integral Bezier curves*.

Most of the standard properties of integral Bezier curves carry over readily to rational Bezier curves, although sometimes there are some minor restrictions on the weights. For example, it follows easily from Equation (5.29) that if $w_0 \ne 0$, then $R(0) = P_0$; similarly, if $w_n \ne 0$, then $R(1) = P_n$. Thus rational Bezier curves interpolate their first and last control points just like ordinary Bezier curves.

If the weights are all nonzero, then it is natural to write

$$R_k^n(t) = \frac{w_k B_k^n(t)}{\sum_{j=0}^{n} w_j B_j^n(t)}, \quad k = 0,\dots,n$$

$$R(t) = \sum_{k=0}^{n} R_k^n(t) P_k \ . \tag{5.30}$$

Thus for a fixed set of nonzero weights, the functions $R_k^n(t)$, $k = 0,\dots,n$, are rational blending functions, and these functions behave much like the standard Bernstein blending functions. Indeed, since the denominator is the same for all values of k, it is easy to show that the rational functions $\{R_k^n(t)\}$ incorporate many of the features of the Bernstein polynomials $\{B_k^n(t)\}$. For this reason rational Bezier curves with non-zero weights share many of the geometric properties of integral Bezier curves (see Exercises 1–5).

If a weight $w_j = 0$, then the mass-point $(w_j P_j, w_j)$ is not just discarded but rather is replaced by a vector $(v_j, 0)$. Thus, in general,

$$P(t) = \sum_{w_k \neq 0} B_k^n(t)(w_k P_k, w_k) + \sum_{w_j = 0} B_j^n(t)(v_j, 0),$$

so first adding in Grassmann space and then projecting into affine space, we arrive at the rational Bezier curve

$$R(t) = \frac{\displaystyle\sum_{w_k \neq 0} B_k^n(t) w_k P_k + \sum_{w_j = 0} B_j^n(t) v_j}{\displaystyle\sum_{w_k \neq 0} w_k B_k^n(t)}.$$

Examples abound in which it is necessary to set some weights to zero in order to represent a rational curve segment exactly in rational Bezier form (see Figure 5.33 discussed later in this section and also Exercises 8–10).

To see how the rational Bezier representation works in practice, let's consider the circle as a rational Bezier curve. Recall that the circle $x^2 + y^2 = 1$ has the rational quadratic parametrization

$$x(t) = \frac{2t}{1 + t^2} \qquad y(t) = \frac{1 - t^2}{1 + t^2}.$$

This parametrization lifts into Grassmann space to the polynomial curve

$$P(t) = \left(2t, 1 - t^2, 1 + t^2 \right).$$

To find a rational quadratic Bezier representation for the circle, we need to find its control points P_0, P_1, P_2 and scalar weights w_0, w_1, w_2.

We begin by representing the mass distribution $1 + t^2$—the denominator of the rational curve—in terms of the Bernstein basis functions. Since the Bernstein polynomials form a polynomial basis, certainly there are constants w_0, w_1, w_2 such that

$$1 + t^2 = w_0 B_0^2(t) + w_1 B_1^2(t) + w_2 B_2^2(t).$$

Moreover, it is easy to show either directly or from Equation (5.9) (with $k = 0,2$ and $n = 2$) that

$$1 + t^2 = B_0^2(t) + B_1^2(t) + 2 B_2^2(t).$$

Thus, we can just read off the weights from the coefficients of the Bernstein basis functions:

$$w_0 = 1 \qquad w_1 = 1 \qquad w_2 = 2.$$

To find the control points is also quite easy. Once more, since the Bernstein polynomials form a polynomial basis, we can certainly write

$$(2t, 1 - t^2) = w_0 P_0 B_0^2(t) + w_1 P_1 B_1^2(t) + w_2 P_2 B_2^2(t).$$

But again either directly or using Equation (5.9), we can represent the numerators of $x(t)$ and $y(t)$ in terms of the Bernstein basis functions:

$$2t = B_1^2(t) + 2B_2^2(t)$$

$$1 - t^2 = B_0^2(t) + B_1^2(t).$$

[0,1]

From these identities we can read off the xy-coordinates of $w_k P_k$:

$$w_0 P_0 = (0,1) \qquad w_1 P_1 = (1,1) \qquad w_2 P_2 = (2,0).$$

Since we already know the weights, we can now solve for the control points:

$$P_0 = (0,1) \qquad P_1 = (1,1) \qquad P_2 = (1,0).$$

This representation of the unit circle in terms of control points and weights is not unique. Clearly we can multiply all the weights by the same nonzero scalar without affecting the curve. Moreover, we could reparametrize the curve by a rational linear change of parameter $t = (au + b)/(cu + d)$, and this change of variables would affect both the control points and the weights but not the underlying curve. If we adopt our usual convention of restricting the parameter t to the interval [0,1], then the rational Bezier curve we have just constructed represents the portion of the circle in the first quadrant (see Figure 5.32).

Other segments of the circle would require a different set of control points and a different set of scalar weights. For example, if we want to represent the semicircle in rational Bezier form over the interval [0,1], we must map [0,1] $\rightarrow$ [−1,1] by sending $t \rightarrow 2t - 1$. This linear change of parameter generates the reparametrized circle

$$x(t) = \frac{2(2t-1)}{4t^2 - 4t + 2} \qquad y(t) = \frac{4t(1-t)}{4t^2 - 4t + 2},$$

which lifts in Grassmann space to the polynomial curve

$$P(t) = \left(2(2t-1), 4t(1-t), 4t^2 - 4t + 2\right).$$

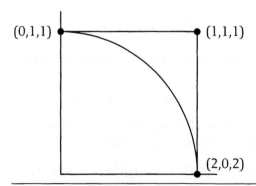

Figure 5.32 A quarter circle as a rational Bezier curve. The control points are represented with three coordinates denoting $(w_k P_k, w_k)$.

Solving as before for the scalar weights, we now find that

$$4t^2 - 4t + 2 = 2B_0^2(t) + 2B_2^2(t),$$

so

$$w_0 = 2, \quad w_1 = 0, \quad w_2 = 2.$$

Moreover, it is easy to verify that

$$2(2t - 1) = -2B_0^2(t) + 2B_2^2(t)$$
$$4t(1 - t) = 2B_1^2(t) \ .$$

Therefore,

$$w_0 P_0 = (-2,0,2) \qquad v_1 = (0,2,0) \qquad w_2 P_2 = (2,0,2).$$

We illustrate this rational Bezier representation for the semicircle in Figure 5.33.

The weights of a rational Bezier curve can be used to control its shape. As the weight w_k increases, the influence of the control point P_k increases and the curve passes closer to P_k; as w_k decreases, the curve is pushed away from P_k. Thus the weights behave like tension parameters (see Figure 5.34).

Typically all the weights are chosen to be positive to avoid singularities, but zero and negative weights are permitted and sometimes, as we have just seen, are even necessary to represent certain curves exactly. Unlike rational Lagrange or rational Hermite interpolation, the effect of increasing the weights on a rational Bezier curve is rather benign because rational Bezier curves are not constrained to interpolate their control points. Negative weights, however, may introduce singularities even if

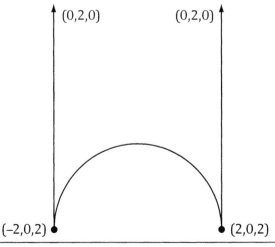

Figure 5.33 A semicircle as a rational Bezier curve. The two control points are represented with three coordinates denoting $(w_k P_k, w_k)$. Notice here the control vector $(0,2,0)$.

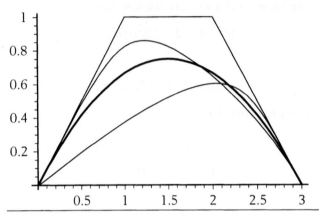

Figure 5.34 Bezier curves with fixed control points, but with different values for the weights. The dark curve is an integral cubic Bezier curve with control points at $P_0 = (0,0)$. $P_1 = (1,1)$, $P_2 = (2,1)$. $P_3 = (3,0)$. The upper and lower curves have the same control points, but in the upper curve the weight at P_1 is increased to 3, while in the lower curve the weight at P_1 is decreased to -0.05.

we restrict the parameter domain to $[0,1]$, so negative weights are generally avoided (see Exercise 12(b)).

Nevertheless, even though a rational Bezier curve is continuous everywhere except at parameter values where the denominator vanishes, and even though the effect on the shape of the curve of increasing the weights is generally benign, in the limit as a single weight approaches infinity a rational Bezier curve collapses to a disjoint collection of points. We already know that when $w_0, w_n \neq 0$ a rational Bezier curve interpolates its first and last control point, so for $t = 0,1$

$$\lim_{w_j \to \infty} R(0) = P_0$$
$$\lim_{w_j \to \infty} R(1) = P_n.$$

But for any other value of t,

$$\lim_{w_j \to \infty} R(t) = \lim_{w_j \to \infty} \frac{\sum\limits_{k=0}^{n} B_k^n(t) w_k P_k}{\sum\limits_{k=0}^{n} w_k B_k^n(t)} = \lim_{w_j \to \infty} \frac{\sum\limits_{k=0}^{n} B_k^n(t) \frac{w_k}{w_j} P_k}{\sum\limits_{k=0}^{n} \frac{w_k}{w_j} B_k^n(t)} = P_j.$$

Thus the limit curve consists of only three points.

If the mass of a Bezier curve in Grassmann space is ever zero, then the projection of the curve into affine space is not continuous. We can avoid these discontinuities by projecting the curve instead into projective space. Therefore, for a rational Bezier curve, just as for a rational Lagrange curve, the control points reside in Grassmann space, but the curve itself may lie in projective space.

Typically, algorithms for integral Bezier curves carry over directly to algorithms for rational Bezier curves because generally we can apply these algorithms separately to the numerator and denominator. For example, we can evaluate points along a rational Bezier curve by applying the de Casteljau algorithm independently to the numerator and denominator. Similarly, change of basis algorithms can be applied separately to the numerator and denominator. Therefore, the algorithms for degree elevation and subdivision can be applied independently to the numerator and denominator. In essence there is nothing new here; we simply apply the algorithm in question to the control points $(w_0 P_0, w_0), ..., (w_n P_n, w_n)$ in Grassmann space and then divide by the weight to get the desired result in affine space.

Exercises

1. Show that for rational Bezier curves reversing both the order of the control points and the order of the weights generates the same rational curve but with the opposite orientation.

2. Show that the rational blending functions defined in Equation (5.30) satisfy the identity

$$\sum_{k=0}^{n} R_k^n(t) \equiv 1.$$

3. Using Equation (5.30) and Exercise 2, show that if all the weights are positive, then a rational Bezier curve lies in the convex hull of its control points.

4. Using Equation (5.30), show that if all the weights are nonzero, then a rational Bezier is nondegenerate provided that there are no indices j,k for which $(w_k P_k, w_k) = c_{jk}(w_j P_j, w_j)$.

5. Suppose that all the weights of a rational Bezier curve are positive.

   a. Show that the rational functions in Equation (5.30) satisfy Descartes' Law of Signs in the interval $(0,1)$.

   b. Conclude that rational Bezier curves with positive weights satisfy the variation diminishing property.

6. Show that for some choices of positive weights the rational blending functions in Equation (5.30) are not unimodal in k.

7. Find control points and weights to represent the quarter circles in the second, third, and fourth quadrants as rational Bezier curves.

8. Find control points and weights to represent the lower half circle as a rational Bezier curve.

9. Apply the subdivision algorithm for rational Bezier curves to the quarter circle given in the text to derive the Bezier control points and weights for the upper half circle.

10. Find Bezier control points and scalar weights for

   a. the ellipse: $x = \dfrac{2at}{1+t^2}$ $y = \dfrac{b(1-t^2)}{1+t^2}$

b. the hyperbola: $x = \dfrac{2at}{1-t^2}$ $y = \dfrac{b(1+t^2)}{1-t^2}$

Which segments of these curves are represented by your choice of control points and weights?

11. Consider a rational Bezier curve with control points $P_0,\dots,P_n$ and nonzero weights $w_0,\dots,w_n$. What does the limit curve look like if two or more weights are allowed to approach infinity simultaneously?

12. Consider the rational cubic curves in Figure 5.34.

   a. Plot the point with $t = .99$ for larger and larger values of the weight at $P_1 = (1,1)$.

      i.   What do you observe?

      ii. Explain what is happening.

   b. Plot the curve for different negative values of the weight at $P_1 = (1,1)$.

      i.   What do you observe?

      ii. Explain what is happening.

13. Consider the conic section $R(t) = P(t)/w(t)$, where

$$P(t) = \sum_{k=0}^{2} B_k^2(t)w_k P_k \text{ and } w(t) = \sum_{k=0}^{2} w_k B_k^2(t).$$

   a. Show that $degree(w(t)) < 2 \Leftrightarrow w_0 - 2w_1 + w_2 = 0$.

   b. Conclude that if $w_0 - 2w_1 + w_2 = 0$, then

   $R(t)$ is a parabola $\Leftrightarrow w(t)$ does not divide $P(t)$.

   c. Prove that if $w_0 - 2w_1 + w_2 \neq 0$, then

      i. $w(t)$ has 2 real roots $\Leftrightarrow w_1^2 - w_0 w_2 > 0$

      ii. $w(t)$ has 1 real root $\Leftrightarrow w_1^2 - w_0 w_2 = 0$

      iii. $w(t)$ has 0 real roots $\Leftrightarrow w_1^2 - w_0 w_2 < 0$

   (Hint: Divide $w(t)$ by $(1-t)^2$.)

   d. Conclude that if $w_0 - 2w_1 + w_2 \neq 0$ and $P(t), w(t)$ have no nontrivial common factor, then

      i. $R(t)$ is a hyperbola $\Leftrightarrow w_1^2 - w_0 w_2 > 0$

      ii. $R(t)$ is a parabola $\Leftrightarrow w_1^2 - w_0 w_2 = 0$

      iii. $R(t)$ is an ellipse $\Leftrightarrow w_1^2 - w_0 w_2 < 0$

   Therefore, in general, the type of a rational quadratic Bezier curve depends only on the weights and not on the location of the control points.

   e. What is the curve $R(t)$ if $P(t)$ and $w(t)$ have a nontrivial common factor?

14. Suppose that $R(t)$, $0 \leq t \leq 1$, is a rational Bezier curve with control points $P_0, \ldots, P_n$ and weights $w_0, \ldots, w_n$. Let $\hat{R}(t)$, $0 \leq t \leq 1$, be the rational Bezier curve with control points $P_0, \ldots, P_n$ and weights $\hat{w}_k = (-1)^k w_k$, $k = 0, \ldots, n$.

    a. Show that $\hat{R}(t) = R(t)/(2t - 1)$.

    b. Conclude that the curve $\hat{R}(t)$, $t \in [0,1]$, is the same as the curve $R(t)$, $t \in \{(-\infty, \infty) - (0,1)\}$. That is, $\hat{R}(t)$ is the complement of $R(t)$.

    c. Find the control points and weights for the complement of the quarter circle depicted in Figure 5.32.

15. Consider a rational Bezier curve $R(t)$, $0 \leq t \leq 1$, with weights $w_0, \ldots, w_n$. Suppose that $w_0, w_n = 0$, $w_1, w_{n-1} \neq 0$ and that the control vectors v_0, v_n are the zero vector. Show that

    a. $\lim_{t \to 0} R(0) = P_1$

    b. $\lim_{t \to 1} R(1) = P_{n-1}$

16. Given a rational Bezier curve with control points $P_0, \ldots, P_n$ and weights $w_0, \ldots, w_n$, use the known formulas for integral Bezier curves to derive explicit formulas for the control points and weights of

    a. the degree elevated rational curve

    b. the subdivided rational curve

17. Implement the following algorithms for rational Bezier curves:

    a. de Casteljau evaluation algorithm

    b. degree elevation algorithm

    c. subdivision algorithm

18. Implement the recursive subdivision algorithm for

    a. rendering a rational Bezier curve

    b. intersecting two rational Bezier curves

19. Let $R(t)$ be a rational Bezier curve with control points $P_0, \ldots, P_n$ and positive weights $w_0, \ldots, w_n$. Show that the arc length of $R(t)$ is bounded below by the length of the chord joining P_0 and P_n and above by the perimeter of its control polygon.

    (Hint: Use the result for integral Bezier curves and the fact that recursive subdivision commutes with projection.)

20. Let $R(t)$ be a rational Bezier curve with control points $P_0, \ldots, P_n$ and weights $w_0, \ldots, w_n$. Define

$$P(t) = \sum_{k=0}^{n} B_k^n(t) w_k P_k \quad \text{and} \quad w(t) = \sum_{k=0}^{n} w_k B_k^n(t),$$

so that $R(t) = \dfrac{P(t)}{w(t)}$. Prove that

    a. $\int_0^1 w(t) dt$ = average of the weights

b. $\dfrac{\int_0^1 P(t)dt}{\int_0^1 w(t)dt}$ = center of mass of $\{(w_0 P_0, w_0),...,(w_n P_n, w_n)\}$

5.7.1 Differentiating Rational Bezier Curves

One algorithm that cannot be applied independently to the numerator and denominator of a rational Bezier curve is the algorithm for differentiating of Bezier curves because the derivative of a quotient is not equal to the quotient of the derivatives. To find the derivatives of a rational Bezier curve, we proceed in the following manner. Let

$$R(t) = \frac{\sum_{k=0}^{n} B_k^n(t) w_k P_k}{\sum_{k=0}^{n} w_k B_k^n(t)}$$

be a rational Bezier curve. Then the numerator and denominator of $R(t)$ behave like ordinary Bezier curves, which we already know how to differentiate. Let

$$P(t) = \sum_{k=0}^{n} B_k^n(t) w_k P_k$$

$$w(t) = \sum_{k=0}^{n} w_k B_k^n(t) .$$

Then $R(t) = P(t)/w(t)$, so multiplying both sides by $w(t)$ we obtain

$$w(t)R(t) = P(t).$$

To find the derivatives of $R(t)$, we proceed recursively using Leibniz's rule:

$$\sum_{i=0}^{r} \binom{r}{i} w^{(i)}(t) R^{(r-i)}(t) = P^{(r)}(t). \tag{5.31}$$

By Equation (5.24), we have

$$P^{(r)}(t) = \frac{n!}{(n-r)!} \sum_{k=0}^{n-r} B_k^{n-r}(t) \sum_{j=0}^{r} (-1)^{r-j} \binom{r}{j} w_{j+k} P_{j+k}$$

$$w^{(i)}(t) = \frac{n!}{(n-i)!} \sum_{k=0}^{n-i} B_k^{n-i}(t) \sum_{j=0}^{i} (-1)^{i-j} \binom{i}{j} w_{j+k}.$$

Hence if we know $R^{(p)}(t)$ for $0 \le p < r$, then we can apply Equation (5.31) to compute $R^{(r)}(t)$. Indeed, we can find $R^{(r)}(t)$ by using Equation (5.31) with $r = 1$ to find $R'(t)$, then with $r = 2$ to find $R''(t)$, and so on, till finally we can apply this equation to compute $R^{(r)}(t)$. Notice too that $P^{(i)}(t), w^{(i)}(t)$ can be computed algorithmically by differentiating i rows of the de Casteljau algorithm for the control points $(w_0 P_0, w_0),...,(w_n P_n, w_n)$.

Suppose now that we are given a rational Bezier curve $R(t)$ with control points $(w_0 P_0, w_0), \ldots, (w_n P_n, w_n)$ and we want to construct another rational Bezier curve $C(t)$ with control points $(v_0 Q_0, v_0), \ldots, (v_n Q_n, v_n)$ that meets $R(t)$ and matches its first r derivatives at its end point. For arbitrary r this problem is difficult, so let's consider only the cases $r = 0, 1$. If the weights are nonzero, then by Equation (5.29) $C(0) = Q_0$ and $R(1) = P_n$, so continuity requires that $Q_0 = P_n$ just as for integral Bezier curves. Moreover, by Equation (5.31)

$$w'(1)R(1) + w(1)R'(1) = P'(1),$$

so substituting for the values of w and P we have

$$n(w_n - w_{n-1})P_n + w_n R'(1) = n(w_n P_n - w_{n-1}P_{n-1}).$$

Solving for $R'(1)$, we obtain

$$R'(1) = n \frac{w_{n-1}}{w_n}(P_n - P_{n-1}).$$

Similarly, applying Equation (5.31) to $C(t)$, we get

$$v'(0)C(0) + v(0)C'(0) = Q'(0).$$

Substituting for the values of v and Q yields

$$n(v_1 - v_0)Q_0 + v_0 C'(0) = n(v_1 Q_1 - v_0 Q_0),$$

and solving for $C'(0)$ gives

$$C'(0) = n \frac{v_1}{v_0}(Q_1 - Q_0).$$

Therefore, $R(t)$ and $C(t)$ will meet with one continuous derivative if and only if

$$C(0) = R(1) \implies Q_0 = P_n$$

$$C'(0) = R'(1) \implies n \frac{v_1}{v_0}(Q_1 - Q_0) = n \frac{w_{n-1}}{w_n}(P_n - P_{n-1}).$$

Substituting $Q_0 = P_n$ and solving for Q_1, we find that the second condition reduces to

$$Q_1 = P_n + \frac{v_0 w_{n-1}}{v_1 w_n}(P_n - P_{n-1}).$$

There are three free parameters in this equation: Q_1, v_0, v_1. Continuity of first derivatives does not specify the values of these parameters; rather it specifies only that these parameters must be in some specific relationship. Thus unlike the polynomial case, the location of Q_1 is not fixed by insisting on continuity of first derivatives. Notice, however, that just as in the polynomial case, Q_1 must still lie somewhere along the line joining P_{n-1} and P_n.

Necessary and sufficient conditions for continuity of the first r derivatives for arbitrary values of r are difficult to obtain because we do not have a simple explicit formula for the rth derivative of a rational Bezier curve. However, sufficient conditions are easy to derive. The rth derivative of a quotient depends only on the first r derivatives of the numerator and denominator. Thus if $R(t) = P(t)/w(t)$ and $C(t) = Q(t)/v(t)$, then sufficient conditions for $R^{(i)}(1) = C^{(i)}(0)$, $i = 0,...,n$, are

$$P^{(i)}(1) = Q^{(i)}(0) \qquad\qquad i = 0,...,n$$

$$w^{(i)}(1) = v^{(i)}(0) \qquad\qquad i = 0,...,n.$$

Since $P(t), Q(t), w(t), v(t)$ can be thought of as integral Bezier curves, we can use the methods of Section 5.6.2 (e.g., Equation (5.27)) to find values for Q_k and v_k that guarantee these continuity conditions are satisfied.

Exercises

1. Let $R(t)$ be a rational Bezier curve with control points $P_0,...,P_n$ and positive weights $w_0,...,w_n$.

    a. Find a bound on $R''(t)$.

    b. Find an analogue of Wang's formula for the number of levels of subdivision required so that the control polygon approximates the rational curve within a fixed tolerance ε.

    c. Develop an algorithm for intersecting two rational Bezier curves with positive weights based on recursive subdivision and Wang's formula.

2. Let $R(t)$ be a rational Bezier curve with control points $P_0,...,P_n$ and nonzero weights $w_0,...,w_n$.

    a. Compute explicit formulas for $R''(t), R''(0), R''(1)$.

    b. Find necessary and sufficient conditions on the control points and the weights for two rational Bezier curves to meet with two continuous derivatives at their end points.

3. Let $R(t)$ be a rational Bezier curve with control points $(w_0 P_0, w_0),...,$ $(w_n P_n, w_n)$.

    a. Using the formulas for the first derivative derived in the text, show that

    $$\lim\nolimits_{w_0 \to \infty} R'(0) = 0 \text{ and } \lim\nolimits_{w_n \to \infty} R'(1) = 0.$$

    b. More generally, show that

    $$\lim\nolimits_{w_0 \to \infty} R^{(k)}(0) = 0 \text{ and } \lim\nolimits_{w_n \to \infty} R^{(k)}(1) = 0, \, k \geq 1.$$

    c. Show that if $t \neq 0,1$, then $\lim\nolimits_{w_j \to \infty} R^{(k)}(t) = 0$ for all $k \geq 1$.

4. The formulas for the unit tangent $U(t)$, the curvature $K(t)$, and the torsion $T(t)$ of a parametric curve $P(t)$ are given in Section 5.6.2, Exercise 12.

    a. Compute the unit tangent, curvature, and torsion of a rational Bezier curve at $t = 0,1$.

    b. Find conditions on the control points and weights of a rational Bezier curve $C(t)$ so that it matches a given rational Bezier curve $B(t)$ with continuous unit tangent, curvature, and torsion.

5.8 Bezier Surfaces

Bezier surface patches come in two standard shapes: rectangular and triangular. We have already encountered both rectangular and triangular patches in the context of Lagrange interpolation in Chapter 2; here we explore the corresponding constructions for Bezier approximation. Just as de Casteljau's algorithm for Bezier curves is simpler than Neville's algorithm for Lagrange polynomials, so too are Bezier surfaces simpler to define and analyze than the corresponding Lagrange surfaces because the absence of nodes in Bezier approximation simplifies the domains of these surfaces as well as many of their associated algorithms.

We begin our study of Bezier surfaces with tensor product Bezier patches and then go on to explore the corresponding triangular patches. We also examine briefly rational Bezier patches, both rectangular and triangular. Rational Bezier patches with more than four sides are investigated in Chapter 8.

5.8.1 Tensor Product Bezier Patches

A rectangular tensor product Bezier patch $B(s,t)$ of bidegree (m,n) is defined by setting

$$B(s,t) = \sum_{i=0}^{m} \sum_{j=0}^{n} B_i^m(s) B_j^n(t) P_{ij} \qquad 0 \le s,t \le 1. \tag{5.32}$$

The functions $B_i^m(s)B_j^n(t)$—where $B_i^m(s)$ and $B_j^n(t)$ are the standard Bernstein basis functions of degree m and n in the parameters s and t—are the tensor product Bernstein basis functions (see Figure 5.35).

The rectangular array of control points $\{P_{ij}\}$ generates a control polyhedron for the tensor product Bezier patch that controls the shape of the Bezier patch in much the same way that the Bezier control polygon controls the shape of a Bezier curve (see Figures 5.36 and 5.37). In particular, dragging a control point pulls the surface patch in the same general direction as the control point.

Let $P_i(t)$ be the Bezier curve with control points $P_{i0},...,P_{in}$. Then

$$P_i(t) = \sum_{j=0}^{n} B_j^n(t) P_{ij} \qquad\qquad i = 0,...,m \tag{5.33}$$

$$B(s,t) = \sum_{i=0}^{m} B_i^m(s) P_i(t) \tag{5.34}$$

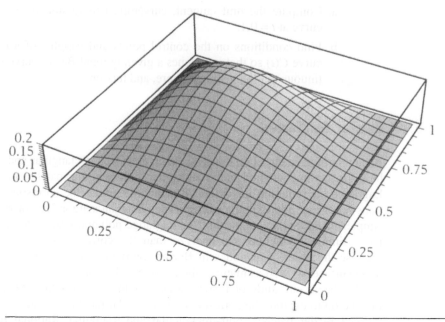

Figure 5.35 The bicubic Bernstein basis function $B_1^3(s)B_2^3(t)$. Compare to the bicubic Lagrange basis function $L_1^3(s)L_2^3(t)$ in Figure 2.20.

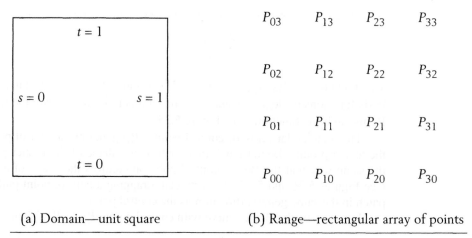

(a) Domain—unit square   (b) Range—rectangular array of points

Figure 5.36 Data for a bicubic tensor product Bezier patch. Notice that the domain is simply the unit square and that, unlike Lagrange interpolation, the domain has no nodes and no grid. Compare to Figure 2.19.

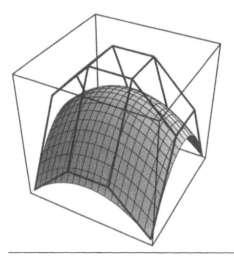

Figure 5.37 A bicubic tensor product Bezier surface with its control polyhedron, formed by connecting control points with adjacent indices. The control points are the same as those for the Lagrange surface in Figure 2.22.

Thus if we fix t, then $B(s,t)$ is the Bezier curve with contro   oints $P_0(t),...,P_m(t)$ (see Figures 5.38 and 5.39).

Equations (5.33) and (5.34) suggest the following evaluation algorithm for tensor product Bezier patches: first use de Casteljau's algorithm $m + 1$ times to compute the points at parameter t along the degree n Bezier curves $P_0(t),...,P_m(t)$; then use de Casteljau's algorithm one more time to compute the point at parameter s along the degree m Bezier curve with control points $P_0(t),...,P_m(t)$ (see Figure 5.40).

Just as in Lagrange tensor product interpolation, there is also an alternative evaluation algorithm for tensor product Bezier patches based on a bilinear recurrence. For simplicity, let us assume that $m = n$. Multiplying together the linear recurrences

$$B_i^n(s) = (1-s)B_i^{n-1}(s) + sB_{i-1}^{n-1}(s)$$

$$B_j^n(t) = (1-t)B_j^{n-1}(t) + tB_{j-1}^{n-1}(t) \ ,$$

generates the bilinear recurrence

$$B_i^n(s)B_j^n(t) = (1-s)(1-t)B_i^{n-1}(s)B_j^{n-1}(t) + s(1-t)B_{i-1}^{n-1}(s)B_j^{n-1}(t)$$

$$+ (1-s)tB_i^{n-1}(s)B_{j-1}^{n-1}(t) + stB_{i-1}^{n-1}(s)B_{j-1}^{n-1}(t) \ .$$

This recurrence can be diagrammed on a square pyramid, with $(n+1)^2$ nodes on the nth level of the pyramid, by placing a 1 at the apex of the pyramid and the four functions $(1-s)(1-t)$, $s(1-t)$, $(1-s)t$, st along the edges connecting the node labeled (i,j) on the nth level respectively to the nodes labeled (i,j), $(i-1,j)$,

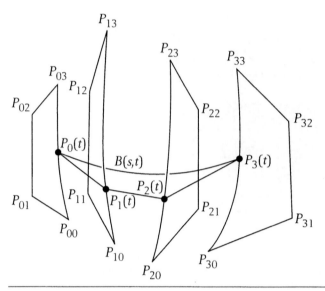

Figure 5.38 A schematic construction for points on a bicubic tensor product Bezier surface $B(s,t)$. First the Bezier curves $P_i(t)$, $i = 0,...,3$ are constructed from the control points $P_{i0}, P_{i1}, P_{i2}, P_{i3}$. Then for a fixed value of t, the Bezier curve $B(s,t)$ is constructed using the points $P_0(t), P_1(t), P_2(t), P_3(t)$ as control points. In general, the Bezier surface $B(s,t)$ does not interpolate its control points. Compare to Figure 2.21   for bicubic Lagrange interpolation.

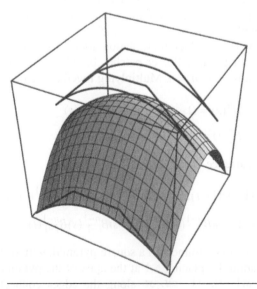

Figure 5.39 The bicubic Bezier patch in Figure 5.37 along with its cubic Bezier control curves. Notice that only the boundary control curves are interpolated by the surface. Compare to the tensor product Lagrange surface in Figure 2.22.

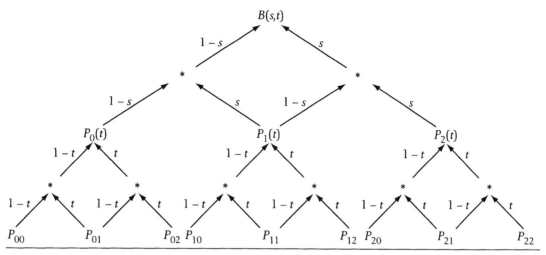

Figure 5.40 De Casteljau's evaluation algorithm for a biquadratic Bezier patch. The three lower triangles represent Bezier curves in the t direction, and the upper triangle blends these curves in the s direction. Compare to Figure 2.23—Neville's algorithm for a biquadratic interpolating patch.

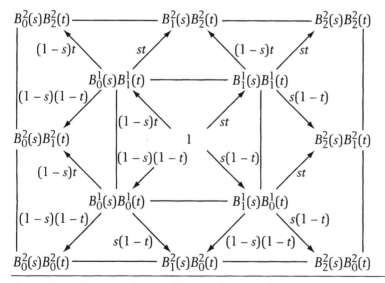

Figure 5.41 The pyramid algorithm for the biquadratic Bernstein basis functions viewed from above. The function $B_1^2(s) B_1^2(t)$ on the base is not shown, since it is obscured by the upper portions of the pyramid.

$(i, j-1)$, $(i-1, j-1)$ on the $(n-1)$st level. It follows by the bilinear recurrence and induction on n that the functions $\{B_i^n(s)B_j^n(t)\}$ emerge on the nth level of the pyramid (see Figure 5.41).

Thus the function $B_i^n(s)B_j^n(t)$ is the sum over all paths between the node (i,j) on the nth level and the 1 at the apex of the pyramid. Therefore, if we place the control points P_{ij} at the base of the pyramid and reverse all the arrows in the diagram, the tensor product patch

$$B(s,t) = \sum_i \sum_j B_i^m(s)B_j^n(t)P_{ij}$$

will emerge at the apex of the pyramid. This is the bilinear evaluation algorithm for tensor product Bezier patches (see Figure 5.42).

Both the de Casteljau algorithm and the pyramid algorithm are $O(n^3)$. Nevertheless, the de Casteljau algorithm is generally faster than the pyramid algorithm. The analysis here is much the same as the comparison between Neville's algorithm and the pyramid algorithm for tensor product Lagrange interpolation given in Section 2.11. When $m = n$, the de Casteljau algorithm employs $n + 2$ triangles: $n + 1$ triangles in the t direction and one additional triangle in the s direction. Each triangle has $n(n+1)/2$ nodes, and each node requires two multiplications. Therefore,

number of multiplications in the de Casteljau algorithm $= n(n+1)(n+2)$.

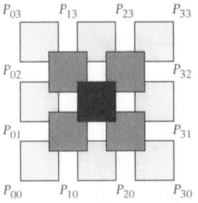

Figure 5.42 A schematic diagram of the bilinear evaluation algorithm for a bicubic tensor product Bezier patch viewed from above. Each panel represents the computation of a point at its center by multiplying the points at its corners with the functions $(1-s)(1-t)$, $s(1-t)$, $(1-s)t$, st and adding the results. The black panel represents the bicubic Bezier patch for the control points at the base of the pyramid; interior control points are obscured by the panels. Notice that the light gray panels represent bilinear Bezier patches and the dark gray panels represent biquadratic Bezier patches. Compare to Figure 2.25, which is the pyramid algorithm for bicubic Lagrange interpolation. The same pyramid is used there, but the algorithm here is much simpler. In the pyramid algorithm for bicubic Lagrange interpolation, the rectangular domain for the interpolation algorithm varies from node to node and level to level, so the labels along the edges also vary from node to node and level to level. For bicubic Bezier patches the same rectangular domain is used at every node and every level, so the labels along the edges are the same from node to node and level to level.

On the other hand, the pyramid algorithm has

$$\sum_{k=1}^{n} k^2 = n(n+1)(2n+1)/6$$

nodes, and each node requires four multiplications. Therefore,

$$\text{number of multiplications in the pyramid algorithm} = \frac{2n(n+1)(2n+1)}{3}.$$

Since $n+2 \le (4n+2)/3$ for $n \ge 4$, the de Casteljau algorithm is generally faster than the pyramid algorithm, though for the most common surfaces, namely, bicubic patches, $n = 3$ and the pyramid algorithm is slightly more efficient.

As in tensor product Lagrange interpolation, the de Casteljau algorithm has another advantage over the pyramid algorithm that is even more substantial. Recall that surfaces are typically rendered by generating points on the surface along isoparameter lines—that is, along lines of constant s or t. If we fix $t = t*$ and vary only s, then we can reuse the computation of the points $P_0(t*),...,P_m(t*)$. Thus along isoparameter lines, de Casteljau's algorithm for tensor product surfaces reduces to the univariate version of de Casteljau's algorithm, which is only $O(n^2)$. No such reduction occurs for the pyramid algorithm along isoparameter lines.

Tensor product Bezier patches inherit many of the characteristic properties of Bezier curves; they are affine invariant, nondegenerate, and lie in the convex hull of their control points (see Exercises 1 and 5). These properties follow easily from Equations (5.33) and (5.34) and the corresponding properties of Bezier curves. Moreover, the boundaries of a tensor product Bezier patch are the Bezier curves determined by their boundary control points, since by Equation (5.34)

$$B(0,t) = P_0(t) = \sum_{j=0}^{n} B_j^n(t) P_{0j}$$

$$B(1,t) = P_m(t) = \sum_{j=0}^{n} B_j^n(t) P_{mj},$$

and symmetric results hold for the boundaries $t = 0$ and $t = 1$. It follows that although tensor product Bezier patches do not generally interpolate their control points, they always interpolate the four corner points $P_{00}, P_{m0}, P_{0n}, P_{mn}$.

One property, however, that does not carry over from curves to surfaces is the variation diminishing property. *There is no known analogue of the variation diminishing property for tensor product Bezier patches.* For example, it is not true that the number of intersections of a Bezier patch with a straight line is always less than or equal to the number of intersections of the same line and the Bezier control polyhedron (see Exercise 22). Nor is it true that a plane always splits a Bezier surface into fewer connected components than it splits the corresponding control polyhedron (see Exercise 23). Geometrically it is difficult to discover an analogue of the variation diminishing property for surfaces because subdivision and degree elevation for surfaces (see below) are not simply vertex slicing procedures for polyhedra. Moreover, algebraically there is no simple analogue of Descartes' Law of Signs in the bivariate setting.

Tensor product Bezier patches do inherit many of the standard algorithms of Bezier curves. Degree elevation and subdivision can be performed independently in each variable. To degree elevate the surface $B(s,t)$ in t, simply degree elevate each of the curves $P_0(t),...,P_m(t)$. Similarly, to subdivide $B(s,t)$ at $t = r$, subdivide each of the curves $P_0(t),...,P_m(t)$ at $t = r$. Symmetric algorithms can be used to degree elevate and subdivide with respect to s instead of t.

To differentiate a tensor product Bezier patch, we can differentiate either de Casteljau's algorithm or the pyramid algorithm. To differentiate the de Casteljau algorithm with respect to s a total of p times, simply differentiate any p of the upper m levels (the s levels) of the de Casteljau algorithm and multiply the result by $m!/(m - p)!$. To differentiate q times with respect to t, differentiate any q of the n levels (the t levels) in each of the m lower triangles (see Figure 5.40) and multiply the results by $n!/(n - q)!$. That this algorithm works is an immediate consequence of the corresponding differentiation algorithm for Bezier curves discussed in Section 5.6.2. Explicit formulas for these derivatives can be generated as well from Equation (5.24) by substituting this explicit formula for the derivatives of a Bezier curve into Equation (5.34).

Differentiation of the pyramid algorithm works in a similar fashion. To find p derivatives with respect to s and q derivatives with respect to t, differentiate any p levels of the algorithm with respect to s—that is, replace $(1 - s)(1 - t) \rightarrow -(1 - t)$, $(1 - s)t \rightarrow -t$, $s(1 - t) \rightarrow (1 - t)$, $st \rightarrow t$ on p levels of the algorithm—then differentiate any q levels (the same or different from the previous p levels) with respect to t, and multiply the result by $(n!)^2/(n - p)!(n - q)!$. This algorithm works because

$$
\left(B_i^m(s)B_j^n(t)\right) = \underbrace{\begin{pmatrix} (1-s)(1-t) & (1-s)t \\ s(1-t) & st \end{pmatrix} \otimes \cdots \otimes \begin{pmatrix} (1-s)(1-t) & (1-s)t \\ s(1-t) & st \end{pmatrix}}_{n \ factors}
$$

$$
= \underbrace{\begin{pmatrix} 1-s \\ s \end{pmatrix} \otimes \cdots \otimes \begin{pmatrix} 1-s \\ s \end{pmatrix}}_{n \ factors} \otimes \underbrace{\begin{pmatrix} 1-t & t \end{pmatrix} \otimes \cdots \otimes \begin{pmatrix} 1-t & t \end{pmatrix}}_{n \ factors}
$$

(see Exercise 15), so we can apply Equation (5.21) and the commutativity of discrete convolution to differentiate the basis functions $B_i^n(s)B_j^n(t)$.

Exercises

1. a. Prove that $\displaystyle\sum_{i=0}^{m} \sum_{j=0}^{n} B_i^m(s)B_j^n(t) \equiv 1$.

    b. Show that every tensor product Bezier patch lies in the convex hull of its control points.

2. Consider a tensor product Bezier patch of bidegree (m,n), where $m < n$. Show that

a. To compute a single point on the surface it is faster to apply de Casteljau's algorithm first in the s direction and then in the t direction.

b. To compute many points along the surface it may be faster to apply de Casteljau's algorithm first in the t direction and then in the s direction.

c. Explain this apparent anomaly.

(Hint: Compare to Section 2.11, Exercise 3.)

3. Complete the analysis of the pyramid algorithm by showing how to implement this algorithm when the degree in s is different from the degree in t.

4. What are the up and down recurrences in the case of tensor product Bezier patches for de Casteljau's algorithm and for the pyramid algorithm?

5. Give necessary and sufficient conditions on the control points for the tensor product Bezier surface to collapse to

a. a single point

b. a line

c. a plane

Justify your answer. (Compare to Section 2.11, Exercise 8.)

6. Let $B[P_{ij}^{mn}](s,t)$ denote the tensor product Bezier patch of bidegree (m,n) with control points $\{P_{ij}\}$. Show that tensor product Bezier patches have the following symmetry properties:

a. $B[P_{ji}^{nm}](t,s) = B[P_{ij}^{mn}](s,t).$

b. $B[P_{i,n-j}^{mn}](s,t) = B[P_{ij}^{mn}](s,1-t).$

c. $B[P_{m-i,j}^{mn}](s,t) = B[P_{ij}^{mn}](1-s,t).$

7. Let $B(s,t) = \sum_i \sum_j B_i^m(s)B_j^n(t)P_{ij}$ be a tensor product Bezier patch, and let

$$B^{(p,q)}(s,t) = \frac{\partial^{p+q}B}{\partial s^p \partial t^q}.$$

a. Show that $B^{(1,0)}(0,0) = m(P_{10} - P_{00})$ and $B^{(0,1)}(0,0) = n(P_{01} - P_{00})$.

b. Conclude that the normal vector at $B(0,0)$ is parallel to
$(P_{10} - P_{00}) \times (P_{01} - P_{00}).$

c. Find the normal vectors at $B(0,1), B(1,0), B(1,1)$.

8. Let $B(s,t) = \sum_i \sum_j B_i^m(s)B_j^n(t)P_{ij}$ be a tensor product Bezier patch, and let $B^{(p,q)}(s,t)$ denote the partial derivatives of $B(s,t)$ as in Exercise 7. Compute explicit expressions for the partial derivatives $B^{(1,0)}(s,t),...,B^{(r,0)}(s,t)$.

9. Suppose we have a tensor product Bezier patch

$$B(s,t) = \sum_i \sum_j B_i^m(s)B_j^n(t)P_{ij}$$

and we want to construct another tensor product Bezier patch

$$C(s,t) = \sum_i \sum_j B_i^m(s) B_j^n(t) Q_{ij}$$

to meet $B(s,t)$ along the boundary $B(1,t)$ and match its first p cross-boundary derivatives. That is, we want to construct a tensor product Bezier patch $C(s,t)$ such that

$$\frac{\partial^r C(0,t)}{\partial s^r} = \frac{\partial^r B(1,t)}{\partial s^r}, \quad r = 0,...,p.$$

a. Show that

    i. $p = 0 \implies Q_{0k} = P_{mk}, \quad k = 0,...,n$

    ii. $p = 1 \implies Q_{1k} = P_{mk} + (P_{mk} - P_{m-1,k}), \quad k = 0,...,n$

b. Derive formulas for the location of the control points Q_{ij} that guarantee

$$\frac{\partial^r C(0,t)}{\partial s^r} = \frac{\partial^r B(1,t)}{\partial s^r}, \quad r = 0,...,p.$$

10. Implement both de Casteljau's algorithm and the pyramid algorithm for tensor product Bezier surfaces. Which algorithm do you prefer? Why? Experiment with tensor product surfaces of different degrees. Determine how changing the location of the control points affects the shape of the surface.

11. Implement the recursive subdivision algorithm for tensor product Bezier patches. Apply this algorithm to

    a. render a tensor product Bezier patch

    b. intersect two tensor product Bezier patches

12. Recall from Section 5.6.3 that we can speed up recursive subdivision for Bezier curves by applying Wang's formula.

    a. Develop an analogue of Wang's formula for tensor product Bezier surfaces.

    b. Implement the recursive subdivision algorithm for rendering a tensor product Bezier surface with and without Wang's formula.

    c. How much does Wang's formula speed up this algorithm?

13. Prove that the control polyhedra generated by recursive subdivision converge to the tensor product Bezier patch.

14. Prove that the control polyhedra generated by degree elevation converge to the tensor product Bezier patch.

15. Define the convolution of two doubly indexed arrays of functions $A(s,t) = \{A_{ij}(s,t)\}$ and $B(s,t) = \{B_{kl}(s,t)\}$ by setting

$$(A \otimes B)_{pq}(s,t) = \sum_{i+k=p} \sum_{j+l=q} A_{ij}(s,t) B_{kl}(s,t).$$

Now introduce the following indexing scheme:

Function	Index
$1-s$	$(0,0)$
$1-t$	$(0,0)$
s	$(1,0)$
t	$(0,1)$
$B_i^m(s)$	$(i,0)$
$B_j^n(t)$	$(0,j)$
$B_i^m(s)B_j^n(t)$	(i,j)

Show that with this indexing scheme

a.
$$\begin{pmatrix} (1-s)(1-t) & (1-s)t \\ s(1-t) & st \end{pmatrix} = \begin{pmatrix} 1-s \\ s \end{pmatrix} \otimes \begin{pmatrix} 1-t & t \end{pmatrix};$$

b.
$$\left(B_i^m(s)B_j^n(t)\right) = \begin{pmatrix} B_0^m(s) \\ \vdots \\ B_m^m(s) \end{pmatrix} \otimes \begin{pmatrix} B_0^n(t) & \cdots & B_n^n(t) \end{pmatrix};$$

c.
$$\left(B_i^m(s)B_j^n(t)\right) = \underbrace{\begin{pmatrix} 1-s \\ s \end{pmatrix} \otimes \cdots \otimes \begin{pmatrix} 1-s \\ s \end{pmatrix}}_{n \ factors} \otimes \underbrace{\begin{pmatrix} 1-t & t \end{pmatrix} \otimes \cdots \otimes \begin{pmatrix} 1-t & t \end{pmatrix}}_{n \ factors}$$

d.
$$\left(B_i^m(s)B_j^n(t)\right) = \underbrace{\begin{pmatrix} (1-s)(1-t) & (1-s)t \\ s(1-t) & st \end{pmatrix} \otimes \cdots \otimes \begin{pmatrix} (1-s)(1-t) & (1-s)t \\ s(1-t) & st \end{pmatrix}}_{n \ factors}$$

e. Use the preceding results to derive the differentiation procedure for the pyramid algorithm described in the text. In particular, show that to compute p derivatives with respect to s and q derivatives with respect to t, we can differentiate any p levels of the pyramid algorithm with respect to s, then differentiate any q levels (the same or different from the previous p levels) with respect to t, and multiply the result by

$$\frac{(n!)^2}{(n-p)!(n-q)!}.$$

16. Describe the properties of the surface you would generate if you replace the Lagrange basis with the Bernstein basis in the Boolean sum construction given in Section 2.15.

17. Let $P_j(s) = \sum_i B_i^m(s)P_{ij}$, $j = 0,...,n$, be a sequence of Bezier curves. Show how to combine Neville's algorithm and the de Casteljau algorithm to generate a surface $C(s,t)$ that interpolates the curves $P_0(s),...,P_n(s)$ at the parameter values $t_0,...,t_n$.

18. Let

$$B(s,t) = \sum_i \sum_j B_i^n(s)B_j^n(t)P_{ij}$$

be a tensor product Bezier patch of bidegree (n,n), and let

$$B_{00}^{n-1}(s,t), B_{10}^{n-1}(s,t), B_{01}^{n-1}(s,t), B_{11}^{n-1}(s,t)$$

denote the four values computed by the pyramid algorithm on the penultimate level just below the apex of the pyramid. Show that

a. $B(s,t) = (1-s)(1-t)B_{00}^{n-1}(s,t)$

$\quad + s(1-t)B_{10}^{n-1}(s,t) + (1-s)tB_{01}^{n-1}(s,t) + stB_{11}^{n-1}(s,t)$

b. $\dfrac{\partial B}{\partial s} = n(1-t)\Big(B_{10}^{n-1}(s,t) - B_{00}^{n-1}(s,t)\Big) + nt\Big(B_{11}^{n-1}(s,t) - B_{01}^{n-1}(s,t)\Big)$

c. $\dfrac{\partial B}{\partial t} = n(1-s)\Big(B_{01}^{n-1}(s,t) - B_{00}^{n-1}(s,t)\Big) + ns\Big(B_{11}^{n-1}(s,t) - B_{10}^{n-1}(s,t)\Big)$

d. Conclude that to compute point values and normal vectors at any point on a tensor product Bezier surface of bidegree (n,n) by the pyramid algorithm costs

$$\frac{2n(n+1)(2n+1)}{3} + 6$$

multiplications and one cross product.

19. Let

$$B(s,t) = \sum_i \sum_j B_i^n(s)B_j^n(t)P_{ij}$$

be a tensor product Bezier patch of bidegree (n,n), and let $B_0^{n-1}(s,t), B_1^{n-1}(s,t)$ denote the two values computed by the de Casteljau algorithm on the penultimate level just below the apex of the algorithm. Show that

a. $B(s,t) = (1-s)B_0^{n-1}(s,t) + sB_1^{n-1}(s,t)$

b. $\dfrac{\partial B}{\partial s} = n\Big(B_1^{n-1}(s,t) - B_0^{n-1}(s,t)\Big)$

Let $P_{00}(t), P_{01}(t),...,P_{n0}(t), P_{n1}(t)$ denote the values computed by the de Casteljau algorithm on the penultimate level for the curves

$$P_i(t) = \sum_{j=0}^{n} B_j^n(t)P_{ij}, \quad i = 0,...,n.$$

   c. Show that $\partial B / \partial t$ can be computed from these values using only an additional $n(n+1)$ multiplications.

   d. Conclude that to compute point values and normal vectors at any point on a bidegree (n,n) tensor product Bezier surface by the de Casteljau algorithm costs a total of $n(n+1)(n+3)+1$ multiplications and one cross product, but that along isoparameter lines de Casteljau's algorithm for computing points and normals is only $O(n^2)$.

20. Show that

   a. The pyramid algorithm is more efficient than the de Casteljau algorithm for computing point values and normal vectors at a single point when $2 \leq n \leq 6$.

   b. The de Casteljau algorithm is more efficient than the pyramid algorithm for computing point values and normal vectors at a single point when $n > 6$.

   c. Along isoparameter lines, the de Casteljau algorithm for computing point values and normal vectors is more efficient than the pyramid algorithm for all values of n.

   (Hint: Compare the results of Exercises 18(d) and 19(d).)

21. a. Show that any level of bilinear interpolation in the pyramid algorithm can be replaced by one level of linear interpolation in s followed by one level of linear interpolation in t.

   b. Draw the diagram of the evaluation algorithm for bicubic Bezier patches where the second level of the pyramid algorithm is replaced by one level of linear interpolation in s followed by one level of linear interpolation in t.

   c. What would the evaluation algorithm look like for tensor product Bezier patches if each level of bilinear interpolation in the pyramid algorithm is replaced by two successive levels of linear interpolation?

   d. Which algorithm is more efficient: the pyramid algorithm or the algorithm where each level of bilinear interpolation is replaced by two successive levels of linear interpolation?

22. Give an example to show that the number of intersections of a tensor product Bezier patch with a straight line may be greater than the number of intersections of the same line and the corresponding Bezier control polyhedron.

23. Give an example to show that a plane may split a tensor product Bezier surface into more connected components than it splits the corresponding control polyhedron.

5.8.2 Triangular Bezier Patches

We can also define Bezier patches over triangular domains. To construct these triangular surfaces, we proceed by simplifying Neville's tetrahedral algorithm for interpolation of a triangular array of control points over a triangular grid (Section 2.12). In Neville's algorithm, for triangular Lagrange interpolation we apply barycentric coordinates at each node of the algorithm to build higher-order interpolants from lower-order interpolants. The domain triangles, and hence too the barycentric coordinates, vary from node to node and depend on the position of the node in the tetrahedron. In de Casteljau's tetrahedral algorithm for triangular Bezier patches, the domain is simplified from a triangular grid to a single triangle (see Figure 5.43), so we apply the same barycentric coordinates at each node of the algorithm (see Figure 5.44).

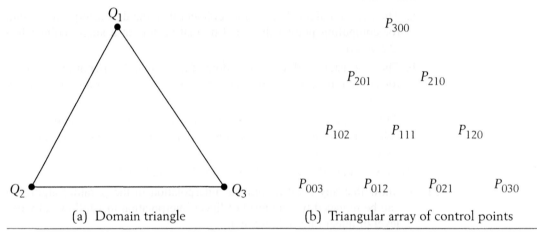

(a) Domain triangle   (b) Triangular array of control points

Figure 5.43 Data for a cubic triangular Bezier patch. The domain is an arbitrary triangle. Compare to Figure 2.26 for triangular Lagrange interpolation.

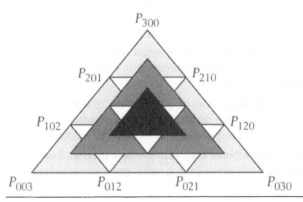

Figure 5.44 Schematic version of de Casteljau's tetrahedral algorithm for a cubic Bezier patch over a triangular domain. Each triangular panel represents the computation of a point at its center calculated by multiplying the points at its vertices by the barycentric coordinates of the domain triangle (Figure 5.43(a)) and adding the results. The light gray triangles represent linear triangular Bezier patches, and the dark gray triangles represent quadratic triangular Bezier patches. Notice that the control point P_{111} is obscured by the panels, and down-pointing triangles are ignored. Compare to Figure 2.28 for triangular Lagrange interpolation.

Just as we did for Bezier curves, we can develop explicit formulas for the basis functions of triangular Bezier patches. One way to do so is to observe that all paths between the apex of the tetrahedron and the control point P_{ijk} at the base of the pyramid are identical to $\beta_1^i \beta_2^j \beta_3^k$, $i + j + k = n$, where $\beta_1, \beta_2, \beta_3$ are the barycentric coordinate functions with respect to the domain triangle $\Delta Q_1 Q_2 Q_3$. Every path has this form because there are n levels in the tetrahedron and to travel between the apex and the point P_{ijk} we must choose the β_1 direction i times, the β_2 direction j times, and the β_3 direction k times. Moreover, there are $\binom{n}{ijk} = n!/i!\,j!\,k!$ paths between the apex and the point P_{ijk} on the nth level of the triangle (see Exercise 1). Thus the basis functions for a degree n triangular Bezier patch are given by

$$B_{ijk}^n(s,t) = \binom{n}{ijk} \beta_1^i(s,t) \beta_2^j(s,t) \beta_3^k(s,t) \qquad i + j + k = n,$$

$$\binom{n}{ijk} = \frac{n!}{i!\,j!\,k!} \tag{5.35}$$

Notice that the functions $B_{ijk}^n(s,t)$ are exactly those functions that appear in the trinomial expansion of $(\beta_1(s,t) + \beta_2(s,t) + \beta_3(s,t))^n$. An example of a triangular patch constructed using these basis functions is provided in Figure 5.45.

The functions $B_{ijk}^n(s,t)$, $i + j + k = n$, are called the *bivariate Bernstein basis functions of degree n*. Notice that by the down recurrence—placing a one at the apex,

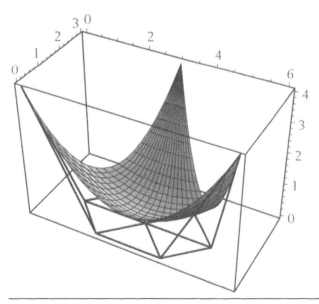

Figure 5.45 A cubic triangular Bezier patch with its control polyhedron. Here the control points lie on a regular triangular mesh in the *xy*-plane, with the corner points raised to height $z = 4$.

reversing all the arrows, and collecting the blending functions at the base of the tetrahedron—these bivariate Bernstein basis functions satisfy the recurrence

$$B_{ijk}^{n+1}(s,t) = \beta_1(s,t)B_{i-1,j,k}^n(s,t) + \beta_2(s,t)B_{i,j-1,k}^n(s,t) + \beta_3(s,t)B_{i,j,k-1}^n(s,t). \quad (5.36)$$

The degree n triangular Bezier patch $T(s,t)$ with control points $\{P_{ijk}\}$, $i + j + k = n$, and domain triangle $\Delta Q_1 Q_2 Q_3$ can be written as

$$T(s,t) = \sum_{i+j+k=n} B_{ijk}^n(s,t)P_{ijk}. \quad (5.37)$$

Typically we shall use the canonical triangle $\Delta = \{(s,t) \mid s,t \geq 0 \text{ and } s + t \leq 1\}$ as our domain. For this triangle the barycentric coordinate functions are $\beta_1 = s$, $\beta_2 = t$, $\beta_3 = 1 - s - t$, so the basis functions are given explicitly by

$$B_{ijk}^n(s,t) = \binom{n}{ijk}s^i t^j (1-s-t)^k \qquad i + j + k = n,$$

$$\binom{n}{ijk} = \frac{n!}{i!\,j!\,k!} \qquad\qquad\qquad\qquad (5.38)$$

To simplify our notation in the remainder of this section, we shall adopt this canonical triangle as our domain and use these canonical basis functions as our blending functions. The proofs, however, do not change much for arbitrary domain triangles and arbitrary barycentric coordinates. Additional identities involving these bivariate Bernstein basis functions can be found at the end of this chapter.

Many of the characteristic properties of Bezier curves extend to triangular Bezier patches. Triangular Bezier patches are affine invariant, nondegenerate, lie in the convex hull of their control points, and interpolate their corner points. These properties follow easily from Equations (5.37) and (5.38) or more directly from de Casteljau's tetrahedral algorithm. Moreover, the boundary curves of triangular Bezier patches are the Bezier curves determined by their boundary control points. For example, along the boundary $s = 0$,

$$B_{ijk}^n(0,t) = 0 \qquad\qquad\qquad i \neq 0$$

$$= \binom{n}{j}t^j(1-t)^{n-j} \qquad i = 0.$$

Substituting into Equation (5.37), we find that

$$T(0,t) = \sum_{i+j+k=n} B_{ijk}^n(0,t)P_{ijk} = \sum_{j=0}^{n} B_j^n(t)P_{0,j,n-j},$$

which is the Bezier curve for the boundary control points $P_{00n},...,P_{0n0}$—that is, the control points for which $i = 0$. Similar results hold along the boundaries $t = 0$ $(j = 0)$ and $s + t = 1$ $(k = 0)$ (see Exercise 20). One property that does not extend from curves to triangular surfaces is the variation diminishing property. Just as with tensor product Bezier patches—and for much the same reasons—*there is no known analogue of the variation diminishing property for triangular Bezier patches.*

Standard algorithms for Bezier curves also extend readily to triangular Bezier patches. For example, the triangular arrays along the three lateral faces of the tetrahedral de Casteljau algorithm are the control points for the three surface patches that subdivide the triangular surface at the point $T(s,t)$ (Exercise 12—see also Section 6.5.1 for a simpler derivation). Moreover, the two-term degree-elevation formula for Bezier curves (Equation (5.17) extends to a three-term degree-elevation formula for triangular Bezier patches (Exercise 10)).

Differentiating of the tetrahedral de Casteljau algorithm for triangular Bezier patches is very similar to differentiating of the pyramid algorithm for tensor product Bezier patches. To find p derivatives with respect to s and q derivatives with respect to t, differentiate any p levels of the algorithm with respect to s—that is, replace $(1 - s - t) \rightarrow -1$, $s \rightarrow 1$, $t \rightarrow 0$ on p levels of the algorithm—then differentiate any q levels (different from the previous p levels) with respect to t, and multiply the result by $n!/(n - p - q)!$. This algorithm works because it follows by induction from Equation (5.36) that

$$\left(B_{ijk}^n(s,t) \right) = \underbrace{\begin{pmatrix} 1 - s - t & t \\ s & 0 \end{pmatrix} \otimes \cdots \otimes \begin{pmatrix} 1 - s - t & t \\ s & 0 \end{pmatrix}}_{n \ \ factors}$$

(see Exercise 18). Therefore, we can apply Equation (5.21) and the commutativity of discrete convolution to differentiate the basis functions $\{B_{ijk}^n(s,t)\}$.

Suppose now that we are given a triangular Bezier surface $P(s,t)$ with control points $\{P_{ijk}\}$, $i + j + k = n$, and domain triangle $\Delta Q_1 Q_2 Q_3$ and we want to construct another triangular Bezier surface $R(s,t)$ with control points $\{R_{ijk}\}$, $i + j + k = n$, and domain triangle $\Delta Q_1 Q_2 \tilde{Q}_3$ that meets $P(s,t)$ continuously and matches its first r derivatives along the boundary parametrized by $Q_1 Q_2$ (see Figure 5.46). For the surfaces to meet continuously, the boundary curves must certainly be identical, so the control points along the common boundary, say, $i = 0$, must match. Thus we must have $P_{0jk} = R_{0jk}$.

What about higher-order smoothness? We could try to compute cross-boundary derivatives to derive formulas for the location of the control points R_{ijk}, $i \neq 0$, but since, in general, the domain triangles need not line up with the coordinate axes, this computation is not so simple as the tensor product case. There is, however, a more straightforward way to proceed based on subdivision. If we apply de Casteljau's tetrahedral algorithm to the patch $P(s,t)$ at the point $\tilde{Q}_3$, then, since this algorithm is also a subdivision procedure, one of the triangular faces of the tetrahedron gives the control points for the surface $P(s,t)$ on the domain $\Delta Q_1 Q_2 \tilde{Q}_3$. Now this triangular patch certainly matches the original patch $P(s,t)$ smoothly, since it is the identical polynomial extended to the domain $\Delta Q_1 Q_2 \tilde{Q}_3$. But it is easy to show using our derivative algorithm that the first r partial derivatives and hence too the first r cross-boundary derivatives along the boundary $i = 0$ depend only on the control points R_{ijk}, $0 \leq i \leq r$ (see Exercise 22). Thus we need only choose these control points to match the control points derived from the subdivision algorithm, and we will generate a surface that meets the original surface smoothly with r common derivatives across the common boundary; no derivative computations are ever required. This trick is

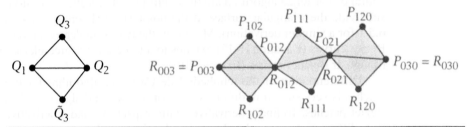

Figure 5.46 Control points for two triangular cubic Bezier patches meeting across their common boundary with continuity of the first cross-boundary derivative: the domain is depicted on the left and the range is on the right. Shaded triangles sharing a common edge must be coplanar. In fact, $R_{102} = \beta_1 P_{003} + \beta_2 P_{012} + \beta_3 P_{102}$, where $\beta_1, \beta_2, \beta_3$ are the barycentric coordinates of $\tilde{Q}_3$ with respect to the domain triangle $\Delta Q_1 Q_2 Q_3$. Similar identities must hold for the control points R_{111} and R_{120}.

essentially the same device we employed to match two Bezier curves smoothly at a common boundary point (see Section 5.6.2). We illustrate this result for continuity of the first cross-boundary derivative in Figure 5.46.

There is another way to construct triangular Bezier patches that is quite similar in design to the de Casteljau algorithm for tensor product Bezier patches. Let $\{P_{ijk}\}$ be a triangular array of control points. For $i = 0,...,n$, define $P_i(t)$ to be the Bezier curve of degree $n - i$ for the control points $P_{i,0,n-i},...,P_{i,n-i,0}$. That is, set

$$P_i(t) = \sum_{k=0}^{n-i} B_k^{n-i}(t) P_{i,k,n-k-i}.$$

Now define a point on the surface $P(s,t)$ to be the value of the degree n Bezier curve $P(s)$ for the control points $P_0(t),...,P_n(t)$ (see Figure 5.47). That is, set

$$P(s,t) = \sum_{i=0}^{n} B_i^n(s) P_i(t) \qquad 0 \le s,t \le 1.$$

It is easy to verify that along the parameter lines $s = 0$, $t = 0$, $t = 1$ this surface interpolates the Bezier curves defined by the boundary control points. Indeed

$$P(0,t) = \sum_{i=0}^{n} B_i^n(0) P_i(t) = P_0(t) = \sum_{k=0}^{n} B_k^n(t) P_{0,k,n-k} \qquad (i = 0)$$

$$P(s,0) = \sum_{i=0}^{n} B_i^n(s) P_i(0) = \sum_{i=0}^{n} B_i^n(s) P_{i,0,n-i} \qquad (j = 0)$$

$$P(s,1) = \sum_{i=0}^{n} B_i^n(s) P_i(1) = \sum_{i=0}^{n} B_i^n(s) P_{i,n-i,0} \ . \qquad (k = 0).$$

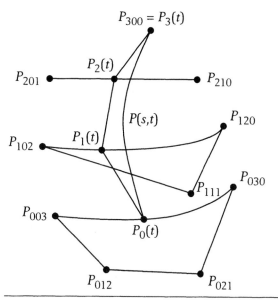

Figure 5.47 A schematic construction for a three-sided Bezier patch $P(s,t)$.

Moreover, along the boundary $s = 1$, the surface collapses to the point $P_n(t) = P_{n00}$. Thus we have constructed a three-sided patch with the same boundaries as the triangular Bezier patch generated by de Casteljau's tetrahedral algorithm. But is this new triangular surface really the same as our original triangular Bezier patch?

Yes and no! Point for point it is the same surface, but not parameter for parameter. That is, this construction generates the same triangular surface, but with a different parametrization. By the way, the analogous construction fails for triangular Lagrange patches; see Section 2.12, Exercise 8.

The original triangular patch is parametrized by

$$T(s,t) = \sum_{i+j+k=n} \binom{n}{ijk} s^i t^j (1-s-t)^k P_{ijk} = \sum_{i=0}^{n} \binom{n}{i} s^i \sum_{j+k=n-i} \binom{n-i}{j} t^j (1-s-t)^k P_{ijk}.$$

Multiplying and dividing the ith term by $(1-s)^{n-i}$, we can rewrite this parametrization as

$$T(s,t) = \sum_{i=0}^{n} \binom{n}{i} s^i (1-s)^{n-i} \sum_{j+k=n-i} \binom{n-i}{j} \left(\frac{t}{1-s}\right)^j \left(\frac{1-s-t}{1-s}\right)^{n-i-j} P_{ijk}.$$

Now if we let $u = t/(1-s)$, then $1-u = (1-s-t)/(1-s)$ and

$$T(s,t) = \sum_{i=0}^{n} \binom{n}{i} s^i (1-s)^{n-i} \sum_{j+k=n-i} \binom{n-i}{j} u^j (1-u)^{n-i-j} P_{ijk}.$$

Setting $P_i(u) = \sum_{j+k=n-i} \binom{n-i}{j} u^j (1-u)^{n-i-j} P_{ijk}$, we arrive at

$$T(s,t) = \sum_{i=0}^{n} B_i^n(s) P_i(u). \tag{5.39}$$

By construction, the curve $P_i(u)$ is the Bezier curve with control points $P_{i,0,n-i}, \dots, P_{i,n-i,0}$. Thus the triangular patch $T(s,t)$ generated by de Casteljau's tetrahedral algorithm is point for point the same as the patch depicted schematically in Figure 5.47. But the parametrization is different, since in the surface $P(s,t)$, we set $u = t$ rather than $u = t/(1-s)$. In fact, our new construction for the triangular patch $P(s,t)$ is really a tensor product construction, since the domain is not the triangle, but the unit square. The transformation $s = s$, $u = t/(1-s)$ maps the canonical triangle into the unit square, although there is a singularity along the edge $s = 1$. This singularity shows up on the surface as a singularity at the point P_{n00}. This surface singularity is not essential; it is an artifact of our artificial tensor product parametrization.

Algorithmically, we can apply Equation (5.39) to compute points along a triangular Bezier surface using only the univariate version of de Casteljau's algorithm (see Figure 5.48). Notice how similar this algorithm is to the algorithm for computing points along a tensor product surface (see Figure 5.40). Indeed, we could make this algorithm look exactly the same as the de Casteljau algorithm for tensor product Bezier surfaces by degree raising the curves $P_i(u)$.

Both the de Casteljau algorithm and the pyramid algorithm are $O(n^3)$. Nevertheless, the triangular version of the de Casteljau algorithm is generally faster than the pyramid algorithm. The triangular de Casteljau algorithm has

$$n(n+1)/2 + \sum_k k(k+1)/2$$

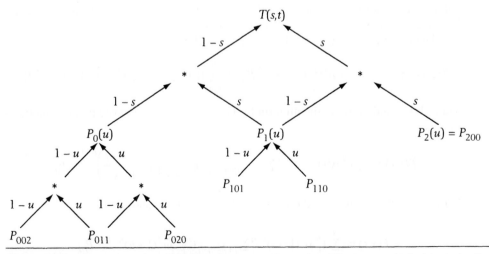

Figure 5.48 A tensor product algorithm for computing points along a quadratic triangular Bezier patch. Here $u = t/(1-s)$. If $u = t$, this triangular version of the de Casteljau algorithm generates the same surface as the pyramid algorithm, but with a different parametrization. Compare to Figure 5.40.

nodes, and each node requires two multiplications. Therefore, this de Casteljau algorithm requires a total of $n(n+1) + \sum_k k(k+1)$ multiplies. But by Exercise 9

$$\sum_{k=1}^{n} \binom{k+1}{2} = \frac{n(n+1)(n+2)}{6}.$$

Therefore,

number of multiplications in the triangular de Casteljau algorithm = $\dfrac{n(n+1)(n+5)}{3}$.

On the other hand, the tetrahedral algorithm has $\sum_k \binom{k+1}{2}$ nodes, and each node requires three multiplications. Therefore,

number of multiplications in the tetrahedral algorithm = $\dfrac{n(n+1)(n+2)}{2}$.

Since $(n+5)/3 \le (n+2)/2$ for $n \ge 4$, the triangular de Casteljau algorithm is generally faster than the tetrahedral algorithm, although for the most common triangular surfaces, namely, cubic patches, $n = 3$ and the tetrahedral algorithm is slightly faster. Near $s = 1$ the triangular de Casteljau algorithm is unstable, but we can overcome this problem by reversing the roles of s and t near $s = 1$.

Moreover, just like the tensor product de Casteljau algorithm for rectangular patches, the triangular de Casteljau algorithm for triangular patches has another advantage over the pyramid algorithm that is even more substantial. If we render the surfaces by generating points on the surface along isoparameter lines—that is, along lines of constant u—then we can reuse the computation of the points $P_0(u), \dots, P_n(u)$. Thus along isoparameter lines, de Casteljau's algorithm for triangular Bezier surfaces reduces to the univariate version of de Casteljau's algorithm, which is only $O(n^2)$. No such reduction occurs for the pyramid algorithm for triangular patches along isoparameter lines.

Finally, notice that we can differentiate the triangular de Casteljau algorithm q times in t by differentiating any q levels (the u levels) in each of the n lower triangles (see Figure 5.48) and multiplying the result by $n!/(n-q)!$. That this algorithm works is an immediate consequence of the corresponding differentiation algorithm for Bezier curves discussed in Section 5.6.2. We cannot, however, differentiate this algorithm p times with respect to s by differentiating p levels of the upper triangle because $u = t/(1-s)$, so s appears as well in the lower triangles. To get around this problem, we can, if we like, simply reverse the roles of s and t.

Exercises

1. Prove that there are $\binom{n}{ijk} = n!/i!\,j!\,k!$ paths between the apex and the point P_{ijk} at the base of a tetrahedron with n levels.

2. Prove that

    a. $\sum_{i+j+k=n} B_{ijk}^n(s,t) = 1$.

    b. $\sum_{i+j+k=n} (-1)^{i+j} B_{ijk}^n(s,t) = (1-2s-st)^n$.

   c. Every triangular Bezier patch lies in the convex hull of its control points. (Hint: Use part (a).)

3. What are the up and down recurrences for the bivariate Bernstein basis functions?

4. a. Prove that the bivariate Bernstein basis functions $\{B_{ijk}^n(s,t)\}$ form a basis for the bivariate polynomials of degree n.

   b. Conclude that triangular Bezier patches are nondegenerate.

   (Hint: Consider $B_{ijk}^n(s,t)/(1-s-t)^n$.)

5. Give necessary and sufficient conditions on the control points for a triangular Bezier surface to collapse to

   a. a single point

   b. a line

   c. a plane

   Justify your answer.

6. Let $P(s,t)$ be a triangular Bezier patch with control points $\{P_{ijk}\}$, $i+j+k=n$.

   a. Develop an algorithm to determine whether or not $P(s,t)$ represents a polynomial patch of degree $n-1$.

   b. If $P(s,t)$ degenerates to a polynomial surface of degree $n-1$, develop an algorithm to find the control points $\{Q_{\alpha\beta\gamma}\}$, $\alpha+\beta+\gamma=n-1$, that represent $P(s,t)$ as a Bezier surface of degree $n-1$ from the control points $\{P_{ijk}\}$, $i+j+k=n$, that represent $P(s,t)$ as a Bezier surface of degree n. (Hint: Compare to Section 5.6.2, Exercise 10.)

7. Let $\{P_{ijk}\}$ be the triangular array of points generated from the de Casteljau algorithm for Bezier curves applied to the control points Q_h, $h=0,...,n$, at some fixed parameter t. What is the triangular Bezier patch generated by the tetrahedral de Casteljau algorithm applied to the control points $\{P_{ijk}\}$?

8. Let $B[P_{ijk}](s,t)$ denote the triangular Bezier patch with control points $\{P_{ijk}\}$. Show that triangular Bezier patches have the following symmetry properties:

   a. $B[P_{jik}](s,t) = B[P_{ijk}](t,s)$

   b. $B[P_{ikj}](s,t) = B[P_{ijk}](s,1-s-t)$

   c. $B[P_{kji}](s,t) = B[P_{ijk}](1-s-t,t)$

9. Prove by induction that

   a. $\displaystyle\sum_{k=1}^{n}\binom{k+1}{2} = \frac{n(n+1)(n+2)}{6}$

   b. $\displaystyle\sum_{k=1}^{n}\binom{p+k}{p+1} = \binom{n+p+1}{p+2}$

10. Show that

a. $sB_{ijk}^n(s,t) = \dfrac{i+1}{n+1} B_{i+1,j,k}^{n+1}(s,t)$

b. $tB_{ijk}^n(s,t) = \dfrac{j+1}{n+1} B_{i,j+1,k}^{n+1}(s,t)$

c. $(1-s-t)B_{ijk}^n(s,t) = \dfrac{k+1}{n+1} B_{i,j,k+1}^{n+1}(s,t)$

d. $B_{ijk}^n(s,t) = \dfrac{i+1}{n+1} B_{i+1,j,k}^{n+1}(s,t) + \dfrac{j+1}{n+1} B_{i,j+1,k}^{n+1}(s,t) + \dfrac{k+1}{n+1} B_{i,j,k+1}^{n+1}(s,t)$

e. Conclude that for triangular Bezier patches the degree-elevation formula is

$$Q_{ijk} = \frac{i}{n+1} P_{i-1,j,k} + \frac{j}{n+1} P_{i,j-1,k} + \frac{k}{n+1} P_{i,j,k-1}.$$

11. Prove that the control polyhedron generated by the degree-elevation formula in Exercise 10 converges to the original triangular Bezier patch.

12. Use a three-color urn model and sampling with replacement to derive the following identities using probabilistic arguments:

a. $B_{ijk}^n(su, sv + t) = \displaystyle\sum_{q=0}^{j} \sum_{r=0}^{k} B_{i,j-q,k-r}^p(u,v) B_{pqr}^n(s,t)$

b. $B_{ijk}^n(tu + s, tv) = \displaystyle\sum_{p=0}^{i} \sum_{r=0}^{k} B_{i-p,j,k-r}^q(u,v) B_{pqr}^n(s,t)$

c. $B_{ijk}^n((1-s-t)u + s, (1-s-t)v + t) = \displaystyle\sum_{q=0}^{j} \sum_{p=0}^{i} B_{i-p,j-q,k}^r(u,v) B_{pqr}^n(s,t)$

d. Use these identities to prove that the triangular arrays along the three lateral faces of the tetrahedral de Casteljau algorithm are the control points for the three surface patches that subdivide the triangular Bezier surface at $T(s,t)$.

13. Implement the de Casteljau subdivision algorithm for triangular Bezier patches described in Exercise 12.

a. Explain why this subdivision algorithm is not an effective tool for rendering triangular Bezier patches. (For a more effective tetrahedral subdivision algorithm, see Section 6.5.1, Exercise 15.)

b. Does this subdivision algorithm converge to the original triangular Bezier patch?

14. Describe how to use the tensor product construction for a triangular Bezier patch to subdivide the patch. Explain why this subdivision procedure is superior to the subdivision procedure generated from the pyramid algorithm.

15. Implement the recursive subdivision algorithm for triangular Bezier patches described in Exercise 14. Apply this algorithm to

    a. render a triangular Bezier patch

    b. intersect two triangular Bezier patches

16. Let $T(s,t) = \sum_{i+j+k=n} B_{ijk}^n(s,t) P_{ijk}$, and let $T^{(p,q)}(s,t) = \dfrac{\partial^{p+q} T}{\partial s^p \partial t^q}$.

    a. Show that

    $$T^{(1,0)}(0,0) = n(P_{1,0,n-1} - P_{00n}) \quad \text{and} \quad T^{(0,1)}(0,0) = n(P_{0,1,n-1} - P_{00n}).$$

    b. Conclude that the normal vector at $T(0,0)$ is parallel to the vector

    $$(P_{0,1,n-1} - P_{00n}) \times (P_{1,0,n-1} - P_{00n}).$$

    c. Find the normal vectors at $T(0,1)$ and $T(1,0)$.

17. Define the convolution of two triply indexed arrays of functions $A(s,t) = \{A_{ijk}(s,t)\}$ and $B(s,t) = \{B_{pqr}(s,t)\}$ by setting

    $$(A \otimes B)_{\alpha\beta\gamma}(s,t) = \sum_{i+p=\alpha} \sum_{j+q=\beta} \sum_{k+r=\gamma} A_{ijk}(s,t) B_{pqr}(s,t).$$

    Now introduce the following indexing scheme:

Function	Index
s	$(1,0,0)$
t	$(0,1,0)$
$1-s-t$	$(0,0,1)$
$B_{ijk}^n(s,t)$	(i,j,k)

    a. Show that with this indexing scheme

    $$\left(B_{ijk}^n(s,t) \right) = \underbrace{\begin{pmatrix} 1-s-t & t \\ s & 0 \end{pmatrix} \otimes \cdots \otimes \begin{pmatrix} 1-s-t & t \\ s & 0 \end{pmatrix}}_{n \ \ factors}$$

    b. Use part (a) to derive the differentiation procedure for the pyramid algorithm presented in the text.

18. Implement both de Casteljau's algorithm and the pyramid algorithm for triangular Bezier surfaces. Which algorithm do you prefer? Why? Experiment with triangular Bezier surfaces of different degrees. Determine how changing the location of the control points affects the shape of the surface.

19. Develop an $O(n^2)$ ladder evaluation algorithm for triangular Bezier patches.

20. Consider a degree n triangular Bezier patch $T(s,t)$ with control points $\{P_{ijk}\}$, $i+j+k = n$.

a. Show that the curve corresponding to the boundary $t = 0$ is given by the degree n Bezier curve for the boundary control points $\{P_{ijk}\}$ with $j = 0$.

b. Show that the curve corresponding to the boundary $s + t = 1$ is given by the degree n Bezier curve for the boundary control points $\{P_{ijk}\}$ with $k = 0$.

21. Show directly from de Casteljau's tetrahedral algorithm that the boundaries of a triangular Bezier patch are the Bezier curves determined by the boundary control points.

22. Let $P(s,t)$ be a triangular Bezier patch with control points $\{P_{ijk}\}$. Show that the first r partial derivatives across the boundary $i = 0$ depend only on the control points P_{ijk}, $0 \le i \le r$.

23. Show that

a. $\dfrac{\partial B^n_{ijk}}{\partial s} = n\left\{ B^{n-1}_{i-1,j,k}(s,t) - B^{n-1}_{i,j,k-1}(s,t) \right\}$

b. $\dfrac{\partial B^n_{ijk}}{\partial t} = n\left\{ B^{n-1}_{i,j-1,k}(s,t) - B^{n-1}_{i,j,k-1}(s,t) \right\}$

c. Conclude that the function $B^n_{ijk}(s,t)$ is unimodal in (s,t) and takes on its maximum value at $(i/n, j/n)$.

d. Prove the recurrence

$$B^n_{ijk}(s,t) = sB^{n-1}_{i-1,j,k}(t) + tB^{n-1}_{i,j-1,k}(t) + (1 - s - t)B^{n-1}_{i,j,k-1}(t).$$

e. Conclude that the functions $\{B^n_{ijk}(s,t)\}$ are unimodal in (i,j,k).

24. Let $\Delta^2 = \{(s,t) \mid 0 \le s,t \text{ and } s+t \le 1\}$. Using the results in Exercise 23(a) and (b), show that

$$\iint_{\Delta^2} B^n_{ijk}(\sigma,\tau)\,d\sigma d\tau = \frac{1}{(n+1)(n+2)}.$$

25. Let

$$T(s,t) = \sum_{i+j+k=n} B^n_{ijk}(s,t)P_{ijk}$$

be a triangular Bezier patch of degree n, and let

$$T^{n-1}_{100}(s,t), T^{n-1}_{010}(s,t), T^{n-1}_{001}(s,t)$$

denote the three values computed by the pyramid algorithm on the penultimate level just below the apex of the pyramid. Show that

a. $T(s,t) = sT^{n-1}_{100}(s,t) + tT^{n-1}_{010}(s,t) + (1 - s - t)T^{n-1}_{001}(s,t)$

b. $\dfrac{\partial T}{\partial s} = n\left(T^{n-1}_{100}(s,t) - T^{n-1}_{001}(s,t) \right)$

c. $\dfrac{\partial T}{\partial t} = n\left(T^{n-1}_{010}(s,t) - T^{n-1}_{001}(s,t) \right)$

d. Conclude that to compute point values and normal vectors at any point on a triangular Bezier patch of degree n by the pyramid algorithm costs

$$\frac{n(n+1)(n+2)}{2} + 2$$

multiplications and one cross product.

e. How does the result in part (d) compare to computing point values and normal vectors at any point using the triangular version of the de Casteljau algorithm?

26. Show that

a. $B_i^n(s) = \sum_{j=0}^{n-i} B_{i,j,n-i-j}^n(s,t)$

b. $B_j^n(t) = \sum_{i=0}^{n-j} B_{i,j,n-i-j}^n(s,t)$

c. $B_i^m(s)B_j^n(t) = \sum_{p=i}^{m+n} \sum_{q=j}^{m+i-j} \frac{\binom{p}{i}\binom{q}{j}\binom{m+n-p-q}{m-i+j-q}}{\binom{m+n}{n}} B_{pqr}^{m+n}(s,t)$

d. Conclude that every tensor product surface of bidegree (m,n) can be represented as a triangular Bezier patch of degree $m + n$.

27. Show that

a. $\sum_{i+j+k=n} B_{ijk}^n(s,t)x^i y^j = \{(1-s-t)+sx+ty)\}^n$

b. $\sum_{i+j+k=n} B_{ijk}^n(s,t)e^{ix}e^{jy} = \{(1-s-t)+se^x+te^y)\}^n$

28. Show that

a. $B_{ijk}^n(s,t) = \sum_{p=i}^{n} \sum_{q=j}^{n-p} (-1)^{i+j+p+q}\binom{p}{i}\binom{q}{j}\binom{n}{pqr}s^p t^q$

b. $\binom{n}{ijk}s^i t^j = \sum_{p=i}^{n} \sum_{q=j}^{n-p} \binom{p}{i}\binom{q}{j}B_{pqr}^n(s,t)$ $0 \le i+j \le n$

29. Show that

$$(sx+ty+1)^n = \sum_{i=0}^{n} \sum_{j=0}^{n-i}(x+1)^i(y+1)^j B_{ijk}^n(s,t).$$

30. Show by example that the number of intersections of a line with a triangular Bezier patch can be greater than the number of intersections of the line with its control polyhedron. Conclude that triangular Bezier patches do not satisfy this version of the variation diminishing property.

5.8.3   **Rational Bezier Patches**

When we studied Lagrange interpolation, we observed that many common surfaces such as the sphere and the torus cannot be represented exactly by polynomial parametrizations. As with nonpolynomial curves, we could approximate these surfaces with polynomials using a bivariate version of the Weierstrass Approximation Theorem. Unfortunately, as with curves, often we would need to use polynomials of quite high degree to generate a good approximation.

The sphere, however, like the circle, has a rational parametrization; in fact, we observed in Chapter 2 that the sphere has different rational parametrizations, and we explicitly provided two distinct parametrizations.

- *Quadratic parametrization of the sphere*

$$x = \frac{2s}{1+s^2+t^2} \qquad y = \frac{2t}{1+s^2+t^2} \qquad z = \frac{1-s^2-t^2}{1+s^2+t^2} \qquad (2.27)$$

- *Biquadratic parametrization of the sphere*

$$x = \frac{2s(1-t^2)}{(1+s^2)(1+t^2)} \qquad y = \frac{2t(1+s^2)}{(1+s^2)(1+t^2)} \qquad z = \frac{(1-s^2)(1-t^2)}{(1+s^2)(1+t^2)} \qquad (2.28)$$

Thus we need to introduce the notion of a rational Bezier surface to represent the sphere exactly.

A *rational Bezier surface* in affine space, like a rational Bezier curve, is the projection of a polynomial Bezier surface from Grassmann space. Thus rational Bezier patches can be introduced by associating a scalar weight w_{ij} (or w_{ijk}) with each control point P_{ij} (or P_{ijk}). This approach leads to the following explicit formulas for rational triangular and rational tensor product Bezier patches.

- *Rational triangular Bezier representation*

$$R(s,t) = \frac{\displaystyle\sum_{i+j+k=n} w_{ijk}P_{ijk}B_{ijk}^n(s,t)}{\displaystyle\sum_{i+j+k=n} w_{ijk}B_{ijk}^n(s,t)} \qquad 0 \le s,t \quad \text{and} \quad s+t \le 1 \qquad (5.40)$$

- *Rational tensor product Bezier representation*

$$R(s,t) = \frac{\displaystyle\sum_{k=0}^{m}\sum_{l=0}^{n} w_{kl}P_{kl}B_k^m(s)B_l^n(t)}{\displaystyle\sum_{k=0}^{m}\sum_{l=0}^{n} w_{kl}B_k^m(s)B_l^n(t)} \qquad 0 \le s,t \le 1 \qquad (5.41)$$

For example, we can represent the sphere in rational tensor product Bezier form by considering the biquadratic parametrization given in Equation (2.28). To find the weights, we express the denominator (the mass) in terms of the Bernstein basis by observing that

$$(1+s^2)(1+t^2) = \left(B_0^2(s) + B_1^2(s) + 2B_2^2(s)\right)\left(B_0^2(t) + B_1^2(t) + 2B_2^2(t)\right)$$

$$= \sum_{i=0}^{2}\sum_{j=0}^{2} w_{ij}B_i^2(s)B_j^2(t) \ .$$

Comparing coefficients, we can read off the weights

$$w_{ij} = \begin{pmatrix} 1 & 1 & 2 \\ 1 & 1 & 2 \\ 2 & 2 & 4 \end{pmatrix}.$$

Similarly, to find the control points, we must express the numerators of x,y,z in terms of the bivariate Bernstein basis. To find the x-coordinates of the control points, observe that

$$2s(1-t^2) = \left(B_1^2(s) + 2B_2^2(s)\right)\left(B_0^2(t) + B_1^2(t)\right)$$

so

$$(w_{ij}P_{ij})_x = \begin{pmatrix} 0 & 0 & 0 \\ 1 & 1 & 0 \\ 2 & 2 & 0 \end{pmatrix}.$$

Similarly,

$$2t(1+s^2) = \left(B_1^2(t) + 2B_2^2(t)\right)\left(B_0^2(s) + B_1^2(s) + 2B_2^2(s)\right)$$

so

$$(w_{ij}P_{ij})y = \begin{pmatrix} 0 & 1 & 2 \\ 0 & 1 & 2 \\ 0 & 2 & 4 \end{pmatrix}.$$

Finally,

$$(1-s^2)(1-t^2) = \left(B_0^2(s) + B_1^2(s)\right)\left(B_0^2(t) + B_1^2(t)\right)$$

so

$$(w_{ij}P_{ij})_z = \begin{pmatrix} 1 & 1 & 0 \\ 1 & 1 & 0 \\ 0 & 0 & 0 \end{pmatrix}.$$

Now using the weights computed above, we can easily solve for the coordinates of the control points.

As with rational Bezier curves, typically all the weights are chosen to be positive to avoid singularities, but zero and negative weights are permitted and sometimes, as we have seen with curves, are even necessary to represent certain surface patches exactly. Like rational Bezier curves, the effect of increasing the weights on a rational Bezier surface is rather benign. Negative weights, however, may introduce singularities even if we restrict the parameter domain, so negative weights are generally avoided.

Rational Bezier surfaces with positive weights share many of the geometric properties of standard Bezier surfaces. They are affine invariant, nondegenerate, lie in the convex hull of their control points, and interpolate the rational boundary curves determined by their boundary control points. These results follow easily from Equations (5.40) and (5.41).

Typically algorithms for polynomial Bezier patches carry over directly to algorithms for rational Bezier patches because generally we can apply these algorithms separately to the numerator and denominator. For example, we can evaluate points along a rational Bezier patch by applying the pyramid algorithm independently to the numerator and denominator. Similarly, change of basis algorithms can be applied separately to the numerator and denominator. Therefore, the algorithms for degree elevation and subdivision can be computed by applying these algorithms on the mass-points $(w_{ij}P_{ij}, w_{ij})$ $\{(w_{ijk}P_{ijk}, w_{ijk})\}$ and then dividing the results by the masses. One algorithm that cannot be applied in this way is the algorithm for differentiating a Bezier surface because the derivative of a quotient is not equal to the quotient of the derivatives, so here we must proceed recursively in a manner similar to our approach in Section 5.7.1 for differentiating rational Bezier curves.

Exercises

1. What is the effect on a rational Bezier surface if one of the mass-points has zero weight?

2. Experiment with altering the weights in a rational Bezier surface.

   a. What are the local and global effects of altering a single weight?

   b. What is the effect of a negative weight?

   c. What happens if all the weights are changed simultaneously?

3. a. Which part of the sphere does the Bezier representation given in the text represent?

   b. Use the results in the text together with de Casteljau's algorithm to render the sphere.

4. a. Find the Bezier control points and the Bezier weights for the sphere given by the quadratic parametrization in Equation (2.27).

   b. Which part of the sphere does this Bezier patch represent?

   c. Use the results of part (a) together with the pyramid algorithm to render the sphere.

5. Recall from Section 2.14, Exercise 5, that the torus with inner radius $d - a$ and outer radius $d + a$ has the biquadratic parametrization

$$x = \frac{d(1+s^2)(1-t^2) + a(1-s^2)(1-t^2)}{(1+s^2)(1+t^2)}$$

$$y = \frac{2d(1+s^2)t + 2a(1-s^2)t}{(1+s^2)(1+t^2)}$$

$$z = \frac{2as(1+t^2)}{(1+s^2)(1+t^2)}$$

    a. Find the Bezier control points and the Bezier weights for the torus given by this biquadratic parametrization.

    b. Use the results of part (a) together with de Casteljau's algorithm to render the torus.

6. Let $x = f(s)$, $z = g(s)$ be a curve in the xz-plane. Recall from Section 2.14, Exercise 6, that the surface of revolution generated by rotating this curve around the z-axis can be represented by the parametric equations

$$x = \frac{(1-t^2)f(s)}{1+t^2}, \qquad y = \frac{2tf(s)}{1+t^2} \qquad z = \frac{(1+t^2)g(s)}{1+t^2}.$$

    a. Use this parametrization for a surface of revolution to generate rational parametrizations for the right circular cylinder and right circular cone by rotating a line about the z-axis.

       i.   Find the triangular Bezier control points and the Bezier weights for the cylinder and cone given by these parametrizations.

      ii.   Use the results of part (i) together with the pyramid algorithm for triangular Bezier patches to render the right circular cylinder and right circular cone.

    b. Use this parametrization for a surface of revolution to generate rational parametrizations for the sphere and the torus by rotating a circle about the z-axis.

       i.   Find the triangular Bezier control points and the Bezier weights for the sphere and the torus given by these parametrizations.

      ii.   Use the results of part (i) together with the pyramid algorithm for triangular Bezier patches to render the sphere and the torus.

7. Let $R(s,t)$ be a rational Bezier surface (triangular or rectangular) with control points $(w_h P_h, w_h)$. Let w_g increase and hold w_h fixed for $h \neq g$.

    a. Show that $\lim_{w_g \to \infty} R(s,t) = R(s,t)$ if (s,t) is a corner point of the domain.

    b. Show that $\lim_{w_g \to \infty} R(s,t) = P_g$ if (s,t) is not a corner point of the domain.

    c. Conclude that the limit surface is a disconnected collection of points.

    d. What does the limit surface look like if several weights are allowed to increase simultaneously?

8. Let $B_g(s,t)$, $g \in G$, be a collection of Bernstein basis functions, either triangular or rectangular, and let $\{w_g\}$, $g \in G$, be a collection of nonzero scalar weights. Define

$$R_g(s,t) = \frac{w_g B_g(s,t)}{\displaystyle\sum_{h \in G} w_h B_h(s,t)}, \quad g \in G.$$

Show that these functions behave like rational Bernstein basis functions. In particular, show that

a. $\sum_{g \in G} R_g(s,t) \equiv 1$

b. $R(s,t) = \sum_{g \in G} R_g(s,t) P_g$

9. Let $R(s,t)$ be a rational Bezier surface (triangular or rectangular) with control points $(w_h P_h, w_h)$. What is the effect on the surface if the weight w_{00} (w_{00n}) is zero and the corresponding control vector is also the zero vector? (Hint: Compare to Section 5.7, Exercise 15.)

10. Let $R(s,t)$ be a rational Bezier surface with control points $(w_h P_h, w_h)$. Define

$$P(s,t) = \sum_h B_h^n(s,t) w_h P_h \text{ and } w(s,t) = \sum_h B_h^n(s,t) w_h, \text{ so that } R(s,t) = \frac{P(s,t)}{w(s,t)}.$$

Prove that

$$\frac{\displaystyle\iint_{domain} P(s,t)\,ds\,dt}{\displaystyle\iint_{domain} w(s,t)\,ds\,dt} = \text{center mass of } \{(w_h P_h, w_h)\}.$$

5.9 Summary

Bernstein/Bezier approximation is an extremely rich theory, which can be approached from many different analytic perspectives: dynamic programming procedures (the de Casteljau algorithm), Bernstein basis functions (explicit expressions and recursive formulas), the binomial theorem (generating functions), probability theory (the binomial distribution), conversion to monomial form (division by $(1 - t)^n$), and discrete convolution. Geometric constructions like subdivision also lead to rich and often unexpected results, including the variation diminishing property (corner cutting), rendering and intersection algorithms (divide and conquer), and methods for smoothly joining together Bezier curves and surfaces (extrapolation).

Although any one of these techniques may be powerful enough to develop the entire Bernstein/Bezier canon, we have purposely avoided consistently adopting any one particular method in order not to impoverish the theory. Rather, in each instance we have tried to select the specific approach most suitable to the problem at hand. Each of these techniques may be a jumping-off point for a new class of curves and surfaces and a new line of investigation. Certain formulas—such as the two-term degree elevation formula (Polya polynomials, Section 5.5.4.2, Exercise 12) and the two-term differentiation formula (B-splines, Chapter 7)—can also be taken as the starting points for the constructions of new approximation schemes.

In the next chapter we shall introduce yet another powerful approach to the analysis of Bernstein/Bezier curves and surfaces: dual functionals, embodied by blossoming or polar forms. At first, this very richness of the theory may seem overwhelming, but it is well to keep in mind when approaching new problems that a variety of attacks are possible and there are many weapons in our arsenal. Below we summarize these analysis techniques, collect in one place the standard properties of and algorithms for Bezier curves and surfaces, and then list as well a collection of useful identities for the univariate and bivariate Bernstein basis functions.

- *Tools for Analyzing Bezier Curves and Surfaces*
    1. De Casteljau algorithm
        - Pascal's triangle and paths arguments
        - Induction + recursion
    2. Bernstein basis functions
        - Properties of Bernstein polynomials $\Rightarrow$ properties of Bezier curves and surfaces
    3. Binomial theorem
        - Generating functions
    4. Binomial distribution
        - Probability theory
    5. Conversion to monomial basis
        - Divide by $(1-t)^n$
    6. Subdivision
        - Divide and conquer
        - Rendering and intersection algorithms
    7. Discrete convolution
        - Commutativity
        - Differentiation
    8. Dual Functionals
        - Blossoming (Chapter 6)
- *Properties of Bezier Curves and Surfaces*
    1. Polynomial
    2. Affine invariant
    3. Convex hull
    4. Symmetry
    5. Interpolation at boundaries

6. Nondegenerate
7. Variation diminishing (curves only)

■ *Algorithms for Bezier Curves and Surfaces*

1. Evaluation
2. Subdivision
3. Differentiation
4. Conversion to and from monomial form
5. Degree elevation
6. Blossoming (Chapter 6)

5.9.1 Identities for the Bernstein Basis Functions

1. Definitions

   a. $B_k^n(t) = \binom{n}{k} t^k (1-t)^{n-k}$ $0 \le k \le n$

   $\binom{n}{k} = \dfrac{n!}{k!(n-k)!}$

   b. $B_{ijk}^n(s,t) = \binom{n}{ijk} s^i t^j (1-s-t)^k$ $i+j+k = n$

   $\binom{n}{ijk} = \dfrac{n!}{i!\,j!\,k!}$

2. Nonnegativity

   a. $B_k^n(t) \ge 0$ $0 \le t \le 1$

   b. $B_{ijk}^n(s,t) \ge 0$ $0 \le s,t$ and $s+t \le 1$

3. Symmetries

   a. $B_k^n(t) = B_{n-k}^n(1-t)$

   b. $B_{ijk}^n(s,t) = B_{jik}^n(t,s)$

   c. $B_{ijk}^n(s,t) = B_{ikj}^n(s,1-s-t)$

   d. $B_{ijk}^n(s,t) = B_{kji}^n(1-s-t,t)$

4. Corner values

   a. $B_k^n(0) = 0$ $k \neq 0$

           $= 1$ $k = 0$

   b. $B_k^n(1) = 0$ $k \neq n$

           $= 1$ $k = n$

   c. $B_{ijk}^n(0,0) = 0$ $k \neq n$

           $= 1$ $k = n$

   d. $B_{ijk}^n(1,0) = 0$ $i \neq n$

           $= 1$ $i = n$

   e. $B_{ijk}^n(0,1) = 0$ $j \neq n$

           $= 1$ $j = n$

5. Boundary values

   a. $B_{ijk}^n(s,0) = 0$ $j \neq 0$

           $= B_i^n(s)$ $j = 0$

   b. $B_{ijk}^n(0,t) = 0$ $i \neq 0$

           $= B_j^n(t)$ $i = 0$

   c. $B_{ijk}^n(s,1-s) = 0$ $k \neq 0$

           $= B_i^n(s)$ $k = 0$

6. Maximum values

   a. $Max\{B_k^n(t)\}$ occurs at $t = k/n$

   b. $Max\{B_{ijk}^n(s,t)\}$ occurs at $s = i/n, \; t = j/n$

7. Partitions of unity

   a. $\displaystyle\sum_{k=0}^{n} B_k^n(t) = 1$

   b. $\sum_{i+j+k=n} B_{ijk}^n(s,t) = 1$

8. Alternating sums

   a. $\displaystyle\sum_{k=0}^{n} (-1)^k B_k^n(t) = (1-2t)^n$

   b. $\sum_{i+j+k=n} (-1)^{i+j} B_{ijk}^n(s,t) = (1-2s-st)^n$

9. Representation of monomials

 a. $\binom{n}{j}t^j = \sum_{k=j}^{n}\binom{k}{j}B_k^n(t)$ $0 \le j \le n$

 b. $\binom{n}{ijk}s^it^j = \sum_{p=i}^{n}\sum_{q=j}^{n-p}\binom{p}{i}\binom{q}{j}B_{pqr}^n(s,t)$ $0 \le i+j \le n$

10. Representation in terms of monomials

 a. $B_k^n(t) = \sum_{j=k}^{n}(-1)^{j-k}\binom{j}{k}\binom{n}{j}t^j$

 b. $B_{ijk}^n(s,t) = \sum_{p=i}^{n}\sum_{q=j}^{n-p}(-1)^{i+j+p+q}\binom{p}{i}\binom{q}{j}\binom{n}{pqr}s^pt^q$

11. Conversion to monomial form

 a. $\dfrac{B_k^n(t)}{(1-t)^n} = \binom{n}{k}u^k$ $u = \dfrac{t}{1-t}$

 b. $\dfrac{B_k^n(t)}{t^n} = \binom{n}{k}u^{n-k}$ $u = \dfrac{1-t}{t}$

 c. $\dfrac{B_{ijk}^n(s,t)}{(1-s-t)^n} = \binom{n}{ijk}u^iv^j$ $u = \dfrac{s}{1-s-t}, \quad v = \dfrac{t}{1-s-t}$

 d. $\dfrac{B_{ijk}^n(s,t)}{s^n} = \binom{n}{ijk}u^kv^j$ $u = \dfrac{1-s-t}{s}, \quad v = \dfrac{t}{s}$

 e. $\dfrac{B_{ijk}^n(s,t)}{t^n} = \binom{n}{ijk}u^iv^k$ $u = \dfrac{s}{t}, \quad v = \dfrac{1-s-t}{t}$

12. Linear independence

 a. $\sum_{k=0}^{n}c_kB_k^n(t) = 0 \iff c_k = 0$ for all k

 b. $\sum_{i+j+k=n}c_{ijk}B_{ijk}^n(s,t) = 0 \iff c_{ijk} = 0$ for all i,j,k

13. Descartes' Law of Signs

 a. Zeros in $(0,1)\left\{\sum_{k=0}^{n}c_kB_k^n(t)\right\} \le$ sign alternations $\{c_0,...,c_n\}$

 b. No known analogous formula exists for bivariate Bernstein bases.

14. Generating functions

    a. $\displaystyle\sum_{k=0}^{n} B_k^n(t)x^k = \{(1-t)+tx)\}^n$

    b. $\displaystyle\sum_{k=0}^{n} B_k^n(t)e^{ky} = \{(1-t)+te^y)\}^n$

    c. $\displaystyle\sum_{i+j+k=n} B_{ijk}^n(s,t)x^i y^j = \{(1-s-t)+sx+ty)\}^n$

    d. $\displaystyle\sum_{i+j+k=n} B_{ijk}^n(s,t)e^{ix} e^{jy} = \{(1-s-t)+se^x+te^y)\}^n$

15. Recursion

    a. $B_k^n(t) = tB_{k-1}^{n-1}(t) + (1-t)B_k^{n-1}(t)$

    b. $B_{ijk}^n(s,t) = sB_{i-1,j,k}^{n-1}(s,t) + tB_{i,j-1,k}^{n-1}(s,t) + (1-s-t)B_{i,j,k-1}^{n-1}(s,t)$

16. Discrete convolution

    a. $\{B_i^n(t),...,B_n^n(t)\} = \underbrace{\{(1-t),t\} \otimes \cdots \otimes \{(1-t),t\}}_{n \ \ factors}$

    b. $\left(B_{ijk}^n(s,t)\right) = \underbrace{\begin{pmatrix} 1-s-t & t \\ s & 0 \end{pmatrix} \otimes \cdots \otimes \begin{pmatrix} 1-s-t & t \\ s & 0 \end{pmatrix}}_{n \ \ factors}$

17. Unimodality

    a. $B(k,t) = B_k^n(t)$ is unimodal in k

    b. $B(i,j,k,t) = B_{ijk}^n(s,t)$ is unimodal in i,j,k

18. Subdivision

a. $B_i^n(rt) = \sum\limits_{k=i}^{n} B_i^k(r) B_k^n(t)$

b. $B_i^n((1-t)r + t) = \sum\limits_{k=0}^{i} B_{i-k}^{n-k}(r) B_k^n(t)$

c. $B_i^n((1-t)r + ts) = \sum\limits_{k=0}^{n} \left\{ \sum\limits_{p+q=i} B_p^{n-k}(r) B_q^k(s) \right\} B_k^n(t)$

d. $B_{ijk}^n(su, sv + t) = \sum\limits_{q=0}^{j} \sum\limits_{r=0}^{k} B_{i,j-q,k-r}^p(u,v) B_{pqr}^n(s,t)$

e. $B_{ijk}^n(tu + s, tv) = \sum\limits_{p=0}^{i} \sum\limits_{r=0}^{k} B_{i-p,j,k-r}^q(u,v) B_{pqr}^n(s,t)$

f. $B_{ijk}^n((1-s-t)u + s, (1-s-t)v + t) = \sum\limits_{q=0}^{j} \sum\limits_{p=0}^{i} B_{i-p,j-q,k}^r(u,v) B_{pqr}^n(s,t)$

g. $B_{ijk}^n((1-s-t)u_1 + sv_1 + tw_1, (1-s-t)u_2 + sv_2 + tw_2)$

$$= \sum\limits_{p+q+r=n} A(u_1, u_2, v_1, v_2, w_1, w_2) B_{pqr}^n(s,t)$$

$A(u_1, u_2, v_1, v_2, w_1, w_2)$

$$= \sum\limits_{a+b+e=i} \sum\limits_{b+d+f=j} B_{abc}^r(u_1, u_2) B_{def}^p(v_1, v_2) B_{ghl}^q(w_1, w_2)$$

19. Partial derivatives

a. $\dfrac{dB_k^n}{dt} = n\left\{ B_{k-1}^{n-1}(t) - B_k^{n-1}(t) \right\}$

b. $\dfrac{d^p B_k^n}{dt^p} = \dfrac{n!}{(n-p)!} \sum\limits_{j=0}^{p} (-1)^{p-j} \binom{p}{j} B_{k-j}^{n-p}(t)$

c. $\dfrac{\partial B_{ijk}^n}{\partial s} = n\left\{ B_{i-1,j,k}^{n-1}(s,t) - B_{i,j,k-1}^{n-1}(s,t) \right\}$

d. $\dfrac{\partial B_{ijk}^n}{\partial t} = n\left\{ B_{i,j-1,k}^{n-1}(s,t) - B_{i,j,k-1}^{n-1}(s,t) \right\}$

e. $\dfrac{\partial^{p+q} B_{ijk}^n}{\partial s^p \partial t^q}$

$$= \dfrac{n!}{(n-p-q)!} \sum\limits_{\alpha=0}^{p} \sum\limits_{\beta=0}^{q} (-1)^{p+q+\alpha+\beta} \binom{p}{\alpha}\binom{q}{\beta} B_{i-\alpha,j-\beta,k+\alpha+\beta-p-q}^{n-p-q}(s,t)$$

20. Directional derivatives

a. $D_{\mathbf{u}}\left\{B_{ijk}^n(s,t)\right\}$

$$= n\left\{u_1 B_{i-1,j,k}^{n-1}(s,t) + u_2 B_{i,j-1,k}^{n-1}(s,t) - (u_1+u_2)B_{i,j,k-1}^{n-1}(s,t)\right\}$$

b. $D_{\mathbf{u}}^m\left\{B_{ijk}^n(s,t)\right\} = \dfrac{n!}{(n-m)!}$

$$\left\{\sum_{\alpha=0}^{i}\sum_{\beta=0}^{j}(-1)^{m-\alpha-\beta}\binom{m}{\alpha\beta\gamma}u_1^\alpha u_2^\beta(u_1+u_2)^{m-\alpha-\beta}B_{i-\alpha,j-\beta,k-\gamma}^{n-m}(s,t)\right\}$$

$D_{\mathbf{u}}^m$ denotes the mth directional derivative in the direction $\mathbf{u} = (u_1, u_2)$.

21. Integrals

a. $\displaystyle\int_0^t B_k^n(\tau)d\tau = \frac{1}{n+1}\sum_{j=k+1}^{n+1}B_j^{n+1}(t)$

b. $\displaystyle\int_t^1 B_k^n(\tau)d\tau = \frac{1}{n+1}\sum_{j=0}^{k}B_j^{n+1}(t)$

c. $\displaystyle\int_0^1 B_k^n(\tau)d\tau = \frac{1}{n+1}$

d. $\displaystyle\int_0^s B_{ijk}^n(\sigma,t)d\sigma = \frac{1}{n+1}\sum_{h=i+1}^{i+k+1}B_{h,j,i+k+1-h}^{n+1}(s,t)$

e. $\displaystyle\int_s^{1-t} B_{ijk}^n(\sigma,t)d\sigma = \frac{1}{n+1}\sum_{h=0}^{i}B_{h,j,i+k+1-h}^{n+1}(s,t)$

f. $\displaystyle\int_0^{1-t} B_{ijk}^n(\sigma,t)d\sigma = \frac{1}{n+1}B_j^{n+1}(t)$

g. $\displaystyle\int_0^t B_{ijk}^n(s,\tau)d\tau = \frac{1}{n+1}\sum_{h=j+1}^{j+k+1}B_{i,h,j+k+1-h}^{n+1}(s,t)$

h. $\displaystyle\int_t^{1-s} B_{ijk}^n(s,\tau)d\tau = \frac{1}{n+1}\sum_{h=0}^{j}B_{i,h,j+k+1-h}^{n+1}(s,t)$

i. $\displaystyle\int_0^{1-s} B_{ijk}^n(s,\tau)d\tau = \frac{1}{n+1}B_i^{n+1}(s)$

j. $\displaystyle\iint_{\Delta^2} B_{ijk}^n(\sigma,\tau)d\sigma d\tau = \frac{1}{(n+1)(n+2)}$

22. Degree elevation

a. $(1-t)B_k^n(t) = \dfrac{n+1-k}{n+1} B_k^{n+1}(t)$

b. $tB_k^n(t) = \dfrac{k+1}{n+1} B_{k+1}^{n+1}(t)$

c. $B_k^n(t) = \dfrac{n+1-k}{n+1} B_k^{n+1}(t) + \dfrac{k+1}{n+1} B_{k+1}^{n+1}(t)$

d. $sB_{ijk}^n(s,t) = \dfrac{i+1}{n+1} B_{i+1,j,k}^{n+1}(s,t)$

e. $tB_k^n(s,t) = \dfrac{j+1}{n+1} B_{i,j+1,k}^{n+1}(s,t)$

f. $(1-s-t)B_{ijk}^n(s,t) = \dfrac{k+1}{n+1} B_{i,j,k+1}^{n+1}(s,t)$

g. $B_{ijk}^n(s,t) = \dfrac{i+1}{n+1} B_{i+1,j,k}^{n+1}(s,t) + \dfrac{j+1}{n+1} B_{i,j+1,k}^{n+1}(s,t) + \dfrac{k+1}{n+1} B_{i,j,k+1}^{n+1}(s,t)$

23. Products and higher-order degree elevation

a. $B_j^m(t)B_k^n(t) = \dfrac{\binom{m}{j}\binom{n}{k}}{\binom{m+n}{j+k}} B_{j+k}^{m+n}(t)$

b. $B_k^n(t) = \displaystyle\sum_{j=0}^{m} \dfrac{\binom{m}{j}\binom{n}{k}}{\binom{m+n}{j+k}} B_{j+k}^{m+n}(t)$

c. $B_{ijk}^m(s,t)B_{pqr}^n(s,t) = \dfrac{\binom{m}{ijk}\binom{n}{pqr}}{\binom{m+n}{i+p\ \ j+q\ \ k+r}} B_{i+p,j+q,k+r}^{m+n}(s,t)$

d. $B_{pqr}^n(s,t) = \displaystyle\sum_{i+j+k=m} \dfrac{\binom{m}{ijk}\binom{n}{pqr}}{\binom{m+n}{i+p\ \ j+q\ \ k+r}} B_{i+p,j+q,k+r}^{m+n}(s,t)$

24. Marsden identities

a. $(x-t)^n = \displaystyle\sum_{k=0}^{n} \dfrac{(-1)^k}{\binom{n}{k}} B_{n-k}^n(x)B_k^n(t)$

b. $(sx+ty+1)^n = \displaystyle\sum_{i=0}^{n} \sum_{j=0}^{n-i} (x+1)^i (y+1)^j B_{ijk}^n(s,t)$

25. De Boor–Fix formulas

a. $\displaystyle\sum_{p=0}^{n} \dfrac{(-1)^{j+p}}{n!\binom{n}{i}} B_i^n(t)^{(p)} B_{n-j}^n(t)^{(n-p)} = \delta_{ij}$

b. $\displaystyle\sum_{p=0}^{i} \sum_{q=0}^{n-p} \dfrac{\binom{i}{p}\binom{j}{q}(n-p-q)!}{n!} \dfrac{\partial^{p+q} B_{klm}^n(0,0)}{\partial s^p \partial t^q} = \delta_{ik}\delta_{jl}$

26. Relationships between univariate and bivariate basis functions

    a. $B_i^n(s) = \sum_{j+k=n-i} B_{ijk}^n(s,t)$

    b. $B_j^n(t) = \sum_{i+k=n-j} B_{ijk}^n(s,t)$

    c. $B_k^n(1-s-t) = \sum_{i+j=n-k} B_{ijk}^n(s,t)$

    d. $B_{ijk}^n(s,t) = \sum\limits_{p=i}^{n} \sum\limits_{q=j}^{n-p} (-1)^{n-p-q} \binom{p}{p}\binom{q}{q}\binom{n}{n-p-q} B_i^p(s)B_j^q(t)$

    e. $B_{ijk}^n(su,tv) = \sum\limits_{p=i}^{n} \sum\limits_{q=j}^{n-p} B_i^p(s)B_j^q(t)B_{pqr}^n(s,t)$

27. Conversion between bivariate and tensor product bases

    a. $B_i^m(s)B_j^n(t) = \sum\limits_{p=i}^{m+n} \sum\limits_{q=j}^{m+i-j} \dfrac{\binom{p}{i}\binom{q}{j}\binom{m+n-p-q}{m-i+j-q}}{\binom{m+n}{n}} B_{pqr}^{m+n}(s,t)$

    b. $B_{ijk}^n(s,t) = \sum\limits_{h=i}^{n} \sum\limits_{l=j}^{n} \left\{ \sum\limits_{p=i}^{n-h+i} \sum\limits_{q=j}^{n-l+j} \dfrac{(-1)^r \binom{n}{pqr}\binom{p}{i}\binom{q}{j}\binom{n-p}{h-i}\binom{n-q}{l-j}}{\binom{n}{h}\binom{n}{l}} \right\} B_h^n(s)B_l^n(t)$

CHAPTER 6

Blossoming

A good labeling scheme can provide a lot of information about an algorithm, but till now we have avoided labeling the interior nodes in de Casteljau's algorithm. Here we shall introduce a labeling scheme for these nodes suggested by a probabilistic interpretation of the de Casteljau diagram. This labeling scheme will lead us to the notion of blossoming, an extremely powerful technique for analyzing the properties of Bezier curves and surfaces. Blossoming is a particularly effective tool for deriving change of basis algorithms. In Sections 6.3 and 6.5 we will see that formulas for degree elevation, subdivision, and conversion from monomial to Bezier form are easily derived from blossoming. We shall also apply blossoming in Chapter 7 to extend the de Casteljau algorithm to a more general evaluation procedure called the de Boor algorithm and in this manner introduce B-spline curves and surfaces.

6.1 Blossoming the de Casteljau Algorithm

In Neville's algorithm for Lagrange interpolation there is a natural labeling scheme for the nodes: the jth node on the kth level above the base is denoted by $P_{j...j+k}(t)$, since this polynomial interpolates the points $P_j,...,P_{j+k}$ at the nodes $t_j,...,t_{j+k}$ (see Figure 2.5). Similarly in de Casteljau's algorithm for Bernstein approximation the polynomial in the jth node on the kth level is the Bezier curve for the control points $P_j,...,P_{j+k}$ (see Figure 5.3). Therefore, you might be inclined to label this curve as $B_{j...j+k}(t)$; that is, you might be inclined to use much the same indices as in Neville's algorithm for the nodes in de Casteljau's algorithm.

But this notation for the nodes would not capture the functions along the edges of the de Casteljau diagram. The labeling scheme for Neville's algorithm is compelling because it captures both the control points and their associated parameter values; it tells us how to label the functions along the edges, and it tells us as well exactly where we are in the diagram. We would like to introduce a labeling scheme for the de Casteljau algorithm that captures these same properties.

When we began our investigation of Bezier curves, we saw that de Casteljau's algorithm is much simpler than Neville's algorithm because the functions along the edges are all the same: $(b-t)/(b-a)$ along all the left edges and $(t-a)/(b-a)$ along all the right edges. One possible paradigm for labeling de Casteljau's algorithm is a random walk: at each node you can proceed with some probability either to the left (a) or to the right (b). Remember that the Bernstein basis functions represent the binomial distribution, so there is a probabilistic flavor already inherent in Bezier curves.

Let us consider then the de Casteljau diagram for a degree n Bezier curve from this probabilistic point of view. Starting at the apex of the triangle, to arrive at the kth position at the base you must make exactly $n - k$ left turns and k right turns. If we think of the label along each arrow as the probability of proceeding left or right, then when $t = a$ the probability of turning left is one and when $t = b$ the probability of turning right is one. Following this line of reasoning, we shall adopt the notation $a^{n-k}b^k$ to denote $n - k$ left turns and k right turns or equivalently the kth position along the base of the triangle.

What about nodes above the base? Suppose we arrive at some internal node by making exactly j left turns and k right turns; then we still have $n - j - k$ choices to make before arriving at the base of the triangle. At a left turn we set $t = a$ and at a right turn we set $t = b$, but there are still $n - j - k$ values of t that need to be decided. Thus we shall denote this node by $a^j b^k t^{n-j-k}$; that is, for an internal node at the kth position on the $(n - j - k)$th level above the base, we adopt the notation $a^j b^k t^{n-j-k}$. The node at the apex of the triangle has the label t^n, since no turns have yet occurred. This notation tells us precisely where we are in the diagram, and it captures as well the labels along the edges. We illustrate this labeling of the nodes for the case $n = 3$ in Figure 6.1.

Notice that Figure 6.1 makes perfect sense if we interpret the values at the nodes as real numbers. Indeed multiplying the linear interpolation identity

$$t = \frac{b-t}{b-a}a + \frac{t-a}{b-a}b$$

by $a^i b^j t^{k-1}$ yields the identity

$$a^i b^j t^k = \frac{b-t}{b-a}a^{i+1}b^j t^{k-1} + \frac{t-a}{b-a}a^i b^{j+1} t^{k-1},$$

which is precisely the computation diagrammed in Figure 6.1.

But does it really make sense to adopt the notation $a^j b^k t^{n-j-k}$ for a node at the kth position on the $(n - j - k)$th level? We argued above that at each turn we must set $t = a$ or $t = b$, so at the $(n - j - k)$th level above the base we still need to set t a total of $n - j - k$ more times. But how can t keep changing value? If we are serious about setting some parameter to the value a or b depending on whether we turn right or left in the diagram, then shouldn't we be setting the value of a new parameter each time? Let's then introduce n new variables $u_1, ..., u_n$, where u_k appears in place of t on the kth level of the diagram, and let's adopt the notation

$$a^j b^k u_1 \cdots u_{n-j-k}$$

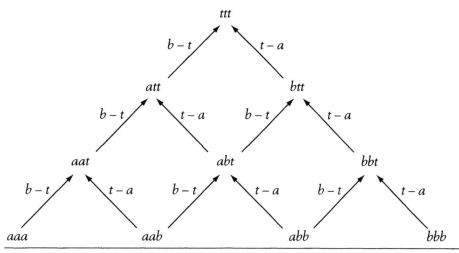

Figure 6.1 The labeling of the nodes for a cubic Bezier curve suggested by a random walk in de Casteljau's diagram. Here we have written *aaa* for a^3, *bbb* for b^3, and so on for reasons that will become clear in the next section.

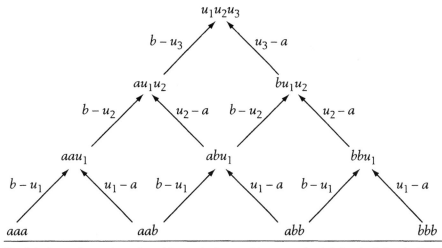

Figure 6.2 The labeling scheme introduced by replacing t with u_k on the kth level of the de Casteljau algorithm. The function that emerges at the apex of the triangle is the blossom of the Bezier curve depicted in Figure 6.1. Observe the close connection to discrete convolution—compare to Figure 5.29.

for a node at the kth position on the $(n - j - k)$th level of this diagram. This convention generates the labeling scheme depicted in Figure 6.2.

Notice again that Figure 6.2 makes perfect sense if we interpret the values at the nodes as multiplication of real numbers because

$$a^i b^j u_1 \cdots u_k = \frac{b - u_k}{b - a} a^{i+1} b^j u_1 \cdots u_{k-1} + \frac{u_k - a}{b - a} a^i b^{j+1} u_1 \cdots u_{k-1},$$

which is precisely the computation diagrammed in Figure 6.2.

The function that emerges at the apex of this triangle is no longer a polynomial in t; rather it is a function of n variables $u_1, \ldots, u_n$. If we place the control points $P_0, \ldots, P_n$ of a Bezier curve $P(t)$ at the base of the diagram, then the function of $u_1, \ldots, u_n$ that emerges at the apex is called the *blossom* of $P(t)$ and is denoted by $p(u_1, \ldots, u_n)$.

The function $p(u_1, \ldots, u_n)$ is symmetric in the variables $u_1, \ldots, u_n$, since the algorithm represents discrete convolution and discrete convolution is commutative (see Section 5.6.1). Indeed, let $P = (P_0, \ldots, P_n)$; then

$$p(u_1, \ldots, u_n) = \left\{ \left(\frac{b - u_1}{b - a}, \frac{u_1 - a}{b - a} \right) \otimes \cdots \otimes \left(\frac{b - u_n}{b - a}, \frac{u_n - a}{b - a} \right) \right\} \bullet P^T,$$

where the functions

$$\left(\frac{b - u_1}{b - a}, \frac{u_1 - a}{b - a} \right) \otimes \cdots \otimes \left(\frac{b - u_n}{b - a}, \frac{u_n - a}{b - a} \right)$$

are the convolution basis functions, P^T denotes the transpose of P, and $\bullet$ signifies matrix multiplication. Also $p(u_1, \ldots, u_n)$ is multiaffine in the variables $u_1, \ldots, u_n$—that is, p preserves affine combinations in each parameter—because each variable appears only to the first power, since variables on the same level of the de Casteljau algorithm never multiply one another. Finally, notice that to blossom the de Casteljau algorithm, we simply replace the parameter t on each level of the algorithm by a different variable u_k. Thus if we replace each variable u_k in the blossom by the parameter t, then we retrieve the original polynomial $P(t)$ because the algorithm reverts back to the de Casteljau algorithm for $P(t)$. Thus along the diagonal, $p(t, \ldots, t) = P(t)$. It turns out that these three properties—symmetry, multiaffine, diagonal—completely characterize the blossom. In the next section we shall define the blossom abstractly in terms of these three properties and then relate this definition back to our concrete construction of the blossom using the de Casteljau algorithm.

6.2   **Existence and Uniqueness of the Blossom**

Polynomial functions are complicated, but linear functions are simple. For example, if $L(t) = at$, then $L(\mu s + \lambda t) = \mu L(s) + \lambda L(t)$. More generally, if $L(t) = at + b$, then it is easy to verify that $L((1 - \lambda)s + \lambda t) = (1 - \lambda)L(s) + \lambda L(t)$, so $L(t)$ is an affine transformation since $L(t)$ preserves affine combinations. On the other hand, if $P(t) = a_n t^n + \cdots + a_0$, $n > 1$, then

$$P(\mu s + \lambda t) \neq \mu P(s) + \lambda P(t)$$

$$P((1 - \lambda)s + \lambda t) \neq (1 - \lambda)P(s) + \lambda P(t).$$

The main idea behind blossoming is to replace a complicated function $P(t)$ in one variable by a simple function $p(u_1,...,u_n)$ of many variables.

The *blossom* of a degree n polynomial $P(t)$ is the unique symmetric multiaffine function $p(u_1,...,u_n)$ that reduces to $P(t)$ along the diagonal. That is, $p(u_1,...,u_n)$ is the unique multivariate polynomial with the following three properties:

  i. *Symmetry*

$$p(u_1,...,u_n) = p(u_{\sigma(1)},...,u_{\sigma(n)}) \text{ for any permutation } \sigma \text{ of } \{1,...,n\}.$$

  ii. *Multiaffine*

$$p(u_1,...,(1-\alpha)u_k + \alpha w_k,...,u_n) = (1-\alpha)p(u_1,...,u_k,...,u_n) + \alpha p(u_1,...,w_k,...,u_n)$$

  iii. *Diagonal*

$$p(\underbrace{t,...,t}_{n}) = P(t)$$

The second property says that $p(u_1,...,u_n)$ is degree 1 in each variable (see Lemma 6.2); the third property connects the blossom back to the original polynomial.

Of course, it remains to establish the existence and uniqueness of a function satisfying these three properties. Shortly, we shall use the constructions in Section 6.1 to establish both existence and uniqueness. But before we proceed, let's compute a few simple examples.

Consider the functions $1, t, t^2, t^3$ as cubic polynomials. It is easy to blossom these monomials, since in each case it is easy to verify that the associated function $p(u_1,u_2,u_3)$ is symmetric, multiaffine, and reduces to the required monomial along the diagonal:

$$P(t) = 1 \implies p(u_1,u_2,u_3) = 1$$

$$P(t) = t \implies p(u_1,u_2,u_3) = \frac{u_1 + u_2 + u_3}{3}$$

$$P(t) = t^2 \implies p(u_1,u_2,u_3) = \frac{u_1 u_2 + u_2 u_3 + u_3 u_1}{3}$$

$$P(t) = t^3 \implies p(u_1,u_2,u_3) = u_1 u_2 u_3.$$

Now we can blossom any cubic polynomial, since

$$P(t) = a_3 t^3 + a_2 t^2 + a_1 t + a_0$$

$$p(u_1,u_2,u_3) = a_3 u_1 u_2 u_3 + a_2 \frac{u_1 u_2 + u_2 u_3 + u_3 u_1}{3} + a_1 \frac{u_1 + u_2 + u_3}{3} + a_0.$$

Similar techniques can be used to blossom polynomials of arbitrary degree by first blossoming t^k, $k = 0,...,n$, (Exercise 1) and then applying linearity (Proposition 6.4). Nevertheless, instead of proceeding in this manner, it is easier to apply the

Bezier constructions given in Section 6.1 to prove both existence and uniqueness. We begin by establishing a crucial connection between symmetric, multiaffine functions and the control points of Bezier curves.

THEOREM
6.1

Let $P(t)$ be a Bezier curve defined over the interval $[a,b]$ with control points $P_0,...,P_n$, and let $p(u_1,...,u_n)$ be a symmetric, multiaffine polynomial satisfying

$$p(\underbrace{t,...,t}_{n}) = P(t).$$

Then $P_k = p(\underbrace{a,...,a}_{n-k},\underbrace{b,...,b}_{k})$.

Proof   Consider Figure 6.1. We can now interpret this diagram in the following fashion:

  i. $a^{n-k}b^k$ denotes $p(\underbrace{a,...,a}_{n-k},\underbrace{b,...,b}_{k})$,

  ii. t^n denotes $p(\underbrace{t,...,t}_{n}) = P(t)$,

  iii. $a^j b^k t^{n-j-k}$ denotes $p(\underbrace{a,...,a}_{j},\underbrace{b,...,b}_{k},\underbrace{t,...,t}_{n-j-k})$.

Then Figure 6.1 shows how to compute $P(t)$ from

$$p(\underbrace{a,...,a}_{n-k},\underbrace{b,...,b}_{k}),\ k = 0,...,n,$$

recursively by applying the multiaffine and symmetry properties at each node of the triangle. But Figure 6.1 is also the de Casteljau algorithm for computing $P(t)$ from the control points P_k, $k = 0,...,n$. Since the Bernstein polynomials $B_0^n(t),...,B_n^n(t)$ form a polynomial basis, the control points of a Bezier curve are unique. It follows then that the values represented by $a^{n-k}b^k$, $k = 0,...,n$, must be the control points of $P(t)$; that is,

$$P_k = p(\underbrace{a,...,a}_{n-k},\underbrace{b,...,b}_{k}).$$

Theorem 6.1 is the central property relating the blossom to Bezier curves. This property is an extremely powerful result—in fact, it is equivalent to the diagonal property (see Section 6.3, Exercise 7)—and we shall have a good deal more to say about it in the next section. However, before we can proceed, we still need to establish the existence and uniqueness of the blossom. For this purpose, we provide an alternative characterization of the multiaffine property.

LEMMA
6.2

Let $p(u_1,...,u_n)$ be a polynomial in which each variable appears to at most the first power. Then $p(u_1,...,u_n)$ is multiaffine.

Proof We must show that

$$p(u_1,...,(1-\alpha)u_k + \alpha w_k,...,u_n) = (1-\alpha)p(u_1,...,u_k,...,u_n) + \alpha p(u_1,...,w_k,...,u_n) \ .$$

Since each variable appears to at most the first power in $p(u_1,...,u_n)$, we can write

$$p(u_1,...,u_k,...,u_n) = q(u_1,...,u_n) + u_k r(u_1,...,u_n),$$

where $q(u_1,...,u_n)$ and $r(u_1,...,u_n)$ are polynomials in which u_k does not appear. Therefore,

$$p(u_1,...,(1-\alpha)u_k + \alpha w_k,...,u_n)$$
$$= \{(1-\alpha)+\alpha\}q(u_1,...,u_n) + \{(1-\alpha)u_k + \alpha w_k\}r(u_1,...,u_n)$$
$$= (1-\alpha)\{q(u_1,...,u_n) + u_k r(u_1,...,u_n)\} + \alpha\{q(u_1,...,u_n) + w_k r(u_1,...,u_n)\}$$
$$= (1-\alpha)p(u_1,...,u_k,...,u_n) + \alpha p(u_1,...,w_k,...,u_n) \ .$$

THEOREM
6.3

Let $P(t)$ be a polynomial of degree n. Then its blossom $p(u_1,...,u_n)$ exists and is unique.

Proof *Existence.* In Figure 6.2 we showed how to blossom any Bezier curve $P(t)$—that is, how to generate a symmetric, multiaffine polynomial that reduces to $P(t)$ along the diagonal—by replacing t with u_k on the kth level of the de Casteljau algorithm. (The multiaffine property follows from Lemma 6.2 because in this construction each parameter u_k appears only to the first power.) Since the Bernstein polynomials $B_0^n(t),...,B_n^n(t)$ form a basis for polynomials of degree n, every polynomial can be written in Bezier form. Therefore, every polynomial has a blossom.

Uniqueness. Again consider Figure 6.2. We can now interpret this diagram in the following fashion:

i. $a^{n-k}b^k$ denotes $p(\underbrace{a,...,a}_{n-k},\underbrace{b,...,b}_{k})$,

ii. $u_1 \cdots u_n$ denotes $p(u_1,...,u_n)$,

iii. $a^j b^k u_1 \cdots u_{n-j-k}$ denotes $p(\underbrace{a,...,a}_{j},\underbrace{b,...,b}_{k},u_1,...,u_{n-j-k})$.

Then Figure 6.2 shows how to compute the blossom value $u_1 \cdots u_n$ from the blossom values $a^{n-k}b^k$, $k = 0,\ldots,n$, recursively by applying the multiaffine and symmetry properties at each node of the triangle. Thus the blossom of $P(t)$ is completely determined by the blossom values $a^{n-k}b^k$, $k = 0,\ldots,n$. But by Theorem 6.1 the blossom values $a^{n-k}b^k$ are the Bezier control points of $P(t)$. Now suppose that the polynomial $P(t)$ has two blossoms $p(u_1,\ldots,u_n)$ and $q(u_1,\ldots,u_n)$. Then by Theorem 6.1

$$p(\underbrace{a,\ldots,a}_{n-k},\underbrace{b,\ldots,b}_{k}) = q(\underbrace{a,\ldots,a}_{n-k},\underbrace{b,\ldots,b}_{k}),$$

since both sides must represent the Bezier control points of $P(t)$ and these control points are unique. But we have seen (Figure 6.2) that any arbitrary blossom value can be computed from the blossom values $a^{n-k}b^k$, $k = 0,\ldots,n$. Therefore,

$$p(u_1,\ldots,u_n) = q(u_1,\ldots,u_n),$$

so the blossom of $P(t)$ is unique.

Why does this approach to blossoming work? Figures 6.1 and 6.2 were originally derived for ordinary multiplication of real numbers. But the only properties of the real numbers that we need are that multiplication is commutative (symmetry) and distributes through addition (multiaffinity). Thus any function with these two features will satisfy these diagrams (see Exercise 10). This observation is the key to understanding the properties of the blossom.

The uniqueness of the blossom plays an important role in the derivations of formulas for the blossom. If we want to establish that some polynomial $p(u_1,\ldots,u_n)$ represents the blossom of $P(t)$, all we need to prove is that the expression $p(u_1,\ldots,u_n)$ is symmetric, multiaffine, and reduces to $P(t)$ along the diagonal. It then follows by the uniqueness of the blossom that the polynomial $p(u_1,\ldots,u_n)$ must be the blossom of $P(t)$. We have seen this trick once before, in Section 4.4, where we introduced axioms for the divided difference. To verify that an expression represented a divided difference, we simply verified that it satisfied the axioms characterizing the divided difference. We can now do the same for the blossom. We illustrate this proof technique in the following proposition. Additional examples are provided in the exercises.

PROPOSITION
6.4

Let $P(t)$ and $Q(t)$ be polynomials of degree n.

    a.   If $R(t) = P(t) + Q(t)$, then $r(u_1,\ldots,u_n) = p(u_1,\ldots,u_n) + q(u_1,\ldots,u_n)$.

    b.   If $S(t) = cP(t)$, then $s(u_1,\ldots,u_n) = cp(u_1,\ldots,u_n)$.

Thus blossoming is a linear operator.

Proof Since both p and q are symmetric functions in the parameters $u_1,...,u_n$, so too are the functions $p + q$ and cp. Similarly since both p and q are multi-affine functions in the parameters $u_1,...,u_n$, the functions $p + q$ and cp are also multiaffine functions in the parameters $u_1,...,u_n$ (see Exercise 11). Moreover, along the diagonal

$$p(t,...,t) + q(t,...,t) = P(t) + Q(t) = R(t)$$
$$cp(t,...,t) = cP(t) = S(t) \ .$$

Thus $p(u_1,...,u_n) + q(u_1,...,u_n)$ is a symmetric, multiaffine function that reduces to $R(t)$ along the diagonal. Hence by the uniqueness of the blossom,

$$r(u_1,...,u_n) = p(u_1,...,u_n) + q(u_1,...,u_n).$$

Similarly, $cp(u_1,...,u_n)$ is a symmetric, multiaffine function that reduces to $S(t)$ along the diagonal. Hence again by the uniqueness of the blossom,

$$s(u_1,...,u_n) = cp(u_1,...,u_n).$$

Exercises

1. Let $M_k^n(t) = \binom{n}{k} t^k$, and consider the function

$$m_k^n(u_1,...,u_n) = \sum u_{i_1} \cdots u_{i_k},$$

where the sum is taken over all subsets $\{i_1,...,i_k\}$ of $\{1,...,n\}$. Show that

 a. $m_k^n(u_1,...,u_n)$ is a symmetric multiaffine function that reduces to $M_k^n(t)$ along the diagonal.

 b. $\left(m_0^n(u_1,...,u_n),...,m_n^n(u_1,...,u_n)\right) = (1,u_1) \otimes \cdots \otimes (1,u_n)$.

 Use the result in part (a) to establish the existence of the blossom.

2. Let $P_k^n(t) = (t - a_k)^n$ and let $p_k^n(u_1,...,u_n) = (u_1 - a_k) \cdots (u_n - a_k)$, $k = 0,...,n$.

 a. Show that $p_k^n(u_1,...,u_n)$ is a symmetric multiaffine function that reduces to $P_k^n(t)$ along the diagonal.

 b. Use the result in part (a) to establish the existence of the blossom.

3. Let $b_i^n(u_1,...,u_n)$ denote the blossom of the Bernstein basis function $B_i^n(t)$. Prove that

 a. $b_i^n(u_1,...,u_n) = \sum_{i_1+\cdots+i_n=i} B_{i_1}^1(u_1) \cdots B_{i_n}^1(u_n)$

 b. $\left(b_0^n(u_1,...,u_n),...,b_n^n(u_1,...,u_n)\right) = \left((1-u_1),u_1\right) \otimes \cdots \otimes \left((1-u_n),u_n\right)$

4. Let $L_0(t),L_1(t)$ be arbitrary linear functions in t, and suppose that

$$\left(F_0(t),...,F_n(t)\right) = \underbrace{\left(L_0(t),L_1(t)\right) \otimes \cdots \otimes \left(L_0(t),L_1(t)\right)}_{n-factors}.$$

Show that

$$\left(f_0(u_1,\ldots,u_n),\ldots,f_n(u_1,\ldots,u_n)\right) = \left(L_0(u_1),L_1(u_1)\right) \otimes \cdots \otimes \left(L_0(u_n),L_1(u_n)\right).$$

5. Prove the converse of Lemma 6.2. That is, prove that if $p(u_1,\ldots,u_n)$ is a multi-affine polynomial, then each variable appears to at most the first power. (Hint: Observe that $u_k = (1-u_k) \bullet 0 + u_k \bullet 1$.)

6. Let $P(t)$ be a polynomial of degree n. Then $P(t)$ is also a polynomial of degree $n + 1$, so the blossoms $p(u_1,\ldots,u_n)$ and $p(u_1,\ldots,u_n,u_{n+1})$ both exist. Show that, in general, $p(u_1,\ldots,u_n,0) \neq p(u_1,\ldots,u_n)$.

7. Let $P(t)$ be a polynomial of degree n, and let $\psi(t) = (u_1 - t)\cdots(u_n - t)$. Prove that

$$p(u_1,\ldots,u_n) = \sum_{k=0}^{n} \frac{(-1)^{n-k}}{n!} P^{(k)}(\tau)\psi^{(n-k)}(\tau).$$

The right-hand side of this expression is called the de Boor–Fix form of the blossom. (Hint: Show that the right-hand side satisfies the three defining properties of the blossom and apply uniqueness.)

8. Let $P(t)$ and $Q(t)$ be polynomials of degree n, and define

$$[P(t),Q(t)]_n = \sum_{k=0}^{n} \frac{(-1)^{n-k}}{n!} P^{(k)}(\tau)Q^{(n-k)}(\tau).$$

Show that

a. $[P(t),Q(t)]_n$ is a bilinear operator.

b. $[P^{(r)}(t),Q(t)]_n = (-1)^r[P(t),Q^{(r)}(t)]_n$.

c. $[P(t),Q(t)]_n$ is a constant independent of the choice of τ.

d. $P(x) = [P(t),(x-t)^n]_n$.

e. $P^{(r)}(x) = \dfrac{n!}{(n-r)!}[P(t),(x-t)^{n-r}]_n$.

f. $p(u_1,\ldots,u_n) = [P(t),(u_1 - t)\cdots(u_n - t)]_n$.

g. Compare part (f) to Exercise 7.

9. Use the de Boor–Fix representation of the blossom given in Exercise 7 to show that the formulas for differentiating and degree elevating the Bernstein polynomials are equivalent—that is, show that either formula implies the other.

10. Let $u_1\cdots u_n$ denote the polynomial $(u_1 - x)\cdots(u_n - x)$.

a. Show that with this interpretation the algorithms represented by Figures 6.1 and 6.2 remain valid.

b. Recall the bracket operator defined in Exercise 8. What do you get if you bracket each polynomial in the algorithms represented by Figures 6.1 and 6.2 with a fixed polynomial $P(x)$?

11. Prove that if both $p(u_1,...,u_n)$ and $q(u_1,...,u_n)$ are symmetric multiaffine functions in the parameters $u_1,...,u_n$, so too are the functions $p+q$ and cp.

12. Let $P(t)$ be a polynomial of degree m, and let $Q(t)$ be a polynomial of degree n. Derive the following blossoming identities:

a. Products: If $R(t) = P(t)Q(t)$, then

$$r(u_1,...,u_{m+n}) = \frac{\sum_\sigma p(u_{\sigma(1)},...,u_{\sigma(m)})q(u_{\sigma(m+1)},...,u_{\sigma(m+n)})}{(m+n)!}.$$

b. Composites: If $S(t) = (P \circ Q)(t)$, then

$$s(u_1,...,u_{mn}) = \frac{\sum_\sigma p\big(q(u_{\sigma(1)},...,u_{\sigma(n)}),...,q(u_{\sigma(mn-n+1)},...,u_{\sigma(mn)})\big)}{(mn)!}.$$

(Hint: Check first that the right-hand side satisfies the blossoming axioms and then invoke uniqueness.)

13. Let $A(t) = \big(A_0(t),...,A_m(t)\big)$ be a sequence of $m+1$ polynomials of degree m with blossoms $a(u_1,...,u_m) = \big(a_0(u_1,...,u_m),...,a_m(u_1,...,u_m)\big)$, and let $B(t) = \big(B_0(t),...,B_n(t)\big)$ be a sequence of $n+1$ polynomials of degree n with blossoms $b(u_1,...,u_n) = \big(b_0(u_1,...,u_n),...,b_n(u_1,...,u_n)\big)$. If $C(t) = A(t) \otimes B(t)$, show that

$$c(u_1,...,u_{m+n}) = \frac{\sum_\sigma a(u_{\sigma(1)},...,u_{\sigma(m)}) \otimes b(u_{\sigma(m+1)},...,u_{\sigma(m+n)})}{(m+n)!}.$$

6.3 Change of Basis Algorithms

Blossoming is a powerful machine for deriving change of basis formulas. Theorem 6.1 is sometimes called the *dual functional property* of the blossom because it shows how to use the blossom to compute the Bezier control points of any polynomial curve. Indeed, just as the divided difference evaluated at the nodes provides the dual functionals for the Newton basis (see Section 4.2), the blossom evaluated at the end points of the parameter domain represents the dual basis for the Bernstein basis. Applying Theorem 6.1, we can convert any polynomial curve to Bezier form by blossoming the curve and evaluating the blossom at the end points of the parameter domain. Here we shall show how to apply blossoming to derive three important results from Chapter 5: subdivision, degree elevation, and conversion from monomial to Bezier form.

Subdivision is very easy using blossoming. Let $P(t)$ be a Bezier curve over the interval $[a,b]$ with control points $P_0,...,P_n$. By the dual functional property,

$$P_k = p(\underbrace{a,...,a}_{n-k},\underbrace{b,...,b}_{k}).$$

To subdivide this Bezier curve at the parameter t, we must find the Bezier control points for the intervals $[a,t]$ and $[t,b]$. Again by the dual functional property for Bezier curves, these control points $Q_0,...,Q_n$ and $R_0,...,R_n$ are

$$Q_k = p(\underbrace{a,...,a}_{n-k},\underbrace{t,...,t}_{k})$$

$$R_k = p(\underbrace{t,...,t}_{n-k},\underbrace{b,...,b}_{k}).$$

But look at Figure 6.1. If we interpret every triple uvw as the blossom value $p(u,v,w)$, then the desired control points Q_k and R_k emerge along the left and right edges of the triangle. Generalized to arbitrary degree, this observation is precisely the de Casteljau subdivision algorithm of Section 5.5.4.2.

Let's try degree elevation. Here we are given the control points for a Bezier curve of degree n, and we must find the control points that represent the same curve as a Bezier curve of degree $n + 1$. Let $P(t)$ be a Bezier curve of degree n over the interval $[0,1]$ with control points $P_0,...,P_n$, and let $Q_0,...,Q_{n+1}$ be the control points that represent $P(t)$ over the interval $[0,1]$ as a Bezier curve of degree $n + 1$. If p_n denotes the blossom of P as a polynomial of degree n and p_{n+1} denotes the blossom of P as a polynomial of degree $n + 1$, then by the dual functional property

$$P_k = p_n(\underbrace{0,...,0}_{n-k},\underbrace{1,...,1}_{k})$$

$$Q_k = p_{n+1}(\underbrace{0,...,0}_{n+1-k},\underbrace{1,...,1}_{k}).$$

Our problem is to find formulas for the control points $Q_0,...,Q_{n+1}$ in terms of the control points $P_0,...,P_n$. From the perspective of blossoming, this problem reduces to finding a formula for the blossom p_{n+1} in terms of the blossom p_n.

**PROPOSITION
6.5**

Degree Elevation

$$p_{n+1}(u_1,...,u_{n+1}) = \frac{\displaystyle\sum_{i=1}^{n+1} p_n(u_1,...,u_{i-1},u_{i+1},...,u_{n+1})}{n+1}.$$

Proof By the uniqueness of the blossom, it is enough to prove that the right-hand side is symmetric, multiaffine, and reduces to $P(t)$ along the diagonal. But the right-hand side is certainly symmetric and multiaffine because p_n is symmetric and multiaffine.   Moreover, by the diagonal property of p_n,

$$P(t) = \frac{\displaystyle\sum_{i=1}^{n+1} p_n(\underbrace{t,...,t}_{n})}{n+1}.$$

Hence the result follows by the uniqueness of the blossom.

We can apply Proposition 6.5 to solve the degree-elevation problem for Bezier curves. By the dual functional property, we already know that

$$Q_k = p_{n+1}(\underbrace{0,...,0}_{n+1-k},\underbrace{1,...,1}_{k}).$$

By Proposition 6.5 omitting a single zero $n + 1 - k$ times and omitting a single one k times yields

$$Q_k = \frac{n+1-k}{n+1} p_n(\underbrace{0,...,0}_{n-k},\underbrace{1,...,1}_{k}) + \frac{k}{n+1} p_n(\underbrace{0,...,0}_{n+1-k},\underbrace{1,...,1}_{k-1})$$

$$= \frac{n+1-k}{n+1} P_k + \frac{k}{n+1} P_{k-1},$$

which is exactly the degree elevation formula derived in Section 5.5.3.

Finally, to convert from monomial to Bezier form, let $M_k^n(t) = \binom{n}{k}t^k$. Then the blossom of the monomial $M_k^n(t)$ is given by the function

$$m_k^n(u_1,...,u_n) = \sum u_{i_1} \cdots u_{i_k}, \tag{6.1}$$

where the sum is taken over all subsets $\{i_1,...,i_k\}$ of $\{1,...,n\}$. This result follows by the uniqueness of the blossom because the right-hand side is clearly symmetric, multiaffine, and reduces to $M_k^n(t)$ along the diagonal (see Exercise 1 of Section 6.2). Now to convert from monomial to Bezier form, consider a polynomial $Q(t) = \sum_k Q_k M_k^n(t)$. By the linearity of the blossom (Proposition 6.4),

$$q(u_1,...,u_n) = \sum_{k=0}^{n} Q_k m_k^n(u_1,...,u_n) .$$

Therefore, by the dual functional property of the blossom, the Bezier control points $P_0,...,P_n$ of $Q(t)$ over the interval $[0,1]$ are given by

$$P_j = q(\underbrace{0,...,0}_{n-j},\underbrace{1,...,1}_{j}) = \sum_{k=0}^{n} Q_k m_k^n(\underbrace{0,...,0}_{n-j},\underbrace{1,...,1}_{j}) .$$

But whenever we substitute 0 for one of the parameters u_{i_h} on the right-hand side of Equation (6.1), the term vanishes. Hence we are concerned only with terms where a 1 is substituted for each parameter u_{i_h}. Since there are k u-parameters and 1 is repeated j times, there are $\binom{j}{k}$ ways to choose k 1s. Thus

$$m_k^n(\underbrace{0,...,0}_{n-j},\underbrace{1,...,1}_{j}) = \binom{j}{k}$$

so

$$P_j = \sum_{k=0}^{j} \binom{j}{k} Q_k,$$

which is exactly the change of basis formula derived in Section 5.5.1.

The derivations of these three change of basis algorithms—subdivision, degree elevation, and conversion from monomial to Bezier form—are very slick. In Chapter 5 we had to invent a new trick every time we wanted to convert from some new basis to Bezier form. We applied probability theory (urn models) to derive subdivision, algebraic identities to achieve degree elevation, and generating functions to convert from monomial to Bezier form. Here there is only one trick: blossoming. Thus blossoming simplifies and unifies the analysis of Bezier curves. It is a very clever idea.

Exercises

1. Use blossoming to derive the following identities:

   a. $\displaystyle\sum_{k=0}^{n} B_k^n(t) = 1$

   b. $\displaystyle\sum_{k=0}^{n} (-1)^k B_k^n(t) = (1-2t)^n$

   c. $\displaystyle\sum_{j=k}^{n} \binom{j}{k} B_j^n(t) = \binom{n}{k} t^k$

2. Apply blossoming and Exercise 12(b) of Section 6.2 to derive the following identities:

   a. $\displaystyle B_i^n(rt) = \sum_{k=i}^{n} B_i^k(r) B_k^n(t)$

   b. $\displaystyle B_i^n((1-t)r+t) = \sum_{k=0}^{i} B_{i-k}^{n-k}(r) B_k^n(t)$

3. Consider the following Marsden identity for the Bernstein basis:

   $$(x-t)^n = \sum_{k=0}^{n} \frac{(-1)^k}{\binom{n}{k}} B_{n-k}^n(x) B_k^n(t).$$

   Prove this Marsden identity using

   a. the binomial theorem (Hint: $x - t = x(1-t) - (1-x)t$.)

   b. blossoming

4. Using the result in Exercise 3 and the bracket operator defined in Section 6.2, Exercise 8, show that

   $$f(\underbrace{0,\ldots,0}_{n-k},\underbrace{1,\ldots,1}_{k}) = [F(t), \frac{(-1)^k}{\binom{n}{k}} B_{n-k}^n(t)]_n.$$

5. Blossoming polynomial identities yields identities for the blossom.

   a. What are the blossom identities corresponding to the identities in Exercises 1–3?

   b. What formula for the blossom do you get when you blossom Taylor's Theorem?

6. Let $P_k^n(t) = \dfrac{(t - t_k)^n}{\prod_{j \neq k}(t_j - t_k)}$, $k = 0,\ldots,n$, and let $P(t) = \sum\limits_{k=0}^{n} P_k^n(t) P_k$.

   Using the result of Section 6.2, Exercise 2, show that
   $$P_k = p(t_0,\ldots,t_{k-1},t_{k+1},\ldots,t_n).$$

7. Let $P(t)$ be a degree n Bezier curve defined over the interval $[a,b]$ with control points $P_0,\ldots,P_n$, and let $p(u_1,\ldots,u_n)$ be a symmetric, multiaffine polynomial satisfying the dual functional property
   $$P_k = p(\underbrace{a,\ldots,a}_{n-k},\underbrace{b,\ldots,b}_{k}) \quad k = 0,\ldots,n.$$

   a. Show that $P(t) = p(\underbrace{t,\ldots,t}_{n})$.

   b. Conclude that in the axioms for the blossom the diagonal property can be replaced by the dual functional property.

   c. In the axioms for the divided difference (see Section 4.4), which axiom corresponds to the diagonal property? What is the corresponding dual functional property?

8. Let $b_j^n(u_1,\ldots,u_n)$ denote the blossom of the Bernstein basis function $B_j^n(t)$ over the interval $[0,1]$. Show that
   $$b_j^n(\underbrace{0,\ldots,0}_{n-k},\underbrace{1,\ldots,1}_{k}) = \delta_{jk}.$$

9. Generalize the degree elevation formula in Proposition 6.5 by showing that
   $$p_{n+k}(u_1,\ldots,u_{n+k}) = \frac{\sum p_n(u_{i_1},\ldots,u_{i_n})}{\binom{n+k}{k}}.$$

6.4 Differentiation and the Homogeneous Blossom

We know how to represent points along polynomial curves in terms of blossom values by invoking the diagonal property, but what about derivatives? Curves take on values that are points in affine space; so too does the blossom. Indeed, only affine combinations of blossom values are permitted. But derivatives are vectors, not points. We need a variant of the blossom that takes on values in a vector space, rather than an affine space. Here we shall construct such a blossom by applying the technique of *homogenization*. Homogenization lifts the domain and the range from affine space to Grassmann space, so that all our computations can be performed in a vector space.

Consider a degree n polynomial $P(t) = \sum_k a_k t^k$. To homogenize $P(t)$, we multiply each term t^k by w^{n-k}. This creates a new polynomial in two variables,
$$P(t,w) = \sum_k a_k t^k w^{n-k}.$$

The point about $P(t,w)$ is that it is homogeneous of degree n; that is, each term has the same total degree n. Homogeneous polynomials are sometimes simpler to manipulate. For example, it is easy to verify that $P(ct,cw) = c^n P(t,w)$, but there is no comparable simple formula for $P(ct)$. We can easily recover $P(t)$ from $P(t,w)$ because $P(t) = P(t,1)$. Constructing $P(t,w)$ from $P(t)$ is called *homogenization*, and recovering $P(t)$ from $P(t,w)$ is called *dehomogenization*.

Now we have two processes that we can apply to polynomials: blossoming and homogenization. Can we homogenize the blossom or blossom the homogenization? Yes. To homogenize the blossom $p(u_1,...,u_n)$, we homogenize with respect to each variable independently. Thus the homogeneous version of $p(u_1,...,u_n)$ is another polynomial $p((u_1,v_1),...,(u_n,v_n))$ that is homogeneous with respect to each pair of variables (u_k,v_k). The original parameters $u_1,...,u_n$ lie in affine space; the homogeneous parameters $(u_1,v_1),...,(u_n,v_n)$ lie in Grassmann space. In every term of $p(u_1,...,u_n)$, each variable u_k appears to at most the first power, so every term of the homogeneous polynomial $p((u_1,v_1),...,(u_n,v_n))$ has as a factor either u_k or v_k but not both. Since $p((u_1,v_1),...,(u_n,v_n))$ is homogeneous of degree 1 in each pair of variables (u_k,v_k),

$$p((u_1,v_1),...,c(u_k,v_k),...,(u_n,v_n)) = cp((u_1,v_1),...,(u_k,v_k),...,(u_n,v_n)).$$

Again we can dehomogenize $p((u_1,v_1),...,(u_n,v_n))$ by setting $v_k = 1$, $k = 1,...,n$. Thus $p((u_1,1),...,(u_n,1)) = p(u_1,...,u_n)$.

We can also blossom the homogenization. We define the blossom of a homogeneous polynomial $P(t,w)$ to be the unique symmetric, multilinear polynomial $p((u_1,v_1),...,(u_n,v_n))$ such that $p((t,w),...,(t,w)) = P(t,w)$. Note that for the homogeneous blossom, we have replaced the multiaffine property by the multilinear property. By *multilinear* we mean that the polynomial $p((u_1,v_1),...,(u_n,v_n))$ is a linear function in each pair of variables (u_k,v_k), $k = 1,...,n$. That is,

$$p((u_1,v_1),...,(u_k,v_k)+(r_k,s_k),...,(u_n,v_n))$$
$$= p((u_1,v_1),...,(u_k,v_k),...,(u_n,v_n)) + p((u_1,v_1),...,(r_k,s_k),...,(u_n,v_n)) \qquad (6.2)$$

$$p((u_1,v_1),...,c(u_k,v_k),...,(u_n,v_n)) = cp((u_1,v_1),...,(u_k,v_k),...,(u_n,v_n))$$

Thus this multilinear blossom must take on values in a vector space, since we can add and multiply by scalars.

Now we seem to have two notions of multilinear functions: polynomials that are homogeneous of degree 1 in each pair of parameters, and functions that satisfy Equation (6.2). Fortunately, in analogy with Lemma 6.2, we have the following result.

LEMMA 6.6	Let $p((u_1,v_1),...,(u_n,v_n))$ be a polynomial in which for each $k = 1,...,n$ either u_k or v_k, but not both, appears in every term to the first power. Then $p((u_1,v_1),...,(u_n,v_n))$ satisfies Equation (6.2)—that is, $p((u_1,v_1),...,(u_n,v_n))$ is multilinear.

Proof Since the proof of this result is so similar to the proof of Lemma 6.2, we leave this result as an exercise (see Exercise 3).

We constructed the multiaffine blossom by blossoming the de Casteljau algorithm, and we can do the same for the multilinear blossom. Begin by homogenizing the de Casteljau algorithm. This amounts to replacing $b-t \rightarrow bw-t$ and $t-a \rightarrow t-aw$ to ensure that each term has the same total degree (see Figure 6.3). To blossom, we now replace the pair (t,w) by the pair (u_k, v_k) on the kth level of the algorithm (see Figure 6.4). This process generates a symmetric, multilinear function that reduces to the homogeneous curve when we replace each pair (u_k, v_k) by (t,w). This function is multilinear rather than multiaffine because it is linear in (u_k, v_k) on the kth level of the algorithm. Thus u_k or v_k, but not both, appears in every term to the first power, so by Lemma 6.6 this function is multilinear.

Notice, by the way, that if we blossom first and then homogenize, we get exactly the same diagram (see Figure 6.5)! Blossoming first gives us Figure 6.2; homogenizing this diagram generates Figure 6.4 once again. Observe that in each of these diagrams, we are applying the multilinear property using the identity

$$(t,w) = \frac{bw-t}{b-a}(a,1) + \frac{t-aw}{b-a}(b,1).$$

We built the homogeneous blossom to deal specifically with differentiation. Consider now what happens in the homogeneous version of de Casteljau's algorithm (Figure 6.3) if we replace $(t,w) \rightarrow (1,0)$ on any level of the algorithm. The effect is to

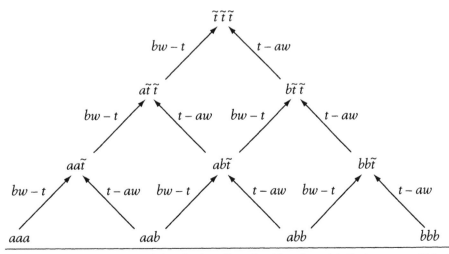

Figure 6.3 The homogeneous version of the de Casteljau algorithm for cubic Bezier curves. This diagram is generated from the de Casteljau algorithm (Figure 6.1) by replacing $b - t \rightarrow bw - t$ and $t - a \rightarrow t - aw$ along the edges of the triangle. We use ~ to denote homogeneous values, so $\tilde{t} = (t, w)$ while $a = (a, 1)$.

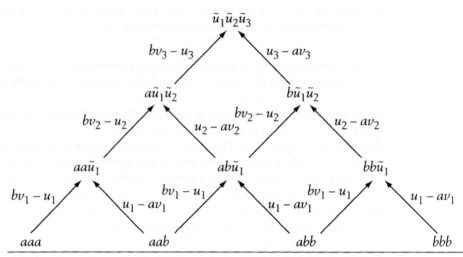

Figure 6.4 The multilinear blossom of a cubic Bezier curve. This diagram is generated from the multi-affine blossom (Figure 6.2) by homogenizing the functions along the edges. It can also be generated by blossoming the homogeneous version of the de Casteljau algorithm (Figure 6.3)—that is, by replacing the pair (t, w) with the pair (u_k, v_k) on the kth level of the algorithm. Thus blossoming and then homogenizing is equivalent to homogenizing and then blossoming. As in Figure 6.3, we use ~ to denote homogeneous values, so $\tilde{u} = (u, v)$ while $a = (a, 1)$.

replace $bw - t \rightarrow -1$ and $t - aw \rightarrow +1$; that is, the effect is to differentiate one level of the algorithm. Now if we replace $(t, w) \rightarrow (t, 1)$ on the remaining levels of the algorithm—that is, if we dehomogenize the remaining levels—then, up to a constant multiple, we get the derivative of the original Bezier curve (see Figure 5.30). In fact, if on r levels of the homogeneous de Casteljau algorithm, we replace $(t, w) \rightarrow (1, 0)$ and on the remaining $n - r$ levels we replace $(t, w) \rightarrow (t, 1)$, then, up to a constant multiple, we obtain the rth derivative of the original Bezier curve.

Thus we can compute derivatives for any polynomial $P(t)$ using the homogeneous blossom. In fact, let $\delta = (1, 0)$ and let $t = (t, 1)$. Then we have just proved that

$$P'(t) = np(\underbrace{t, ..., t}_{n-1}, \delta)$$

$$P^{(r)}(t) = \frac{n!}{(n-r)!} p(\underbrace{t, ..., t}_{n-r}, \underbrace{\delta, ..., \delta}_{r}).$$

These formulas that tell us how to compute the derivative of a polynomial curve in terms of its multilinear blossom turn out to be central to our understanding of how to differentiate B-splines. We shall apply these formulas in Chapter 7 to differentiate B-spline curves.

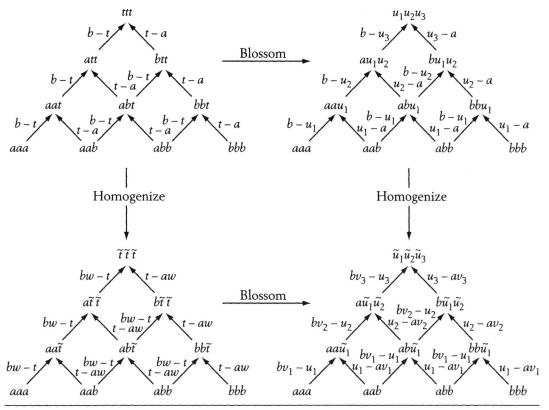

Figure 6.5   Blossoming and homogenizing commute. We illustrate this phenomenon with four versions of the de Casteljau algorithm for cubic Bezier curves: the standard version (upper left), the homogeneous version (lower left), the blossomed version (upper right), and the homogeneous blossom (lower right). Again we use ~ to denote homogeneous values, so $\tilde{t} = (t, w)$ and $a = (a, 1)$.

Exercises

1. Let $P(t,w)$ and $Q(t,w)$ be two homogeneous polynomials of degree n. Show that

    a. $\dfrac{P(ct,cw)}{Q(ct,cw)} = \dfrac{P(t,w)}{Q(t,w)}$

    b. $P(t,w) = w^n P(t/w, 1)$

2. Let $P(t)$ be a polynomial of degree m and let $Q(t)$ be a polynomial of degree n. Show that

    a. $R(t) = P(t) + Q(t) \Rightarrow R(t,w) = P(t,w) + Q(t,w)$ if and only if $m = n$

    b. $S(t) = P(t)Q(t) \Rightarrow S(t,w) = P(t,w)Q(t,w)$.

3. Prove Lemma 6.6. That is, prove that if $p((u_1,v_1),...,(u_n,v_n))$ is a polynomial in which for each $k = 1,...,n$ either u_k or v_k, but not both, appears in every term to the first power, then $p((u_1,v_1),...,(u_n,v_n))$ is multilinear. (Hint: Use the same techniques as in the proof of Lemma 6.2.)

4. Prove the converse of Lemma 6.6. That is, prove that if the function $p((u_1,v_1),...,(u_n,v_n))$ is a multilinear polynomial, then for each $k = 1,...,n$ either u_k or v_k, but not both, appears in every term to the first power.

5. Suppose that $P(t)$ is a polynomial of degree n. Let p' denote the blossom of $P'(t)$ and let $p^{(r)}$ denote the blossom of $P^{(r)}(t)$. Show that

   a. $p'(u_1,...,u_{n-1}) = np(u_1,...,u_{n-1},\delta)$

   b. $p^{(r)}(u_1,...,u_{n-r}) = \dfrac{n!}{(n-r)!}p(u_1,...,u_{n-r},\underbrace{\delta,...,\delta}_{r})$

6. Using the bracket operator defined in Section 6.2, Exercise 8, show that

$$P^{(r)}(x) = \frac{n!}{(n-r)!}p(\underbrace{x,...,x}_{n-r},\underbrace{\delta,...,\delta}_{r}) = [P(t),\frac{d^r}{dx^r}(x-t)^n]_n .$$

7. Let m_j denote the multiplicity of the parameter u_j in the sequence $u_1,...,u_n$. Prove that

   a. $\dfrac{\partial p(u_1,...,u_n)}{\partial u_j} = m_j p(u_1,...,u_{j-1},\delta,u_{j+1},...,u_n)$

   b. $\dfrac{\partial p(u_1,...,u_n)}{\partial u_j} = \dfrac{m_j}{n} p'(u_1,...,u_{j-1},u_{j+1},...,u_n)$

   Compare this result to Section 4.3, Exercise 3(a). (Hint: Consider the basis $P_k^n(t) = (t-a_k)^n$, $k = 0,...,n$.)

8. Prove that the multilinear blossom of a homogeneous polynomial $P(t,w)$ is unique.

9. Let $P(t)$ be a polynomial of degree n with homogenization $P(t,w)$ and blossom $p(u_1,...,u_n)$. Suppose that $q((u_1,v_1),...,(u_n,v_n))$ is a symmetric, multilinear function satisfying the dehomogenization property $q((u_1,1),...,(u_n,1)) = p(u_1,...,u_n)$.

   a. Show that $q((t,w),...,(t,w)) = P(t,w)$.

   b. Conclude that in the axioms for the multilinear blossom the diagonal property can be replaced by the dehomogenization property.

10. Let $L_0(t),L_1(t)$ be arbitrary polynomials of degree 1 in t, and suppose that

$$(F_0(t),...,F_n(t)) = \underbrace{\{L_0(t),L_1(t)\} \otimes \cdots \otimes \{L_0(t),L_1(t)\}}_{n-factors}.$$

Show that

$$(F_0(t,w),...,F_n(t,w)) = \underbrace{\{L_0(t,w),L_1(t,w)\} \otimes \cdots \otimes \{L_0(t,w),L_1(t,w)\}}_{n-factors}.$$

11. Let $Q(t) = \sum_j Q_j t^j$ be a polynomial curve of degree n.

   a. Use Taylor's Theorem to show that

   $$Q_j = \binom{n}{j} q(\underbrace{0,...,0}_{n-j}, \underbrace{\delta,...,\delta}_{j}).$$

   b. Apply the result of part (a) to develop an algorithm to convert from Bezier to monomial form.

12. Let $M_k^n(t,w) = \binom{n}{k} t^k w^{n-k}$ and consider the function

   $$m_k^n\big((u_1,v_1),...,(u_n,v_n)\big) = \sum_\sigma u_{\sigma(1)} \cdots u_{\sigma(k)} v_{\sigma(k+1)} \cdots v_{\sigma(n)},$$

   where the sum is taken over all permutations σ of $\{1,...,n\}$.

   a. Show that

   $$m_k^n\big((u_1,v_1),...,(u_n,v_n)\big)$$

   is a symmetric multilinear function that reduces to $M_k^n(t,w)$ along the diagonal.

   b. Use the result in part (a) to establish the existence of the multilinear blossom.

   c. Verify that Figure 6.6 commutes.

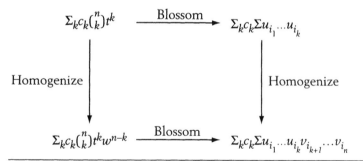

Figure 6.6 Blossoming and homogenizing commute.

6.5 **Blossoming Bezier Patches**

Blossoming can be extended to polynomials in several variables. Since we are mostly concerned with surfaces, we shall confine our attention here to polynomials in two variables, but the same constructions work quite generally for polynomials in an arbitrary number of parameters. We have encountered two distinct kinds of Bezier patches—triangular and rectangular—and these two surface types are associated with different variants of the blossom, so we shall deal here with each one separately.

6.5.1 Blossoming Triangular Bezier Patches

Consider first de Casteljau's tetrahedral algorithm for triangular Bezier patches. Recall that for curves we labeled the nodes of the de Casteljau algorithm with the end points of the parameter interval; for surfaces we shall label the nodes with the corner points of the triangular domain. Let Δabc be the domain triangle, where $a = (a_1, a_2)$, $b = (b_1, b_2)$, $c = (c_1, c_2)$, and let $t = (t_1, t_2)$ denote an arbitrary point in the parameter domain.

The de Casteljau pyramid algorithm for a degree n triangular Bezier patch is represented by a tetrahedral array with edges pointing into the nodes from three directions. Suppose we start at the apex of the tetrahedron and want to reach the control point P_{ijk} at the base (see Figure 5.44). Then we must take i steps along one direction, j steps in a second direction, and k steps in the third direction. We can use the vertices a,b,c of the domain triangle to encode these three directions. This convention leads to the notation $a^i b^j c^k$ for the position ijk along the base of the triangle. Nodes located above the base we shall denote by

$$a^\alpha b^\beta c^\gamma t^{n-\alpha-\beta-\gamma},$$

since there are still $n - \alpha - \beta - \gamma$ levels to consider. The node at the apex of the tetrahedron has the label t^n, since no edges have been traversed. This notation tells us precisely where we are in the tetrahedron, and it captures as well the labels along the edges of the tetrahedron, which are just the barycentric coordinates of the point $t = (t_1, t_2)$ relative to the domain triangle Δabc. We illustrate this labeling of the nodes for the case $n = 3$ in Figure 6.7.

As with curves we blossom this bivariate de Casteljau algorithm by replacing $t = (t_1, t_2)$ with $u_k = (u_{k_1}, u_{k_2})$ on the kth level of the diagram, and replacing

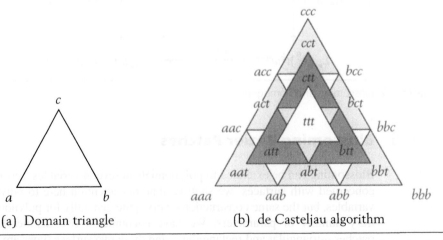

(a) Domain triangle   (b) de Casteljau algorithm

Figure 6.7 The labeling for the nodes in de Casteljau's pyramid algorithm for a cubic triangular Bezier patch. The vertices of the domain triangle (a) provides the labels for the directions in de Casteljau's algorithm (b).

$$a^{\alpha}b^{\beta}c^{\gamma}t^{n-\alpha-\beta-\gamma} \text{ with } a^{\alpha}b^{\beta}c^{\gamma}u_1 \cdots u_{n-\alpha-\beta-\gamma}.$$

This procedure generates Figure 6.8, where the labels along the edges on the kth level of the tetrahedron are the barycentric coordinates of the points $u_k = (u_{k_1}, u_{k_2})$ relative to Δabc.

The function that emerges at the apex of this tetrahedron depends on the n variables $u_1, \ldots, u_n$. If we place the control points $\{P_{ijk}\}$ of the triangular Bezier surface $P(t_1, t_2)$ at the base of the diagram, then the function that emerges at the apex is called the *blossom* of $P(t_1, t_2)$ and, as in the univariate setting, is denoted by $p(u_1, \ldots, u_n)$. Here, however, the variables $u_1, \ldots, u_n$ are points in the affine plane, so each variable has two coordinates.

The function $p(u_1, \ldots, u_n)$ is symmetric in the variables $u_1, \ldots, u_n$, since the coefficients of the control points are convolutions and discrete convolution is commutative (see Section 5.6.1). Also $p(u_1, \ldots, u_n)$ is multiaffine in the variables $u_1, \ldots, u_n$ because each coordinate of each variable appears only to the first power in $p(u_1, \ldots, u_n)$, since the barycentric coordinates that label the edges of the diagram are linear functions in the parameters. Finally, if we replace each variable $u_k = (u_{k_1}, u_{k_2})$ in $p(u_1, \ldots, u_n)$ by $t = (t_1, t_2)$, the algorithm reverts back to the de Casteljau algorithm for $P(t_1, t_2)$. Thus along the diagonal, $p(t, \ldots, t) = P(t)$. Hence the function $p(u_1, \ldots, u_n)$ is symmetric, multiaffine, and reduces to $P(t)$ along the diagonal—that is, $p(u_1, \ldots, u_n)$ is the blossom of $P(t)$. This argument establishes the existence of the bivariate blossom; uniqueness follows by a demonstration similar to the one given in Theorem 6.3 (see Exercise 2).

The dual functional property also holds for triangular Bezier patches. That is, the control points $\{P_{ijk}\}$ for a Bezier patch $P(t)$ over the domain triangle Δabc are given by

$$P_{ijk} = p(\underbrace{a, \ldots, a}_{i}, \underbrace{b, \ldots, b}_{j}, \underbrace{c, \ldots, c}_{k}).$$

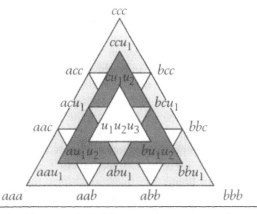

Figure 6.8   The blossom of a cubic triangular Bezier patch.

The proof follows from the uniqueness of the Bezier coefficients and is much the same as the proof of Theorem 6.1 (see Exercise 1).

As in the univariate setting, the dual functional property simplifies the derivation of various change of basis procedures. For example, suppose we want to split a triangular patch into the three patches formed by joining the vertices of $\triangle abc$ to some point $t = (t_1, t_2)$ in the parameter domain. By the dual functional property, the control points for these three patches are given by

$$Q_{ijk} = p(\underbrace{a,...,a}_{i}, \underbrace{b,...,b}_{j}, \underbrace{t,...,t}_{k})$$

$$R_{ijk} = p(\underbrace{b,...,b}_{i}, \underbrace{c,...,c}_{j}, \underbrace{t,...,t}_{k})$$

$$S_{ijk} = p(\underbrace{a,...,a}_{i}, \underbrace{c,...,c}_{j}, \underbrace{t,...,t}_{k}) \; .$$

Therefore, it follows immediately from Figure 6.7 that we can subdivide a triangular Bezier patch into three patches by taking the values off the three lateral faces of the de Casteljau tetrahedron. This algorithm is the generalization to triangular patches of the de Casteljau subdivision algorithm for Bezier curves.

We can also homogenize bivariate polynomials in much the same way that we homogenize univariate polynomials. Let

$$P(t_1, t_2) = \sum_{ij} c_{ij} t_1^i t_2^j$$

be a polynomial of degree n. To homogenize $P(t_1, t_2)$, we multiply each term $t_1^i t_2^j$ by w^{n-i-j} to obtain

$$P(t_1, t_2, w) = \sum_{ij} c_{ij} t_1^i t_2^j w^{n-i-j} \; .$$

To homogenize de Casteljau's tetrahedral algorithm, we simply homogenize the linear barycentric coordinate functions along the edges of the tetrahedron. As in the univariate setting, we can homogenize the blossom or blossom the homogenization by first blossoming and then homogenizing or first homogenizing and then blossoming de Casteljau's tetrahedral algorithm. Again blossoming and homogenization commute (see Exercise 3). Notice too that in the bivariate setting, the homogeneous blossom is once again multilinear rather than multiaffine—that is, both the domain and the range are lifted from affine space to Grassmann space.

The homogeneous blossom can be used to compute the partial derivatives of a triangular Bezier patch. Let $\delta_1 = (1,0,0)$, $\delta_2 = (0,1,0)$, and $t = (t_1, t_2, 1)$. Consider now what happens in the homogeneous version of de Casteljau's algorithm if we replace $t \to \delta_1$ on any level of the algorithm. Since the labels are barycentric coordinates and since barycentric coordinates are linear functions, the effect is to replace each barycentric coordinate function with the coefficient of t_1; that is, the effect is to differentiate one level of the algorithm with respect to t_1. If now we replace

$(t_1,t_2,w) \to (t_1,t_2,1)$ on the remaining levels of the algorithm—that is, if we dehomogenize the remaining levels—then, up to a constant multiple, we get the partial derivative with respect to t_1 of the original Bezier surface. Similarly, if we replace $t \to \delta_2$ on any level of the algorithm and dehomogenize the remaining levels, then, up to a constant multiple, we obtain the partial derivative with respect to t_2. Therefore, it follows that

$$\frac{\partial P}{\partial t_1} = np(\underbrace{t,...,t}_{n-1},\delta_1)$$

$$\frac{\partial P}{\partial t_2} = np(\underbrace{t,...,t}_{n-1},\delta_2).$$

More generally by an analogous argument,

$$\frac{\partial^{i+j} P}{\partial t_1^i \partial t_2^j} = \frac{n!}{(n-i-j)!} p(\underbrace{t,...,t}_{n-i-j},\underbrace{\delta_1,...,\delta_1}_{i},\underbrace{\delta_2,...,\delta_2}_{j}) .$$

Exercises

1. Prove the dual functional property of the bivariate blossom for triangular Bezier patches. That is, prove that if $P(t_1,t_2)$ is a Bezier surface with control points $\{P_{ijk}\}$ over the domain triangle Δabc, then

$$P_{ijk} = p(\underbrace{a,...,a}_{i},\underbrace{b,...,b}_{j},\underbrace{c,...,c}_{k}).$$

2. Using Exercise 1, prove that the blossom of a bivariate polynomial $P(t_1,t_2)$ is unique.

3. Prove that for bivariate polynomials, blossoming and homogenizing commute.

4. Let $P_{ijk}^n(s,t) = (s+t-a_{ijk})^n$. Show that

$$p_{ijk}^n\big((u_1,v_1),...,(u_n,v_n)\big) = (u_1+v_1-a_{ijk})\cdots(u_n+v_n-a_{ijk}).$$

5. Consider the monomial $M_{ijk}^n(s,t) = \binom{n}{ijk}s^i t^j$.
   a. Show that

$$m_{ijk}^n\big((u_1,v_1),...,(u_n,v_n)\big) = \sum u_{\alpha_1}\cdots u_{\alpha_i} v_{\beta_1}\cdots v_{\beta_j},$$

   where the sum is taken over all subsets $\{\alpha_1,...,\alpha_i,\beta_1,...,\beta_j\}$ of the integers $\{1,...,n\}$.
   b. Show that

$$\left\{m_{ijk}^n\big((u_1,v_1),...,(u_n,v_n)\big)\right\} = \begin{pmatrix} 1 & v_1 \\ u_1 & 0 \end{pmatrix} \otimes \cdots \otimes \begin{pmatrix} 1 & v_n \\ u_n & 0 \end{pmatrix}.$$

c. Using the result in part (a), show that

$$(^n_{ijk})s^i t^j = \sum_{p=i}^{n} \sum_{q=j}^{n-p} (^p_i)(^q_j) B^n_{pqr}(s,t) \qquad i+j+k=n\,.$$

d. Using the result in part (c), develop a formula for converting from bivariate monomial to bivariate Bezier form.

6. Let $b^n_{ijk}((u_1,v_1),...,(u_n,v_n))$ denote the blossom of the degree n Bernstein basis function $B^n_{ijk}(t_1,t_2)$. Prove that

a. $b^n_{ijk}((u_1,v_1),...,(u_n,v_n)) =$

$$\sum_{i_1+\cdots+i_n=i}\sum_{j_1+\cdots+j_n=j} B^1_{i_1 j_1 k_1}(u_1,v_1)\cdots B^1_{i_n j_n k_n}(u_n,v_n)$$

b. $\left\{ b^n_{ijk}((u_1,v_1),...,(u_n,v_n)) \right\} = \begin{pmatrix} 1-u_1-v_1 & v_1 \\ u_1 & 0 \end{pmatrix} \otimes \cdots \otimes \begin{pmatrix} 1-u_n-v_n & v_n \\ u_n & 0 \end{pmatrix}$

7. Let $L_{100}(s,t), L_{010}(s,t), L_{001}(s,t)$ be arbitrary linear functions in s,t, and suppose that

$$\left(F_{ijk}(s,t) \right) = \underbrace{\begin{pmatrix} L_{001}(s,t) & L_{010}(s,t) \\ L_{100}(s,t) & 0 \end{pmatrix} \otimes \cdots \otimes \begin{pmatrix} L_{001}(s,t) & L_{010}(s,t) \\ L_{100}(s,t) & 0 \end{pmatrix}}_{n \ linear \ factors}$$

Show that

a. $\left(F_{ijk}(s,t,w) \right) =$

$$\underbrace{\begin{pmatrix} L_{001}(s,t,w) & L_{010}(s,t,w) \\ L_{100}(s,t,w) & 0 \end{pmatrix} \otimes \cdots \otimes \begin{pmatrix} L_{001}(s,t,w) & L_{010}(s,t,w) \\ L_{100}(s,t,w) & 0 \end{pmatrix}}_{n \ linear \ factors}$$

b. $\left(f_{ijk}((u_1,v_1),...,(u_n,v_n)) \right)$

$$\begin{pmatrix} L_{001}(u_1,v_1) & L_{010}(u_1,v_1) \\ L_{100}(u_1,v_1) & 0 \end{pmatrix} \otimes \cdots \otimes \begin{pmatrix} L_{001}(u_n,v_n) & L_{010}(u_n,v_n) \\ L_{100}(u_n,v_n) & 0 \end{pmatrix}$$

8. Let $P(s)$ be a univariate polynomial of degree n. Then $P(s,t) = P(s)$ is also a bivariate polynomial of degree n. Prove that

$$p((u_1,v_1),...,(u_n,v_n)) = p(u_1,...,u_n)\,.$$

9. Using the result of Exercise 8, prove that

a. $B^n_i(s) = \sum_{j+k=n-i} B^n_{ijk}(s,t)$

b. $B^n_j(t) = \sum_{i+k=n-j} B^n_{ijk}(s,t)$

10. Use blossoming to derive the following identities:

   a. $\sum_{i+j+k=n} B_{ijk}^n(s,t) = 1$

   b. $\sum_{i+j+k=n}(-1)^{i+j} B_{ijk}^n(s,t) = (1-2s-2t)^n$

   c. $\sum_{i+j=n-k} B_{ijk}^n(s,t) = B_k^n(1-s-t)$

   d. $\sum_{i=0}^{n} \sum_{j=0}^{n-i} (x+1)^i (y+1)^j B_{ijk}^n(s,t) = (sx+ty+1)^n$

   e. $B_{ijk}^n(su,tv) = \sum_{p=i}^{n} \sum_{q=j}^{n-p} B_i^p(u)B_j^q(v)B_{pqr}^n(s,t)$

11. Let $P(s,t)$ be a bivariate polynomial of total degree n, and let $p^{(i,j)}(u_1,...,u_{n-i-j})$ denote the blossom of $P^{(i,j)}(s,t)$. Prove that

   a. $P^{(i,j)}(s,t) = \dfrac{n!}{(n-i-j)!} p(\underbrace{(s,t),...,(s,t)}_{n-i-j},\underbrace{\delta_1,...,\delta_1}_{i},\underbrace{\delta_2,...,\delta_2}_{j})$

   b. $p^{(i,j)}(u_1,...,u_{n-i-j}) = \dfrac{n!}{(n-i-j)!} p(u_1,...,u_{n-i-j},\underbrace{\delta_1,...,\delta_1}_{i},\underbrace{\delta_2,...,\delta_2}_{j})$

12. Let $b_{ijk}^n\big((u_1,v_1),...,(u_n,v_n)\big)$ denote the blossom of the degree n Bernstein basis function $B_{ijk}^n(s,t)$. Prove that

$$b_{ijk}^n(\underbrace{(0,0),...,(0,0)}_{p},\underbrace{(1,0),...,(1,0)}_{q},\underbrace{(0,1),...,(0,1)}_{r}) = \delta_{ip}\delta_{jq}\delta_{kr}.$$

13. Let $Q(s,t) = \sum_j Q_{ij} s^i t^j$ be a polynomial surface of total degree n.

   a. Use Taylor's Theorem in two variables to prove that

$$Q_{ij} = \binom{n}{ijk} p(\underbrace{\delta_1,...,\delta_1}_{i},\underbrace{\delta_2,...,\delta_2}_{j},\underbrace{(s,t),...,(s,t)}_{k}), \quad i+j+k=n.$$

   b. Apply the result of part (a) to show that

$$B_{ijk}^n(s,t) = \sum_{p=i}^{n} \sum_{q=j}^{n-p} (-1)^{i+j+p+q} \binom{p}{i}\binom{q}{j}\binom{n}{pqr} s^p t^q$$

   c. Apply the result of part (b) to develop a formula to convert from bivariate Bezier to bivariate monomial form.

14. Consider the degree elevation formula in Proposition 6.5.

   a. Show that if we let $u_j = (u_{j_1}, u_{j_2})$, then this identity is valid for the bivariate blossom.

   b. Use this bivariate degree elevation formula to derive the degree elevation formula for triangular Bezier surfaces given in Section 5.8.2, Exercise 10(e).

15. Consider the subdivision algorithm for triangular Bezier patches described in the text.

    a. Explain why the control polyhedra generated by iterating this subdivision algorithm will not, in general, converge to the original triangular patch. (Hint: Consider the boundaries of the patch.)

    b. Use blossoming to subdivide a Bezier patch into four subpatches as in Figure 6.9.

    c. Verify experimentally that iterating the algorithm in part (b) generates control polyhedra that converge to the original triangular patch.

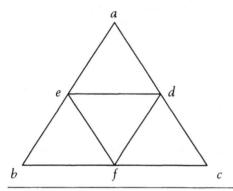

Figure 6.9   Subdivision of a Bezier patch into four subpatches.

16. Recall from Section 5.6.3 that we can speed up recursive subdivision for Bezier curves by applying Wang's formula.

    a. Develop an analogue of Wang's formula for triangular Bezier surfaces for the subdivision algorithm in Exercise 15(b).

    b. Implement the recursive subdivision algorithm for rendering a triangular Bezier surface with and without Wang's formula.

    c. How much does Wang's formula speed up this algorithm?

17. Let $P(r,s,t)$ and $Q(r,s,t)$ be homogeneous polynomials of total degree n, and define

$$[P(r,s,t),Q(r,s,t)]_n = \frac{1}{n!}\sum_{i+j+k=n}\frac{P^{(i,j,k)}(\rho,\sigma,\tau)Q^{(i,j,k)}(\rho,\sigma,\tau)}{i!\,j!\,k!}.$$

Show that

    a. $[P(r,s,t),Q(r,s,t)]_n$ is a bilinear operator.

    b. $[P(r,s,t),Q(r,s,t)]_n$ is a constant independent of the choice of (ρ,σ,τ).

    c. $P(x,y,z) = [P(r,s,t),(rx+sy+tz)^n]_n$.

d. $p\big((u_1,v_1,w_1),\ldots,(u_n,v_n,w_n)\big) =$

$$[P(r,s,t),(ru_1 + sv_1 + tw_1)\cdots(ru_n + sv_n + tw_n)]_n.$$

(Compare to Section 6.2, Exercise 8.)

18. What formula do you get for the homogeneous blossom when you homogenize and then blossom the bivariate form of Taylor's Theorem?

6.5.2 Blossoming Tensor Product Bezier Patches

Look at Figure 5.40, the de Casteljau algorithm for a tensor product Bezier patch. Unlike the tetrahedral algorithm for a triangular Bezier patch, here the s and t parameters always appear on separate levels of the diagram, never on the same level. The trick for blossoming and homogenizing a tensor product surface is to treat each variable independently.

To blossom the de Casteljau algorithm for a tensor product surface $P(s,t)$ of bidegree (m,n), we blossom the t levels exactly as we blossom Bezier curves, and then we do the same for the s levels. To keep track of the fact that s and t are distinct variables, we use different blossom parameters to replace s and t: $u_1,\ldots,u_m$ for s and $v_1,\ldots,v_n$ for t. Thus, we shall denote the blossom of $P(s,t)$ by $p(u_1,\ldots,u_m;v_1,\ldots,v_n)$, where the semicolon separates the s and t blossom parameters. Algorithmically we can generate the blossom of $P(s,t)$ by replacing t with v_j on the jth level of the de Casteljau algorithm in t and then replacing s with u_k on the kth level of the de Casteljau algorithm in s (see Figure 6.10).

This blossom $p(u_1,\ldots,u_m;v_1,\ldots,v_n)$ is bisymmetric—that is, it is symmetric in the u and v parameters independently. This means that we can exchange any two u parameters or any two v parameters, but we cannot exchange a u parameter with a v parameter. Bisymmetry follows directly from the commutativity of discrete convolution. This blossom is also multiaffine in the u and v parameters, since these parameters appear only to the first power. Finally, by construction, $p(u_1,\ldots,u_m;v_1,\ldots,v_n)$ reduces to $P(s,t)$ along the diagonal—that is, $p(s,\ldots,s;t,\ldots,t) = P(s,t)$—because if we replace each u parameter by s and each v parameter by t, then the algorithm reverts back to the original de Casteljau evaluation algorithm for $P(s,t)$. Thus the blossom $p(u_1,\ldots,u_m;v_1,\ldots,v_n)$ satisfies the following properties:

i. *Bisymmetry*

$$p(u_1,\ldots,u_m;v_1,\ldots,v_n) = p(u_{\sigma(1)},\ldots,u_{\sigma(m)};v_{\tau(1)},\ldots,v_{\tau(n)}).$$

ii. *Multiaffine*

$$p(u_1,\ldots,u_m;v_1,\ldots,v_n) \text{ is multiaffine in each variable.}$$

iii. *Diagonal*

$$p(\underbrace{s,\ldots,s}_{m};\underbrace{t,\ldots,t}_{n}) = P(s,t).$$

The argument in the previous paragraph establishes the existence of the blossom of bidegree (m,n); uniqueness can be proved by invoking the linear independence of the bivariate tensor product Bernstein basis functions (see Exercise 2).

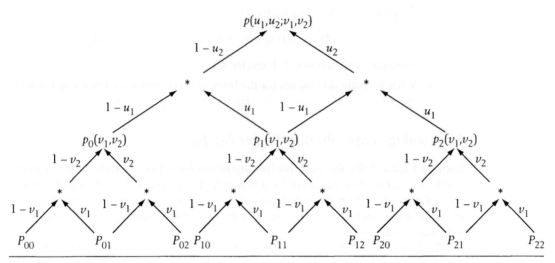

Figure 6.10   The blossom $p(u_1,u_2;v_1,v_2)$ of the biquadratic tensor product Bezier patch $P(s,t) =$

$$\sum_{i=0}^{2} \sum_{j=0}^{2} B_i^2(s)B_j^2(t)P_{ij}.$$ Notice the internal nodes labeled with the blossoms $p_i(v_1,v_2)$ of the

Bezier curves $P_i(t) = \sum_{k=0}^{2} B_k^2(t)P_{ik}.$

The dual functional property also holds for tensor product patches. Suppose that the domain of $P(s,t)$ is $[a,b]\times[c,d]$. Then

$$P_{ij} = p(\underbrace{a,...,a}_{m-i},\underbrace{b,...,b}_{i};\underbrace{c,...,c}_{n-j},\underbrace{d,...,d}_{j})\cdot$$

As usual the proof follows by the uniqueness of the Bezier coefficients of a tensor product Bezier patch and is much the same as the proof for Bezier curves (see Exercise 1).

To homogenize a polynomial $P(s,t)$ of bidegree (m,n), we homogenize in each variable independently. Thus every term of the homogenization $P(s,w;t,\omega)$ is of exact bidegree (m,n). In particular, to homogenize, we multiply the term $s^i t^j$ by $w^{m-i}\omega^{n-j}$. Hence

$$P(s,t) = \sum_{i=0}^{m} \sum_{j=0}^{n} c_{ij} s^i t^j$$

$$P(s,w;t,\omega) = \sum_{i=0}^{m} \sum_{j=0}^{n} c_{ij} s^i w^{m-i} t^j \omega^{n-j}\cdot$$

As with blossoming we can homogenize the de Casteljau algorithm by homogenizing each level of the algorithm with respect to the appropriate variable.

Blossoming and homogenization commute just as in the univariate setting (see Exercise 3). To homogenize the bidegree blossom, we again homogenize with respect to each variable independently. Hence we write

$$p\big((u_1,w_1),...,(u_m,w_m);(v_1,\omega_1),...,(v_n,\omega_n)\big)$$

for the homogeneous blossom of $p(u_1,...,u_m;v_1,...,v_n)$. Partial derivatives are calculated in the usual way by substituting $s \to \delta_1 = (1,0)$ or $t \to \delta_2 = (0,1)$ on any level of the de Casteljau algorithm. Thus

$$\frac{\partial P}{\partial s} = mp(\underbrace{s,...,s}_{m-1},\delta_1;\underbrace{t,...,t}_{n})\ ,$$

$$\frac{\partial P}{\partial t} = np(\underbrace{s,...,s}_{m};\underbrace{t,...,t}_{n-1},\delta_2)\ ,$$

and, more generally,

$$\frac{\partial^{i+j}P}{\partial s^i \partial t^j} = \frac{m!}{(m-i)!}\frac{n!}{(n-j)!}\,p(\underbrace{s,...,s}_{m-i},\underbrace{\delta_1,...,\delta_1}_{i};\underbrace{t,...,t}_{n-j},\underbrace{\delta_2,...,\delta_2}_{j})\ .$$

Finally, to clarify the distinction between the different variants of bivariate homogenization and blossoming, let's consider a simple concrete example. We can think of the polynomial $P(s,t) = s^2 + st + t$ either as a bivariate polynomial of total degree 2 or as a bivariate polynomial of bidegree (2,2). Thus we have the following formulas:

- *Polynomial*

$$P(s,t) = s^2 + st + t$$

- *Homogenizations*

$$P(s,t,w) = s^2 + st + tw$$

$$P(s,w;t,\omega) = s^2\omega^2 + swt\omega + w^2 t\omega$$

- *Blossoms*

$$p\big((u_1,v_1),(u_2,v_2)\big) = u_1u_2 + \frac{u_1v_2 + u_2v_1}{2} + \frac{v_1 + v_2}{2}$$

$$p(u_1,u_2;v_1,v_2) = u_1u_2 + \frac{(u_1+u_2)(v_1+v_2)}{4} + \frac{v_1+v_2}{2}$$

$$p\big((u_1,v_1,w_1),(u_2,v_2,w_2)\big) = u_1u_2 + \frac{u_1v_2 + u_2v_1}{2} + \frac{v_1w_2 + v_2w_1}{2}$$

$$p\big((u_1,w_1),(u_2,w_2);(v_1,\omega_1),(v_2,\omega_2)\big)$$
$$= u_1u_2\omega_1\omega_2 + \frac{(u_1w_2 + u_2w_1)(v_1\omega_2 + v_2\omega_1)}{4} + \frac{w_1w_2v_1\omega_2 + w_1w_2v_2\omega_1}{2}$$

Exercises

1. Prove the dual functional property of the bivariate blossom for rectangular Bezier patches. That is, prove that if $P(s,t)$ is a tensor product Bezier surface of bidegree (m,n) with control points $\{P_{ij}\}$ over the domain $[a,b] \times [c,d]$, then

$$P_{ij} = p(\underbrace{a,...,a}_{m-i},\underbrace{b,...,b}_{i};\underbrace{c,...,c}_{n-j},\underbrace{d,...,d}_{j}).$$

2. Prove that the blossom of a polynomial $P(s,t)$ of bidegree (m,n) is unique.

3. Prove that for polynomials of bidegree (m,n) homogenization and blossoming commute.

4. Here we provide an alternative construction for the blossom of a tensor product Bezier patch based on the pyramid algorithm. Consider the pyramid algorithm for a tensor product Bezier patch $P(s,t)$ of bidegree (n,n). Let

$$\tilde{p}\big((u_1,v_1),...,(u_n,v_n)\big)$$

denote the function generated by replacing the parameters (s,t) by a different parameter pair (u_k,v_k) on each level of the pyramid.

a. Show that the function $\tilde{p}\big((u_1,v_1),...,(u_n,v_n)\big)$ is bisymmetric, multiaffine in the u and v parameters, and that

$$\tilde{p}\big((u_1,v_1),...,(u_n,v_n)\big)$$

reduces to $P(s,t)$ along the diagonal.

b. Conclude that $\tilde{p}\big((u_1,v_1),...,(u_n,v_n)\big)$ is equal to the blossom $p(u_1,...,u_n;v_1,...,v_n)$ of $P(s,t)$.

c. Let $\{P_{\mu\nu}\}$ be the control points and let $Q_{ij} = (i,j)$, $i,j = 0,1$, be the vertices of the domain rectangle of $P(s,t)$. Show that $P_{\mu\nu}$ is equal to

$$\tilde{p}\left(\underbrace{Q_{00},...,Q_{00}}_{d-i-j-k},\underbrace{Q_{10},...,Q_{10}}_{i},\underbrace{Q_{01},...,Q_{01}}_{j},\underbrace{Q_{11},...,Q_{11}}_{k}\right) \quad \mu = i+k, \ \nu = j+k.$$

5. Let $P(s,t) = s^3 + st^2 + st + s$. Compute the following blossom values:

a. $p\big((u_1,v_1),(u_2,v_2),(u_3,v_3)\big)$

b. $p(u_1,u_2,u_3;v_1,v_2,v_3)$

c. $p\big((u_1,v_1,w_1),(u_2,v_2,w_2),(u_3,v_3,w_3)\big)$

d. $p\big((u_1,w_1),(u_2,w_2),(u_3,w_3);(v_1,\omega_1),(v_2,\omega_2),(v_3,\omega_3)\big)$

6. Let $P_{ij}(s,t) = s^i t^j$, where $i \le m$ and $j \le n$. Then $P_{ij}(s,t)$ is both a polynomial of total degree $m + n$ and a polynomial of bidegree (m,n). Show that

a. $p\big((u_1,v_1),...,(u_{m+n},v_{m+n})\big) = \Sigma \dfrac{u_{\alpha_1}\cdots u_{\alpha_i}v_{\beta_1}\cdots v_{\beta_j}}{\binom{m+n}{i\ j\ k}}$,

where the sum is taken over all sets of distinct indices $\{\alpha_1,...,\alpha_i,\beta_1,...,\beta_j\}$.

b. $p(u_1,...,u_m;v_1,...,v_n) = \Sigma \dfrac{u_{\alpha_1}\cdots u_{\alpha_i}v_{\beta_1}\cdots v_{\beta_j}}{\binom{m}{i}\binom{n}{j}}$,

where the sum is taken over all indices $\{\alpha_1,...,\alpha_i\}$ and $\{\beta_1,...,\beta_j\}$ such that $\{\alpha_1,...,\alpha_i\}$ are distinct and $\{\beta_1,...,\beta_j\}$ are distinct.

7. Let $P(s)$ be a univariate polynomial of degree n. Then $P(s,t) = P(s)$ is both a bivariate polynomial of total degree n and a bivariate polynomial of bidegree (n,n). Prove that

a. $p\big((u_1,v_1),...,(u_n,v_n)\big) = p(u_1,...,u_n)$

b. $p(u_1,...,u_n;v_1,...,v_n) = p(u_1,...,u_n)$

c. Using parts (a) and (b), conclude that if $P(s,t) = s^i$, $i \le m$, or $P(s,t) = t^j$, $j \le m$, then $p(u_1,...,u_n;v_1,...,v_n) = p\big((u_1,v_1),...,(u_n,v_n)\big)$.

8. Let $Q(s,t) = \Sigma_j Q_{ij}s^i t^j$ be a bivariate polynomial of bidegree (m,n).

a. Use Taylor's Theorem in two variables to prove that

$$Q_{ij} = \binom{m}{i}\binom{n}{j}p(\underbrace{s,...,s}_{m-i},\underbrace{\delta_1,...,\delta_1}_{i};\underbrace{t,...,t}_{n-j},\underbrace{\delta_2,...,\delta_2}_{j}) .$$

b. Apply the result of part (a) to develop an algorithm to convert from bivariate Bezier to bivariate monomial form.

9. Let $b_{ij}^{mn}\big((u_1,...,u_m;v_1,...,v_n)\big)$ denote the blossom of the bidegree (m,n) Bernstein basis function $B_i^m(s)B_j^n(t)$. Prove that

a. $b_{ij}^{mn}\big((u_1,...,u_m;v_1,...,v_n)\big)$

$= \Big(\Sigma_{i_1+\cdots+i_m=i}B_{i_1}^1(u_1)\cdots B_{i_m}^1(u_m)\Big)\Big(\Sigma_{j_1+\cdots+j_n=j}B_{j_1}^1(v_1)\cdots B_{j_n}^1(v_n)\Big).$

b. $\Big\{b_{ij}^{mn}\big((u_1,...,u_m;v_1,...,v_n)\big)\Big\} = U_m \otimes V_n$

i. $U_m = \begin{pmatrix}1-u_1\\u_1\end{pmatrix}\otimes\cdots\otimes\begin{pmatrix}1-u_m\\u_m\end{pmatrix}$

ii. $V_n = \begin{pmatrix}1-v_1 & v_1\end{pmatrix}\otimes\cdots\otimes\begin{pmatrix}1-v_n & v_n\end{pmatrix}$

6.6 **Summary**

Blossoming is easy. Blossoming is slick. Blossoming is fun. Blossoming is powerful. But there is a lot of material in this chapter on blossoming to try to absorb all at
once. Here then is a short summary of the main points that you need to remember:

- *Primary Properties*
  - i. Symmetry

    $$p(u_1,...,u_n) = p(u_{\sigma(1)},...,u_{\sigma(n)}) \text{ for any permutation } \sigma \text{ of } \{1,...,n\}.$$

  - ii. Multiaffine $p(u_1,...,(1-\alpha)u_k + \alpha w_k,...,u_n) =$

    $$(1-\alpha)p(u_1,...,u_k,...,u_n) + \alpha p(u_1,...,w_k,...,u_n)$$

  - iii. Diagonal

    $$p(\underbrace{t,...,t}_{n}) = P(t)$$

  - iv. Dual functional

    $$P_k = p(\underbrace{a,...,a}_{n-k},\underbrace{b,...,b}_{k}) \quad P_{ijk} = p(\underbrace{a,...,a}_{i},\underbrace{b,...,b}_{j},\underbrace{c,...,c}_{k})$$

    $$P_{ij} = p(\underbrace{a,...,a}_{m-i},\underbrace{b,...,b}_{i};\underbrace{c,...,c}_{n-j},\underbrace{d,...,d}_{j})$$

  - v. Existence
  - vi. Uniqueness

  The first three properties are the blossoming axioms—the three properties
  that uniquely characterize the blossom. Property (iv) is the key connection
  between blossoming and Bezier curves and surfaces. It leads to all the applications listed below. Property (iv) is so strong that it can replace the diagonal property as one of the blossoming axioms. Property (v), existence, is
  essential; otherwise there is nothing to talk about. But uniqueness, Property
  (vi), is also critical. It is uniqueness that allows us to identify formulas that
  represent the blossom simply by verifying the axioms. Many proofs of blossoming identities are based on uniqueness.

- *Key de Casteljau Constructions*
  1. Blossoming the de Casteljau algorithm
     - Blossom each level of the de Casteljau algorithm independently.
     - Replace $t \to u_k$ on the *k*th level of the de Casteljau algorithm.

   2. Homogenizing the de Casteljau algorithm
      - Homogenize each level of the de Casteljau algorithm independently.
      - Replace $t - a \rightarrow t - aw$ and $b - t \rightarrow bw - t$ on every level.

- *Central Ideas*
   1. Blossoming $\leftrightarrow$ powerful tool for generating change of basis algorithms.
   2. Homogenization $\leftrightarrow$ differentiation.
   3. Blossoming and homogenization commute.

- *Principal Applications = Change of Basis Algorithms*
   1. Subdivision
   2. Differentiation
   3. Degree elevation
   4. Conversion between Bezier and Monomial form

- *Essential Tools for Analyzing the Blossom*
   1. Blossoming axioms
      - Uniqueness
   2. de Casteljau algorithm
      - Recursion
   3. Marsden identity
      - Power basis
   4. Elementary symmetric functions
      - Monomial basis
   5. Convolution
      - Bernstein basis

Finally, it is difficult to remember all the interesting identities for the blossom that we have encountered in the text and in the exercises. For quick recall, we have collected these formulas below in the following section on blossoming identities.

6.6.1 Blossoming Identities

1. *Symmetry*

$$p(u_1,...,u_i,...,u_j,...,u_n) = p(u_1,...,u_j,...,u_i,...,u_n)$$

2. *Multiaffine*

$$p(u_1,...,(1-\alpha)u_k + \alpha w_k,...,u_n) =$$

$$(1-\alpha)p(u_1,...,u_k,...,u_n) + \alpha p(u_1,...,w_k,...,u_n)$$

3. *Diagonal*

$$p(\underbrace{t,...,t}_{n}) = P(t)$$

4. *Dual functionals*

a. Bezier coefficients

$$P_k = p(\underbrace{a,...,a}_{n-k},\underbrace{b,...,b}_{k})$$

$$P_{ijk} = p(\underbrace{a,...,a}_{i},\underbrace{b,...,b}_{j},\underbrace{c,...,c}_{k})$$

$$P_{ij} = p(\underbrace{a,...,a}_{m-i},\underbrace{b,...,b}_{i};\underbrace{c,...,c}_{n-j},\underbrace{d,...,d}_{j})$$

b. Monomial coefficients

$$P_k = p(\underbrace{a,...,a}_{n-k},\underbrace{\delta,...,\delta}_{k})$$

$$P_{ijk} = p(\underbrace{a,...,a}_{i},\underbrace{\delta_1,...,\delta_1}_{j},\underbrace{\delta_2,...,\delta_2}_{k})$$

$$P_{ij} = p(\underbrace{a,...,a}_{m-i},\underbrace{\delta_1,...,\delta_1}_{i};\underbrace{c,...,c}_{n-j},\underbrace{\delta_2,...,\delta_2}_{j})$$

c. Power coefficients

$$P_k = p(t_0,...,t_{k-1},t_{k+1},...,t_n)$$

$$P_{ij} = p(s_0,...,s_{i-1},s_{i+1},...,s_m;t_0,...,t_{j-1},t_{j+1},...,t_n)$$

5. *Linearity*

$$(p+q)(u_1,...,u_n) = p(u_1,...,u_n) + q(u_1,...,u_n)$$

$$(cp)(u_1,...,u_n) = cp(u_1,...,u_n)$$

6. *Products*

$$(pq)(u_1,...,u_{m+n}) = \frac{\sum_\sigma p(u_{\sigma(1)},...,u_{\sigma(m)})q(u_{\sigma(m+1)},...,u_{\sigma(m+n)})}{(m+n)!}$$

7. *Composites*

$$(p\circ q)(u_1,...,u_{mn}) = \frac{\sum_\sigma p\left(q(u_{\sigma(1)},...,u_{\sigma(n)}),...,q(u_{\sigma(mn-n+1)},...,u_{\sigma(mn)})\right)}{(mn)!}$$

8. *Degree elevation*

$$p_{n+1}(u_1,...,u_{n+1}) = \frac{\sum_{i=1}^{n+1} p_n(u_1,...,u_{i-1},u_{i+1},...,u_{n+1})}{n+1}$$

$$p_{n+k}(u_1,...,u_{n+k}) = \frac{\sum p_n(u_{i_1},...,u_{i_n})}{\binom{n+k}{k}}$$

9. *Derivatives*

$$P'(t) = np(t, \ldots, t, \delta)$$

$$P^{(r)}(t) = \frac{n!}{(n-r)!} p(t, \ldots, t, \underbrace{\delta, \ldots, \delta}_{r})$$

$$\frac{\partial P}{\partial t_1} = np(\underbrace{t, \ldots, t}_{n-1}, \delta_1)$$

$$\frac{\partial P}{\partial t_2} = np(\underbrace{t, \ldots, t}_{n-1}, \delta_2)$$

$$\frac{\partial^{i+j} P}{\partial t_1^i \partial t_2^j} = \frac{n!}{(n-i-j)!} p(\underbrace{t, \ldots, t}_{n-i-j}, \underbrace{\delta_1, \ldots, \delta_1}_{i}, \underbrace{\delta_2, \ldots, \delta_2}_{j})$$

$$\frac{\partial P}{\partial s} = mp(\underbrace{s, \ldots, s}_{m-1}, \delta_1; \underbrace{t, \ldots, t}_{n})$$

$$\frac{\partial P}{\partial t} = np(\underbrace{s, \ldots, s}_{m}; \underbrace{t, \ldots, t}_{n-1}, \delta_2)$$

$$\frac{\partial^{i+j} P}{\partial s^i \partial t^j} = \frac{m!}{(m-i)!} \frac{n!}{(n-j)!} p(\underbrace{s, \ldots, s}_{m-i}, \underbrace{\delta_1, \ldots, \delta_1}_{i}; \underbrace{t, \ldots, t}_{n-j}, \underbrace{\delta_2, \ldots, \delta_2}_{j})$$

10. *Convolutions*

$$\left(F_0^n(t), \ldots, F_n^n(t) \right) = \underbrace{\left(L_0(t), L_1(t) \right) \otimes \cdots \otimes \left(L_0(t), L_1(t) \right)}_{n \ linear \ factors}$$

$$\left(f_0^n(u_1, \ldots, u_n), \ldots, f_n^n(u_1, \ldots, u_n) \right) = \left(L_0(u_1), L_1(u_1) \right) \otimes \cdots \otimes \left(L_0(u_n), L_1(u_n) \right)$$

$$\left(F_{ijk}(s,t) \right) =$$

$$\underbrace{\begin{pmatrix} L_{001}(s,t) & L_{010}(s,t) \\ L_{100}(s,t) & 0 \end{pmatrix} \otimes \cdots \otimes \begin{pmatrix} L_{001}(s,t) & L_{010}(s,t) \\ L_{100}(s,t) & 0 \end{pmatrix}}_{n \ linear \ factors}$$

$$\left(f_{ijk}\big((u_1, v_1) \ldots (u_n, v_n)\big) \right) =$$

$$\begin{pmatrix} L_{001}(u_1, v_1) & L_{010}(u_1, v_1) \\ L_{100}(u_1, v_1) & 0 \end{pmatrix} \otimes \cdots \otimes \begin{pmatrix} L_{001}(u_n, v_n) & L_{010}(u_n, v_n) \\ L_{100}(u_n, v_n) & 0 \end{pmatrix}$$

11. *Special univariate bases*

    a. $B_j^n(t) = \binom{n}{j} t^j (1-t)^{n-j}$

       $b_j^n(u_1,...,u_n) = \sum_{i_1+\cdots i_n = j} B_{i_1}^1(u_1) \cdots B_{i_n}^1(u_n)$

    b. $M_j^n(t) = \binom{n}{j} t^j$

       $m_j^n(u_1,...,u_n) = \sum u_{i_1} \cdots u_{i_j}$

    c. $P_j^n(t) = (t - a_j)^n$

       $p_j^n(u_1,...,u_n) = (u_1 - a_j) \cdots (u_n - a_j)$

12. *Special bivariate bases*

    a. $B_{ijk}^n(t) = \binom{n}{ijk} s^i t^j (1-s-t)^{n-i-j}$

       $b_{ijk}^n\big((u_1,v_1),...,(u_n,v_n)\big) =$

       $\sum_{i_1+\cdots+i_n=i} \sum_{j_1+\cdots+j_n=j} B_{i_1 j_1 k_1}^1(u_1,v_1) \cdots B_{i_n j_n k_n}^1(u_n,v_n)$

    b. $M_{ijk}^n(t) = \binom{n}{ijk} s^i t^j$

       $m_{ijk}^n\big((u_1,v_1),...,(u_n,v_n)\big) = \sum_{\alpha_i \neq \beta_j} u_{\alpha_1} \cdots u_{\alpha_i} v_{\beta_1} \cdots v_{\beta_j}$

    c. $P_{ijk}^n(t) = (s + t - a_{ijk})^n$

       $p_{ijk}^n\big((u_1,v_1),...,(u_n,v_n)\big) = (u_1 + v_1 - a_{ijk}) \cdots (u_n + v_n - a_{ijk})$

13. *Special tensor product bases*

    a. $B_{ij}^{mn}(t) = \binom{m}{i} \binom{n}{j} s^i t^j$

       $b_{ij}^{mn}\big((u_1,...,u_m; v_1,...,v_n)\big)$

          $= \Big(\sum_{i_1+\cdots+i_m=i} B_{i_1}^1(u_1) \cdots B_{i_m}^1(u_m)\Big) \Big(\sum_{j_1+\cdots+j_n=j} B_{j_1}^1(v_1) \cdots B_{j_n}^1(v_n)\Big)$

    b. $M_{ij}^{mn}(t) = \binom{m}{i} \binom{n}{j} s^i t^j$

       $m_{ij}^{mn}\big((u_1,...,u_m; v_1,...,v_n)\big) = \sum u_{\alpha_1} \cdots u_{\alpha_i} v_{\beta_1} \cdots v_{\beta_j}$

    c. $P_{ij}^{mn}(t) = (s - a_i)^m (t - b_j)^n$

       $p_{ij}^{mn}\big((u_1,...,u_m; v_1,...,v_n)\big) = (u_1 - a_i) \cdots (u_m - a_i)(v_1 - b_j) \cdots (v_n - b_j)$

14. *De Boor–Fix representation*

    a. Univariate

    $$p(u_1,\ldots,u_n) = \sum_{k=0}^{n} \frac{(-1)^{n-k}}{n!} P^{(k)}(\tau)\psi^{(n-k)}(\tau)$$

    $$\psi(t) = (u_1 - t)\cdots(u_n - t)$$

    b. Bivariate (homogeneous)

    $$p\big((u_1,v_1,w_1),\ldots,(u_n,v_n,w_n)\big) = \frac{1}{n!} \sum_{i+j+k=n} \frac{P^{(i,j,k)}(\rho,\sigma,\tau)\psi^{(i,j,k)}(\rho,\sigma,\tau)}{i!\,j!\,k!}$$

    $$\psi(r,s,t) = (ru_1 + sv_1 + tw_1)\cdots(ru_n + sv_n + tw_n)$$

15. *Blossom of the derivatives*

    $$p'(u_1,\ldots,u_{n-1}) = np(u_1,\ldots,u_{n-1},\delta)$$

    $$p^{(r)}(u_1,\ldots,u_{n-r}) = \frac{n!}{(n-r)!} p(u_1,\ldots,u_{n-r},\underbrace{\delta,\ldots,\delta}_{r})$$

16. *Partial derivatives of the blossom*

    $$\frac{\partial p(u_1,\ldots,u_n)}{\partial u_j} = m_j p(u_1,\ldots,u_{j-1},\delta,u_{j+1},\ldots,u_n)$$

    $$m_j = \text{multiplicity } of\ u_j$$

    $$\frac{\partial p(u_1,\ldots,u_n)}{\partial u_j} = \frac{m_j}{n} p'(u_1,\ldots,u_{j-1},u_{j+1},\ldots,u_n)$$

CHAPTER 7

B-Spline Approximation and the de Boor Algorithm

B-splines are generalizations of Bernstein polynomials and share many of their analytic and geometric properties. A spline is a piecewise polynomial whose pieces fit together smoothly at the joins. B-spline curves and surfaces have two advantages over polynomial curves and surfaces. For a large collection of control points, a Bezier curve or surface approximates the control polygon or polyhedron with a single polynomial of high degree. But high-degree polynomials take a long time to compute and are numerically unstable. Splines provide low-degree approximations, which are faster to compute and numerically more tractable. We could, of course, manufacture splines by forming piecewise Bezier curves and surfaces. To do so, however, we would need to constrain the location of the control points so that the Bezier segments would meet smoothly at their joins. B-splines provide an approximation scheme where such constraints on the location of the control points are not necessary; B-spline curves and surfaces meet smoothly at their joins for completely arbitrary collections of control points. Thus B-splines provide a simpler, numerically more stable approach to approximating large amounts of data. For these reasons B-splines have become extremely popular in large-scale industrial applications.

7.1 The de Boor Algorithm

We are going to introduce B-spline curves by invoking blossoming to generalize the de Casteljau algorithm for Bezier curves. Consider again the de Casteljau algorithm depicted as a blossoming recurrence in Figure 6.1 (see page 309).

Examining Figure 6.1 from the bottom up, we see that what makes this recurrence work is that the blossom variables in adjacent nodes on the same level differ by only a single parameter. This juxtaposition allows us to invoke the multiaffine and symmetry properties to compute new blossom values from old blossom values as we proceed up the de Casteljau triangle. Thus to generalize the de Casteljau construction all we need to do is to ensure that this adjacency property holds throughout the triangle. The basic step is illustrated in Figure 7.1.

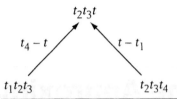

Figure 7.1 The basic step in the blossoming recurrence for a cubic polynomial: computing $t_2 t_3 t$ from $t_1 t_2 t_3$ and $t_2 t_3 t_4$ using linear interpolation together with the multiaffine and symmetry properties of the blossom. As usual each triple *uvw* stands for the blossom evaluated at (u, v, w).

The computation in Figure 7.1 is valid because from linear interpolation, we have

$$t = \frac{t_4 - t}{t_4 - t_1} t_1 + \frac{t - t_1}{t_4 - t_1} t_4 \, . \tag{7.1}$$

Therefore, for any symmetric multiaffine function p,

$$p(t_2, t_3, t) = p\left(t_2, t_3, \frac{t_4 - t}{t_4 - t_1} t_1 + \frac{t - t_1}{t_4 - t_1} t_4 \right) = \frac{t_4 - t}{t_4 - t_1} p(t_1, t_2, t_3) + \frac{t - t_1}{t_4 - t_1} p(t_2, t_3, t_4),$$

which is precisely the identity depicted in Figure 7.1.

Now to generalize the de Casteljau algorithm, we start at the base of the triangle not with the parameters $a^{n-k} b^k$, $k = 0, \dots, n$, but with some arbitrary parameters $t_{j+1} \dots t_{j+n}$, $j = 0, \dots, n$. Since adjacent nodes on the same level still differ by only a single parameter, this approach leads to the recurrence depicted in Figure 7.2. This recurrence is the *de Boor algorithm* for a single segment of a B-spline curve.

The computation illustrated in Figure 7.2 is just the computation depicted in Figure 7.1 repeated again and again as we proceed up the triangle. Notice that by Equation (7.1) this computation also makes sense if we treat each triple *uvw* as a product of real numbers, instead of as a sequence of blossom parameters. This observation provides a simple mnemonic for remembering the de Boor algorithm. Another simple mnemonic that emerges from the diagram is that if you follow along in the direction of any arrow, the labels you encounter along the edges do not change. Of course, this observation is true only for the numerators; the denominators, which are suppressed in the diagram, change from level to level. The label exiting $t_{j+1} \cdots t_{j+n}$ to the left is $t - t_j$, and the label exiting to the right is $t_{j+n+1} - t$. Knowing these labels allows us to label the numerators for the entire diagram. The denominators can be retrieved easily from the numerators, since the labels entering each node must sum to one.

The de Boor algorithm evaluates points along any degree n polynomial curve $P(t)$ by starting with $n + 1$ blossom values $p(t_1, \dots, t_n), \dots, p(t_{n+1}, \dots, t_{2n})$ and running a recurrence to compute $p(t, \dots, t)$. The parameters $t_1, \dots, t_{2n}$ are called *knots*. The only restriction on the knots is that the denominators in the de Boor algorithm must not vanish. This requirement is equivalent to the constraint $t_{j+n} \neq t_i$ whenever

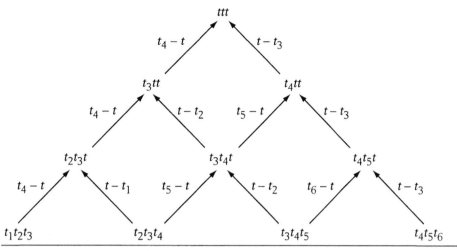

Figure 7.2 The de Boor algorithm for a single segment of a cubic B-spline curve. Again each triple *uvw* stands for the blossom evaluated at (*u,v,w*).

$1 \leq j \leq i \leq n$. Any knot sequence that satisfies this constraint is called a *progressive sequence*. For reasons that will become clear in Section 7.3, we shall generally assume that $t_j \leq t_{j+1}$ and shall restrict our attention only to the segment of the curve for which $t_n \leq t \leq t_{n+1}$.

Now consider the functions

$b_k^n(t)$ = the sum of the products along all paths from the kth position at the base to the apex of the de Boor algorithm.

Since the labels along the edges of the de Boor algorithm are linear functions that depend on the knots, the functions $b_0^n(t),...,b_n^n(t)$ are polynomials of degree n that also depend on the knots, but once the knots are fixed these functions are also fixed. Moreover, for any polynomial of degree n, it follows from the de Boor algorithm that

$$P(t) = \sum_{k=0}^{n} b_k^n(t) p(t_{k+1},...,t_{k+n}) \cdot \qquad (7.2)$$

Thus the polynomials $b_0^n(t),...,b_n^n(t)$ span the space of polynomials of degree n. But any set of $n+1$ polynomials of degree n that span the space of polynomials of degree n must form a basis for the space of polynomials of degree n. Hence the functions $b_0^n(t),...,b_n^n(t)$ are a basis for the polynomials of degree n. We call a polynomial basis that corresponds to a progressive knot sequence a *progressive basis*. If the knot sequence is increasing, then locally these polynomials represent the B-spline basis functions about which we will have a good deal more to say later in this chapter. For now, the important fact to observe is that if $P_0,...,P_n$ are the control points of $P(t)$ relative to the progressive basis $b_0^n(t),...,b_n^n(t)$, then by Equation (7.2)

$$P_k = p(t_{k+1},...,t_{k+n}) \tag{7.3}$$

because the control points relative to a fixed polynomial basis are unique. Equation (7.3) is the dual functional property for the B-splines; this equation is the basic fact connecting blossoming to B-splines and is the generalization to B-splines of the dual functional property for Bezier curves. Again we shall have a good deal more to say about this result in subsequent sections.

The de Boor algorithm for a B-spline segment has an elegant geometric interpretation, similar to the geometric interpretation of the de Casteljau algorithm for Bezier curves. Each step of the de Boor algorithm represents a linear interpolation. Labeling the points with their blossom values and joining the nodes with straight lines generates the trellis in Figure 7.3. Observe that just like the de Casteljau algorithm, the de Boor algorithm represents a corner-cutting procedure. Notice, however, that unlike Bezier curves, the curves generated by the de Boor algorithm do not necessarily interpolate their first and last control points.

The de Boor algorithm is easy to blossom. Since adjacent nodes differ by only a single parameter, we can introduce a new parameter at each level of the de Boor algorithm by replacing t with u_k on the kth level above the base of the triangle. Thus, we can compute an arbitrary blossom value $p(u_1,...,u_n)$ for any degree n polynomial $P(t)$ by starting with $n + 1$ blossom values $p(t_1,...,t_n),...,p(t_{n+1},...,t_{2n})$, running the de Boor algorithm, and replacing t by u_k on the kth level of the algorithm. This blossoming algorithm is illustrated in Figure 7.4. Notice that this ploy is the same maneuver we applied to blossom the de Casteljau algorithm.

In order to verify that B-splines generate smooth piecewise polynomial curves, we shall need to understand how to differentiate a segment of a B-spline curve. In Section 6.4 we showed that for any degree n polynomial $P(t)$,

$$P'(t) = np(\underbrace{t,...,t}_{n-1},\delta)$$

$$P^{(k)}(t) = \frac{n!}{(n-k)!} p(\underbrace{t,...,t}_{n-k},\underbrace{\delta,...,\delta}_{k}).$$

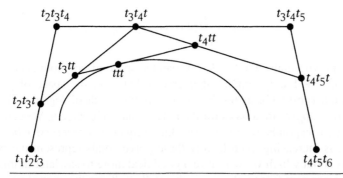

Figure 7.3 Geometric construction algorithm for a point on a segment of a cubic B-spline curve. All the points are labeled with blossom values. Compare to the geometric interpretation of the de Casteljau algorithm in Figure 5.4.

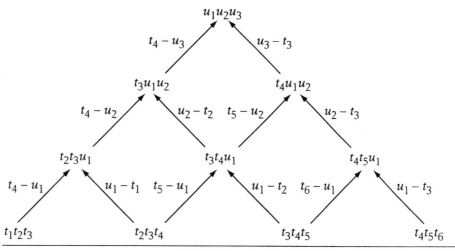

Figure 7.4 Blossoming the de Boor algorithm for a single segment of a cubic B-spline curve. This diagram is derived from Figure 7.2 by replacing t with u_k on the kth level of the algorithm. Again each triple uvw stands for the blossom evaluated at (u,v,w).

Thus to differentiate one segment of a B-spline curve, we must blossom and homogenize the de Boor algorithm and then evaluate at t's and δ's. We already know how to blossom the de Boor algorithm (see Figure 7.4). To homogenize, we simply homogenize each level of the algorithm; that is, we replace

$$t_j - u_k \rightarrow t_j v_k - u_k \text{ and } u_k - t_i \rightarrow u_k - t_i v_k$$

on the kth level of the algorithm. Setting $(u_k,v_k) = (t,1)$ gives us back our original labels, and setting $(u_k,v_k) = \delta = (1,0)$ replaces

$$t_j v_k - u_k \rightarrow -1 = \frac{d(t_j - t)}{dt}$$

$$u_k - t_i v_k \rightarrow +1 = \frac{d(t - t_i)}{dt}$$

on the kth level of the algorithm. Thus the net effect on the de Boor algorithm of evaluating

$$p(\underbrace{t,...,t}_{n-k},\underbrace{\delta,...,\delta}_{k})$$

is to differentiate k levels of the algorithm and leave $n - k$ levels unchanged. We illustrate this differentiation algorithm in Figure 7.5. Notice again that this approach is akin to the procedure we used to differentiate the de Casteljau algorithm for Bezier curves. We shall use this observation in Section 7.3 to prove that two adjacent B-spline segments meet smoothly at their join.

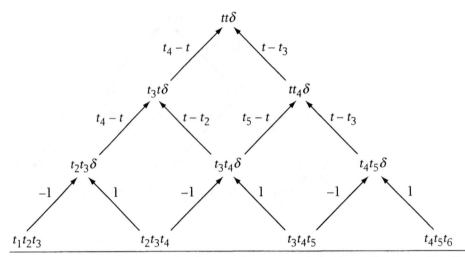

Figure 7.5 Differentiating the de Boor algorithm for a single segment of a cubic B-spline curve. This diagram is derived from Figure 7.2 by differentiating one level of the algorithm. To get the actual derivative, we need to multiply the output at the apex by $n = 3$. As usual, each triple uvw stands for the blossom evaluated at (u, v, w), $t_k = (t_k, 1)$, $t = (t, 1)$, and $\delta = (1, 0)$.

Another way to understand the computation in Figure 7.5 is to observe that

$$\delta = (1,0) = \frac{1}{t_k - t_j}(t_k,1) - \frac{1}{t_k - t_j}(t_j,1) \ . \tag{7.4}$$

Thus the differentiation algorithm is simply a consequence of Equations (7.1) and (7.4) coupled with the symmetry and multilinearity properties of the homogeneous blossom. Notice that the denominators $t_k - t_j$ are suppressed in Figure 7.5.

Exercises

1. Show that the de Boor algorithm and its variants (Figures 7.2, 7.4, and 7.5) remain valid if we interpret the nodes as univariate polynomials whose roots are the specified parameter values. That is, let $t_{j+1} \cdots t_{j+n}$ denote the polynomial with roots $t_{j+1}, \ldots, t_{j+n}$, $j = 0, \ldots, n$.

    a. Show that the de Boor algorithm (Figure 7.2) can be used to compute the polynomial $(t - x)^n$ from the polynomials

    $$\{(t_{j+1} - x) \cdots (t_{j+n} - x)\}, \ \ j = 0, \ldots, n.$$

    b. Show that the blossomed de Boor algorithm (Figure 7.4) can be used to compute the polynomial $(u_1 - x) \cdots (u_n - x)$ from the polynomials $\{(t_{j+1} - x) \cdots (t_{j+n} - x)\}, j = 0, \ldots, n$.

    c. Show that the differentiated de Boor algorithm (Figure 7.5) can be used to compute the polynomial $(t - x)^{n-k}$ from the polynomials

$$\{(t_{j+1} - x)\cdots(t_{j+n} - x)\}, \quad j = 0,...,n.$$

d. Apply the Third Principle of Duality from Section 5.5 to find the coefficients $\{Q_j\}$ of a polynomial $P(t)$ relative to the basis

$$D_j(t) = \{(t_{j+1} - x)\cdots(t_{j+n} - x)\}, \quad j = 0,...,n$$

given the coefficients $\{P_k\}$ relative to the basis

$$B_k(t) = \{(u_{k+1} - x)\cdots(u_{k+n} - x)\}, \quad k = 0,...,n.$$

(Hint: You will need two triangles.)

e. Use the result of part (d) to develop an $O(n^2)$ algorithm to find the Bezier control points of a degree n polynomial curve given $n + 1$ points on the curve along with their associated parameter values. (Hint: Convert from Lagrange to Bezier form using part (d).)

2. Let $b_0^d(t \mid t_{j+1},...,t_{j+2d}),...,b_d^d(t \mid t_{j+1},...,t_{j+2d})$ denote the progressive basis of degree d for the progressive knot sequence $t_{j+1},...,t_{j+2d}$.

a. Show that Figure 7.6 represents the down recurrence for the progressive basis functions.

b. Conclude that the progressive basis functions $\{b_k^n(t \mid t_1,...,t_{2n})\}$ satisfy the recurrence

$$b_k^n(t \mid t_1,...,t_{2n}) =$$

$$\frac{t_{k+n+1} - t}{t_{k+n+1} - t_{k+1}} b_k^{n-1}(t \mid t_2,...,t_{2n-2}) + \frac{t - t_k}{t_{k+n} - t_k} b_{k-1}^{n-1}(t \mid t_2,...,t_{2n-2})$$

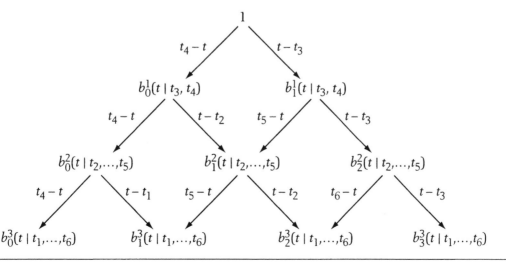

Figure 7.6 The down recurrence for the cubic progressive basis functions.

3. The Bernstein polynomials can be represented by an urn model employing sampling with replacement. Here we develop an urn model for a progressive basis with uniform knots.

   Consider an urn containing w white balls and b black balls. One ball at a time is drawn at random from the urn, its color inspected, then returned to the urn, and $w + b$ balls of the *opposite* color are added to the urn. Let $t = w/(w + b)$ be the probability of selecting a white ball on the first trial, and define

   $f_k^{n-1}(t) =$ probability of selecting a black ball after selecting exactly k white balls in the first $n - 1$ trials

   $s_k^{n-1}(t) =$ probability of selecting a white ball after selecting exactly k white balls in the first $n - 1$ trials

   $b_k^n(t) =$ probability of selecting exactly k white balls in the first n trials

   Using probabilistic arguments, show that

   a. $f_k^{n-1}(t) = \dfrac{k+1-t}{n}$

   b. $s_k^{n-1}(t) = \dfrac{t+n-1-k}{n}$

   c. $b_k^n(t) = f_k^{n-1}(t)b_k^{n-1}(t) + s_{k-1}^{n-1}(t)b_{k-1}^{n-1}(t)$

   By matching the recurrence in part (c) to the recurrence in Exercise 2(b), conclude that

   d. $b_0^d(t),\dots,b_d^d(t)$ is the progressive basis over the interval $(0,1)$ for the knots $t_k = k - d$, $k = 1,\dots,2d$.

   (Compare to Polya's urn model in Section 5.5.4.2, Exercise 12.)

4. Here we extend the urn model in the previous exercise to progressive bases corresponding to arbitrary progressive knot sequences $t_1,\dots,t_{2n}$. Again we study urn models where we add balls only of the opposite color.

   Consider an urn containing w white balls and b black balls. One ball at a time is drawn at random from the urn, its color inspected, and then returned to the urn. If the ball was the jth black ball chosen, then c_j additional white balls are added to the urn; if the ball was the kth white ball chosen, then d_k additional black balls are added to the urn. Thus in each case only balls of the *opposite* color to the color of the chosen ball are added to the urn, and in every case the number of balls added to the urn depends only on the number of balls of the *same* color that have previously been selected. Let $t = w/(w+ b)$ be the probability of selecting a white ball on the first trial, and let

$$c_j = \frac{t_{n+1-j} - t_{n-j}}{t_{n+1} - t_n}(w + b)$$

$$d_k = \frac{t_{n+k+1} - t_{n+k}}{t_{n+1} - t_n}(w + b)$$

$f_k^{n-1}(t) =$ probability of selecting a black ball after selecting exactly k white balls in the first $n - 1$ trials

$s_k^{n-1}(t) =$ probability of selecting a white ball after selecting exactly k white balls in the first $n - 1$ trials

$b_k^n(t) =$ probability of selecting exactly k white balls in the first n trials

Using probabilistic arguments, show that

a. $s_k^{n-1}\left(\dfrac{t - t_n}{t_{n+1} - t_n}\right) = \dfrac{t - t_{k+1}}{t_{n+k+1} - t_{k+1}}$

b. $f_k^{n-1}\left(\dfrac{t - t_n}{t_{n+1} - t_n}\right) = \dfrac{t_{k+n+1} - t}{t_{k+n+1} - t_{k+1}}$

c. $b_k^n\left(\dfrac{t - t_n}{t_{n+1} - t_n}\right) =$

$$f_k^{n-1}\left(\frac{t - t_n}{t_{n+1} - t_n}\right)b_k^{n-1}\left(\frac{t - t_n}{t_{n+1} - t_n}\right) + s_{k-1}^{n-1}\left(\frac{t - t_n}{t_{n+1} - t_n}\right)b_{k-1}^{n-1}\left(\frac{t - t_n}{t_{n+1} - t_n}\right)$$

By matching the recurrence in part (c) to the recurrence in Exercise 2(b), conclude that

d. $b_0^n\left(\dfrac{t - t_n}{t_{n+1} - t_n}\right),\dots,b_n^n\left(\dfrac{t - t_n}{t_{n+1} - t_n}\right)$ is the progressive basis for the knots $t_1,\dots,t_{2n}$.

(Compare to the generalized Polya urn model in Section 5.5.4.2, Exercise 13.)

7.2 Progressive Polynomial Bases Generated by Progressive Knot Sequences

Every progressive knot sequence determines a progressive polynomial basis. Each such basis has an evaluation algorithm that is a special case of the de Boor algorithm. Here we look at three important examples of progressive polynomial bases: the Bernstein basis, the monomial basis, and the Newton dual basis. The power basis is yet another example of a progressive polynomial basis; this basis is examined briefly in Exercise 1.

The de Casteljau algorithm is the special case of the de Boor algorithm where $t_1 = \dots = t_n = a$ and $t_{n+1} = \dots = t_{2n} = b$ (compare Figures 6.1 and 7.2). Thus Bezier curves are special types of B-spline segments.

If we homogenize the de Casteljau algorithm with respect to the knots and then replace b by $\delta = (1,0)$, we get an evaluation algorithm for the monomial basis

$$M_k^n(t) = \binom{n}{k}(t-a)^k$$

at $t = a$ (see Figure 7.7).

Since, up to constant multiples, the monomial coefficients at $t = a$ are the derivatives of the polynomial at $t = a$, the monomial coefficients are the values of the blossom at $a^{n-k}\delta^k$, $k = 0,...,n$. Therefore, the knot sequence for the monomial basis at $t = a$ is $t_1 = \cdots = t_n = a$ and $t_{n+1} = \cdots = t_{2n} = \delta$. Here the monomial basis functions are $\binom{n}{k}(t-a)^k$ instead of the standard Taylor basis functions $(t-a)^k / k!$ because there are $\binom{n}{k}$ paths from the kth position at the base to the apex of the triangle. Notice that in Figure 7.7 we have used the homogeneous version of the blossom. The computation now follows from the identity

$$t = (t,1) = (t-a)(1,0) + (a,1) = (t-a)\delta + a \tag{7.5}$$

and the multilinear property of the homogeneous blossom.

We close with one final example that is less familiar, but nevertheless plays an important role in algorithms for B-spline curves. Consider a sequence of knots $t_1,...,t_{2n}$, where $t_1,...,t_n$ are not multiples of δ and $t_{n+1} = \cdots = t_{2n} = \delta$. The corresponding polynomial basis is called the *Newton dual basis* because, in a way we shall make precise in Section 7.7.2, Exercise 5, his basis is dual to the Newton basis we studied in Chapter 4. The Newton dual basis appears in differentiation and knot insertion algorithms, which we will discuss in Section 7.6.4.2. Notice that the mono-

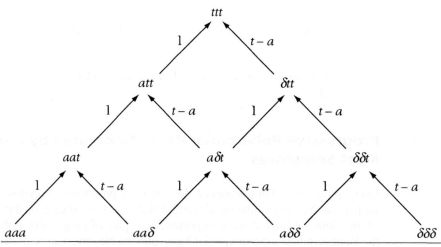

Figure 7.7 An evaluation algorithm for the cubic monomial basis at $t = a$. Here we use the homogeneous version of the blossom. As usual, each triple *uvw* stands for the blossom evaluated at (u,v,w), $t = (t,1)$, $a = (a,1)$, and $\delta = (1,0)$. Notice that by Equation (7.5), there is no normalization in the labels along the edges.

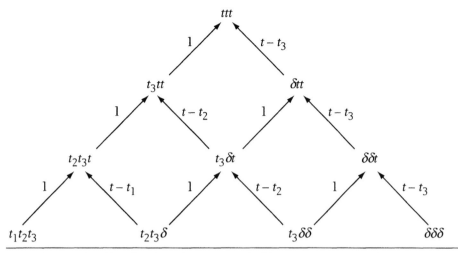

Figure 7.8 An evaluation algorithm for the cubic Newton dual basis. Again we use the homogeneous version of the blossom. As usual, each triple uvw stands for the blossom evaluated at (u, v, w), $t_k = (t_k, 1)$, $t = (t, 1)$, and $\delta = (1, 0)$. As in the evaluation algorithm for the monomial basis, no normalization is required in the labels along the edges.

mial basis at $t = a$ is just a special case of the Newton dual basis where $t_1 = \cdots = t_n = a$. The computation in Figure 7.8 follows from Equation (7.5) with the parameter a replaced by t_k.

Exercises

1. a. Prove by induction on n that the progressive basis corresponding to the knot sequence $t_1, ..., t_n, t_0, ..., t_{n-1}$ is the power basis

$$P_k^n(t) = \frac{(t - t_k)^n}{\prod_{j \neq k}(t_k - t_j)}, \quad k = 0, ..., n.$$

(Compare to Section 6.3, Exercise 6.)

   b. What is the urn model corresponding to the power basis? How are the $\{c_j\}$ and $\{d_k\}$ parameters related? (See Section 7.1, Exercise 4.)

2. Let $D_k^n(t)$, $k = 0, ..., n$, denote the Newton dual basis of degree n relative to the knots $t_1, ..., t_n$. Show that

   a. $D_0^n(t) = 1$.

   b. $D_1^n(t) = nt - \sum_k t_k$.

   c. $D_n^n(t) = (t - t_n)^n$.

   d. $D_k^n(t)$ is a polynomial of degree k that depends only on the knots $t_k, ..., t_n$.

3. Let $E_k^n(t)$, $k = 0,...,n$, be the progressive basis corresponding to the knot sequence

$$\underbrace{0,...,0}_{n},t_1,...,t_n \, ,$$

and let $E(t) = \sum_k E_k^n(t) P_k$.

a. Construct the de Boor algorithm for $E(t)$.

b. Show that $E(0) = P_0$.

7.3 B-Spline Curves

So far we have shown how to apply the de Boor algorithm to compute points only along a single polynomial segment, but our goal is to produce smooth piecewise polynomial curves. To achieve this end, we must somehow string together B-spline segments.

A B-spline segment of degree n is defined by $2n$ knots and $n + 1$ control points. Given knots $t_1,...,t_{2n}$ and control points $P_0,...,P_n$, we place the control points at the base of the de Boor algorithm (Figure 7.2) and the knots in the linear functions along the edges. Now suppose we are given some additional knots $t_{2n+1}, t_{2n+2},...$ and some additional control points $P_{n+1}, P_{n+2},...$. Then we can define some additional polynomial segments simply by shifting all the indices in the de Boor algorithm by a constant. For example, if we are given one additional knot and one additional control point, then we can form the polynomial segment illustrated in Figure 7.9.

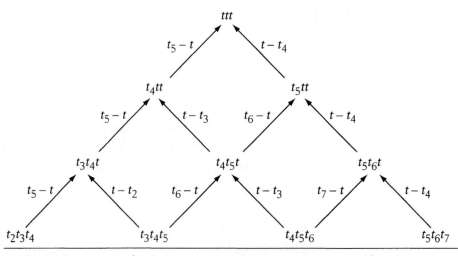

Figure 7.9 A cubic B-spline segment for the knots $t_2,...,t_7$. This diagram is generated from Figure 7.2 by shifting all the indices by one.

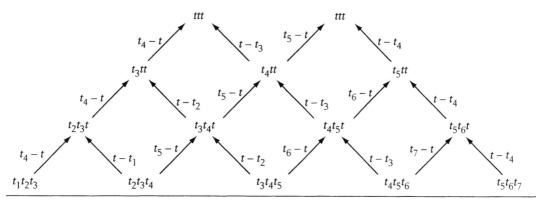

Figure 7.10 The de Boor algorithm for two segments of a cubic B-spline curve. Notice that the symbols *ttt* at the two apexes represent the values of two distinct polynomial curves over two distinct knot intervals.

How is this new segment related to the original segment? By construction, the two diagrams for the two de Boor algorithms fit together as in Figure 7.10; that is, they share common control points as well as nodes and edges with common labels.

This overlapping of the two diagrams is very suggestive. Notice, however, that we have overloaded our notation here. The two overlapping triangles represent two distinct polynomial curves $P_1(t)$ and $P_2(t)$. The symbols *ttt* at the two apexes are not identical; rather they represent $p_1(t,t,t) = P_1(t)$ and $p_2(t,t,t) = P_2(t)$. Similarly, the notation $t_{k+1}t_{k+2}t_{k+3}$ is overloaded; here $t_{k+1}t_{k+2}t_{k+3}$ represents both $p_1(t_{k+1},t_{k+2},t_{k+3})$ and $p_2(t_{k+1},t_{k+2},t_{k+3})$. To construct B-spline curves, we start with a fixed collection of control points $\{P_k\}$. It is these control points that we actually place at the base of the diagram for the de Boor algorithm and use to compute the polynomials $P_1(t)$ and $P_2(t)$. It then follows from Equation (7.3) that

$$p_1(t_{k+1},t_{k+2},t_{k+3}) = P_k = p_2(t_{k+1},t_{k+2},t_{k+3});$$

it is this equality that permits us to overload our notation.

We shall now show that the curve segments that these overlapping diagrams represent also fit together smoothly. By convention, the first apex represents the curve segment for $t_3 \leq t \leq t_4$; shifting indices by one, we see that the second apex must represent the curve segment for $t_4 \leq t \leq t_5$. The question then is this: How smoothly do these two curve segments meet at $t = t_4$?

To begin, we must show that these curves meet continuously at $t = t_4$. Consider the first segment in Figure 7.10. At $t = t_4$, the label $t_4 - t = 0$; therefore, since the labels entering any node sum to one, the nodes labeled *ttt* and t_4tt have the same value at $t = t_4$. But by the same argument the nodes labeled *ttt* and t_4tt have the same value at $t = t_4$ for the second segment. Since *ttt* represents the value of the polynomial in each segment, it follows that these two polynomials agree at $t = t_4$.

An identical argument works for derivatives. Recall from Section 7.1 that to differentiate a B-spline segment, we simply differentiate the labels on any level. If we differentiate the labels on the first level above the control points in Figure 7.10, then exactly the same argument we used to show that the two polynomials agree at $t = t_4$ shows that their derivatives match at $t = t_4$. Differentiating the second level as well shows that second derivatives also match at $t = t_4$. The third derivatives, however, need not agree since differentiating the third level destroys the label $t_4 - t$, so we cannot conclude that the node at the apex now labeled $\delta\delta\delta$ has the same value at $t = t_4$ as the interior node labeled $t_4\delta\delta$.

Exactly the same arguments work for B-spline segments of degree n, except now $n - 1$ derivatives agree at the knots. Stringing together B-spline segments of degree n in this manner generates smooth curves that join at the knots with $n - 1$ continuous derivatives. Notice that this construction is completely independent of either the location of the control points or the values of the knots. We have assumed only that the knots are strictly increasing, that is, that $t_j < t_{j+1}$, so that adjacent knots define nonempty intervals.

What happens if for some knots $t_j = t_{j+1}$? Consider again Figure 7.10. If $t_4 = t_5$, then in the second segment the denominators of the labels on the arrows pointing into the apex vanish. That is, there is a singularity in the algorithm. Fortunately, however, if $t_4 = t_5$, then the segment this algorithm represents is essentially null so we can ignore this part of the diagram. Proceeding directly to the next segment—the segment for the interval $[t_5, t_6]$—we see that the diagram for this segment still overlaps the diagram for the segment $[t_3, t_4]$, but there are fewer common nodes and edges. In fact, there are no common nodes on the second level and the only common node on the first level is $t_4 t_5 t$ (see Figure 7.11). Now exactly the same arguments that we used before show that the segment over $[t_3, t_4]$ meets the segment over $[t_5, t_6]$ with one continuous derivative at $t = t_4 = t_5$, since they share the value at the node $t_4 t_5 t$.

In general, increasing the multiplicity of a knot shifts the diagram for the next segment to the right and decreases the differentiability of the B-spline curve at the knot. Thus we have the following result; the proof is just an elaboration of the arguments in the preceding paragraph.

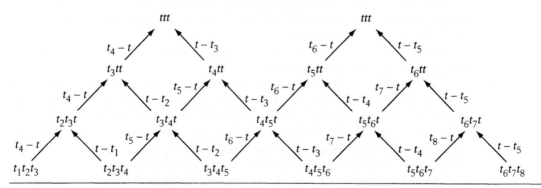

Figure 7.11   The de Boor algorithm for two adjacent segments of a cubic B-spline curve when $t_4 = t_5$.

THEOREM 7.1	At a knot of multiplicity μ, a B-spline curve of degree n has $n - \mu$ continuous derivatives.

Exercises

1. Prove Theorem 7.1.
2. Extend Figure 7.10 to three B-spline segments.
3. Draw the de Boor algorithm for two adjacent segments of a cubic B-spline curve when $t_4 = t_5 = t_6$.

7.4 Elementary Properties of B-Spline Curves

A B-spline curve is a piecewise polynomial curve specified by an arbitrary collection of control points $\{P_j\}$ and a nondecreasing sequence of knots $\{t_k\}$, where each individual polynomial segment is defined by the de Boor algorithm. By construction, the kth segment of a degree n B-spline curve

- lies over the parameter interval $[t_k, t_{k+1}]$,
- has $n + 1$ control points—$P_{k-n}, ..., P_k$,
- depends on $2n$ knots—$t_{k-n+1}, ..., t_{k+n}$.

The labels on the de Boor algorithm are specified as follows: $t - t_k$ labels the edge exiting P_k to the left and $t_{k+n+1} - t$ labels the edge exiting P_k to the right (see Figure 7.12). The remainder of the edges can be labeled by observing that if you follow along in the direction of any arrow, the labels you encounter along the edges (in the numerator) do not change (see Figure 7.2).

Here is a list of the elementary properties of B-spline curves:

1. Piecewise polynomial
2. Continuity of order $C^{n-\mu}$ at knots of multiplicity μ on curves of degree n
3. Local control

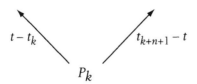

Figure 7.12 The (unnormalized) labels in the de Boor algorithm on the arrows exiting from the node containing the control point P_k. The entire de Boor algorithm can be recovered from this diagram because in the direction of any arrow, the labels encountered along the edges (in the numerator) do not change (see Figure 7.2).

4. Affine invariance
5. Local convex hull
6. Locally nondegenerate
7. Interpolation of control points at knots where the multiplicity μ is equal to the degree n

Many of these features are analogous to the elementary properties of Bezier curves. We shall derive each of these properties directly from the de Boor algorithm:

- *Piecewise polynomial.* Between any two adjacent knots, the de Boor algorithm computes a polynomial curve. Therefore, B-spline curves are piecewise polynomials with break points at the knots.

- *Continuity of order $C^{n-\mu}$ at knots of multiplicity μ.* This property is just Theorem 7.1.

- *Local control.* Since the polynomial segment over the parameter interval $[t_k, t_{k+1}]$ is defined by the control points $P_{k-n}, ..., P_k$, the control point P_k influences only the $n + 1$ curve segments over the parameter interval $[t_k, t_{k+n+1}]$. Hence moving a single control point has only a local effect on the B-spline curve, in contrast to Bezier curves where each control point influences the entire curve. Similarly, since the curve segment over the parameter interval $[t_k, t_{k+1}]$ depends only on the $2n$ knots $t_{k-n+1}, ..., t_{k+n}$, the knot t_k influences solely the $2n$ curve segments over the parameter interval $[t_{k-n}, t_{k+n}]$, so altering the location of a knot also has only a local influence on the curve.

- *Affine invariance.* Since the nodes in the de Boor algorithm are computed from affine combinations of lower-level nodes, it follows by induction on the level of the node that each node in the de Boor algorithm represents a point in affine space. Therefore, the B-spline curve at the apex of the de Boor algorithm is a collection of points in affine space.

- *Local convex hull.* Each segment of a B-spline curve lies in the convex hull of its control points because the functions along the edges of the de Boor algorithm sum to one and are nonnegative in the parameter domain. To see that this is so, consider the segment of a B-spline curve of degree n over the knot interval $[t_n, t_{n+1}]$. For this segment the labels along the edges depend only on the knots $t_1 \le \cdots \le t_{2n}$. Hence here all the denominators of the functions along the edges are of the form $t_{n+j} - t_i$, $1 \le i, j \le n$, so the denominators are certainly positive. Moreover, the numerators are all of the form $t_{n+i} - t$ or $t - t_i$, $i = 1, ..., n$. Since $t_n \le t \le t_{n+1}$, it follows that the numerators too are nonnegative. Because the labels along the edges are nonnegative and sum to one, it follows by induction on the level of the node that each node in the local de Boor algorithm lies in the convex hull of the local control points. Hence B-spline curves have the local convex hull property.

- *Locally nondegenerate.* Suppose that locally the B-spline curve collapses to a single point P. Then locally the B-spline curve is given by the constant polynomial $P(t) = P$. Therefore, by the dual functional property for B-spline curve segments (Equation (7.3)), the control points of $P(t)$ are given by $P_k = p(t_{k+1},...,t_{k+n}) = P$. Hence if the B-spline curve collapses locally to a single point, all the control points of the curve segment must lie at that point.

■ *Interpolation of control points at knots where the multiplicity μ is equal to the degree n.* Suppose that $t_{k+1} = \cdots = t_{k+n}$, and consider the B-spline segment $P_{k+n}(t)$ over the knot interval $[t_{k+n}, t_{k+n+1}]$. By the dual functional property, the initial control point of this segment is

$$P_k = p_{k+n}(t_{k+1},...,t_{k+n}) = p_{k+n}(t_{k+n},...,t_{k+n}) = P_{k+n}(t_{k+n}),$$

so the B-spline curve interpolates its kth control point when $t_{k+1} = \cdots = t_{k+n}$.

Unlike Bezier curves, B-spline curves do not generally interpolate their first or last control points. This property makes it difficult to connect two arbitrary B-spline curves. To overcome this problem, n-fold knots are often placed at the start and the end of the knot vector to force interpolation of the initial and final control points.

There is another way to force interpolation of control points. If we set $P_k = \cdots = P_{k+n}$, then by the local convex hull property the B-spline segment over the knot interval $[t_{k+n}, t_{k+n+1}]$ will collapse to the point P_k because all its control points are located at P_k. Thus the B-spline curve will certainly interpolate P_k. But introducing multiple control points is not as benign as introducing multiple knots. If

$$P_k = \cdots = P_{k+n},$$

then the segment over $[t_{k+n+1}, t_{k+n+2}]$ collapses to a line, since all but one of the control points for this segment are identical. Additional nearby segments will exhibit similar degeneracies because many of their control points will be identical. Multiple knots reduce the differentiability of the curve at the knot, but knot multiplicities do not introduce additional degeneracies in nearby segments. Therefore, to force interpolation it is more desirable to introduce multiple knots than multiple control points.

Exercises

1. Show that translating each control point of a B-spline curve by a vector v translates the entire B-spline curve by v.

2. Let $S(t)$ be a B-spline curve with control points $\{P_k\}$ and knots $\{t_k\}$. Form a new B-spline curve $R(t)$ by replacing each knot t_k by the knot $\tau_k = at_k + b$ for some fixed constants $a > 0$ and b. Show that changing all the knots in this way has no effect on the shape of the B-spline curve. In particular, using the de Boor algorithm, show that $R(at + b) = S(t)$. What happens if we choose $a < 0$? (Compare to Section 2.2, Exercise 4.)

3. Construct B-spline curves that interpolate a fixed control point P_k by introducing

   a. multiple knots

   b. repeated control points

   Compare your results. Which curves do you prefer? Why?

4. Use the de Boor algorithm to prove directly, without appealing to the dual functional property, that a degree n B-spline curve with $t_{k+1} = \cdots = t_{k+n}$ interpolates its kth control point P_k.

5. Use the de Boor algorithm to construct a B-spline curve. Experiment with how moving the control points or altering the location of the knots changes the shape of the curve.

7.5 All Splines Are B-Splines

The de Boor algorithm constructs a spline from a knot sequence and a control polygon. But suppose we are given a spline—that is, a piecewise polynomial curve where the pieces meet smoothly at the joins. Is there a B-spline curve that matches this spline? Is every spline curve a B-spline curve? Remarkably the answer is yes. Thus by investigating B-spline curves we study all spline curves.

To prove this result, we begin with a lemma showing how the blossom values of two polynomials are related when their derivatives agree at a point.

LEMMA 7.2

Let $P(t)$ and $Q(t)$ be two polynomials of degree n. Then the following statements are equivalent:

1. $P^{(j)}(\tau) = Q^{(j)}(\tau),\ 0 \le j \le k$.

2. $p(\underbrace{\tau,...,\tau}_{n-j},u_1,...,u_j) = q(\underbrace{\tau,...,\tau}_{n-j},u_1,...,u_j)$

   for any parameters $u_1,...,u_j, 0 \le j \le k$.

Proof $1 \Rightarrow 2$. Suppose that $P^{(j)}(\tau) = Q^{(j)}(\tau),\ 0 \le j \le k$. In Section 6.4 we showed that

$$P^{(j)}(\tau) = \frac{n!}{(n-j)!}\, p(\underbrace{\tau,...,\tau}_{n-j},\underbrace{\delta,...,\delta}_{j})$$

$$Q^{(j)}(\tau) = \frac{n!}{(n-j)!}\, q(\underbrace{\tau,...,\tau}_{n-j},\underbrace{\delta,...,\delta}_{j}).$$

Therefore,

$$P^{(j)}(\tau) = Q^{(j)}(\tau) \iff p(\underbrace{\tau,...,\tau}_{n-j},\underbrace{\delta,...,\delta}_{j}) = q(\underbrace{\tau,...,\tau}_{n-j},\underbrace{\delta,...,\delta}_{j}). \tag{7.6}$$

Now starting with the $k + 1$ blossom values

$$p(\tau,...,\tau), p(\tau,...,\tau,\delta),..., p(\underbrace{\tau,...,\tau}_{n-k},\underbrace{\delta,...,\delta}_{k})$$

along the base of a triangle and running the blossomed version of the homogeneous de Boor algorithm, we see that the $k + 1$ blossom values

$$p(\tau,...,\tau), p(\tau,...,\tau,u_1),..., p(\underbrace{\tau,...,\tau}_{n-k},u_1,...,u_k)$$

emerge along the left edge of the triangle (see Figure 7.13). But the same algorithm (Figure 7.13) starting with the $k+1$ blossom values

$$q(\tau,...,\tau), q(\tau,...,\tau,\delta),..., q(\underbrace{\tau,...,\tau}_{n-k},\underbrace{\delta,...,\delta}_{k})$$

generates the $k+1$ blossom values

$$q(\tau,...,\tau), q(\tau,...,\tau,u_1),..., q(\underbrace{\tau,...,\tau}_{n-k},u_1,...,u_k).$$

Since by assumption the input to these two algorithms is the same, the output must also be the same. Therefore,

$$p(\underbrace{\tau,...,\tau}_{n-j},u_1,...,u_j) = q(\underbrace{\tau,...,\tau}_{n-j},u_1,...,u_j)$$

for any parameters $u_1,...,u_j$, $0 \le j \le k$.

$2 \Rightarrow 1$. Conversely suppose that

$$p(\underbrace{\tau,...,\tau}_{n-j},u_1,...,u_j) = q(\underbrace{\tau,...,\tau}_{n-j},u_1,...,u_j)$$

for any parameters $u_1,...,u_j$, $0 \le j \le k$. Then setting $u_i = \delta$ for $i = 1,...,k$, we obtain

$$p(\underbrace{\tau,...,\tau}_{n-j},\underbrace{\delta,...,\delta}_{j}) = q(\underbrace{\tau,...,\tau}_{n-j},\underbrace{\delta,...,\delta}_{j}) \cdot$$

Hence by Equation (7.6), $P^{(j)}(\tau) = Q^{(j)}(\tau)$, $0 \le j \le k$.

THEOREM 7.3

Every spline curve is a B-spline curve.

Proof Let $S(t)$ be a spline curve defined over the knot intervals $[t_k, t_{k+1}]$ by the degree n polynomials $S_k(t)$, $k = n,...,m$. Suppose further that for each index k,

$$S_k^{(j)}(t_{k+1}) = S_{k+1}^{(j)}(t_{k+1}), \quad j = 0,...,n-1.$$

Our goal is to find a collection of points $\{P_k\}$ such that the B-spline curve of degree n for the control points $\{P_k\}$ and the knot sequence $\{t_k\}$ exactly reproduces the spline $S(t)$. Consider the polynomial $S_n(t)$ for the first interval $[t_n, t_{n+1}]$. Let $P_j = s_n(t_{j+1},...,t_{j+n})$, $j = 0,...,n$. Then by Equation 7.3, the de Boor algorithm for the knots $t_1,...,t_{2n}$ and the control points $P_0,...,P_n$ generates the polynomial $S_n(t)$. Similarly, if we set

$$Q_j = s_{n+1}(t_{j+2},...,t_{j+n+1}) \quad j = 0,...,n,$$

then the de Boor algorithm for the knots $t_2,\ldots,t_{2n+1}$ and the control points $Q_0,\ldots,Q_n$ generates the polynomial $S_{n+1}(t)$. To show that these two polynomials generate a B-spline curve, we need to show that $Q_j = P_{j+1}$ or equivalently that

$$s_{n+1}(t_{j+2},\ldots,t_{j+n+1}) = s_n(t_{j+2},\ldots,t_{j+n+1}), \quad j = 0,\ldots,n-1;$$

that is, we need to show that the two algorithms share n control points. But by assumption,

$$S_n^{(j)}(t_{n+1}) = S_{n+1}^{(j)}(t_{n+1}), \quad j = 0,\ldots,n-1.$$

Therefore, by Lemma 7.2,

$$s_{n+1}(t_{j+2},\ldots,t_{j+n+1}) = s_n(t_{j+2},\ldots,t_{j+n+1}), \quad j = 0,\ldots,n-1,$$

since t_{n+1} is always one of the blossom parameters. Thus these two segments form a B-spline curve. Now in the same manner using the de Boor algorithm, we can generate more and more polynomial segments that match the segments of the given spline $S(t)$. By the same argument, these segments will share common control points; therefore, these segments form a B-spline curve. Since the entire spline can be generated in this manner from the de Boor algorithm, it follows that every spline curve is a B-spline curve.

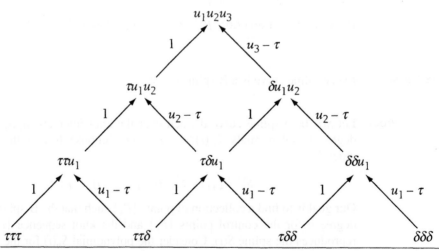

Figure 7.13 Computing the blossom values $\tau\tau\tau, \tau\tau u_1, \tau u_1 u_2, u_1 u_2 u_3$ from the blossom values $\tau\tau\tau, \tau\tau\delta,$ $\tau\delta\delta, \delta\delta\delta$. Notice that the first k blossom values in the first set depend only on the first k blossom values in the second set. Here we use the homogeneous blossom and the identity $u_k = (u_k - \tau)\delta + \tau$, so no normalization is necessary in the labels along the edges.

Theorem 7.3 can be generalized to splines $S(t)$ where the spline segments meet with arbitrary smoothness. That is, we need only suppose that

$$S_k^{(j)}(t_{k+1}) = S_{k+1}^{(j)}(t_{k+1}), \quad j = 0,\ldots,n - \mu_{k+1}.$$

In this case, we use a knot sequence where each knot t_k has multiplicity μ_k. The remainder of the argument is much the same; the details are left as an exercise.

Exercises

1. Prove that Theorem 7.3 remains valid if the segments $S_k(t)$ of the spline $S(t)$ satisfy

$$S_k^{(j)}(t_{k+1}) = S_{k+1}^{(j)}(t_{k+1}), \quad j = 0,\ldots,n - \mu_{k+1}.$$

2. Prove that every polynomial curve is a B-spline curve for any arbitrary choice of knots.

7.6 **Knot Insertion Algorithms**

Knot insertion is one of the main innovations of CAGD in the study of B-spline curves and surfaces. Knot insertion is to B-spline curves what subdivision is to Bezier curves. Given a knot sequence and a control polygon, the idea behind knot insertion is to construct a new knot sequence and a new control polygon that generates the same B-spline curve as the original knot sequence and original control polygon. The motivation is to create a control polygon with additional control points that more closely approximates the curve than the original control polygon. This new control polygon could then be used for rendering and intersection algorithms, as well as for providing additional control over the shape of the curve.

To make the notion of knot insertion more precise, consider a degree n B-spline curve $S(t)$ defined by a knot sequence $\{t_j\}$ and a collection of control points $\{P_k\}$. A knot sequence $\{u_i\}$ is said to be a *refinement* of $\{t_j\}$ if $\{u_i\} \supset \{t_j\}$. Given a refinement $\{u_i\}$ of $\{t_j\}$, the knot insertion problem is to find a collection of control points $\{Q_h\}$ such that the degree n B-spline curve generated by the knot sequence $\{u_i\}$ and the control points $\{Q_h\}$ is identical to the original degree n B-spline curve $S(t)$ generated by the knot sequence $\{t_j\}$ and control points $\{P_k\}$. This problem is called *knot insertion* because additional knots have been inserted into the knot sequence $\{t_j\}$ to form the refined knot sequence $\{u_i\}$.

The knot insertion problem always has a solution. To understand why, we need to recall precisely what it means for $S(t)$ to be a spline of degree n over the knot sequence $\{t_j\}$. A spline $S(t)$ of degree n over the knot sequence $\{t_j\}$ is a piecewise polynomial curve whose pieces are degree n polynomials in the knot intervals $[t_j, t_{j+1}]$. Moreover, adjacent segments must fit together at the knot t_j with at least $n - \mu_j$ continuous derivatives, where μ_j is the multiplicity of the knot t_j in the knot sequence. If $\{u_i\}$ is a refinement of $\{t_j\}$, then $S(t)$ is also a spline curve over the knot sequence $\{u_i\}$. Indeed if u_i is not one of the original knots t_j, then $S(t)$ is represented by a single polynomial on both sides of u_i; therefore, adjacent segments fit together with infinitely many continuous derivatives at u_i. Now by Theorem 7.3, since every spline is a B-spline, there must exist control points $\{Q_h\}$ that represent $S(t)$ as a B-spline curve over the knot sequence $\{u_i\}$.

The problem then is to develop an algorithm to generate the new control points $\{Q_h\}$ from the original control points $\{P_k\}$, the original knots t_j, and the new knots $\{u_i\}$. We will present three such algorithms: Boehm's algorithm, the Oslo algorithm, and a factored knot insertion algorithm. All three of these algorithms are based on blossoming and the de Boor algorithm, and all make use of the fundamental dual functional property for B-splines that if the spline $S(t)$ is represented in a subinterval of $[u_h, u_{h+n}]$ by the polynomial $P(t)$, then by Equation (7.3), $Q_h = p(u_{h+1}, ..., u_{h+n})$.

Exercise

1. Let $S_n\{t_j\}$ denote the collection of all splines of degree n with knots at $\{t_j\}$. Show that

   a. $S_n\{t_j\}$ is a vector space

   b. $\{u_i\} \supset \{t_j\} \Rightarrow S_n\{u_i\} \supset S_n\{t_j\}$

   Conclude that if $\{t_j\}$ and $\{u_i\}$ are nested knot sequences, then $S_n\{t_j\}$ and $S_n\{u_i\}$ are nested vector spaces.

7.6.1   Boehm's Knot Insertion Algorithm

Boehm's knot insertion algorithm inserts one knot at a time. Consider a cubic B-spline curve $S(t)$ defined over a knot sequence $\{t_j\}$. Suppose that we want to insert the knot u between t_k and t_{k+1}; that is, we want to replace the old knot sequence $..., t_{k-1}, t_k, t_{k+1}, t_{k+2}, ...$ with the new knot sequence $..., t_{k-1}, t_k, u, t_{k+1}, t_{k+2}, ...$ By Equation (7.3) the control points of $S(t)$ can be computed by evaluating the (local) blossom of $S(t)$ at consecutive knots. Thus if $S(t)=P(t)$ in the knot interval $[t_k, t_{k+1}]$, then we have

- *Old control points*

$$p(t_{k-2}, t_{k-1}, t_k), p(t_{k-1}, t_k, t_{k+1}), p(t_k, t_{k+1}, t_{k+2}), p(t_{k+1}, t_{k+2}, t_{k+3})$$

- *New control points*

$$p(t_{k-2}, t_{k-1}, t_k), p(t_{k-1}, t_k, u), p(t_k, u, t_{k+1}), p(u, t_{k+1}, t_{k+2}), p(t_{k+1}, t_{k+2}, t_{k+3})$$

Moreover, all the other control points with respect to the two knot sequences are identical because the new knot u does not appear in any other sequence of three consecutive knots. Thus to find the new control points, we need to replace

$$p(t_{k-1}, t_k, t_{k+1}), p(t_k, t_{k+1}, t_{k+2}) \rightarrow p(t_{k-1}, t_k, u), p(t_k, u, t_{k+1}), p(u, t_{k+1}, t_{k+2}) \cdot$$

That is, two of the original control points must be replaced by three new control points. Notice that we need to increase the number of control points by one because by inserting a new knot we have increased the number of knot intervals by one.

Boehm's algorithm uses the de Boor algorithm to compute the new control points from the original control points. If we replace t by u in the de Boor algorithm (Figure 7.2), then the new control points emerge on the first level of the de Boor algorithm (see Figure 7.14).

If we wish to insert the knot u as a multiple knot, we could run Boehm's algorithm several times—once for each time we wish to insert u—but it is more efficient instead to run the de Boor algorithm for several levels. For cubic curves, if we want to insert u as a double knot between t_k and t_{k+1}, then the old knot sequence is $\dots, t_{k-1}, t_k, t_{k+1}, t_{k+2}, \dots$ and the new knot sequence is $\dots, t_{k-1}, t_k, u, u, t_{k+1}, t_{k+2}, \dots$. Taking three consecutive knots at a time, we have

- *Old control points*

$$p(t_{k-2}, t_{k-1}, t_k), p(t_{k-1}, t_k, t_{k+1}), p(t_k, t_{k+1}, t_{k+2}), p(t_{k+1}, t_{k+2}, t_{k+3})$$

- *New control points*

$$p(t_{k-2}, t_{k-1}, t_k), p(t_{k-1}, t_k, u), p(t_k, u, u), p(u, u, t_{k+1}),$$

$$p(u, t_{k+1}, t_{k+2}), p(t_{k+1}, t_{k+2}, t_{k+3})$$

Thus two of the original control points must be replaced by four new control points— that is,

$$p(t_{k-1}, t_k, t_{k+1}), p(t_k, t_{k+1}, t_{k+2}) \rightarrow p(t_{k-1}, t_k, u), p(t_k, u, u), p(u, u, t_{k+1}), p(u, t_{k+1}, t_{k+2}).$$

If we replace t by u in the de Boor algorithm, the new control points we seek now emerge along the lateral edges and on the second level of the de Boor algorithm (see Figure 7.15).

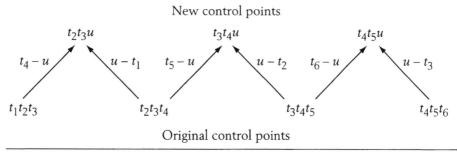

Figure 7.14 Boehm's knot insertion algorithm for a cubic B-spline curve. This algorithm computes just one level of the de Boor algorithm (compare to Figure 7.2).

New control points

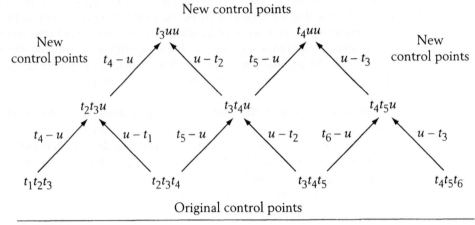

Original control points

Figure 7.15   Boehm's knot insertion algorithm for inserting a double knot into a cubic B-spline curve. Note that this algorithm computes two levels of the de Boor algorithm (compare to Figure 7.2).

Finally, for cubic curves, if we wish to insert u as a triple knot between t_k and t_{k+1}, then

- *Old control points*

$$p(t_{k-2},t_{k-1},t_k), p(t_{k-1},t_k,t_{k+1}), p(t_k,t_{k+1},t_{k+2}), p(t_{k+1},t_{k+2},t_{k+3})$$

- *New control points*

$$p(t_{k-2},t_{k-1},t_k), p(t_{k-1},t_k,u), p(t_k,u,u), p(u,u,u),$$
$$p(u,u,t_{k+1}), p(u,t_{k+1},t_{k+2}), p(t_{k+1},t_{k+2},t_{k+3})$$

Thus two of the original control points must be replaced by five new control points:

$$p(t_{k-1},t_k,t_{k+1}), p(t_k,t_{k+1},t_{k+2}) \rightarrow$$
$$p(t_{k-1},t_k,u), p(t_k,u,u), p(u,u,u), p(u,u,t_{k+1}), p(u,t_{k+1},t_{k+2}).$$

Now if we replace t by u in the de Boor algorithm, the control points we seek emerge along the two lateral edges of the complete de Boor algorithm (see Figure 7.16).

Analogous results hold for B-spline curves of arbitrary degree. We can summarize Boehm's approach to knot insertion as follows:

Boehm's Knot Insertion Algorithm

To insert a knot u a total of p times between the knots t_k and t_{k+1}:

1. Run the de Boor algorithm for the kth B-spline segment to the pth level and take the new control points consecutively off the left lateral edge, the pth level, and the right lateral edge of the triangle.

2. Discard the old control points of the kth segment.

3. Keep all of the other original control points.

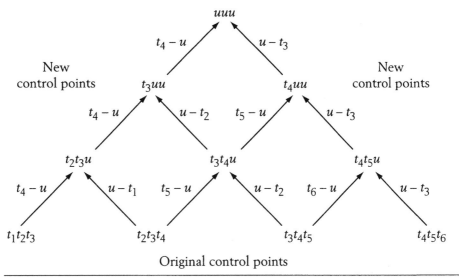

Figure 7.16 Boehm's knot insertion algorithm for inserting a triple knot into a cubic B-spline curve. Note that this algorithm computes the complete de Boor evaluation algorithm (compare to Figure 7.2).

7.6.2 The Oslo Algorithm

In Boehm's knot insertion algorithm the basic step is inserting one new knot; in the Oslo algorithm the fundamental step is computing one new control point. Boehm's knot insertion algorithm inserts one knot at a time; the Oslo algorithm inserts many knots simultaneously.

To insert new knots, we need to compute new control points. By Equation (7.3) these new control points are the (local) blossom of the spline evaluated at the new knots. Thus we need a method to evaluate the (local) blossom of the spline at arbitrary parameter values. But the blossom evaluated at arbitrary parameter values is precisely the output of the blossomed version of the de Boor algorithm (Figure 7.4). The Oslo algorithm simply applies the blossomed version of the de Boor algorithm to compute each of the required new control points.

The blossomed version of the de Boor algorithm is the original version of the Oslo algorithm. But this version of the Oslo algorithm is much less efficient than Boehm's knot insertion algorithm. Consider, for example, inserting four distinct knots u_1, u_2, u_3, u_4 into a single segment $[t_k, t_{k+1}]$ of a cubic B-spline curve. This operation requires the computation of six new control points, corresponding to the six blossom values

$$p(t_{k-1}, t_k, u_1), p(t_k, u_1, u_2), p(u_1, u_2, u_3), p(u_2, u_3, u_4), p(u_3, u_4, t_{k+1}), p(u_4, t_{k+1}, t_{k+2}).$$

To find these new control points, we can run either Boehm's algorithm four times or the Oslo algorithm six times. Each time we run Boehm's algorithm we must compute 3 affine combinations. So to insert 4 new knots with Boehm's algorithm requires a total of $4 \times 3 = 12$ affine combinations. On the other hand, to compute a single new control point using the Oslo algorithm, we need to run the blossomed version of the de Boor algorithm, which requires us to perform 6 affine combinations. So to compute 6 new control points using the Oslo algorithm, we would need to compute a total of $6 \times 6 = 36$ affine combinations. Clearly, then, Boehm's algorithm is more efficient than this version of the Oslo algorithm.

There is, however, a more efficient version of the Oslo algorithm. Consider again a cubic B-spline curve $S(t)$ defined over a knot sequence $\{t_j\}$. Suppose that we want to insert four knots u_1, u_2, u_3, u_4 between the knots t_k and t_{k+1}; that is, we want to replace the old knot sequence $\dots, t_{k-1}, t_k, t_{k+1}, t_{k+2}, \dots$ with the new knot sequence $\dots, t_{k-1}, t_k, u_1, u_2, u_3, u_4, t_{k+1}, t_{k+2}, \dots$ If $S(t) = P(t)$ in the knot interval $[t_k, t_{k+1}]$, then by Equation (7.3), we have

- *Old control points*

  $$p(t_{k-2}, t_{k-1}, t_k), p(t_{k-1}, t_k, t_{k+1}), p(t_k, t_{k+1}, t_{k+2}), p(t_{k+1}, t_{k+2}, t_{k+3})$$

- *New control points*

  $$p(t_{k-2}, t_{k-1}, t_k), p(t_{k-1}, t_k, u_1), p(t_k, u_1, u_2), p(u_1, u_2, u_3),$$
  $$p(u_2, u_3, u_4), p(u_3, u_4, t_{k+1}), p(u_4, t_{k+1}, t_{k+2}), p(t_{k+1}, t_{k+2}, t_{k+3})$$

We can find all the new control points by running the blossomed version of the de Boor algorithm just twice—once to compute $p(u_1, u_2, u_3)$ and once to compute $p(u_2, u_3, u_4)$. The other new control points emerge off the left and right lateral edges of the two triangles (see Figure 7.17). Notice that in the second triangle the knots must be introduced in reverse order; that is, the knot u_4 appears on the lowest level, the knot u_3 on the second level, and the knot u_2 on the top level.

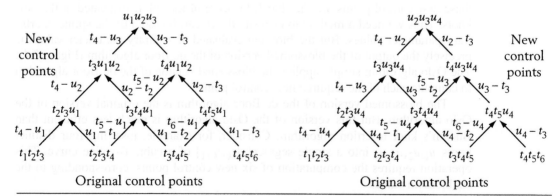

Figure 7.17 An efficient version of the Oslo algorithm for knot insertion in cubic B-spline curves. Two blossoming variants of the de Boor algorithm are invoked. Notice that in the second triangle the knots are introduced in reverse order. The new control points emerge off the lateral edges of the triangles (compare to Figure 7.4).

This version of the Oslo algorithm for cubic curves is just as efficient as Boehm's algorithm. To evaluate each triangle, we must perform a total of 6 affine combinations. Thus to evaluate both triangles requires $2 \times 6 = 12$ affine combinations, which is exactly the same count as for Boehm's knot insertion algorithm.

To insert $n + 1$ knots $u_1, \ldots, u_{n+1}$ in a single interval of a degree n B-spline curve, the Oslo algorithm again employs two triangles: the first triangle computes the blossom $p(u_1, \ldots, u_n)$, and the second triangle computes the blossom $p(u_2, \ldots, u_{n+1})$ from the original control points using the blossomed version of the de Boor algorithm. As in the cubic case, the other control points emerge off the left and right lateral edges of the two triangles, and in the second triangle the knots $u_2, \ldots, u_{n+1}$ must be introduced in reverse order. For degree n B-spline curves, both Boehm's algorithm and the efficient version of the Oslo algorithm require $n(n + 1)$ affine combinations to insert $n + 1$ knots in a single interval. Boehm's algorithm proceeds by inserting one knot at a time; the Oslo algorithm by inserting $n + 1$ knots all at once.

Exercises

1. Prove that for degree n B-spline curves, both Boehm's algorithm and the efficient version of the Oslo algorithm require $n(n + 1)$ affine combinations to insert $n + 1$ knots in a single interval.

2. Explain how the Oslo algorithm can be formulated to insert fewer than $n + 1$ knots in a single interval of a degree n B-spline curve just as efficiently as Boehm's algorithm.

3. Implement both Boehm's knot insertion algorithm and the Oslo algorithm. Which algorithm do you prefer? Why?

4. Consider what happens to the control polygon of a B-spline curve with knots $\{t_k\}$ as the knot spacing gets arbitrarily small. Let $h = \max(t_{k+1} - t_k)$.

   a. Use the Oslo algorithm to show that the control points get closer together as h decreases.

   b. Use part (a) and the fact that a B-spline curve lies in the local convex hull of its control points to conclude that the control polygon converges to the B-spline curve as the knot spacing $h \to 0$.

5. Use the result of Exercise 4 and the fact that knot insertion is a corner-cutting procedure to prove that B-spline curves are variation diminishing. (Compare to Theorem 7.4 in Section 7.6.3.)

6. Based on knot insertion and Exercise 4:

   a. implement a rendering algorithm for B-spline curves

   b. implement an intersection algorithm for B-spline curves

7. *Sablonniere's Tetrahedral Algorithm*

   Suppose we want to convert a cubic polynomial $P(t)$ from the progressive basis with respect to the knots $t_1, \ldots, t_6$ to the progressive basis with respect to the knots $u_1, \ldots, u_6$. That is, given the coefficients $p(t_1, t_2, t_3), \ldots, p(t_4, t_5, t_6)$, we want to find the coefficients $p(u_1,u_2,u_3),\dots,p(u_4,u_5,u_6)$. We could proceed by applying the blossomed version of the de Boor algorithm (Figure 7.4), which is essentially identical to the original, inefficient version of the Oslo algorithm, to find each one of these new coefficients. If we were to stack these four triangles one atop the other, the triangles would form a triangular prism. Sablonniere's algorithm reduces the amount of computation by collapsing this prism into a tetrahedron.

a. Show how to arrange the four triangles in Figure 7.18 into a tetrahedral computation for a change of basis algorithm.

b. Explain how to generalize the tetrahedral computation in part (a) to polynomials of arbitrary degree.

c. Apply the Oslo algorithm—the blossomed version of the de Boor algorithm—to perform the same change of basis with only two triangles. (Hint: Use the output along one of the lateral edges of the first trangle as the input to the second triangle.)

d. Prove that Sablonniere's tetrahedral algorithm is less efficient than the Oslo algorithm. In particular, show that Sablonniere's tetrahedral algorithm is $O(n^3)$, whereas the Oslo algorithm is $O(n^2)$.

e. Explain why Sablonniere's algorithm might be more stable numerically than the Oslo algorithm.

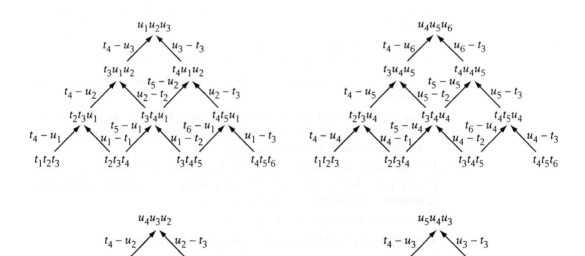

Figure 7.18 The four triangles used in Sablonniere's tetrahedral algorithm to convert between two cubic progressive polynomial bases.

7.6.3   Change of Basis Algorithms via Knot Insertion

Knot insertion algorithms are change of basis procedures for B-spline curves. Given the B-spline coefficients of a spline curve with respect to a knot vector $\{t_k\}$, a knot insertion procedure finds the B-spline coefficients of the same spline curve with respect to another knot vector $\{u_j\} \supset \{t_k\}$. We shall exhibit several applications of knot insertion here, including conversion to piecewise Bezier form, conversion between the monomial and Bezier form, Bezier subdivision, and algorithms for differentiating B-spline curves.

7.6.3.1   Conversion to Piecewise Bezier Form

Bezier curves are special types of B-spline segments. In Chapter 5 we developed fast algorithms for analyzing Bezier segments, including simple procedures for rendering and intersecting Bezier curves. One of the simplest ways to analyze B-spline curves is to convert them to piecewise Bezier form and then to perform the analysis on the Bezier segments, using the Bezier algorithms we have already developed. We shall also see that this approach leads to a straightforward proof of the variation diminishing property for B-spline curves.

We can apply knot insertion to convert from B-spline to piecewise Bezier form. Consider a cubic B-spline segment $P(t)$ defined over a knot sequence $t_1,...,t_6$. Relative to this knot sequence, we are interested only in the segment of $P(t)$ for which $t_3 \leq t \leq t_4$, and we want to convert from this B-spline representation of $P(t)$ to a Bezier representation of $P(t)$. For the interval $t_3 \leq t \leq t_4$, the Bezier knots are simply t_3,t_3,t_3,t_4,t_4,t_4. By Equation (7.3) the control points of $P(t)$ can be computed by evaluating the blossom of $P(t)$ at consecutive knots. Thus we have

- *B-spline control points*

    $p(t_1,t_2,t_3), p(t_2,t_3,t_4), p(t_3,t_4,t_5), p(t_4,t_5,t_6)$

- *Bezier control points*

    $p(t_3,t_3,t_3), p(t_3,t_3,t_4), p(t_3,t_4,t_4), p(t_4,t_4,t_4)$

To get from the B-spline control points to the Bezier control points, we need to incorporate t_3 and t_4 as triple knots. Thus, we need to insert the knots t_3,t_3,t_4,t_4 between the knots t_3 and t_4. To do so, we can apply Boehm's knot insertion algorithm (see Figure 7.19). Notice that in this case we only proceed up two levels in the triangle because even though we require t_3 and t_4 to be triple knots, we only need to insert each of these knots twice.

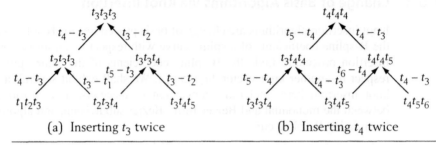

(a) Inserting t_3 twice   (b) Inserting t_4 twice

Figure 7.19 Boehm's knot insertion algorithm for converting one segment of a cubic B-spline curve to Bezier form. In (a) we insert t_3 twice and in (b) we insert t_4 twice. Some of the output from (a) is used as input to (b). Since some computations are redundant, we can omit $t_4t_5t_6$ from the input to (a) and $t_3t_3t_3$ from the input to (b). Notice that in (a) the Bezier control points emerge along the right edge, but in (b) they emerge along the left edge of the triangle.

THEOREM 7.4   B-spline curves are variation diminishing.

Proof   The de Boor algorithm is a corner-cutting procedure (see Figure 7.3). Since knot insertion is the blossomed version of the de Boor algorithm, knot insertion is also a corner-cutting procedure. Therefore, the piecewise Bezier control polygon is variation diminishing with respect to the original B-spline control polygon. But by Corollary 5.4 each Bezier segment is variation diminishing with respect to its Bezier control polygon. Hence the entire B-spline curve must be variation diminishing with respect to the original B-spline control polygon.

Exercises

1. Implement the knot insertion algorithm to convert from B-spline to piecewise Bezier form.

    a. Draw both the B-spline and the Bezier control polygons for each B-spline curve.

    b. Use this conversion algorithm to render B-spline curves.

    c. Apply this conversion procedure to intersect pairs of B-spline curves.

2. Prove that the arc length of a B-spline curve is never greater than the perimeter of its control polygon. (Hint: Convert to piecewise Bezier form.)

7.6.3.2 Bezier Subdivision and Conversion between Bezier and Monomial Form

Several standard algorithms for Bezier curves can be derived from and interpreted as knot insertion procedures. For example, the standard subdivision algorithm for degree n Bezier curves at $t = r$ can be viewed as a procedure that converts from the knot sequence

$$\underbrace{0,...,0}_{n},\underbrace{1,...,1}_{n}$$

to the knot sequence

$$\underbrace{0,...,0}_{n},\underbrace{r,...,r}_{n},\underbrace{1,...,1}_{n}\,.$$

Thus, the standard Bezier subdivision algorithm is simply n-fold knot insertion at $t = r$. For example, for a cubic Bezier curve $P(t)$, we have

- *Original Bezier control points*

  $p(0,0,0), p(0,0,1), p(0,1,1), p(1,1,1)$

- *Bezier control points after subdivision at $t = r$*

  Left segment— $p(0,0,0), p(0,0,r), p(0,r,r), p(r,r,r)$

  Right segment— $p(r,r,r), p(r,r,1), p(r,1,1), p(1,1,1)$

We illustrate subdivision as knot insertion for cubic Bezier curves in Figure 7.20.

Conversion between degree n Bezier and degree n monomial form can also be viewed as n-fold knot insertion. To convert from Bezier to monomial form, we must

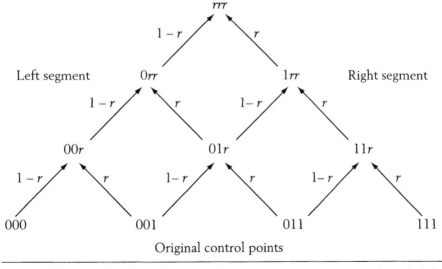

Figure 7.20 Bezier subdivision as knot insertion at $t = r$. The new control points emerge along the left and right lateral edges of the triangle (compare to Figure 5.23).

perform a change of basis from the progressive basis represented by the knot sequence

$$\underbrace{0,...,0}_{n},\underbrace{1,...,1}_{n}$$

to the progressive basis represented by the knot sequence

$$\underbrace{0,...,0}_{n},\underbrace{\delta,...,\delta}_{n}.$$

(Here, as usual, to avoid normalization problems, we use the monomial basis $\{\binom{n}{k}t^{k}\}$, $k = 0,...,n$, rather than the monomial basis $\{t^{k}\}$, $k = 0,...,n$.) We can interpret this change of basis as n-fold knot insertion at $\delta = (1,0)$. Conversely, to convert from monomial to Bezier form, we must convert from the progressive basis represented by the knot sequence

$$\underbrace{0,...,0}_{n},\underbrace{\delta,...,\delta}_{n}$$

to the progressive basis represented by the knot sequence

$$\underbrace{0,...,0}_{n},\underbrace{1,...,1}_{n}.$$

We can do this conversion by performing n-fold knot insertion at $t = 1$. We illustrate these two procedures for cubic curves in Figure 7.21.

Notice that the right edge of both triangles in Figure 7.21 contains the coefficients with respect to the monomial basis centered at $t = 1$. Thus as a bonus the change of basis algorithm that converts from monomial to Bezier form can also be used to convert from one monomial basis to another monomial basis. In fact, this

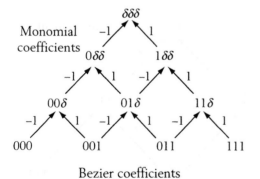

(a) Bezier to monomial form

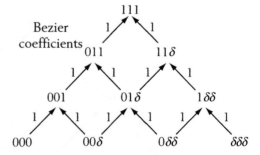

(b) Monomial to Bezier form

Figure 7.21 Conversion between cubic Bezier and cubic monomial form. (a) To convert from Bezier to monomial form, we perform triple knot insertion at $\delta = (1,0)$. (b) To convert from monomial to Bezier form, we perform triple knot insertion at $t = 1$. Notice that the labels along the edges in these diagrams do not need to be normalized because $1 = (1,1) = (0,1) + (1,0) = 0 + \delta$. (Compare to Figures 5.18 and 5.19.)

algorithm is the standard synthetic division algorithm for converting from the monomial basis centered at $t = 0$ to the monomial basis centered at $t = 1$. If we think of $\delta = (1,0)$ as a knot at infinity, then the monomial form is simply a special case of the Bezier form. From this perspective, Bezier subdivision, conversion from Bezier to monomial form, conversion from monomial to Bezier form, and synthetic division are all one and the same algorithm.

Exercise

1. Use knot insertion to develop change of basis algorithms between the power basis and

   a. the Bernstein basis

   b. the monomial basis

   (Hint: See Section 7.2, Exercise 1.)

7.6.4   Differentiation and Knot Insertion

Evaluation, blossoming, and knot insertion algorithms for B-spline curves are intimately related; each of these procedures is just a variant of the de Boor algorithm. Differentiation too is another simple variant of the de Boor algorithm (see Figure 7.5), so differentiation and knot insertion are also closely connected. Here we will show how to interpret standard differentiation algorithms for B-spline curves as knot insertion procedures. We will then go on to show how to use these differentiation algorithms to generate another fast knot insertion procedure.

7.6.4.1   Differentiation as Knot Insertion

Let's first revisit differentiation for B-spline curves from the perspective of knot insertion. Figure 7.5 depicts the differentiation algorithm for one segment of a cubic B-spline curve. If we isolate the lowest level of this diagram, what we see is exactly Boehm's knot insertion algorithm at $u = \delta$ (see Figure 7.22).

Thus the degree $n - 1$ B-spline coefficients for the first derivative of a B-spline curve of degree n can be computed by knot insertion at $u = \delta$. Similarly, if we want to differentiate a degree n B-spline curve r times, the degree $n - r$ B-spline coefficients

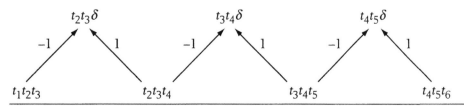

Figure 7.22 The first level of the differentiation algorithm for a cubic B-spline curve (see Figure 7.5). Compare to Boehm's knot insertion algorithm (Figure 7.14).

are given by inserting an r-fold knot at δ and taking the coefficients off the top (rth) level of the diagram.

7.6.4.2   Boehm's Derivative Algorithm

The standard derivative algorithm finds the rth derivative of a B-spline curve at an arbitrary value of t by differentiating r levels of the de Boor algorithm. Boehm's derivative algorithm computes all the derivatives of a B-spline curve at a single parameter $t = a$ simultaneously by converting a B-spline segment to monomial form. In Section 6.4 we showed that for any degree n polynomial $P(t)$,

$$P^{(k)}(a) = \frac{n!}{(n-k)!} \, p(\underbrace{a,...,a}_{n-k},\underbrace{\delta,...,\delta}_{k}) \cdot$$

But the values

$$p(\underbrace{a,...,a}_{n-k},\underbrace{\delta,...,\delta}_{k}), \; k = 0,...,n,$$

are just the monomial coefficients of $P(t)$ at $t = a$. This identity between derivatives and monomial coefficients is a simple consequence of Taylor's Theorem. Thus, up to constant multiples, to find the derivatives of a B-spline segment at a fixed parameter value, we need only convert from B-spline to monomial form. We can do this conversion by performing n-fold knot insertion at δ followed by n-fold knot insertion at a. Computing the derivatives of a B-spline curve at a single point in this manner is Boehm's derivative algorithm. We illustrate this algorithm for cubic curves in Figure 7.23.

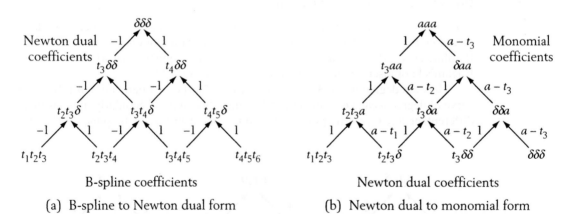

Figure 7.23 Boehm's derivative algorithm for cubic curves. In (a) we perform triple knot insertion at δ, converting from B-spline to Newton dual form. In (b) we perform triple knot insertion at a, converting from Newton dual to monomial form. Up to constant multiples, the derivatives at $t = a$ are the monomial coefficients, which emerge along the right lateral edge of the second triangle. Observe that the labels along the edges in (a) must be normalized by the same denominators that appear in the de Boor algorithm, but the labels along the edges in (b) do not need to be normalized.

Notice that the first step of Boehm's derivative algorithm converts a B-spline segment into Newton dual form. This step depends only on the choice of segment and not on the parameter within the segment. Thus, we can reuse this part of the computation to find the derivatives at another parameter $t = b$ within the same B-spline segment.

Exercise

1. We could compute all the derivatives of a B-spline curve at $t = a$ by first performing n-fold knot insertion at a and then performing n-fold knot insertion at δ. Give two reasons why it is more efficient to insert an n-fold knot at δ first, before inserting an n-fold knot at a.

7.6.4.3 Knot Insertion from Differentiation

Both Boehm's knot insertion algorithm and the Oslo algorithm are based on the de Boor evaluation algorithm. Here we shall show how to perform fast knot insertion based on the differentiation algorithm.

Let's begin by differentiating a degree n B-spline curve n times. Figure 7.23(a) illustrates this differentiation algorithm for cubic curves. Notice that the values along the left edge of this diagram are the coefficients of the spline segment $P(t)$ with respect to the Newton dual basis. Thus starting with the B-spline coefficients, we can compute the Newton dual coefficients by applying the differentiation algorithm. We can also turn this around. Starting with the Newton dual coefficients, we can retrieve the B-spline coefficients essentially by running the diagram in reverse with new labels along the edges. If we run only one level in reverse, we obtain the algorithm illustrated in Figure 7.24.

Now notice two things: First, we have computed one new blossom value $p(u_2,u_3,u_4)$ from the original Newton dual coefficients. Second, starting with the Newton dual coefficients relative to the knot sequence $u_1,u_2,u_3,\delta,\delta,\delta$, we have computed the Newton dual coefficients relative to the new knot sequence $u_2,u_3,u_4,\delta,\delta,\delta$. We can now replace the knot sequence $u_1,u_2,u_3,\delta,\delta,\delta$ by $u_2,u_3,u_4,\delta,\delta,\delta$ and iterate the same algorithm. At every step we compute one new blossom value as well as a new set of Newton dual coefficients. The structure of this algorithm resembles Boehm's knot insertion algorithm, but with two significant dif-

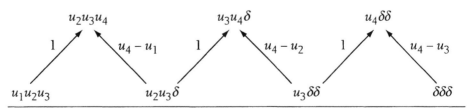

Figure 7.24 Running the differentiation algorithm in reverse—the cubic case. Here again it is not necessary to normalize the labels along the edges.

ferences. Unlike Boehm's algorithm, we compute only a single new coefficient rather than many new coefficients at a time. Computationally, however, each stage of this algorithm is more efficient than Boehm's algorithm because this algorithm uses only multiplication and no division. We call this algorithm *factored knot insertion,* since we have factored knot insertion through the Newton dual basis.

The benefits of this factored approach are heightened when the knots we wish to insert are evenly spaced. Let $\Delta = u_{k+1} - u_k$ denote the knot spacing. By introducing appropriate multiples of the knot spacing, we can now insert knots without any multiplication whatsoever. We illustrate this algorithm for cubic curves in Figure 7.25.

This algorithm should remind you of computing new, evenly spaced values along a polynomial curve via forward differencing (see Section 4.5) because there too only addition is involved; no multiplication is required after the initial start-up step. Thus like fast forward differencing, we can accomplish fast knot insertion by performing an initial start-up step followed by a fast marching algorithm. Notice too the similarity between the start-up step in Figure 7.23(a) and computation of divided differences in Figure 4.2. Moreover, a slight modification to the structure of the marching algorithm for factored knot insertion makes it identical in structure to the marching algorithm for forward differencing. We illustrate this marching algorithm for inserting knots into cubic B-spline curves in Figure 7.26.

Exercises

1. Implement the factored knot insertion algorithm for cubic B-spline curves for

   a. arbitrary knots

   b. evenly spaced knots

   In each case, compare the speed of this algorithm to the speed of your implementation of Boehm's knot insertion procedure.

2. Derive the factored knot insertion algorithm for degree n B-spline curves with evenly spaced knots. By what factors of the knot spacing must you multiply each of the coefficients of the Newton dual basis so that no multiplication is needed in the marching step?

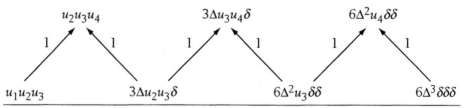

Figure 7.25 Inserting new knots without any multiplication—the cubic case. Since the knots are evenly spaced, $u_{j+k} = u_j + k\Delta$, so all the arrows represent addition; no normalization is required.

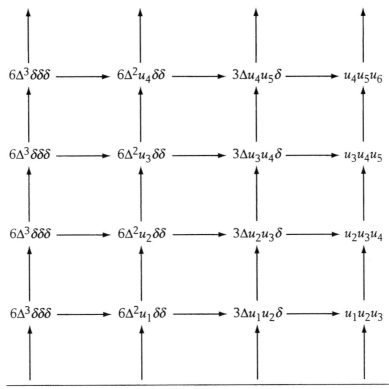

Figure 7.26 Knot insertion for evenly spaced knots via forward differencing for a cubic B-spline segment. The value at each node is computed by adding the values at the nodes that point into it. The values in the leftmost column are all identical, and the values in the rightmost column are the new blossom values for equally spaced knots. (Compare to the forward differencing algorithm in Figure 4.5.)

7.7 The B-Spline Basis Functions

Just like Bezier curves, B-spline curves can be represented in terms of basis functions. Let $S(t)$ be a B-spline curve of degree n with control points $\{P_k\}$ and knot sequence $\{t_k\}$. It follows from the de Boor algorithm that there exist piecewise polynomials $\{N_{k,n}(t)\}$ such that

$$S(t) = \sum_k N_{k,n}(t)P_k \, . \tag{7.7}$$

The functions $\{N_{k,n}(t)\}$ are called the *B-spline basis functions* or simply the *B-splines*. Just as the Bernstein basis functions can be used to analyze Bezier curves and surfaces, the B-splines can be used to elucidate the properties of B-spline curves and surfaces, so it is to these basis functions that we now turn our attention.

We can compute the B-splines $\{N_{k,n}(t)\}$ from the de Boor algorithm in two ways. If we set $P_j = 0, j \neq k, P_k = 1$, and run the de Boor algorithm, then by Equation

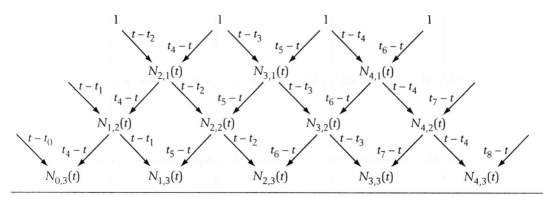

Figure 7.27   The down recurrence for the B-splines.

(7.7) $S(t) = N_{k,n}(t)$. This algorithm is the up recurrence for the B-spline basis functions. From the up recurrence it follows that the piecewise polynomial $N_{k,n}(t)$ represents the sum over all paths from P_k at the base to the various apexes of the de Boor triangles, where the apex we select depends on the knot interval of the parameter t. Therefore, if we place a 1 at each apex and reverse all the arrows in the de Boor algorithm, then the B-splines $\{N_{k,n}(t)\}$ emerge at the base of the diagram. This algorithm is the down recurrence for the B-splines (see Figure 7.27). Again the apex at which we begin the computation depends on the knot interval of the parameter t.

To ensure the validity of Equation (7.7), we must index the B-splines so that the basis function $N_{k,n}(t)$ resides in the same node where the control point P_k would reside in the up recurrence. This indexing scheme makes the down recurrence in Figure 7.27 particularly easy to remember. Lower-degree B-splines emerge at interior nodes of the diagram. The arrow entering the node $N_{k,n}(t)$ from the left has numerator $t - t_k$, and the arrow entering from the right has numerator $t_{k+n+1} - t$; denominators are recovered in the usual fashion by ensuring that the labels on the two arrows that exit each node in the down recurrence sum to one. In addition, if you follow along in the direction of any arrow, then

1. the labels (in the numerators) you encounter along the edges do not change
2. the first index of the B-splines you encounter does not change for right-pointing arrows and decreases by one for left-pointing arrows; the second index, the degree, decreases as you ascend the diagram

The last step of the down recurrence is summarized in Figure 7.28; all of Figure 7.27—the entire down recurrence—can be recovered from this diagram, using the two simple rules listed above.

It follows from Figure 7.28 that the B-splines satisfy the recurrence

$$N_{k,0}(t) = 1 \qquad\qquad t_k \leq t < t_{k+1}$$

$$N_{k,n}(t) = \frac{t - t_k}{t_{k+n} - t_k} N_{k,n-1}(t) + \frac{t_{k+n+1} - t}{t_{k+n+1} - t_{k+1}} N_{k+1,n-1}(t). \tag{7.8}$$

This recurrence is often taken as the definition of the B-splines.

Using this recurrence, we illustrate some B-splines of degrees $n = 0,1,2,3$ with simple knots at the integers in Figures 7.29 and 7.30.

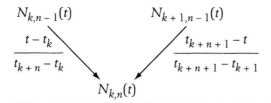

Figure 7.28 The last step of the down recurrence for the B-splines (compare to Figure 7.12).

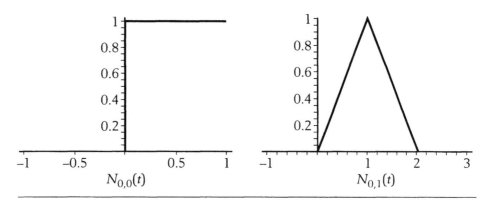

Figure 7.29 B-spline basis functions of degree 0 and degree 1 with simple knots at the integers.

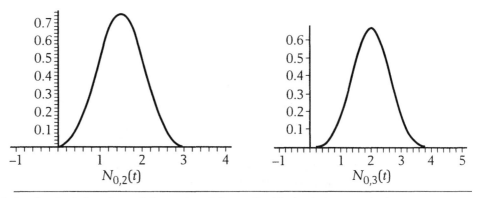

Figure 7.30 B-spline basis functions of degree 2 and degree 3 with simple knots at the integers.

Exercise

1. Draw the shapes of all the different B-splines of degrees 1,2,3 with multiple knots at the integers.

7.7.1   Elementary Properties of the B-Spline Basis Functions

In this section we shall study the elementary properties of the B-spline basis functions. These characteristics of the B-splines $\{N_{k,n}(t)\}$ both mirror and are mirrored in the elementary features of B-spline curves derived in Section 7.4. Below we list these features and then derive each of these properties from the corresponding properties of B-spline curves.

1. Piecewise polynomial
2. Continuity of order $C^{n-\mu}$ at knots of multiplicity μ
3. Compact support
4. Partition of unity
5. Nonnegativity
6. Spline basis
7. Unimodality

- *Piecewise polynomial.* From the up recurrence we know that the B-spline basis functions are B-spline curves. Therefore, the B-splines must be piecewise polynomials.

- *Continuity.* Again since by construction the B-spline basis functions are B-spline curves, the B-splines must have continuity of order $n - \mu$ at knots of multiplicity μ.

- *Compact support.* By the de Boor algorithm, the only B-splines that are nonzero over the parameter interval $[t_k, t_{k+1}]$ are $N_{k-n,n}(t),...,N_{k,n}(t)$. Hence the B-spline $N_{k,n}(t)$ is nonzero only for values of t in the parameter interval $[t_k, t_{k+n+1}]$—that is, $support\{N_{k,n}(t)\} = [t_k, t_{k+n+1}]$. Therefore, from now on, whenever we want to make explicit the knots on which $N_{k,n}(t)$ depends, we shall write $N_{k,n}(t \mid t_k,...,t_{k+n+1})$. By Equation (7.7), the compact support of the B-splines is equivalent to the local control property for the control points of B-spline curves.

- *Partition of unity.* The B-splines form a partition of unity. This result can be proved from the down recurrence (Equation (7.8)) by induction on n. This property can also be derived from the de Boor algorithm by setting $P_k = 1$ for all k and observing that since at every stage of the algorithm we are taking affine combinations of the nodes the value at every interior node is also equal to one. Hence the value at any apex must be one. Therefore,

$$1 = S(t) = \sum_k N_{k,n}(t)P_k = \sum_k N_{k,n}(t).$$

The partition of unity property of the B-spline basis functions is equivalent to the affine invariance of B-spline curves.

■ *Nonnegativity.* Recall that for any parameter interval the labels along the edges of the de Boor algorithm are nonnegative. Since the B-spline $N_{k,n}(t)$ represents the sum over all paths from the kth position at the base to the various apexes of the de Boor triangles, it follows that the B-splines too are nonnegative. The partition of unity and nonnegativity of the B-spline basis functions are equivalent to the affine invariance and the convex hull properties of B-spline curves.

■ *Spline basis.* To prove that the B-splines $\{N_{k,n}(t)\}$ with knots $\{t_j\}$ form a basis for the space of all splines $S(t)$ with knots $\{t_j\}$, we need to show that the B-splines span this space and are linearly independent. But by Theorem 7.3, every spline is a B-spline curve; that is, every spline $S(t)$ with knots $\{t_j\}$ can be generated from the de Boor algorithm for some set of control points $\{P_k\}$. Therefore, by Equation (7.7), $S(t) = \sum_k N_{k,n}(t)P_k$, so the B-splines $\{N_{k,n}(t)\}$ do indeed span the space of all splines with knots $\{t_j\}$. To prove that the B-splines are linearly independent, we must show that if $\sum_k c_k N_{k,n}(t) \equiv 0$, then $c_k = 0$ for all k. Let's restrict our attention to the parameter interval $[t_i, t_{i+1}]$. Over this interval $N_{i-n,n}(t),...,N_{i,n}(t)$ are the only nonzero B-splines, so over this interval

$$\sum_k c_k N_{k,n}(t) = \sum_{h=0}^{n} c_{i-n+h} N_{i-n+h,n}(t).$$

Moreover, over the interval $[t_i, t_{i+1}]$, the B-splines $N_{i-n,n}(t),...,N_{i,n}(t)$ are polynomials, and by Section 7.1 these polynomials are just the progressive basis functions $b_0^n(t),...,b_n^n(t)$, which form a polynomial basis. Therefore, if

$$0 \equiv \sum_{h=0}^{n} c_{i-n+h} N_{i-n+h,n}(t) = \sum_{h=0}^{n} c_{i-n+h} b_h^n(t),$$

then $c_{i-n+h} = 0$ for all h. Hence the B-splines are indeed linearly independent. The linear independence of the B-splines is equivalent to the nondegeneracy of B-spline curves.

■ *Unimodality.* Recall that a function is said to be unimodal if it has only one local maximum. The B-splines $\{N_{k,n}(t)\}$ are unimodal in t. To understand why, consider the graph of the function $N_{k,n}(t)$—that is, the curve $S(t) = (t, N_{k,n}(t))$. The function $F(t) = t$ is a polynomial and hence certainly a spline (see Section 7.5, Exercise 2). Since the B-splines form a basis for the space of all splines, there must be constants $\{c_j\}$ such that $t = \sum_j c_j N_{j,n}(t)$. (We shall derive explicit expressions for the constants $\{c_j\}$ in Section 7.7.2, but for now all we need to know is that such constants exist.) Therefore, $S(t) = \sum_j (c_j, \delta_{j,k}) N_{j,n}(t)$. Thus the control points for the graph of $N_{k,n}(t)$ all lie along the t-axis except for the point at $(c_k, 1)$. Therefore, the control polygon for the graph of $N_{k,n}(t)$ has only one local maximum (see Figure 7.31). But the graph of $N_{k,n}(t)$ is a B-spline curve, and by

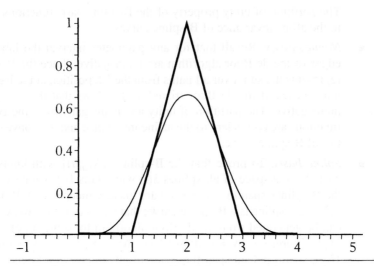

Figure 7.31 Graph of the cubic B-spline $N_{0,3}(t)$ (light) together with its control polygon (dark).

Theorem 7.4, B-spline curves are variation diminishing. Therefore, the graph of $N_{k,n}(t)$ can oscillate no more than its control polygon. Hence $N_{k,n}(t)$ has only one local maximum.

Exercises

1. Let $S(t)$ be a spline of degree n with knots $\{t_k\}$ whose support lies in $[t_n, t_{k+n+1}]$.

   a. Prove that there is a constant c such that $S(t) = cN_{k,n}(t)$.

   b. Conclude from part (a) that the B-splines $\{N_{k,n}(t)\}$ have minimal support. That is, show that if $S(t)$ is a spline of degree n with knots $\{t_k\}$ whose support lies in a closed subinterval of $[t_n, t_{k+n+1}]$, then $S(t)$ is identically zero.

2. Prove that *Sign Alternations*$\left\{\sum_k c_k N_{k,n}(t)\right\} \le$ *Sign Alternations*$\{c_k\}$.

3. Let $\tau_{k+i} = at_{k+i} + b$, $i = 0,\dots,n+1$, for some fixed constants $a > 0$ and b. Show that

$$N_k^n(at + b \mid \tau_k, \dots, \tau_{k+n+1}) = N_k^n(t \mid t_k, \dots, t_{k+n+1}).$$

   Compare this result to Section 7.4, Exercise 2.

4. Show, by example, that the B-splines $\{N_{k,n}(t)\}$ are not necessarily unimodal in k for $n \ge 6$.

5. The B-splines $\{N_{k,n}(t)\}$ are called the *de Boor normalized B-splines*. The *Schoenberg normalized B-splines* $\{M_{k,n}(t)\}$ are defined by setting

$$M_{k,n}(t) = \frac{N_{k,n}(t)}{t_{k+n+1} - t_k} = \frac{N_{k,n}(t)}{Support}$$

a. Prove that $M_{k,n}(t) = \dfrac{t - t_k}{t_{k+n+1} - t_k} M_{k,n-1}(t) + \dfrac{t_{k+n+1} - t}{t_{k+n+1} - t_k} M_{k+1,n-1}(t).$

b. Conclude from part (a) that the Schoenberg normalized B-splines $\{M_{k,n}(t)\}$ are unimodal in k.

7.7.2 Blossoming and Dual Functionals

For polynomial curves generated by the de Boor algorithm, the control points are given by the blossom evaluated at consecutive knots. Thus if $P(t)$ is a degree n progressive polynomial curve specified by control points $P_0,...,P_n$ and knots $t_1,...,t_{2n}$, then by Equation (7.3),

$$P_k = p(t_{k+1},...,t_{k+n}) \ .$$

We would like to extend this result to B-spline curves, but first we must define exactly what we mean by the blossom of a spline evaluated at the knots.

Consider then a B-spline curve $S(t) = \sum_k P_k N_{k,n}(t)$ with knots $\{t_j\}$. Let $S_k(t)$ be the degree n polynomial that represents $S(t)$ over the knot interval $[t_k, t_{k+1}]$—that is, $S(t) = S_k(t)$ for $t_k \le t \le t_{k+1}$. Over the interval $[t_k, t_{k+1}]$, the B-spline curve depends only on the n control points $P_{k-n},...,P_k$ and the $2n$ knots $t_{k-n+1},...,t_{k+n}$. Moreover, by Equation (7.3),

$$P_i = s_k(t_{i+1},...,t_{i+n}) \qquad k-n \le i \le k \ .$$

Therefore,

$$s_j(t_{i+1},...,t_{i+n}) = s_k(t_{i+1},...,t_{i+n})$$

provided that $j-n \le i \le j$ and $k-n \le i \le k$. Now we define the blossom of $S(t)$ evaluated at the knots $t_{i+1},...,t_{i+n}$ by setting

$$s(t_{i+1},...,t_{i+n}) = s_k(t_{i+1},...,t_{i+n}),$$

where k is any index such that $k-n \le i \le k$—that is, k is the index of any knot interval influenced by the control point P_i. It follows that

$$P_i = s(t_{i+1},...,t_{i+n}) \qquad\qquad (7.9)$$
$$S(t) = \sum_i s(t_{i+1},...,t_{i+n}) N_{i,n}(t)$$

Equation (7.9) is the dual functional property of the B-splines and is the basic fact connecting blossoming to B-spline curves.

We can apply Equation (7.9) to find the B-spline coefficients for various spline functions. For example, in our discussion of the unimodality of the B-splines in Section 7.7.1, we observed that there must be constants $\{c_i\}$ such that

$$t = \sum_i c_i N_{i,n}(t) \ .$$

These constants $\{c_i\}$ are called the *nodes* of the B-splines; it is at these values that we must place the B-spline coefficients in order to generate the graph of a spline curve. Using blossoming, these nodes are easy to locate. Let $P(t) = t$. Then by Equation (7.9),

$$c_i = p(t_{i+1},...,t_{i+n}) = \frac{t_{i+1} + \cdots + t_{i+n}}{n} \, .$$

What happens if we evaluate the blossom $n_{i,d}(u_1,...,u_d)$ of a B-spline basis function $N_{i,d}(t)$ at the knots $\{t_k\}$? Since $N_{i,d}(t)$ is a spline curve, and since the blossom evaluated at consecutive knots gives the B-spline coefficients of a spline curve, it follows from the linear independence of the B-splines that

$$n_{i,d}(t_{j+1},...t_{j+d}) = 0 \quad i \neq j$$
$$= 1 \quad i = j \cdot \qquad (7.10)$$

A linear operator that vanishes on all but one of a fixed set of basis functions and yields the value one on a single basis function is called a *dual functional* (see Section 4.2). Since the blossom is a linear operator, Equation (7.10) is just another way of saying that the blossom evaluated at consecutive knots represents the dual functionals for the B-splines $\{N_{i,d}(t)\}$.

Exercises

1. Use Equation (7.9) to prove that $\sum_k N_{k,n}(t) = 1$.

2. Prove that

$$\binom{n}{j} t^j = \sum_i \left\{ \sum t_{i_1} \cdots t_{i_j} \right\} N_{i,n}(t),$$

where the sum in brackets is taken over all subsets $\{i_1,...,i_j\}$ of $\{i + 1,..., i + n\}$.

3. a. Apply Equations (7.9) and (7.10) to prove that

$$N_{k,n}(t \mid t_k,...,t_{k+n+1}) = \frac{\tau - t_k}{t_{k+n} - t_k} N_{k,n}(t \mid t_k,...,\tau,...,t_{k+n})$$

$$+ \frac{t_{k+n+1} - \tau}{t_{k+n+1} - t_{k+1}} N_{k+1,n}(t \mid t_{k+1},...,\tau,...,t_{k+n+1}).$$

   b. Explain why this result is equivalent to Boehm's knot insertion algorithm.

4. a. Apply blossoming to derive Marsden's identity:

$$(x - t)^d = \sum_k (t_{k+1} - t) \cdots (t_{k+d} - t) N_{k,d}(x) \, .$$

   b. Show that Marsden's identity is equivalent to Equation (7.10).

5. Two bases $\{B_k(x)\}$ and $\{D_k(t)\}$ for polynomials of degree n are said to be *dual* if they satisfy a local version of Marsden's identity (Exercise 4). That is, $\{B_k(x)\}$ and $\{D_k(t)\}$ are called *dual bases* if $(x - t)^n = \sum_k D_k(t) B_k(x)$.

    a. Find the dual basis to each of the following progressive bases:

       i.   Bernstein basis

      ii.   Monomial basis

     iii.   Newton dual basis

     iv.   Power basis (see Section 7.2, Exercise 1)

    b. Let $r_{k1},\ldots r_{kn}$ be the roots of $D_k(t)$. Show that

$$b_j(r_{k1},\ldots r_{kn}) = \frac{(-1)^n n! \delta_{jk}}{D_k^{(n)}(t)}.$$

6. Show that the B-splines $\{N_{k,n}(t)\}$ for the knot sequence $\{t_k\}$ are the unique functions satisfying the following axioms:

    a. $N_{k,n}(t)$ is a piecewise polynomial with continuity of order $C^{n-\mu}$ at knots of multiplicity μ.

    b. $\{N_{k,n}(t)\}$ have minimal support.

    c. $\sum_k N_{k,n}(t) = 1$.

    (Hint: Use Exercise 1 and Section 7.7.1, Exercise 1.)

7.7.3 Differentiating and Integrating the B-Splines

We know how to differentiate a degree n B-spline curve; we simply differentiate one level of the de Boor algorithm and multiply the result by n. A B-spline $N_{k,n}(t)$ is just a B-spline curve where the control points are given by $P_j = \delta_{jk}$. Therefore, differentiating the bottom level of the de Boor algorithm and summing over all paths from the kth position at the base to the various apexes yields the derivative of $N_{k,n}(t)$. But the sum over all paths from the kth position on the base to the different apexes is the same as the sum over all paths from the different apexes to the kth position on the base. That is, we can run the down recurrence instead of the up recurrence, remembering to differentiate the bottom level. Since we run the down recurrence for $n-1$ levels, the functions that emerge on the $(n-1)$st level are just the degree $n-1$ B-splines. Moreover, only two arrows point into the kth position on the nth level (see Figure 7.28). Hence for differentiating $N_k^n(t)$ we have Figure 7.32.

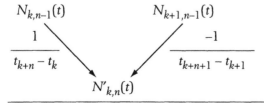

Figure 7.32 The down recurrence for differentiating a B-spline. Remember that the final result needs to be multiplied by the degree n.

Therefore, we conclude that

$$\frac{dN_{k,n}(t)}{dt} = n\left(\frac{N_{k,n-1}(t)}{t_{k+n}-t_k} - \frac{N_{k+1,n-1}(t)}{t_{k+n+1}-t_{k+1}}\right). \tag{7.11}$$

Since $t_{k+n}-t_k$ represents the support of $N_{k,n-1}(t)$ and $t_{k+n+1}-t_{k+1}$ represents the support of $N_{k+1,n-1}(t)$, we shall sometimes write

$$\frac{dN_{k,n}(t)}{dt} = n\left(\frac{N_{k,n-1}(t)}{Support} - \frac{N_{k+1,n-1}(t)}{Support}\right). \tag{7.12}$$

What about the antiderivative of a B-spline? Since $N_{k,n}(t)$ is a piecewise polynomial of degree n with continuity of order $n-\mu$ at knots of multiplicity μ, we would naturally expect that the antiderivative of $N_{k,n}(t)$ should be a piecewise polynomial of degree $n+1$ with continuity of order $n+1-\mu$ at knots of multiplicity μ. Moreover, for values of $t > t_{k+n+1}$, the antiderivative of $N_{k,n}(t)$ should be a constant equal to the area under the curve $N_{k,n}(t)$ because $Support\{N_{k,n}(t)\} \subseteq [t_k,t_{k+n+1}]$. Thus it appears that the antiderivative of $N_{k,n}(t)$ should be a B-spline of degree $n+1$ with the same knots as $N_{k,n}(t)$ but with one additional knot at infinity.

To determine if this analysis is correct, let $N_{k,n}(t \mid t_k,...,t_{k+n+1})$ denote the B-spline with knots $t_k,...,t_{k+n+1}$. To avoid calculating with infinities, we shall represent a knot at infinity by a homogeneous knot at $\delta = (1,0)$. Then, up to a constant multiple, we expect that the antiderivative of $N_{k,n}(t \mid t_k,...,t_{k+n+1})$ is given by $N_{k,n+1}(t \mid t_k,...,t_{k+n+1},\delta)$. Let's see if this works.

Generally, differentiating the de Boor algorithm leads to a two-term derivative formula, but not always. Recall that the de Boor algorithm is derived from a blossoming recurrence, where the input is the blossom evaluated at consecutive knots. If we start with the multilinear blossom, then the knots may take on homogeneous values. Now suppose, in particular, that $(t_{k+n+2},w_{k+n+2}) = (1,0) = \delta$. Then the label $t_{k+n+2}-t \to t_{k+n+2} - tw_{k+n+2} \to 1$, so one of the coefficients in the down recurrence for $N_{k,n+1}(t \mid t_k,...,t_{k+n+1},\delta)$ is a constant (see Figure 7.33(a)).

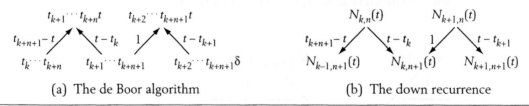

(a) The de Boor algorithm   (b) The down recurrence

Figure 7.33 One level of (a) the de Boor algorithm and (b) the down recurrence with the knot $(t_{k+n+2},w_{k+n+2}) = (1,0) = \delta$. Notice the label 1 along one of the edges of the diagram. Since the derivative of a constant is zero, differentiating this diagram leads to a one-term derivative formula for $N_{k,n+1}(t)$.

Thus the down recurrence becomes

$$N_{k,n+1}(t \mid t_k,...,t_{k+n+1},\delta)$$

$$= \frac{t - t_k}{t_{k+n+1} - t_k} N_{k,n}(t \mid t_k,...,t_{k+n+1}) + N_{k+1,n}(t \mid t_{k+1},...,t_{k+n+1},\delta).$$

Now differentiating the last level of the de Boor algorithm and multiplying the result by $n + 1$ leads to the one-term differentiation formula:

$$\frac{dN_{k,n+1}(t \mid t_k,...,t_{k+n+1},\delta)}{dt} = (n+1)\frac{N_{k,n}(t \mid t_k,...,t_{k+n+1})}{t_{k+n+1} - t_k}.$$

Hence

$$\int N_{k,n}(t \mid t_k,...,t_{k+n+1})dt = \frac{t_{k+n+1} - t_k}{n+1} N_{k,n+1}(t \mid t_k,...,t_{k+n+1},\delta)$$

or equivalently

$$\int_{Support} \frac{N_{k,n}(t)}{Support} dt = \frac{N_{k,n+1}(t \mid t_k,...,t_{k+n+1},\delta)}{n+1}. \tag{7.13}$$

Exercises

1. Let $S(t) = \sum_k N_{k,n}(t)P_k$ be a B-spline curve with knots $\{t_k\}$. Show that

$$S'(t) = n\sum_k N_{k,n-1}(t)\left(\frac{P_k - P_{k-1}}{t_{k+n} - t_k}\right).$$

2. Derive the following recurrence between the derivatives of the B-splines:

$$\left(\frac{n-1}{n}\right)N'_{k,n}(t) = \left(\frac{t - t_k}{t_{k+n} - t_k}\right)N'_{k,n-1}(t) + \left(\frac{t_{k+n+1} - t}{t_{k+n+1} - t_{k+1}}\right)N'_{k+1,n-1}(t).$$

3. Using Equation (7.12) prove that

   a. $$\int_{Support} \frac{N_{k,n-1}(t)}{Support} dt = \int_{Support} \frac{N_{k+1,n-1}(t)}{Support} dt$$

   b. $$\int_{Support} \frac{N_{k,n-1}(t)}{Support} dt \text{ is independent of } k$$

4. Prove that

   $$\int_{Support} \frac{N_{k,n}(t)}{Support} dt = \frac{1}{n+1}$$

   in two different ways.

   a. Insert knots far from the support of $N_{k,n}(t)$ to form a local Bernstein basis. Then apply Exercise 3(b).

    b. Show that $N_{k,n+1}(t_{k+n+1} \mid t_k,...,t_{k+n+1},\delta) = 1$. Then apply Equation (7.13).

5. Let $S(t)$ be a B-spline curve with control points $\{P_k\}$.

    a. Using the results of Exercises 1 and 4, show that

$$\int_{-\infty}^{\infty} \mid S'(t) \mid dt \le \Sigma_k \mid P_k - P_{k-1} \mid .$$

    b. Conclude that *arc length S(t) ≤ perimeter of its control polygon.*

7.7.4   B-Splines and Divided Difference

Two remarkable formulas link B-splines with divided differences:

$$N_{k,n}(t) = \{(t_{k+n+1} - t_k)(x - t)_+^n\}[t_k,...,t_{k+n+1}] \tag{7.14}$$

$$F[t_k,...,t_{k+n+1}] = \int_{t_k}^{t_{k+n+1}} \left\{ \frac{N_{k,n}(t)}{t_{k+n+1} - t_k} \right\} \left\{ \frac{F^{(n+1)}(t)}{n!} \right\} dt$$

$$= \int_{Support} \left\{ \frac{N_{k,n}(t)}{Support} \right\} \left\{ \frac{F^{(n+1)}(t)}{n!} \right\} dt \quad . \tag{7.15}$$

The first formula constructs the B-splines from the divided difference operator; the second builds the divided difference operator by integration with the B-splines. Note that in Equation (7.14) the divided difference is with respect to x, so t is treated as a constant on the right-hand side.

    Equation (7.14) is important because it allows us to use the divided difference identities derived in Chapter 4 to derive identities for the B-splines. Till now we have used blossoming almost exclusively to derive B-spline identities, but the blossoming approach works only when the identities involve spline bases—that is, the knot sequences of the B-splines must be compatible—or when the blossom is easy to compute. This is not always the case—see Exercises 1–3 and 7–9 for examples. Divided differences have no such restrictions.

    Equation (7.15) is important because it allows us to use the properties of the B-splines to derive properties of the divided difference operator. In fact, sometimes this formula is taken as the definition of the divided difference.

    To understand Equation (7.14), we need to define our notation. Let

$$(x - t)_+^n = (x - t)^n \qquad x > t$$
$$= 0 \qquad x \le t.$$

To derive Equation (7.14), we begin by showing that the function

$$\{(x - t)_+^n\} \; [t_k,...,t_{k+n+1}]$$

has the same support as the B-spline $N_{k,n}(t \mid t_k,...,t_{k+n+1})$.

LEMMA 7.5

$Support\{(x-t)_+^n[t_k,...,t_{k+n+1}]\} \subseteq [t_k, t_{k+n+1}].$

Proof We consider two cases: $t < t_k$ and $t > t_{k+n+1}$. If $t < t_k$, then the functions $(x-t)_+^n$ and $(x-t)^n$ agree at the nodes $t_k,...,t_{k+n+1}$. Hence

$$(x-t)_+^n[t_k,...,t_{k+n+1}] = (x-t)^n[t_k,...,t_{k+n+1}] = 0,$$

since divided difference annihilates polynomials of low degree (see Theorem 4.3, 12(a)). On the other hand, if $t > t_{k+n+1}$, then on the nodes $t_k,..., t_{k+n+1}$, we have $(x-t)_+^n = 0$, so it follows immediately that

$$(x-t)_+^n t_k,...,t_{k+n+1}] = 0.$$

Next we establish that $(x-t)_+^n[t_k,...,t_{k+n+1}]$ is a piecewise polynomial with knots at the nodes $t_k,...,t_{k+n+1}$.

LEMMA 7.6

The function $(x-t)_+^n[t_k,...,t_{k+n+1}]$ is a piecewise polynomial of degree n with knots at $t_k,...,t_{k+n+1}$.

Proof Let $t_i < t < t_{i+1}$, and let $P(x) = a_{n+1}x^{n+1} + \cdots + a_1x + a_0$ be the unique polynomial of degree $n+1$ that interpolates the same values as $(x-t)_+^n$ at the nodes $t_k,...,t_{k+n+1}$. Then

$$P(t_k) = a_{n+1}t_k^{n+1} + \cdots + a_1t_k + a_0 = (t_k - t)_+^n = 0$$
$$\vdots \qquad \vdots \qquad \vdots$$
$$P(t_i) = a_{n+1}t_i^{n+1} + \cdots + a_1t_i + a_0 = (t_i - t)_+^n = 0$$
$$P(t_{i+1}) = a_{n+1}t_{i+1}^{n+1} + \cdots + a_1t_{i+1} + a_0 = (t_{i+1} - t)_+^n = (t_{i+1} - t)^n \qquad (7.16)$$
$$\vdots \qquad \vdots \qquad \vdots$$
$$P(t_{k+n+1}) = a_{n+1}t_{k+n+1}^{n+1} + \cdots + a_1t_{k+n+1} + a_0 = (t_{k+n+1} - t)_+^n = (t_{k+n+1} - t)^n$$

Solving for the unknown coefficients $a_0,...,a_{n+1}$, we see that $P(x)$ is a polynomial of degree $n+1$ in x with coefficients that are polynomials of degree n in t. Since the divided difference is the highest-order coefficient of the polynomial interpolant,

$$(x-t)_+^n[t_k,...,t_{k+n+1}] = a_{n+1}$$

is a polynomial of degree n in t for $t_i < t < t_{i+1}$. To show that t_i is a knot of $(x-t)_+^n[t_k,...,t_{k+n+1}]$, let $t_{i-1} < t_- < t_i$ and $t_i < t_+ < t_{i+1}$. In addition, let $Q(x)$ be the unique polynomial of degree $n+1$ that interpolates the same values as the function

$$(x-t_-)_+^n - (x-t_+)_+^n$$

at the nodes $t_k,...,t_{k+n+1}$. Now observe that the functions $(x-t_-)^n_+$ and $(x-t_+)^n_+$ agree—that is, are the same polynomial in the variables t_- and t_+—at all the nodes except t_i; therefore, the $n+1$ nodes $t_k,...,t_{i-1}$, $t_{i+1},...,t_{k+n+1}$ are the $n+1$ roots of $Q(x)$. Moreover,

$$Q(t_i) = (t_i - t_-)^n_+ - (t_i - t_+)^n_+ = (t_i - t_-)^n.$$

Therefore,

$$Q(x) = \frac{(t_i - t_-)^n (x - t_k)\cdots(x - t_{i-1})(x - t_{i+1})\cdots(x - t_{k+n+1})}{(t_i - t_k)\cdots(t_i - t_{i-1})(t_i - t_{i+1})\cdots(t_i - t_{k+n+1})}.$$

It follows by the cancellation property of the divided difference (see Theorem 4.3) that

$$(x-t_-)^n_+[t_k,...,t_{k+n+1}] - (x-t_+)^n_+[t_k,...,t_{k+n+1}]$$

$$= \{(x-t_-)^n_+ - (x-t_+)^n_+\}[t_k,...,t_{k+n+1}]$$

$$= Q[t_k,...,t_{k+n+1}]$$

$$= \frac{(t_i - t_-)^n (x - t_k)\cdots(x - t_{i-1})(x - t_{i+1})\cdots(x - t_{k+n+1})}{(t_i - t_k)\cdots(t_i - t_{i-1})(t_i - t_{i+1})\cdots(t_i - t_{k+n+1})}[t_k,...,t_{k+n+1}]$$

$$= \frac{(t_i - t_-)^n}{(t_i - t_k)\cdots(t_i - t_{i-1})(t_i - t_{i+1})\cdots(t_i - t_{k+n+1})}[t_i]$$

$$= \frac{(t_i - t_-)^n}{(t_i - t_k)\cdots(t_i - t_{i-1})(t_i - t_{i+1})\cdots(t_i - t_{k+n+1})},$$

where the last equality follows because the divided difference is taken with respect to x and the expression

$$\frac{(t_i - t_-)^n}{(t_i - t_k)\cdots(t_i - t_{i-1})(t_i - t_{i+1})\cdots(t_i - t_{k+n+1})}$$

is a constant when viewed as a function of x. Thus

$$(x-t_-)^n_+[t_k,...,t_{k+n+1}] - (x-t_+)^n_+[t_k,...,t_{k+n+1}] = constant(t_i - t_-)^n,$$

so the value and the first $n-1$ derivatives of the two functions

$$(x-t_-)^n_+[t_k,...,t_{k+n+1}] \text{ and } (x-t_+)^n_+[t_k,...,t_{k+n+1}]$$

agree at $t = t_i$. Thus the piecewise polynomial $(x-t)^n_+[t_k,...,t_{k+n+1}]$ has a simple knot at $t = t_i$.

In the proof of Lemma 7.6 we have implicitly assumed that t_i appears only once in the sequence $t_k,...,t_{k+n+1}$—that is, t_i is a simple knot. If t_i is repeated μ times, then the same proof applies except that now we must set

$$Q(x) = \frac{\prod_{j \neq i}(x - t_j)}{\prod_{j \neq i}(t_i - t_j)} \, R(x),$$

where $R(x)$ is the Taylor expansion of $(x - t_-)_+^n$ of order $\mu - 1$ at $t_- = t_i$. The rest of the proof remains the same, and in this case we conclude that $(x - t)_+^n[t_k,...,t_{k+n+1}]$ has a knot of multiplicity μ at $t = t_i$. In any event, the preceding proof of Lemma 7.6 is quite abstract; for a somewhat more concrete proof, see Exercise 6.

With these results in hand, we are finally ready to prove Equation (7.14).

**PROPOSITION
7.7**

$$N_{k,n}(t) = \{(t_{k+n+1} - t_k)(x - t)_+^n\}[t_k,...,t_{k+n+1}].$$

Proof To simplify our notation, let

$$G_{k,n}(t) = \{(t_{k+n+1} - t_k)(x - t)_+^n\}[t_k,...,t_{k+n+1}].$$

Our goal now is to show that $G_{k,n}(t) = N_{k,n}(t)$. By Lemma 7.6, the function $G_{k,n}(t)$ is a piecewise polynomial with knots at the nodes $t_k,...,t_{k+n+1}$. Since the B-splines form a basis for the splines, it follows that we can represent the function $G_{k,n}(t)$ in terms of the B-splines $\{N_{j,n}(t)\}$. Moreover, by Lemma 7.5, $G_{k,n}(t)$ has the same support as $N_{k,n}(t)$. Therefore, it follows by the linear independence of the B-splines that there is a constant c_k such that

$$G_{k,n}(t) = c_k N_{k,n}(t).$$

We can compute the constant c_k from blossoming. Let $g_{k,n}(u_1,...,u_n)$ denote the blossom of $G_{k,n}(t)$. Then

$$g_{k,n}(u_1,...,u_n) = \{(t_{k+n+1} - t_k)(x - t)_+^0(x - u_1)\cdots(x - u_n)\}[t_k,...,t_{k+n+1}],$$

since the right-hand side is symmetric, multiaffine, and reduces to $G_{k,n}(t)$ along the diagonal. Now let $t_k < t < t_{k+n+1}$. By the dual functional property of the blossom (Equation (7.9)) and the cancellation property of the divided difference (Theorem 4.3),

$$c_k = g_{k,n}(t_{k+1},...,t_{k+n})$$
$$= \{(t_{k+n+1} - t_k)(x - t)_+^0(x - t_{k+1})\cdots(x - t_{k+n})\}[t_k,...,t_{k+n+1}]$$
$$= \{(t_{k+n+1} - t_k)(x - t)_+^0\}[t_k, t_{k+n+1}]$$
$$= \frac{(t_{k+n+1} - t_k)(t_{k+n+1} - t)_+^0 - (t_{k+n+1} - t_k)(t_k - t)_+^0}{t_{k+n+1} - t_k}$$
$$= 1,$$

since $t_k < t < t_{k+n+1}$. Hence $G_{k,n}(t) = N_{k,n}(t)$.

We had to work fairly hard to establish Equation (7.14); Equation (7.15) is much easier to derive. Our main tools here are integration by parts and the two-term differentiation formula for the B-splines.

**PROPOSITION
7.8**

$$F[t_k,...,t_{k+n+1}] = \int_{Support} \left\{ \frac{N_{k,n}(t)}{Support} \right\} \left\{ \frac{F^{(n+1)}(t)}{n!} \right\} dt$$

Proof We proceed by induction on n. We can easily verify this result is true for $n = 0$, since

$$F[t_k,t_{k+1}] = \frac{F(t_{k+1}) - F(t_k)}{t_{k+1} - t_k} = \int_{t_k}^{t_{k+1}} \frac{F'(t)dt}{t_{k+1} - t_k}.$$

Now assume this result is true for $n - 1$ and apply integration by parts. Then

$$\int_{Support} \left\{ \frac{N_{k,n}(t)}{Support} \right\} \left\{ \frac{F^{(n+1)}(t)}{n!} \right\} dt$$

$$= \left\{ \frac{F^{(n)}(t)N_{k,n}(t)}{n!(t_{k+n+1} - t_k)} \right\}_{t_k}^{t_{k+n+1}} - \frac{1}{t_{k+n+1} - t_k} \int_{t_k}^{t_{k+n+1}} \left\{ \frac{dN_{k,n}(t)}{dt} \right\} \left\{ \frac{F^{(n)}(t)}{n!} \right\} dt.$$

Since $N_{k,n}(t)$ vanishes on the end points of its support,

$$\left\{ \frac{F^{(n)}(t)N_{k,n}(t)}{n!(t_{k+n+1} - t_k)} \right\}_{t_k}^{t_{k+n+1}} = 0.$$

Moreover, by Equation (7.12)

$$\frac{dN_{k,n}(t)}{dt} = n \left(\frac{N_{k,n-1}(t)}{t_{k+n} - t_k} - \frac{N_{k+1,n-1}(t)}{t_{k+n+1} - t_{k+1}} \right).$$

Therefore,

$$
\int_{Support} \left\{ \frac{N_{k,n}(t)}{Support} \right\} \left\{ \frac{F^{(n+1)}(t)}{n!} \right\} dt
$$

$$
= \frac{-1}{t_{k+n+1} - t_k} \int_{t_k}^{t_{k+n+1}} \left\{ \frac{dN_{k,n}(t)}{dt} \right\} \left\{ \frac{F^{(n)}(t)}{n!} \right\} dt
$$

$$
= \frac{-1}{t_{k+n+1} - t_k} \int_{t_k}^{t_{k+n+1}} n \left\{ \frac{N_{k,n-1}(t)}{t_{k+n} - t_k} - \frac{N_{k+1,n-1}(t)}{t_{k+n+1} - t_{k+1}} \right\} \left\{ \frac{F^{(n)}(t)}{n!} \right\} dt
$$

$$
= \frac{1}{t_{k+n+1} - t_k} \int_{t_{k+1}}^{t_{k+n+1}} \left\{ \frac{N_{k+1,n-1}(t)}{t_{k+n+1} - t_{k+1}} \right\} \left\{ \frac{F^{(n)}(t)}{(n-1)!} \right\} dt
$$

$$
- \frac{1}{t_{k+n+1} - t_k} \int_{t_k}^{t_{k+n}} \left\{ \frac{N_{k,n-1}(t)}{t_{k+n} - t_k} \right\} \left\{ \frac{F^{(n)}(t)}{(n-1)!} \right\} dt
$$

$$
= \frac{F[t_{k+1},\ldots,t_{k+n+1}] - F[t_k,\ldots,t_{k+n}]}{t_{k+n+1} - t_k}
$$

$$
= F[t_k,\ldots,t_{k+n+1}] \; .
$$

Exercises

1. Use Equation (7.14) together with the cancellation property of the divided difference to prove that

$$
N_{k,n}(t_{k+j} \mid t_k,\ldots,t_{k+n+1}) = N_{k,n-1}(t_{k+j} \mid t_k,\ldots,t_{k+j-1},t_{k+j+1},\ldots,t_{k+n+1}).
$$

2. Use Equation (7.14) together with Section 4.3, Exercise 3(a), to prove that

$$
\frac{\partial N_k^n(t \mid t_k,\ldots,t_j,\ldots,t_{k+n+1})}{\partial t_j}
$$

$$
= \mu_j \left(\frac{N_{k+1}^n(t \mid t_{k+1},\ldots,t_j,t_j,\ldots,t_{k+n+1})}{t_{k+n+1} - t_{k+1}} - \frac{N_k^n(t \mid t_k,\ldots,t_j,t_j,\ldots,t_{k+n})}{t_{k+n} - t_k} \right),
$$

where μ_j is the multiplicity of $t_j \neq t_k, t_{k+n+1}$.

3. Use Equation (7.14) together with Section 4.3, Exercise 3(b), to prove that

$$
N_{k,n}(t \mid t_k,\ldots,t_{k+n+1}) = \frac{\sum_j N_{k,n+1}(t \mid t_k,\ldots,t_{k+j},t_{k+j},\ldots,t_{k+n+1})}{n+1} \; .
$$

4. Use Equation (7.14) together with Leibniz's rule for divided difference to give an alternative proof of the divided difference recurrence

$$N_{k,n}(t) = \frac{t - t_k}{t_{k+n} - t_k} N_{k,n-1}(t) + \frac{t_{k+n+1} - t}{t_{k+n+1} - t_{k+1}} N_{k+1,n-1}(t).$$

5. Use Equation (7.14) together with the appropriate property of the divided difference to prove that

$$t_{j+1} = \cdots = t_{j+n} \Rightarrow N_{j,n}(t_{j+1}) = 1.$$

6. Here we give an alternative proof of Lemma 7.6. Since, by definition, the divided difference is the highest-order coefficient of the polynomial interpolant, we can find

$$(x - t)_+^n [t_k, \ldots, t_{k+n+1}]$$

by solving system (7.16) for the coefficient a_{n+1}.

a. Using Cramer's rule, show that for $t_i < t < t_{i+1}$

$$\{(x-t)_+^n\}[t_k, \ldots, t_{k+n+1}] = a_{n+1} = \frac{\begin{vmatrix} 0 & t_k^n & \cdots & 1 \\ \vdots & \vdots & \vdots & \vdots \\ 0 & \vdots & \vdots & \vdots \\ (t_{i+1} - t)^n & \vdots & \vdots & \vdots \\ \vdots & \vdots & \vdots & \vdots \\ (t_{k+n+1} - t)^n & t_{k+n+1}^n & \cdots & 1 \end{vmatrix}}{\begin{vmatrix} t_k^{n+1} & t_k^n & \cdots & 1 \\ \vdots & \vdots & \vdots & \vdots \\ t_{k+n+1}^{n+1} & t_{k+n+1}^n & \cdots & 1 \end{vmatrix}}.$$

b. Similarly show that for $t_{i-1} < t < t_i$,

$$\{(x-t)_+^n\}[t_k, \ldots, t_{k+n+1}] = \frac{\begin{vmatrix} 0 & t_k^n & \cdots & 1 \\ \vdots & \vdots & \vdots & \vdots \\ (t_i - t)^n & \vdots & \vdots & \vdots \\ (t_{i+1} - t)^n & \vdots & \vdots & \vdots \\ \vdots & \vdots & \vdots & \vdots \\ (t_{k+n+1} - t)^n & t_{k+n+1}^n & \cdots & 1 \end{vmatrix}}{\begin{vmatrix} t_k^{n+1} & t_k^n & \cdots & 1 \\ \vdots & \vdots & \vdots & \vdots \\ t_{k+n+1}^{n+1} & t_{k+n+1}^n & \cdots & 1 \end{vmatrix}}.$$

c. Let $Q_i(t)$ be the difference between the two polynomials in parts (a) and (b)—that is, $Q_i(t)$ is the difference between $(x-t)^n_+[t_k,...,t_{k+n+1}]$ to the left and to the right of t_i. Show that $Q_i(t) = constant(t_i - t)^n$.

d. Conclude from part (c) that the function value and the first $n - 1$ derivatives of

$$(x-t)^n_+[t_k,...,t_{k+n+1}]$$

agree to the left and to the right of t_i. Hence the function

$$(x-t)^n_+[t_k,...,t_{k+n+1}]$$

is a piecewise polynomial of degree n with knots at $t_k,...,t_{k+n+1}$.

e. How would you need to alter this proof if t_i is repeated μ times in the sequence $t_k,...,t_{k+n+1}$?

7. a. Use Equation (7.14) together with Proposition 4.6 to prove that for any τ

$$N_{k,n}(t \mid t_k,...,t_{k+n+1}) = \frac{\tau - t_k}{t_{k+n} - t_k} N_{k,n}(t \mid t_k,...,\tau,...,t_{k+n})$$

$$+ \frac{t_{k+n+1} - \tau}{t_{k+n+1} - t_{k+1}} N_{k+1,n}(t \mid t_{k+1},...,\tau,...,t_{k+n+1}).$$

b. By letting $\tau = t$ in part (a) and invoking Exercise 1, derive the B-spline recurrence

$$N_{k,n}(t) = \frac{t - t_k}{t_{k+n} - t_k} N_{k,n-1}(t) + \frac{t_{k+n+1} - t}{t_{k+n+1} - t_{k+1}} N_{k+1,n-1}(t).$$

8. Generalize the B-spline recurrence by showing that for any $1 \le i < j \le n$

$$N_{k,n}(t \mid t_k,...,t_{k+n+1}) = \frac{t - t_{k+i}}{t_{k+j} - t_{k+i}} N_{k,n-1}(t \mid t_k,...,\hat{t}_{k+j},...,t_{k+n+1})$$

$$+ \frac{t_{k+j} - t}{t_{k+j} - t_{k+i}} N_{k,n-1}(t \mid t_k,...,\hat{t}_{k+i},...,t_{k+n+1})$$

where $\hat{t}_\alpha$ means that t_α is omitted from the sequence. (Hint: Use Equation (7.14) together with Proposition 4.6 and Exercise 1. Compare to Exercise 7.)

9. Differentiate Equation (7.14) and then apply the divided difference recurrence to prove that

a. $\dfrac{dN_{k,n}(t)}{dt} = n\left(\dfrac{N_{k,n-1}(t)}{t_{k+n} - t_k} - \dfrac{N_{k+1,n-1}(t)}{t_{k+n+1} - t_{k+1}}\right)$

b. $\dfrac{dN_{k,n}(t)}{dt} =$

$$n\left(\frac{N_{k,n-1}(t \mid t_k,...,\hat{t}_{k+j},...,t_{k+n+1})}{t_{k+j} - t_{k+i}} - \frac{N_{k,n-1}(t \mid t_k,...,\hat{t}_{k+i},...,t_{k+n+1})}{t_{k+j} - t_{k+i}} \right)$$

10. Consider the identities for the divided difference listed at the end of Chapter 4. Using Equation (7.15) and the properties of B-splines, derive as many of these identities as you can.

7.7.5   A Geometric Characterization of the B-Splines

So far we have seen two distinct ways to construct the B-splines: by the de Boor recurrence (Equation (7.8)) and by divided differences (Equation (7.14)). Here we shall give an alternative, geometric characterization of the B-splines.

Let $P_0,...,P_n$ be any $n + 1$ affinely independent points in n dimensions such that $P_{i1} = t_i$, and let $\Delta^n(P)$ denote the n-simplex with vertices $P_0,...,P_n$. We are going to show that

$$N_{0,n-1}(t \mid t_0,...,t_n) = \frac{(t_n - t_0)Vol_{n-1}(\sigma \in \Delta^n(P) \mid \sigma_1 = t)}{nVol_n(\Delta^n(P))} . \qquad (7.17)$$

Equation (7.17) has the following geometric interpretation: Form any n-simplex such that the first coordinates of its vertices $P_0,...,P_n$ lie over the knots $t_0,...,t_n$. Then the $n - 1$ dimensional volume of the cross section of this simplex lying over the parameter t is, up to a constant multiple, the B-spline $N_{0,n-1}(t \mid t_0,...,t_n)$ (see Figure 7.34).

To derive Equation (7.17), we begin by recalling the Hermite-Genocchi formula for the divided difference (see Section 4.3, Exercise 9):

$$F[t_0,...,t_n] = \int_{\Delta^n} F^{(n)}\bigl(t_0 + v_1(t_1 - t_0) + \cdots + v_n(t_n - t_0)\bigr)dv_1 \cdots dv_n ,$$

where $\Delta^n = \{(v_1,...,v_n) \mid v_j \geq 0 \text{ and } \sum_j v_j \leq 1\}$ is the standard n-dimensional simplex. On the other hand, by Equation (7.15),

$$F[t_0,...,t_n] = \int_{t_0}^{t_n} \left\{ \frac{N_{0,n-1}(t \mid t_0,...,t_n)}{t_n - t_0} \right\} \left\{ \frac{F^{(n)}(t)}{(n-1)!} \right\} dt .$$

Therefore, it follows that

$$\int_{\Delta^n} F^{(n)}\bigl(t_0 + v_1(t_1 - t_0) + \cdots + v_n(t_n - t_0)\bigr)dv_1 \cdots dv_n$$

$$= \int_{t_0}^{t_n} \left\{ \frac{N_{0,n-1}(t \mid t_0,...,t_n)}{t_n - t_0} \right\} \left\{ \frac{F^{(n)}(t)}{(n-1)!} \right\} dt ,$$

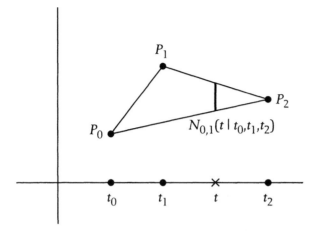

Figure 7.34 Geometric characterization of the B-spline $N_{0,1}(t \mid t_0,t_1,t_2)$ as a cross section of a triangle, normalized by the area of the triangle and the length of the support. Notice that this cross section is a piecewise linear function with support $[t_0,t_2]$ and a knot at t_1, just like the B-spline $N_{0,1}(t \mid t_0,t_1,t_2)$.

or equivalently, replacing $F^{(n)}(t)$ by $G(t)$,

$$\int_{\Delta^n} G\big(t_0 + v_1(t_1 - t_0) + \cdots + v_n(t_n - t_0)\big)dv_1 \cdots dv_n$$

$$= \int_{t_0}^{t_n} \left\{ \frac{N_{0,n-1}(t \mid t_0,\ldots,t_n)}{(n-1)!(t_n - t_0)} \right\} G(t)dt. \tag{7.18}$$

We are now going to perform a change of variables to simplify the left-hand side of Equation (7.18). Let S be the unique affine transformation that maps the vertices of Δ^n to the vertices of $\Delta^n(P)$ by sending the vertex E_i to the vertex P_i, where E_i is the point whose rectangular coordinates $(v_1,\ldots,v_n)$ have a 1 in the ith position and a zero everywhere else. Then

$$S(v_1,\ldots,v_n) = P_0 + v_1(P_1 - P_0) + \cdots + v_n(P_n - P_0),$$

or equivalently

$$s_i = P_{0i} + v_1(P_{1i} - P_{0i}) + \cdots + v_n(P_{ni} - P_{0i}), \, i = 1,\ldots,n.$$

Notice, in particular, that by the choice of $P_0,\ldots,P_n$

$$s_1 = t_0 + v_1(t_1 - t_0) + \cdots + v_n(t_n - t_0).$$

The Jacobian of this transformation between the variables $s = (s_1, \ldots, s_n)$ and the variables $v = (v_1, \ldots, v_n)$ is given by

$$\frac{\partial(s_1, \ldots, s_n)}{\partial(v_1, \ldots, v_n)} = \det \begin{pmatrix} P_{11} - P_{01} & P_{21} - P_{01} & \cdots & P_{n1} - P_{01} \\ P_{12} - P_{02} & P_{22} - P_{02} & \cdots & P_{n2} - P_{02} \\ \vdots & \vdots & \vdots & \vdots \\ P_{1n} - P_{0n} & P_{12n} - P_{0n} & \cdots & P_{nn} - P_{0n} \end{pmatrix}$$

$$= \det(P_1 - P_0, \ldots, P_n - P_0)$$

$$= n! Vol_n\left(\Delta^n(P)\right).$$

Hence by this change of variables

$$\int_{\Delta^n} G\left(t_0 + v_1(t_1 - t_0) + \cdots + v_n(t_n - t_0)\right) dv_1 \cdots dv_n$$

$$= \frac{1}{n! Vol_n(\Delta_n)} \int_{\Delta^n(P)} G(s_1) ds_1 \cdots ds_n$$

$$= \frac{1}{n! Vol_n(\Delta_n)} \int_{t_0}^{t_n} \left\{ \int_{\sigma \in \Delta^n(P) | \sigma_1 = s_1} ds_2 \cdots ds_n \right\} G(s_1) ds_1. \qquad (7.19)$$

Now it follows from Equations (7.18) and (7.19) that

$$\int_{t_0}^{t_n} \left\{ \frac{N_{0,n-1}(t \mid t_0, \ldots, t_n)}{(n-1)!(t_n - t_0)} \right\} G(t) dt = \frac{1}{n! Vol_n(\Delta_n)} \int_{t_0}^{t_n} \left\{ \int_{\sigma \in \Delta^n(P) | \sigma_1 = s_1} ds_2 \cdots ds_n \right\} G(s_1) ds_1.$$

Replacing s_1 by t and comparing the integrands, we find that

$$\frac{N_{0,n-1}(t \mid t_0, \ldots, t_n)}{(n-1)!(t_n - t_0)} = \frac{1}{n! Vol_n(\Delta_n)} \int_{\sigma \in \Delta^n(P) | \sigma_1 = t} ds_2 \cdots ds_n$$

$$= \frac{Vol_{n-1}(\sigma \in \Delta^n(P) \mid \sigma_1 = t)}{n! Vol_n(\Delta^n(P))}.$$

Therefore,

$$N_{0,n-1}(t \mid t_0, \ldots, t_n) = \frac{(t_n - t_0) Vol_{n-1}(\sigma \in \Delta^n(P) \mid \sigma_1 = t)}{n Vol_n(\Delta^n(P))}.$$

Although we have not taken this geometric approach to defining the B-splines, all the many properties of and procedures for the B-splines, such as knot insertion algorithms, can be derived directly from Equation (7.17). Therefore, this formula for the univariate B-splines is often taken as the starting point for the extension of the theory of B-splines to the multivariate setting, where other approaches do not apply. For multivariate B-splines in k dimensions, Vol_{n-1} is replaced by Vol_{n-k}.

Exercises

1. Use the triangle with vertices $(t_0,1),(t_1,2),(t_2,1)$ to calculate the values of the B-spline $N_{0,1}(t \mid t_0,t_1,t_2)$ from Equation (7.17).

2. a. Let S be an affine transformation on affine n-space that preserves the value of the first coordinate. Show that the first column of the matrix representing S is $(1,0,...,0)^T$.

   b. Using part (a), show that the value of

$$\frac{(t_n - t_0)Vol_{n-1}(\sigma \in \Delta^n(P) \mid \sigma_1 = t)}{nVol_n(\Delta^n(P))}$$

   is independent of the choice of the simplex $\Delta^n(P)$ provided that $P_{i1} = t_i$, $i = 0,...,n$.

7.8 Uniform B-Splines

The simplest, most common knot vectors have evenly spaced knots—for example, knot vectors where the knots are located at the integers $\{...,-2,-1,0,1,2,...\}$. When the knots are simple knots (knots of multiplicity one) located at the integers, the knot sequence is called a *uniform knot sequence* and the associated B-splines are called *uniform B-splines*. These uniform B-splines have some especially nice properties, which we are now going to investigate.

For uniform B-splines it follows easily from the de Boor algorithm that all the B-splines of the same degree are translates of a single B-spline so that

$$N_{k,n}(t) = N_{0,n}(t - k) \tag{7.20}$$

(see Exercise 1). Thus, all uniform B-splines of degree n have a support of size $n + 1$. Since the size of the support appears in several B-spline identities, many of these identities simplify for uniform B-splines. For example, for uniform B-splines, it follows from Equations (7.8), (7.11), (7.13), and Exercise 4 of Section 7.7.3 that

$$N_{k,n+1}(t) = \frac{t-k}{n+1}N_{k,n}(t) + \frac{k+n+2-t}{n+1}N_{k+1,n}(t) \tag{7.21}$$

$$\frac{dN_{k,n}(t)}{dt} = N_{k,n-1}(t) - N_{k+1,n-1}(t) \tag{7.22}$$

$$\int N_{k,n}(t)dt = \tilde{N}_{k,n+1}(t) \tag{7.23}$$

$$\int_{Support} N_{k,n}(t)dt = 1. \tag{7.24}$$

Notice that in the antiderivative formula (Equation (7.23)) the B-spline $\tilde{N}_{k,n+1}(t)$ on the right-hand side is not quite uniform, since its last knot is at $\delta = (1,0)$.

Exercises

1. Using the de Boor algorithm, show that when the knots are located at the integers $N_{k,n}(t) = N_{0,n}(t-k)$.

2. Prove that the uniform B-splines $\{N_{k,n}(t)\}$ are unimodal in k.

3. Prove that for uniform B-splines:

   a. $\int_k^{k+n+1} N_{k,n}(t)dt = 1$.

   b. $\dfrac{\int_k^{k+n+1} N_{k,n}(t)F^{(n+1)}(t)dt}{(n+1)!} = F[k,\dots,k+n+1]$.

7.8.1   Continuous Convolution and Uniform B-Splines

Uniform B-splines can be generated from continuous convolutions. The continuous convolution of two functions of a continuous variable is an extension of discrete convolution for two functions of a discrete parameter (i.e., sequences), where summation is replaced by integration. Let $f(t)$ and $g(t)$ be integrable functions defined for all values of t. The continuous convolution $(f*g)(t)$ is defined by setting

$$(f*g)(t) = \int_{-\infty}^{\infty} f(t-x)g(x)dx .$$

We shall show shortly that just as the Bernstein basis can be generated from the discrete n-fold convolution of the sequence $\{1-t,t\}$, the uniform B-splines can be built up by continuous n-fold convolution of the characteristic function of the unit interval.

In the following proposition and its corollaries, $N_{k,n}(t)$ always denotes a uniform B-spline with knots at the integers.

PROPOSITION 7.9 $N_{k,n}(t) = \int_0^1 N_{k,n-1}(t-x)dx \qquad n \geq 1.$ $\qquad\qquad$ (7.25)

Proof   Let

$$L_{k,n}(t) = \int_0^1 N_{k,n-1}(t-x)dx.$$

We shall show by induction on n that $L_{k,n}(t) = N_{k,n}(t)$. For $n=1$, we have

$$L_{k,1}(t) = \int_0^1 N_{k,0}(t-x)dx .$$

But

$$N_{k,0}(t) = 1 \qquad k \leq t < k+1$$
$$= 0 \qquad \text{otherwise.}$$

Therefore,

$$N_{k,0}(t-x) = 1 \qquad k \le t - x < k+1$$
$$= 0 \qquad \text{otherwise.}$$

Hence

$$\int_0^1 N_{k,0}(t-x)dx = \int_0^{t-k} dx = t - k \qquad\qquad k \le t \le k+1$$
$$= \int_{t-k-1}^1 dx = k+2-t \qquad k+1 \le t \le k+2$$
$$= 0 \qquad\qquad\qquad \text{otherwise.}$$

Moreover, by Equation (7.21)

$$N_{k,1}(t) = (t-k)N_{k,0}(t) + (k+2-t)N_{k+1,0}(t) = t - k \qquad k \le t \le k+1$$
$$= k+2-t \qquad k+1 \le t \le k+2$$
$$= 0 \qquad\qquad \text{otherwise.}$$

Hence $L_{k,1}(t) = N_{k,1}(t)$. For the inductive step, observe that by Equation (7.22) and the inductive hypothesis

$$\frac{dL_{k,n}(t)}{dt} = \int_0^1 \left\{ \frac{dN_{k,n-1}(t-x)}{dt} \right\} dx$$
$$= \int_0^1 N_{k,n-2}(t-x)dx - \int_0^1 N_{k+1,n-2}(t-x)dx$$
$$= N_{k,n-1}(t) - N_{k+1,n-1}(t)$$
$$= \frac{dN_{k,n}(t)}{dt} .$$

Therefore, $L_{k,n}(t)$ and $N_{k,n}(t)$ differ by at most a constant. But

$$L_{k,n}(k) = \int_0^1 N_{k,n-1}(k-x)dx = 0 = N_{k,n}(k),$$

since $N_{k,n-1}(k-x) = 0$ for $0 \le x \le 1$. Hence $L_{k,n}(t) = N_{k,n}(t)$.

Let $\chi_{[0,1)}$ denote the characteristic function of $[0,1]$. That is, let

$$\chi_{[0,1)} = 1 \qquad 0 \le t < 1$$
$$= 0 \qquad \text{otherwise.}$$

COROLLARY
7.10

$$N_{k,n} = N_{k,n-1} * \chi_{[0,1)} .$$ (7.26)

Proof From Proposition 7.9

$$N_{k,n}(t) = \int_0^1 N_{k,n-1}(t-x)dx$$

$$= \int_{-\infty}^{\infty} N_{k,n-1}(t-x)\chi_{[0,1)}(x)dx$$

$$= (N_{k,n-1} * \chi_{[0,1)})(t) .$$

COROLLARY
7.11

$$N_{0,n} = \underbrace{\chi_{[0,1)} * \cdots * \chi_{[0,1)}}_{n+1 \ factors} .$$

Proof This result follows immediately from Corollary 7.10 by induction on n.

Exercises

1. Prove that
   a. $f * g = g * f$
   b. $(f * g) * h = f * (g * h)$
   c. $f * (g + h) = f * g + f * h$
2. Let $h_a(x) = h(x - a)$. Prove that $f_a * g = f * g_a$.
3. Prove that for uniform B-splines

   $$N_{k,n} = \chi_{[k,k+1)} * \underbrace{\chi_{[0,1)} * \cdots * \chi_{[0,1)}}_{n \ factors} .$$

4. Let $P(t)$ be the control polygon generated by the control points $\{P_k\}$—that is, $P(t)$ is the piecewise linear curve with $P(k+1) = P_k$—and let $S_n(t)$ be the uniform B-spline curve of degree n generated by these same control points. Show that
   a. $S_0(t) = P_k \qquad k \le t < k+1$
   b. $S_1(t) = P(t)$
   c. $S_n(t) = (P * N_{0,n-2})(t) \qquad n \ge 2$

7.8.2   Chaikin's Knot Insertion Algorithm

Boehm's algorithm and the Oslo algorithm allow us to insert a finite number of new knots at arbitrary locations into the knot sequence of a B-spline curve. But suppose we have a B-spline curve with uniform knots and we want to insert new knots but still keep the knot spacing uniform. For example, suppose we have knots at the integers $\{...,-2,-1,0,1,2,...\}$ and we want to insert new knots at the half integers, creating the new uniform knot sequence $\{...,-2,-1.5,-1,-0.5,0,0.5,1,1.5,2,...\}$. How should we proceed?

Knot insertion is, evidently, a special type of change of basis procedure. Let $\{N_{k,n}(t)\}$ denote the uniform B-splines over the integers, and let $\{H_{j,n}(t)\}$ denote the uniform B-splines over the half integers. Given a B-spline curve $S(t)$ with control points $\{P_k\}$ relative to the B-splines $\{N_{k,n}(t)\}$, we seek control points $\{Q_j\}$ relative to the B-splines $\{H_{j,n}(t)\}$ so that

$$S(t) = \sum_k P_k N_{k,n}(t) = \sum_j Q_j H_{j,n}(t).$$

Since the knot sequences for $\{N_{k,n}(t)\}$ and $\{H_{j,n}(t)\}$ are nested, such control points $\{Q_j\}$ must exist; the problem is how to find them.

Let us begin by considering some simple cases. For $n = 0$, $N_{k,0}(t) = 1$ on the half open unit interval $[k,k+1)$ and is zero everywhere else. Similarly, $H_{k,0}(t) = 1$ on the half-unit interval $[k,k+0.5)$ and $H_{k+0.5,0}(t) = 1$ on the half-unit interval $[k+0.5,k+1)$. Thus

$$N_{k,0}(t) = H_{k,0}(t) + H_{k+0.5,0}(t).$$

Therefore, if $\{P_k\}$ are the control points of a spline $S(t)$ relative to the basis $\{N_{k,0}(t)\}$, then evidently $\{...,P_j,P_j,...\}$ are the control points of $S(t)$ relative to the basis $\{H_{j,0}(t)\}$. That is, when we halve the parameter interval, we must double the control points.

Next let's try the case $n = 1$—that is, linear B-splines. Consider the curve

$$S(t) = \sum_k P_k N_{k,1}(t).$$

Since $N_{j,1}(k+1) = 0$ for $j \neq k$, it follows that $N_{k,1}(k+1) = 1$. Therefore, the spline $S(t)$ is a piecewise linear curve that interpolates the control point P_k at the parameter $k+1$. That is, the spline $S(t)$ is identical to the control polygon generated by the points $\{P_k\}$. Similarly, if we write

$$S(t) = \sum_j Q_j H_{j,1}(t),$$

then $S(t)$ will interpolate the control point Q_k at the parameter $k+0.5$ and the control point $Q_{k+0.5}$ at the parameter $k+1$. Hence we must set

$$Q_k = \frac{P_{k-1} + P_k}{2}$$
$$Q_{k+0.5} = P_k \ ,$$

where the formula for Q_k follows by the linearity of $S(t)$. Thus we can compute the points $\{Q_j\}$ from the points $\{P_k\}$ by doubling and averaging. This algorithm is depicted in Figure 7.35.

Does this pattern persist? Can we find the control points for a quadratic B-spline relative to knots at the half integers by first doubling and then twice averaging the

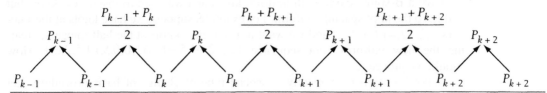

Figure 7.35 Algorithm for inserting knots at the half integers in a degree 1 B-spline curve. The original control points $\{P_k\}$ are doubled and averaged to generate the new control points $\{Q_j\}$.

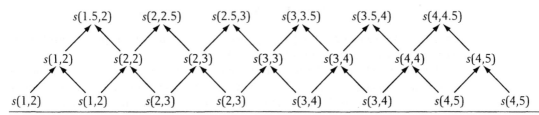

Figure 7.36 Chaikin's algorithm for inserting knots at the half integers for a uniform quadratic B-spline curve. Doubling the control points for the spline $S(t)$ with the knots at the integers and averaging twice yields the control points for the same spline with the knots at the half integers.

control points relative to the knots at the integers? Let's see what blossoming has to say about this question. By the dual functional property for a quadratic spline $S(t)$,

$$P_k = s(k+1, k+2)$$
$$Q_k = s(k+0.5, k+1)$$
$$Q_{k+0.5} = s(k+1, k+1.5) .$$

Now let's try doubling and then averaging twice. Sure enough, Figure 7.36 shows that this technique does indeed generate the points $\{Q_j\}$ from the points $\{P_k\}$. This algorithm of doubling and averaging twice was first introduced by Chaikin and is known as *Chaikin's algorithm*.

Exercises

1. This exercise extends Chaikin's knot insertion algorithm to quadratic B-splines with knots in geometric progression. Let $\{N_{k,2}(t)\}$ denote the quadratic B-splines with knots at $t_{2k} = \beta^{2k}$, and let $\{H_{k,2}(t)\}$ denote the quadratic B-splines with knots at $t_k = \beta^k$. Let $\{P_k\}$ denote the control points of a quadratic B-spline curve $S(t)$ relative to the B-splines $\{N_{k,2}(t)\}$, and let $\{Q_j\}$ denote the control points of the same curve $S(t)$ relative to the B-splines $\{H_{j,2}(t)\}$. Use blossoming to prove that the following algorithm can be used to generate the new points $\{Q_j\}$ from the original points $\{P_k\}$. Start by doubling the original control points $\{P_k\}$. Then take successive weighted averages of adjacent points in the following manner:

$$Q_j^0 = P_k \qquad\qquad j = 2k, 2k+1$$

$$Q_j^m = \frac{\beta^m Q_{j-1}^{m-1} + Q_j^{m-1}}{1 + \beta^m} \qquad m = 1, 2.$$

The points $\{Q_j^2\}$ that emerge on the second level of the algorithm are the control points $\{Q_j\}$ of the quadratic B-spline curve $S(t)$ relative to the B-splines $\{H_{j,2}(t)\}$ (see Figure 7.37).

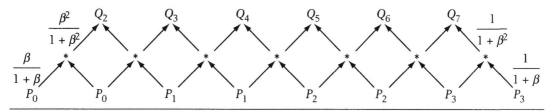

Figure 7.37 Algorithm for inserting knots in geometric progression in a quadratic B-spline curve. The algorithm begins at the base of the diagram where the original control points $\{P_k\}$ are doubled. Then two successive weighted averages are computed to generate the new control points $\{Q_j\}$ at the top of the diagram. All right-pointing arrows on the first level are labeled $\beta/(1 + \beta)$ and on the second level $\beta^2/(1 + \beta^2)$; all left-pointing arrows on the first level are labeled $1/(1 + \beta)$ and on the second level $1/(1 + \beta^2)$.

2. A knot sequence is said to be in affine progression if there are constants a, b such that for all k, $t_{k+1} = at_k + b$. Generalize Exercise 1 to knots in affine progression.

3. Implement a rendering algorithm for uniform quadratic B-spline curves based on Chaikin's knot insertion procedure.

4. Implement an intersection algorithm for uniform quadratic B-spline curves based on Chaikin's knot insertion procedure.

7.8.3   The Lane-Riesenfeld Knot Insertion Algorithm

Unfortunately, the blossoming approach to deriving knot insertion no longer works for cubic splines with uniform knots. (Try it!) But the pattern is still correct. To find the control points relative to the half integers for a spline of degree n, we need only double the control points relative to the integers and then average n times. Below we shall prove this assertion by appealing to the convolution formula for uniform B-splines.

To begin our analysis, recall that since the half integers are a refinement of the integers, we can certainly write

$$N_{k,n}(t) = \sum_j M_{jk}^n H_{j,n}(t).$$

Hence, by the First Principle of Duality in Section 5.5,

$$Q_j = \sum_k M^n_{jk} P_k .$$

Therefore, our knot insertion problem reduces to finding the coefficients $\{M^n_{jk}\}$ that express the B-splines relative to the integers in terms of the B-splines relative to the half integers. But recall that by the de Boor algorithm

$$N_{k,n}(t) = N_{0,n}(t-k) \tag{7.27}$$

$$H_{j,n}(t) = N_{0,n}(2t-j). \tag{7.28}$$

Therefore, it is enough to represent $N_{0,n}(t)$ in terms of $N_{0,n}(2t-j)$. We shall accomplish this goal by applying Corollary 7.11, but first we need some simple facts about continuous convolution.

LEMMA
7.12

Let f,g be arbitrary integrable functions. Then

   a. $f(2t) * g(2t) = \dfrac{(f*g)(2t)}{2}$

   b. $f(t-i) * g(t-j) = (f*g)(t-i-j)$

   c. $f(2t-i) * g(2t-j) = \dfrac{(f*g)(2t-i-j)}{2}$

Proof   To prove the first two identities, apply a change of variables:

   a. $\{f(2x) * g(2x)\}(t) = \displaystyle\int_{-\infty}^{\infty} f(2t-2x)g(2x)dx$

$$= \frac{\displaystyle\int_{-\infty}^{\infty} f(2t-u)g(u)du}{2} \qquad (u = 2x)$$

$$= \frac{(f*g)(2t)}{2}$$

   b. $\{f(x-i) * g(x-j)\}(t) = \displaystyle\int_{-\infty}^{\infty} f(t-x-i)g(x-j)dx$

$$= \int_{-\infty}^{\infty} f(t-i-j-u)g(u)dx \qquad (u = x-j)$$

$$= (f*g)(t-i-j)$$

The last result now follows from the first two, since

   c. $\{f(2x-i) * g(2x-j)\}(t) = \{f(2x)*g(2x)\}(t-i/2-j/2)$

$$= \frac{(f*g)(2t-i-j)}{2}$$

PROPOSITION
7.13

$$N_{0,n}(t) = \sum_{i=0}^{n+1} \frac{\binom{n+1}{i}}{2^n} N_{0,n}(2t - i).$$

Proof To simplify our notation, let $\chi = \chi_{[0,1)}$. Then by Corollary 7.11,

$$N_{0,n}(t) = \underbrace{\chi(t) * \cdots * \chi(t)}_{n+1 \ \text{factors}}.$$

Moreover, it is easy to see that $\chi(t) = \chi(2t) + \chi(2t - 1)$, since

$$\chi(2t) = 1 \qquad 0 \le t < 0.5$$
$$= 0 \qquad \text{otherwise}$$

and

$$\chi(2t - 1) = 1 \qquad 0.5 \le t < 1$$
$$= 0 \qquad \text{otherwise}.$$

Therefore, by Lemma 7.12,

$$\underbrace{\chi(t) * \cdots * \chi(t)}_{n+1 \ \text{factors}} = \underbrace{\left(\chi(2t) + \chi(2t-1) \right) * \cdots * \left(\chi(2t) + \chi(2t-1) \right)}_{n+1 \ \text{factors}}$$

$$= \sum_{i=0}^{n+1} \frac{\binom{n+1}{i}}{2^n} N_{0,n}(2t - i).$$

PROPOSITION
7.14

$$N_{k,n}(t) = \sum_{j=2k}^{n+2k+1} \frac{\binom{n+1}{j-2k}}{2^n} H_{j,n}(t).$$

Proof By Equations (7.27) and (7.28),

$$N_{k,n}(t) = N_{0,n}(t - k)$$

$$H_{j,n}(t) = N_{0,n}(2t - j).$$

Therefore, by Proposition 7.13

$$N_{k,n}(t) = N_{0,n}(t - k) = \sum_{i=0}^{n+1} \frac{\binom{n+1}{i}}{2^n} N_{0,n}(2t - 2k - i)$$

$$= \sum_{i=0}^{n+1} \frac{\binom{n+1}{i}}{2^n} H_{2k+i,n}(t) = \sum_{j=2k}^{n+2k+1} \frac{\binom{n+1}{j-2k}}{2^n} H_{j,n}(t).$$

From Proposition 7.14 we can derive both an explicit formula and a recursive formula for the control points introduced by knot insertion at the half integers.

COROLLARY 7.15   Let $\{P_k\}$ denote the control points of a B-spline curve of degree n relative to the B-splines with knots at the integers. Then the control points $\{Q_j^n\}$ for the same curve relative to the B-splines with knots at the half integers are given explicitly by

$$Q_j^n = \sum_{k=[j/2]}^{\left[\frac{j-n-1}{2}\right]} \frac{\binom{n+1}{j-2k}}{2^n} P_k .$$

Proof   Recall that by the First Principle of Duality in Section 5.5,

$$N_{k,n}(t) = \sum_j M_{jk}^n H_{j,n}(t) \Leftrightarrow Q_j^n = \sum_k M_{jk}^n P_k .$$

But by Proposition 7.14 we know that

$$M_{jk}^n = \frac{\binom{n+1}{j-2k}}{2^n} .$$

Therefore,

$$Q_j^n = \sum_{k=[j/2]}^{\left[\frac{j-n-1}{2}\right]} \frac{\binom{n+1}{j-2k}}{2^n} P_k .$$

COROLLARY 7.16   Let $\{P_k\}$ denote the control points of a B-spline curve of degree n relative to the B-splines with knots at the integers. Then the control points $\{Q_j^n\}$ for the same curve relative to the B-splines with knots at the half integers are given recursively by

$$Q_{2i}^0 = Q_{2i+1}^0 = P_i$$

$$Q_j^n = \frac{Q_{j-1}^{n-1} + Q_j^{n-1}}{2} .$$

Proof   The case $n = 0$ is established in Section 7.8.2. For $n > 0$ we know by Corollary 7.15 that

$$Q_j^n = \sum_k M_{jk}^n P_k, \text{ where } M_{jk}^n = \frac{\binom{n+1}{j-2k}}{2^n} .$$

But from Pascal's triangle,

$$M_{j,k}^n = \frac{M_{j-1,k}^{n-1} + M_{j,k}^{n-1}}{2} \ .$$

Therefore,

$$Q_j^n = \Sigma_k M_{jk}^n P_k = \sum_k \left(\frac{M_{j-1,k}^{n-1} + M_{j,k}^{n-1}}{2} \right) P_k = \frac{Q_{j-1}^{n-1} + Q_j^{n-1}}{2} \ .$$

The recursion formula in Corollary 7.16 leads directly to the algorithm we seek for inserting knots at the half integers in a degree n B-spline curve. Start by repeating each of the original control points $\{P_k\}$. Then take successive averages of adjacent points. The points that emerge at the nth level of the diagram are the control points of the same B-spline curve with knots at the half integers (see Figure 7.38). This generalization of Chaikin's algorithm for quadratic B-splines was first proved by Lane and Riesenfeld and is known as the *Lane-Riesenfeld algorithm*.

Iterating the Lane-Riesenfeld algorithm generates a sequence of control polygons that converges to the original B-spline curve. Indeed it is easy to show that the maximum distance between adjacent control points is halved after each iteration of the algorithm. Since each B-spline segment lies in the convex hull of its control points, it follows that these control polygons must converge to the B-spline curve. Therefore, we can apply the Lane-Riesenfeld knot insertion procedure to render and intersect uniform B-spline curves.

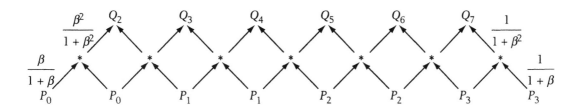

Figure 7.38 The Lane-Riesenfeld algorithm for inserting knots at the half integers in a uniform cubic B-spline curve. The algorithm begins at the base of the diagram where the original control points are doubled. Then three successive averages are computed to generate the new control points at the top of the diagram. Observe that the algorithms for inserting knots at the half integers for uniform linear and uniform quadratic B-splines are also contained in this diagram (compare to Figures 7.35 and 7.36). Notice too the binomial coefficients that multiply the control points.

Exercises

1. The Lane-Riesenfeld algorithm can be extended to inserting knots at the one-third and two-third integers— $\{k, k+1/3, k+2/3\}$ —by tripling the control points and then taking successive averages of three consecutive points. Prove that if we start with the control points for a degree n B-spline curve with knots at the integers, the points that emerge at the nth level of this algorithm are the control points of the same B-spline curve with knots at the one-third and two-third integers.

2. This exercise provides an alternative derivation of the Lane-Riesenfeld knot insertion algorithm. Let $P(t)$ be the control polygon generated by the control points $\{P_k\}$, and let $S_n(t)$ be the uniform B-spline curve of degree n generated by these same control points.

   a. Using Proposition 7.13 and Section 7.8.1, Exercise 4, show that

   $$S_n(t) = \sum_{i=0}^{n-1} \frac{\binom{n-1}{i}}{2^{n-2}} P(t) * N_{0,n-2}(2t - i) .$$

   b. Apply part (a) together with Equation (7.28) and Section 7.8.1, Exercise 2, to conclude that

   $$S_n(t) = \sum_{i=0}^{n-1} \frac{\binom{n-1}{i}}{2^{n-2}} (P_{i/2} * H_{0,n-2})(t), \text{ where } P_{i/2}(t) = P(t - i/2).$$

   c. Let $R_{i,k}^n$ denote the kth control point of $P_{i/2} * H_{0,n-2}$. Show that

   $$R_{i,k}^n = P\left(\frac{k-i}{2}\right).$$

   d. Use part (c) to derive the Lane-Riesenfeld recurrence for the control points Q_j^n of $S_n(t)$ relative to the B-splines with the knots at the half integers.

3. This exercise provides an alternative proof of the Lane-Riesenfeld knot insertion algorithm without resorting to convolution. The proof is based solely on the de Boor recurrence for uniform B-splines (Equation (7.21)). As in the text, let $\{N_{k,n}(t)\}$ denote the uniform B-splines over the integers, and let $\{H_{j,n}(t)\}$ denote the uniform B-splines over the half integers. Then there are constants $\{M_{jk}^n\}$ such that

   $$N_{k,n}(t) = \sum_j M_{jk}^n H_{j,n}(t). \qquad (*)$$

   We shall develop a recurrence for the constants $\{M_{jk}^n\}$.

   a. Substitute (*) into the de Boor recurrence (Equation (7.21)) for $N_{k,n+1}(t)$, and substitute the de Boor recurrence for $H_{j,n+1}(t)$ into (*) for $N_{k,n+1}(t)$ to obtain two different expressions for $N_{k,n+1}(t)$.

b. Comparing the coefficients $H_{j,n}(t)$ in part (a), conclude that

$$\frac{t-2k}{2n+2} M_{j,k}^n + \frac{2k+2n+4-t}{2n+2} M_{j-2,k}^n$$

$$= \frac{t-2k-j}{n+1} M_{j,k}^{n+1} + \frac{2k+j+n+1-t}{n+1} M_{j-2,k}^{n+1} .$$

c. Comparing the coefficients of t in part (b), conclude that

$$\frac{M_{j,k}^n - M_{j-2,k}^n}{2} = M_{j,k}^{n+1} + M_{j-2,k}^{n+1} .$$

Notice that this recurrence is independent of k.

d. Show by induction on j that

$$M_{0,k}^{n+1} = \frac{M_{0,k}^n}{2}$$

$$M_{j-1,k}^{n+1} = \frac{M_{j-2,k}^n + M_{j-1,k}^n}{2}$$

satisfies the recurrence in part (c).

e. Use the result in part (d) to derive the Lane-Riesenfeld algorithm.

4. This exercise extends the Lane-Riesenfeld knot insertion algorithm to B-splines with knots in geometric progression (see Section 7.8.2, Exercise 1). Again the proof is based solely on the de Boor recurrence, but this time for nonuniform B-splines (Equation (7.8)). Let $\{N_{k,n}(t)\}$ denote the B-splines with knots at $t_{2k} = \beta^{2k}$, and let $\{H_{k,n}(t)\}$ denote the B-splines with knots at $t_k = \beta^k$.

a. Show that there are constants $\{M_{jk}^n\}$ such that

$$N_{k,n}(t) = \sum_j M_{jk}^n H_{j,n}(t). \qquad (*)$$

b. Using the strategy developed in Exercise 3, show that the constants $\{M_{jk}^n\}$ satisfy the recurrence

$$\frac{-1}{1+\beta^{n+1}} M_{j,k}^n + \frac{\beta^{2n+2}}{1+\beta^{n+1}} M_{j-2,k}^n = -M_{j,k}^{n+1} + \beta^{n+1} M_{j-1,k}^{n+1} .$$

Observe that this recurrence is independent of k.

c. Show by induction on j that

$$M_{j-1,k}^{n+1} = \frac{\beta^{n+1} M_{j-2,k}^n + M_{j-1,k}^n}{1+\beta^{n+1}}$$

satisfies the recurrence in part (b).

d. Let $\{P_k\}$ denote the control points of a degree n B-spline curve relative to the B-splines $\{N_{k,n}(t)\}$, and let $\{Q_j^n\}$ denote the control points of the same curve relative to the B-splines $\{H_{j,n}(t)\}$. Use the result in part (c) to prove that the following algorithm can be used to generate the new points $\{Q_j^n\}$ from the original points $\{P_k\}$. Start by doubling the original control points $\{P_k\}$. Then take successive weighted averages of adjacent points in the following manner:

$$Q_j^0 = P_k \qquad\qquad\qquad j = 2k, 2k+1$$

$$Q_j^m = \frac{\beta^m Q_{j-1}^{m-1} + Q_j^{m-1}}{1 + \beta^m} \qquad\qquad m = 1,\ldots,n.$$

The points $\{Q_j^n\}$ that emerge at the nth level of the algorithm are the control points of the degree n B-spline curve relative to the B-splines $\{H_{j,n}(t)\}$. (See Figure 7.37 for the case $n = 2$.)

5. A knot sequence is said to be in affine progression if there are constants a,b such that for all k, $t_{k+1} = at_k + b$. Generalize Exercise 4 to knots in affine progression.

6. Implement a rendering algorithm for uniform B-spline curves based on the Lane-Riesenfeld knot insertion procedure.

7. Implement an intersection algorithm for uniform B-spline curves based on the Lane-Riesenfeld knot insertion procedure.

8. Explain how the blossoming proof of Chaikin's algorithm breaks down for cubic B-spline curves with uniform knots.

7.9 Rational B-Splines

When we studied Lagrange and Bezier curves and surfaces, we observed that there are some common curve and surface types, such as conic sections and quadric surfaces, that cannot be represented exactly by polynomials. Since splines are piecewise polynomials, the same limitations hold for splines. To overcome this deficiency, we shall construct rational splines—that is, functions that are the ratios of two splines. These rational functions greatly expand the range of curves and surfaces with exact B-spline representations.

The construction of rational B-spline curves mimics the construction of rational Bezier curves. With each control point P_k we associate a scalar weight w_k. We then define the *rational B-spline curve* $R(t)$ to be the projection from Grassmann space of the B-spline curve

$$S(t) = \sum_k N_{k,n}(t)(w_k P_k, w_k) .$$

The B-spline curve $S(t)$ in Grassmann space projects to the rational B-spline curve

$$R(t) = \frac{\sum_k N_{k,n}(t) w_k P_k}{\sum_k w_k N_{k,n}(t)} \qquad (7.29)$$

in affine space. For Equation (7.29) to make sense we shall always assume that at least one weight is nonzero. The curves in Equation (7.29) are often called *NURBS*, which is an abbreviation for *nonuniform rational B-spline*—a rational B-spline with nonuniform knots.

Just as in the construction of rational Bezier curves, there is associated with each point $R(t)$ on a rational B-spline curve a scalar weight

$$w(t) = \sum_k w_k N_{k,n}(t).$$

Thus a rational B-spline curve is more than just a continuous collection of points in affine space; there is also a scalar field, a mass distribution, associated with each point on a rational B-spline curve.

If $w(t_0) = 0$, then the projection from Grassmann space to affine space is not continuous at $t = t_0$. We can avoid these discontinuities in the usual fashion by projecting the curve into projective space rather than into affine space. Thus, as with rational Bezier curves, the control structures—the mass-points and vectors—always reside in Grassmann space, but the curves themselves may lie in projective space.

If a weight $w_j = 0$, then, as in the Bezier setting, the mass-point $(w_j P_j, w_j)$ is not just discarded but rather is replaced by a vector $(v_j, 0)$. Thus, in general,

$$P(t) = \sum_{w_k \neq 0} N_{k,n}(t)(w_k P_k, w_k) + \sum_{w_j = 0} N_{j,n}(t)(v_j, 0),$$

so first adding in Grassmann space and then projecting into affine space, we arrive at the general rational B-spline curve

$$R(t) = \frac{\displaystyle\sum_{w_k \neq 0} N_{k,n}(t) w_k P_k + \sum_{w_j = 0} N_{j,n}(t) v_j}{\displaystyle\sum_{w_k \neq 0} w_k N_{k,n}(t)}.$$

If all the weights are nonzero, then it is natural, as in the rational Lagrange and rational Bezier settings, to write

$$R_{k,n}(t) = \frac{w_k N_{k,n}(t)}{\sum_j w_j N_{j,n}(t)} \qquad (7.30)$$

$$R(t) = \sum_k R_{k,n}(t) P_k.$$

Thus, for a fixed set of nonzero weights, the functions $\{R_{k,n}(t)\}$ are piecewise rational blending functions, and these functions behave much like the standard B-spline blending functions. Indeed, since the denominator of $R_{k,n}(t)$ is the same for all values of k, it is easy to show that the piecewise rational functions $\{R_{k,n}(t)\}$ incorporate many of the features of the B-splines $\{N_{k,n}(t)\}$. For this reason, rational B-spline curves with nonzero weights share many of the geometric properties of integral B-spline curves. For example, rational B-spline curves are piecewise rational curves with continuity of order $n - \mu$ at knots of multiplicity μ. Moreover, rational B-spline curves are affine invariant and lie in the local convex hull of their control points provided that the weights are nonnegative (see Exercises 1–5).

Algorithms for B-spline curves also extend to algorithms for rational B-spline curves because generally we can apply these algorithms separately to the numerator and denominator. For example, we can evaluate points along a rational B-spline curve by applying the de Boor algorithm independently to the numerator and denominator. Similarly, knot insertion algorithms can be applied separately to the numerator and denominator. Therefore, Boehm's algorithm, the Oslo algorithm, the factored knot insertion algorithm, Chaikin's algorithm, and the Lane-Riesenfeld algorithm for uniform B-splines can be applied independently to the numerator and denominator—simply apply the algorithm in question to the control points $\{(w_k P_k, w_k)\}$ in Grassmann space and then divide by the weight to get the desired result in affine space. The only exception to this rule is the algorithm for differentiation because the derivative of a rational function is not equal to the derivative of the numerator divided by the derivative of the denominator. Differentiation of rational B-spline curves is handled in a manner similar to differentiation for rational Bezier curves (see Section 5.7.1).

In Section 7.6.3.1 we observed that one of the simplest ways to analyze an ordinary B-spline curve is to apply knot insertion to convert the curve to piecewise Bezier form and then perform the analysis on the Bezier segments. Similarly, since knot insertion works as well in the rational setting, rational B-spline curves can be converted to piecewise rational Bezier form, and we can then apply the analysis algorithms we have already developed for rational Bezier curves. Just as in the integral case, this method is that standard approach to rendering and intersecting rational B-spline curves (see Exercise 9).

The weights of a rational B-spline curve can be used to control shape, and the results are again similar to the Bezier setting. As the weight w_k increases, the influence of the control point P_k increases and the curve passes closer to P_k; as w_k decreases, the curve is pushed away from P_k. Typically all the weights are chosen to be positive to avoid singularities, but as with rational Bezier curves, zero and negative weights are permitted and sometimes are even necessary to represent specific curves.

Exercises

1. Show that if $t_{k+1} = \cdots = t_{k+n}$, then the rational B-spline curve of degree n interpolates the control point P_k.

2. a. Show that the rational blending functions defined in Equation (7.30) satisfy the identity $\sum_k R_{k,n}(t) \equiv 1$.

   b. Conclude from part (a) that if all the weights are positive, then a rational B-spline curve lies in the local convex hull of its control points.

3. Using Equation (7.30), show that if all the weights are nonzero, then a rational B-spline is nondegenerate provided that there are no indices j,k for which $(w_k P_k, w_k) = c_{jk}(w_j P_j, w_j)$.

4. Prove that for any choice of knots $\{t_k\}$ it is always possible to choose weights $\{w_k\}$ so that the rational blending functions $\{R_{k,n}(t)\}$ defined in Equation (7.30) are unimodal in k.

5. Prove that rational B-spline curves have continuity of order $n - \mu$ at knots of multiplicity μ.

6. Consider the circle:

$$x = \frac{2t}{1+t^2} \qquad y = \frac{1-t^2}{1+t^2} .$$

Find control points and weights to represent the following segments of the circle as rational B-spline curves with knots at the integers:

a. the quarter circle that lies in the first quadrant

b. the upper half circle

7. Find control points and weights to represent the following curves as a rational B-splines with knots at the integers:

a. The ellipse: $\qquad x = \frac{2at}{1+t^2} \qquad y = \frac{b(1-t^2)}{1+t^2}$

b. The hyperbola: $\qquad x = \frac{2at}{1-t^2} \qquad y = \frac{b(1+t^2)}{1-t^2}$

c. Which segments of these curves are represented by your choice of control points and weights?

8. Implement the following algorithms for rational B-spline curves:

a. De Boor evaluation algorithm

b. Boehm's knot insertion algorithm

c. Factored knot insertion algorithm

d. Lane-Riesenfeld knot insertion algorithm for uniform B-splines

9. a. Apply Boehm's knot insertion algorithm to convert rational B-spline curves to piecewise rational Bezier form.

b. Use the algorithm developed in part (a) to intersect two rational B-spline curves.

10. Consider a rational B-spline curve of degree n with control points $\{P_k\}$ and weights $\{w_k\}$.

a. What does the limit curve look like if

i. one of the weights goes to infinity while the other weights are left fixed?

ii. two or more weights are allowed to approach infinity simultaneously?

b. What happens in part (a) if one of the knots has multiplicity n?

11. Let $R(t)$ be a rational B-spline curve of degree n with control points $P_0,...,P_m$ and weights $w_0,...,w_m$. Define

$$P(t) = \sum_{k=0}^{m} N_{k,n}(t) w_k P_k \text{ and } w(t) = \sum_{k=0}^{m} w_k N_{k,n}(t),$$

so that

$$R(t) = \frac{P(t)}{w(t)}.$$

Prove that for a uniform knot sequence with knots at the integers

a. $\int_0^{n+m+1} w(t)dt = average \ of \ the \ weights$

b. $\dfrac{\int_0^{m+n+1} P(t)dt}{\int_0^{m+n+1} w(t)dt} = center \ of \ mass \ of \ \{(w_0 P_0, w_0),...,(w_n P_n, w_n)\}.$

(Compare to Section 5.7, Exercise 20.)

7.10 Catmull-Rom Splines

Suppose we have a large number of data points. Lagrange polynomials generate smooth interpolations of high degree; B-splines generate smooth approximations of low degree. Our goal here is to provide a simple construction for smooth interpolants of low degree. Catmull-Rom splines combine Lagrange interpolation with B-spline approximation to generate splines of low degree that interpolate the data points. Here is how it is done.

We want to construct a low-degree spline curve $C(t)$ to interpolate a set of data points $\{P_k\}$ at an arbitrary collection of parameter values $\{t_k\}$—that is, we want to build a smooth, low-degree, piecewise polynomial curve $C(t)$ such that $C(t_k) = P_k$. For any arbitrary value of n, let $P_{k...k+n}(t)$ denote the unique polynomial of degree n that interpolates the points $P_k,...,P_{k+n}$ at the nodes $t_k,...,t_{k+n}$, and let $N_{k,n-1}(t)$ be the B-spline basis function of degree $n-1$ for the knots $\{t_j\}$ with support $[t_k, t_{k+n}]$. The *Catmull-Rom spline* $C_n(t)$ is defined by setting

$$C_n(t) = \sum_k N_{k,n-1}(t) P_{k...k+n}(t). \tag{7.31}$$

Since the interpolants $P_{k...k+n}(t)$ are polynomials of degree n and the B-splines $N_{k,n-1}(t)$ are C^{n-2} piecewise polynomials of degree $n-1$, the Catmull-Rom spline $C_n(t)$ is a C^{n-2} piecewise polynomial of degree $2n-1$. We are going to show that the curve $C_n(t)$ interpolates the points $\{P_k\}$ at the knots $\{t_k\}$ and has order of continuity $n-1$.

PROPOSITION 7.17	Given a collection of points $\{P_j\}$ and nodes $\{t_j\}$, let

$$C_n(t) = \Sigma_k N_{k,n-1}(t)P_{k\cdots k+n}(t).$$

Then for all j, $C_n(t_j) = P_j$. That is, the Catmull-Rom spline interpolates the data points $\{P_j\}$ at the knots $\{t_j\}$.

Proof At the knot t_j the only nonzero B-splines are $N_{j-n+1,n-1}(t),...,N_{j,n-1}(t)$. Therefore,

$$C_n(t_j) = \Sigma_k N_{k,n-1}(t_j)P_{k\cdots k+n}(t_j)$$

$$= \sum_{i=0}^{n-1} N_{j-n+1+i,n-1}(t_j)P_{j-n+1+i\cdots j+1+i}(t_j)$$

$$= \sum_{i=0}^{n-1} N_{j-n+1+i,n-1}(t_j)P_j \ .$$

But the B-splines form a partition of unity, so

$$\sum_{i=0}^{n-1} N_{j-n+1+i,n-1}(t_j) = 1.$$

Therefore, for all j, $C_n(t_j) = P_j$.

Interpolation is easy to prove; continuity of order C^{n-1} is a bit harder to establish. We will derive this property from the following rather remarkable result.

LEMMA 7.18	$\Sigma_k N_{k,n-1}(t)P_{k\cdots k+n}(t) = \Sigma_k N_{k,n}(t)P_{k+1\cdots k+n}(t).$

Proof We shall apply both Neville's algorithm (Equation (2.7))

$$P_{k\cdots k+n}(t) = \frac{t_{k+n}-t}{t_{k+n}-t_k}P_{k\cdots k+n-1}(t) + \frac{t-t_k}{t_{k+n}-t_k}P_{k+1\cdots k+n}(t)$$

and the de Boor recurrence (Equation (7.8))

$$N_{k,n}(t) = \frac{t-t_k}{t_{k+n}-t_k}N_{k,n-1}(t) + \frac{t_{k+n+1}-t}{t_{k+n+1}-t_{k+1}}N_{k+1,n-1}(t).$$

From these two identities it follows that

$$\Sigma_k N_{k,n-1}(t)P_{k\cdots k+n}(t)$$

$$= \Sigma_k N_{k,n-1}(t)\left\{\frac{t_{k+n}-t}{t_{k+n}-t_k}P_{k\cdots k+n-1}(t) + \frac{t-t_k}{t_{k+n}-t_k}P_{k+1\cdots k+n}(t)\right\}$$

$$= \Sigma_k \left\{\frac{t_{k+n+1}-t}{t_{k+n+1}-t_{k+1}}N_{k+1,n-1}(t) + \frac{t-t_k}{t_{k+n}-t_k}N_{k,n-1}(t)\right\}P_{k+1\cdots k+n}(t)$$

$$= \Sigma_k N_{k,n}(t)P_{k+1\cdots k+n}(t).$$

**COROLLARY
7.19**

The Catmull-Rom spline $C_n(t) = \sum_k N_{k,n-1}(t)P_{k\cdots k+n}(t)$ has continuity of order $n-1$ at the knots.

Proof By Lemma 7.18: $C_n(t) = \sum_k N_{k,n}(t)P_{k+1\cdots k+n}(t)$. Since the interpolants $P_{k+1\cdots k+n}(t)$ are polynomials of degree $n-1$ and the B-splines $N_{k,n}(t)$ are C^{n-1} piecewise polynomials of degree n, the Catmull-Rom spline $C_n(t)$ is a C^{n-1} piecewise polynomial of degree $2n-1$.

Neville's algorithm for Lagrange interpolation can be combined with the de Boor algorithm for B-spline approximation to generate a recursive evaluation algorithm for Catmull-Rom splines. Place the control points at the base of a triangle and run Neville's algorithm for n levels. Then continue for the next $n-1$ levels with the de Boor algorithm. The values of points on the Catmull-Rom spline will emerge at the apexes because, by construction, the functions $P_{k\cdots k+n}(t)$ occupy the location of the B-spline control points for the de Boor algorithm of degree $n-1$ (see Figure 7.39). Notice that we can also view this algorithm as $n-1$ levels from Neville's algorithm followed by n levels from the de Boor algorithm. This observation accounts for Lemma 7.18.

Catmull-Rom splines inherit many of the characteristic features of Lagrange polynomials and B-spline curves. Here are the most prominent of these properties:

1. Piecewise polynomial of degree $2n-1$
2. Interpolates the point P_k at the knot t_k
3. Continuity at the knots of order C^{n-1}
4. Local control
5. Affine invariance

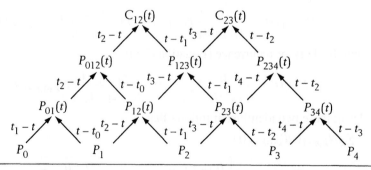

Figure 7.39 A recursive evaluation algorithm for cubic Catmull-Rom splines manufactured from Neville's algorithm for Lagrange interpolation and the de Boor algorithm for B-spline approximation. The two lower levels are taken from Neville's algorithm, and the upper level is taken from the de Boor algorithm. Notice that we can also view this algorithm as one level of Neville's algorithm followed by two levels of the de Boor algorithm. Here $C_{i,i+1}(t)$ denotes the polynomial that represents the cubic Catmull-Rom spline $C_3(t)$ in the interval $[t_i, t_{i+1}]$; therefore, $C_{i,i+1}(t)$ interpolates the points P_i and P_{i+1} at the knots t_i and t_{i+1}.

6. Nondegenerate

7. Exactly reproduces polynomials of degree n

The first property is immediate from the definition of Catmull-Rom splines in Equation (7.31), and we have already established the second and third properties in Proposition 7.17 and Corollary 7.19. Local control is a simple consequence of Equation (7.31) together with the compact support of the B-splines. Indeed, by construction, the point P_k will affect the Catmull-Rom spline only over the interval $[t_{k-n}, t_{k+n}]$. Affine invariance follows directly from the affine invariance of Lagrange interpolation and B-spline approximation, and Catmull-Rom splines are clearly nondegenerate since they interpolate the data points. Finally, Catmull-Rom splines reproduce polynomials of degree n because if the data points $\{P_k\}$ lie on a degree n polynomial $P(t)$ at the nodes $\{t_k\}$, then by the uniqueness of the polynomial interpolant $P_{k\ldots k+n}(t) = P(t)$ for all k. Therefore, since the B-splines form a partition of unity

$$C_n(t) = \sum_k N_{k,n-1}(t)P_{k\ldots k+n}(t) = \sum_k N_{k,n-1}(t)P(t) = P(t) \, .$$

Thus Catmull-Rom splines have many desirable features. Notice, however, that unlike B-spline curves, Catmull-Rom splines do not lie in the convex hull of their control points nor are they variation diminishing because Lagrange polynomials fail to have these properties.

Moreover, there is no knot insertion algorithm for Catmull-Rom splines. Suppose we were to try to insert a new knot at $t = u$. Since Catmull-Rom splines are interpolating splines, we would have to add a new control point at $C_n(u)$. But it is easy to construct examples where the Catmull-Rom spline with the new knot u and the new control point $C_n(u)$ is not the same curve as the original Catmull-Rom spline, even though both the old spline and the new spline interpolate the same data points at all the knots including the new knot u (see Figure 7.40).

Cubic Catmull-Rom splines and piecewise cubic Hermite interpolants both generate C^1 piecewise cubic curves that interpolate the data points. These curves, however, are generally not the same (see Exercise 6). The Catmull-Rom construction does not allow us to choose the tangents at the control points; rather it chooses these tangents for us to ensure that the curve has one continuous derivative. By the uniqueness of Hermite interpolation, the Hermite interpolant will reproduce the Catmull-Rom curve if and only if we choose the Hermite data off the Catmull-Rom spline.

Exercises

1. Show how to construct Catmull-Rom splines to interpolate Hermite data.

2. Explain how to extend the Catmull-Rom construction to rational spline curves.

3. Given a collection of data points $\{P_k\}$ and nodes $\{t_k\}$, define a Catmull-Rom spline $C_{p,q}(t)$ of type (p,q) by setting $C_{p,q}(t) = \sum_k N_{k,p}(t)P_{k\ldots k+q}(t)$. Prove that if $p < q$, then $C_{p,q}(t_j) = P_j$ for all j.

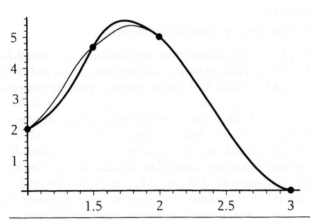

Figure 7.40 Knot insertion fails for Catmull-Rom splines. The first curve is a cubic Catmull-Rom spline with two polynomial segments. The control points are located at (0,0),(1,2),(2,5),(3,0),(4,5) and knots are situated at the parameter values $t = 1,3,4$. Thus the spline interpolates the points (1,2),(2,5),(3,0) at the knots $t = 1,3,4$. At $t = 2$ this curve passes through the point (3/2,14/3). The second curve has an additional control point at (3/2,14/3) and an additional knot at $t = 2$. Observe that for $0 < t < 2$ the two Catmull-Rom splines are not identical, even though they interpolate the same data at the identical knots.

4. Show that the blending functions for Catmull-Rom splines are given by the splines

$$C_{k,2n-1}(t) = \sum_{j=0}^{n-1} N_{k-j,n-1}(t \mid t_{k-j},...,t_{k-j+n})L_j^n(t \mid t_{k-j},...,t_{k-j+n}) \cdot$$

That is, show that the Catmull-Rom spline $C_n(t)$ for the points $\{P_k\}$ and nodes $\{t_k\}$ is given by

$$C_n(t) = \sum_k C_{k,2n-1}(t)P_k \cdot$$

5. Let $C_n(t)$ be a Catmull-Rom spline with control points $\{P_k\}$ and knots $\{t_k\}$. Form a new Catmull-Rom spline $D_n(t)$ by replacing each knot t_k by the knot $\tau_k = at_k + b$ for some fixed constants $a > 0$ and b. Show that changing all the knots in this way has no effect on the shape of the Catmull-Rom spline. In particular, show that $D_n(at + b) = C_n(t)$. What happens if we choose $a < 0$? (Compare to Section 2.2, Exercise 4, and Section 7.4, Exercise 2.)

6. Cubic Catmull-Rom splines and piecewise cubic Hermite interpolants both generate C^1 piecewise cubic curves that interpolate the data points. Draw some curves to illustrate how cubic Catmull-Rom splines generally differ from piecewise cubic Hermite interpolants.

7.11 **Tensor Product B-Spline Surfaces**

A rectangular tensor product B-spline patch $S(s,t)$ of bidegree (m,n) is defined by setting

$$S(s,t) = \sum_i \sum_j N_{i,m}(s) N_{j,n}(t) P_{ij} .$$

The functions $N_{i,m}(s) N_{j,n}(t)$ are the tensor product B-spline basis functions, and the rectangular array of control points $\{P_{ij}\}$ forms the vertices of the control polyhedron for the tensor product B-spline patch. This construction is much the same as the Lagrange and Bezier tensor product constructions. Notice, however, that a knot s_i or t_j of the basis functions becomes a knot line $s = s_i$ or $t = t_j$ on the tensor product patch. Tensor product patches join smoothly along curves defined by knot lines.

As usual in tensor product constructions, we can let $P_i(t)$ be the B-spline curve with control points $P_{i0},...,P_{i\alpha}$. Then

$$P_i(t) = \sum_j N_{j,n}(t) P_{ij}$$

$$S(s,t) = \sum_i N_{i,m}(s) P_i(t).$$

Thus we can evaluate a tensor product B-spline surface by first using the de Boor algorithm in t to evaluate the curves $\{P_i(t)\}$ and then applying the de Boor algorithm in s with the control points $\{P_i(t)\}$. Again this approach is much the same technique we used in Sections 2.11 and 5.8.1 to evaluate Lagrange and Bezier tensor product surfaces.

Tensor product B-spline patches inherit many of the characteristic properties of B-spline curves. They are piecewise polynomials with continuity of order $C^{n-\mu}$ at knot lines of multiplicity μ. In addition, they are affine invariant, nondegenerate, and lie in the local convex hull of their control points. Typically these surfaces do not interpolate any specific curves, but if the knot vectors in s and t have knots of multiplicity n at the start and end, then the boundaries of a tensor product B-spline patch are the B-spline curves determined by their boundary control points.

Tensor product B-spline patches also inherit the standard knot insertion algorithms of B-spline curves. Thus Boehm's algorithm, the Oslo algorithm, factored knot insertion, and the Lane-Riesenfeld algorithm for uniform knots all extend to tensor product patches in a straightforward manner. We simply treat the s and t directions independently. Thus to insert knot lines in the t direction, we just use the standard curve algorithms to insert knots into the curves $\{P_i(t)\}$. Symmetric algorithms can be applied to insert knot lines in the s direction.

Similarly, to differentiate a tensor product B-spline patch, we can differentiate the de Boor algorithm in the usual manner. To differentiate the de Boor algorithm with respect to s a total of p times, simply differentiate any p of the upper m levels (the s levels) of the de Boor algorithm and multiply the result by $m!/(m-p)!$. To differentiate q times with respect to t, differentiate any q of the n levels (the t levels) in each of the lower triangles and multiply the results by $n!/(n-q)!$. That this algorithm works is an immediate consequence of the corresponding differentiation algorithm for B-spline curves.

Rational tensor product B-spline patches can be introduced by associating a scalar weight w_{ij} with each control point P_{ij}. Thus we define a rational tensor product Bezier patch by setting

$$R(s,t) = \frac{\sum_i \sum_j N_{i,m}(s)N_{j,n}(t)w_{ij}P_{ij}}{\sum_i \sum_j w_{ij}N_{i,m}(s)N_{j,n}(t)}.$$

This construction allows us to represent surfaces such as the sphere, which have no exact polynomial representation, as tensor product B-spline patches. Again, most of the properties and algorithms of rational B-spline curves are inherited by rational tensor product B-spline surfaces.

Exercises

1. a. Prove that $\sum_i \sum_j N_{i,m}(s)N_{j,n}(t) \equiv 1$.

    b. Show that every tensor product B-spline patch lies in the local convex hull of its control points.

2. Consider a tensor product B-spline patch of bidegree (m,n), where $m < n$. Show that

    a. to compute a single point on the surface it is faster to apply the de Boor algorithm first in the s direction and then in the t direction.

    b. to compute many points along the surface it may be faster to apply the de Boor algorithm first in the t direction and then in the s direction.

    c. Explain this apparent anomaly. (Compare to Section 5.8.1, Exercise 2.)

3. Implement the de Boor evaluation algorithm for tensor product B-spline surfaces, and then use the de Boor algorithm to render points on a tensor product B-spline surface.

4. Implement the following knot insertion algorithms for tensor product B-spline surfaces:

    a. Boehm's algorithm

    b. Oslo algorithm

    c. Factored knot insertion

    d. Lane-Riesenfeld algorithm for uniform B-splines

5. Apply Boehm's knot insertion algorithm to convert a tensor product B-spline surface to piecewise tensor product Bezier form.

6. Apply the algorithm in Exercise 5 to

    a. render tensor product B-spline patches

    b. intersect two tensor product B-spline patches

7. Explain how to extend Catmull-Rom splines to tensor product surfaces. Construct explicit formulas and implement dynamic programming algorithms for tensor product Catmull-Rom splines.

8. What is the effect on a rational B-spline surface if one of the mass-points has zero weight?

9. Experiment with altering the weights in a rational B-spline surface.

    a. What are the local and global effects of altering a single weight?

    b. What is the effect of a negative weight?

    c. What happens if all the weights are changed simultaneously?

10. Consider the rational biquadratic parametrization of the sphere given by

$$x = \frac{2s(1-t^2)}{(1+s^2)(1+t^2)} \qquad y = \frac{2t(1+s^2)}{(1+s^2)(1+t^2)} \qquad z = \frac{(1-s^2)(1-t^2)}{(1+s^2)(1+t^2)}.$$

    a. Find control points and weights to represent the following segments of the sphere as rational B-spline surfaces with knots at the integers:

      i.   The portion of the sphere that lies in the first octant

      ii. The upper half of the sphere

    b. Use the results in part (a) together with the de Boor algorithm to render the sphere.

    c. Use the results in part (a) together with Chaikin's algorithm to render the sphere.

11. Recall from Section 2.14, Exercise 5, that the torus with inner radius $d - a$ and outer radius $d + a$ has the biquadratic parametrization

$$x = \frac{d(1+s^2)(1-t^2) + a(1-s^2)(1-t^2)}{(1+s^2)(1+t^2)}$$

$$y = \frac{2d(1+s^2)t + 2a(1-s^2)t}{(1+s^2)(1+t^2)}$$

$$z = \frac{2as(1+t^2)}{(1+s^2)(1+t^2)}.$$

    a. Find control points and weights to represent a portion of the torus as a rational B-spline surface with knots at the integers.

    b. Use the results in part (a) together with the de Boor algorithm to render the torus.

    c. Use the results in part (a) together with the Lane-Riesenfeld algorithm to render the torus.

12. Consider a rational B-spline surface of bidegree (m,n) with control points $\{P_{jk}\}$ and weights $\{w_{jk}\}$. What does the limit surface look like if

    a. one of the weights goes to infinity while the other weights are left fixed?

    b. two or more weights are allowed to approach infinity simultaneously?

7.12   **Pyramid Algorithms and Triangular B-Patches**

There is also a pyramid algorithm for tensor product B-spline surfaces, just like for tensor product Lagrange and tensor product Bezier patches. Let

$$P(s,t) = \sum_i \sum_j N_{i,n}(s) N_{j,n}(t) P_{ij}$$

be a tensor product B-spline surface of bidegree (n,n), and consider its blossom $p(u_1,\ldots,u_n; v_1,\ldots,v_n)$. By the multiaffine property of the blossom

$$p(u_1,\ldots,u_{n-1},s; v_1,\ldots,v_{n-1},t)$$

$$= p(u_1,\ldots,u_{n-1}, \frac{\mu_2-s}{\mu_2-\mu_1}\mu_1 + \frac{s-\mu_1}{\mu_2-\mu_1}\mu_2; v_1,\ldots,v_{n-1}, \frac{v_2-t}{v_2-v_1}v_1 + \frac{t-v_1}{v_2-v_1}v_2)$$

$$= \frac{\mu_2-s}{\mu_2-\mu_1}\frac{v_2-t}{v_2-v_1} p(u_1,\ldots,u_{n-1},\mu_1; v_1,\ldots,v_{n-1},v_1)$$

$$+ \frac{\mu_2-s}{\mu_2-\mu_1}\frac{t-v_1}{v_2-v_1} p(u_1,\ldots,u_{n-1},\mu_1; v_1,\ldots,v_{n-1},v_2)$$

$$+ \frac{s-\mu_1}{\mu_2-\mu_1}\frac{v_2-t}{v_2-v_1} p(u_1,\ldots,u_{n-1},\mu_2; v_1,\ldots,v_{n-1},v_1)$$

$$+ \frac{s-\mu_1}{\mu_2-\mu_1}\frac{t-v_1}{v_2-v_1} p(u_1,\ldots,u_{n-1},\mu_2; v_1,\ldots,v_{n-1},v_2) \ . \tag{7.32}$$

Substituting the knots of the B-splines $N_{i,n}(s)$ and $N_{j,n}(t)$ for the blossom parameters u,v,μ,v, leads to a bilinear recurrence for computing the points on the surface $P(s,t)$ from the control points P_{ij}. This recurrence can be diagrammed on a square pyramid, with n^2 nodes on the nth level of the diagram. We illustrate this recurrence in Figure 7.41 for a biquadratic B-spline patch and in Figure 7.42 for a bicubic B-spline patch. Notice that like the pyramid algorithm for tensor product Lagrange interpolation, but unlike the pyramid algorithm for tensor product Bezier approximation, the labels along the edges change from node to node and level to level.

To find p derivatives with respect to s and q derivatives with respect to t, differentiate any p levels of the pyramid algorithm with respect to s and then differentiate any q levels (the same or different from the previous p levels) with respect to t, and multiply the result by $(n!)^2/(n-p)!(n-q)!$. This algorithm works because we know from blossoming that

$$\frac{\partial P}{\partial s} = n p(\underbrace{s,\ldots,s}_{n-1},\delta_1; \underbrace{t,\ldots,t}_{n}) \ ,$$

$$\frac{\partial P}{\partial t} = n p(\underbrace{s,\ldots,s}_{n}; \underbrace{t,\ldots,t}_{n-1},\delta_2) \ .$$

But substituting the parameters $\delta_1 = (1,0)$ or $\delta_2 = (0,1)$ into the blossom is equivalent to differentiating one level of the pyramid with respect to s or t.

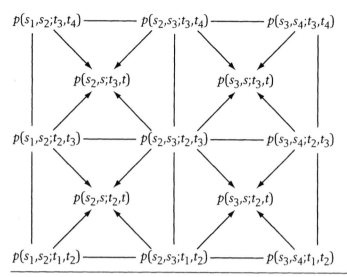

Figure 7.41 The base and the first level of the local pyramid algorithm for a biquadratic tensor product B-spline patch $P(s,t)$ viewed from above. The nine control points represented by the blossom of $P(s,t)$ evaluated at the knots are at the base of the diagram. From these nine control points the four blossom values $p(s_2,s;t_2,t), p(s_3,s;t_2,t), p(s_2,s;t_3,t), p(s_3,s;t_3,t)$ are computed at the first stage of the algorithm. At the next stage (not shown) the blossom value $p(s,s;t,t) = P(s,t)$ is computed from the four blossom values $p(s_2,s;t_2,t), p(s_3,s;t_2,t), p(s_2,s;t_3,t), p(s_3,s;t_3,t)$ using the bilinear recurrence in Equation (7.32).

The local recurrences for tensor product B-spline patches pictured in Figures 7.41 and 7.42 can be pasted together, much like the local triangular de Boor recurrences for B-spline curve segments. Pyramids for adjacent patches share common nodes and common labels along their edges. Therefore, these pyramids can be glued together along the horizontal or vertical axes to form a sequence of overlapping pyramids analogous to the sequence of overlapping triangles that express the de Boor recurrence for B-spline curves. These overlapping pyramids are illustrated schematically in Figure 7.43. Thus the diagrams we adopted to represent B-spline curves extends in a natural way to tensor product B-spline surfaces. For example, surface patches represented by adjacent overlapping pyramids fit together smoothly along their common knot lines.

Both the de Boor algorithm and the pyramid algorithm for tensor product B-spline surfaces are $O(n^3)$. Nevertheless, the de Boor algorithm is generally faster than the pyramid algorithm. The comparative analysis of the relative speeds of these two algorithms is much the same as the comparison between the de Casteljau algorithm and the pyramid algorithm for tensor product Bezier patches presented in Section 5.8.1, so we shall not repeat this analysis here.

Since it is generally slower than the de Boor algorithm, why then have we presented the pyramid algorithm for tensor product B-spline surfaces? Both Lagrange and Bezier surfaces come in two types: rectangular and triangular. So far we have

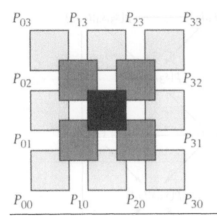

P_{03} P_{13} P_{23} P_{33}

P_{02} P_{32}

P_{01} P_{31}

P_{00} P_{10} P_{20} P_{30}

Figure 7.42 A schematic diagram of the local bilinear evaluation algorithm for a bicubic tensor product B-spline patch viewed from above. Each panel represents the computation of a point at its center by multiplying the points at its corners with the coefficients in Equation (7.32) and adding the results. The black panel represents the bicubic B-spline patch corresponding to the control points at the base of the pyramid; interior control points are obscured by the panels. Notice that the light gray panels represent bilinear B-spline patches and the dark gray panels represent biquadratic B-spline patches. Compare to Figure 5.42, which is the pyramid algorithm for bicubic Bezier approximation. The same pyramid is used there, but the algorithm here is more complicated, since the labels along the edges vary from node to node and from level to level.

constructed only rectangular tensor product B-spline surfaces. Just as Neville's algorithm for tensor product Lagrange interpolation does not extend to triangular Lagrange interpolation, the de Boor algorithm does not extend to triangular B-spline patches. But now that we have a pyramid algorithm for tensor product B-spline surfaces, perhaps, just like in the Lagrange and Bezier settings, we can extend this pyramid approach to an algorithm for generating triangular B-spline surfaces. Let's try and see what happens.

Consider a polynomial surface $P(s,t)$ of total degree n. This surface has a bivariate blossom $p((u_1,v_1),...,(u_n,v_n))$. Using the multiaffine property, we can write a linear recurrence for this blossom by observing that

$$
\begin{aligned}
& p\big((u_1,v_1),\ldots,(u_{n-1},v_{n-1}),(s,t)\big) \\
& \quad = p\big((u_1,v_1),\ldots,(u_{n-1},v_{n-1}),\beta_1(s,t)(\mu_1,v_1)+\beta_2(s,t)(\mu_2,v_2)+\beta_3(s,t)(\mu_3,v_3)\big) \\
& \quad = \beta_1(s,t)p\big((u_1,v_1),\ldots,(u_{n-1},v_{n-1}),(\mu_1,v_1)\big) \\
& \qquad + \beta_2(s,t)p\big((u_1,v_1),\ldots,(u_{n-1},v_{n-1}),(\mu_2,v_2)\big) \\
& \qquad + \beta_3(s,t)p\big((u_1,v_1),\ldots,(u_{n-1},v_{n-1}),(\mu_3,v_3)\big)\,,
\end{aligned}
$$

(7.33)

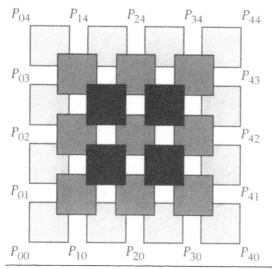

P_{04} P_{14} P_{24} P_{34} P_{44}

P_{03} P_{43}

P_{02} P_{42}

P_{01} P_{41}

P_{00} P_{10} P_{20} P_{30} P_{40}

Figure 7.43 A schematic diagram of the bilinear evaluation algorithm for four adjacent patches of a bicubic tensor product B-spline surface viewed from above. Each panel represents the computation of a point at its center by multiplying the points at its corners with the appropriate bilinear functions and adding the results. The black panels represent the patches of the bicubic B-spline surface corresponding to the control points at their base; interior control points are obscured by the panels. Compare to Figure 7.10, the de Boor algorithm for cubic B-spline curves.

where $\beta_1,(s,t),\beta_2,(s,t),\beta_3,(s,t)$ are the barycentric coordinates of the point (s,t) with respect to the triangle with vertices $(\mu_1,\nu_1),(\mu_2,\nu_2),(\mu_3,\nu_3)$. Now, as in the local tensor product case, we can iterate Equation (7.33) to generate a tetrahedral evaluation algorithm for the polynomial patch $P(s,t)$ that resembles the tetrahedral pyramid algorithms for triangular Lagrange and triangular Bezier patches.

This recurrence can be diagrammed on a tetrahedron with $n(n+1)/2$ nodes on the nth level of the diagram. We illustrate this recurrence in Figure 7.44 for a quadratic surface patch and in Figure 7.45 for a cubic surface patch. Notice that like the pyramid algorithm for triangular Lagrange interpolation, but unlike the pyramid algorithm for triangular Bezier approximation, the labels along the edges vary from node to node and level to level because the barycentric coordinates at different nodes are computed with respect to different triangles. Moreover, unlike the de Boor algorithm for B-spline curves, if you follow along in the direction of any arrow, the labels you encounter along the edges (in the numerator) change from level to level. What is invariant instead is that if an arrow is labeled with the barycentric coordinates of a (t,u) pair, then the next arrow in the same direction is labeled with the barycentric coordinates of another (t,u) pair, albeit with different subscripts.

Now what have we got? Certainly we can form one polynomial patch in this manner. A patch generated by this tetrahedral algorithm is called a *B-patch*. Any degree n polynomial surface can be generated as a B-patch from three sets of

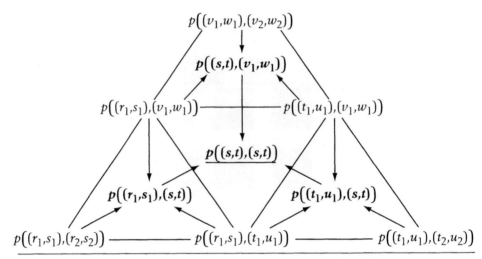

Figure 7.44 The tetrahedral algorithm for a quadratic B-patch $P(s,t)$ viewed from above. The six control points (light) represented by the blossom of $P(s,t)$ evaluated at different parameter values are at the base of the diagram. From these six control points the three blossom values $p((r_1,s_1),(s,t))$, $p((t_1,u_1),(s,t))$, $p((v_1,w_1),(s,t))$ (bold) are computed at the first stage of the algorithm. At the next stage the blossom value $P(s,t) = p((s,t),(s,t))$ (bold and underlined) is computed from the three blossom values $p((r_1,s_1),(s,t))$, $p((t_1,u_1),(s,t))$, $p((v_1,w_1),(s,t))$ using the linear recurrence in Equation (7.33).

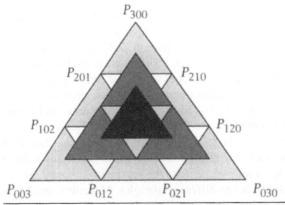

Figure 7.45 Schematic version of the tetrahedral algorithm for a cubic B-patch. Each triangular panel represents the computation of a point at its center calculated by multiplying the points at its vertices by the barycentric coordinates of a different domain triangle and adding the results. The light gray triangles represent linear triangular patches, and the dark gray triangles represent quadratic triangular patches. Notice that some interior control points are obscured by the panels, and down-pointing triangles are ignored. Compare to Figure 5.44 for triangular Bezier approximation.

nodes—$(r_1,s_1),\ldots,(r_n,s_n)$, $(t_1,u_1),\ldots,(t_n,u_n)$, $(v_1,w_1),\ldots,(v_n,w_n)$—provided that the nodes $(r_i,s_i),(t_j,u_j),(v_k,w_k)$ are affinely independent for all $i+j+k \le n+2$. A collection of nodes with this property is called a *knot-net*. Triangular Bezier patches are a special case of B-patches, where the knot-net is given by three affinely independent nodes $(r,s),(t,u),(v,w)$, each repeated n times. Algorithms for blossoming, homogenizing, and differentiating B-patches based on this tetrahedral algorithm are similar to the corresponding algorithms for triangular Bezier patches (see Exercise 4).

But a spline surface is not just a single polynomial patch; a spline surface is a collection of polynomial patches joined together smoothly along common edges. In the tensor product case, pyramids for adjacent patches share common nodes and common labels along their edges. Therefore, these pyramids can be glued together along the horizontal or vertical axes to form a sequence of overlapping pyramids. Moreover, and this observation is the key point, surface patches represented by adjacent overlapping pyramids fit together smoothly along their common knot lines. While we could try to paste together tetrahedra that share common nodes and common labels along their edges, the surface patches represented by such adjacent overlapping tetrahedra would not fit together smoothly along common knot lines. In fact, it is not even clear what a common knot line would be or over what domains these B-patches would fit together. By choosing clouds of knots near the vertices of a single domain triangle, Dahmen, Micchelli, and Seidel succeeded in creating an interesting kind of multivariate B-spline surface using B-patches, but the details of this approach are beyond the scope of this text. The main idea to take away from this discussion is that generating triangular B-spline surfaces is much more difficult than generating tensor product B-spline surfaces. The triangular Bezier construction generalizes readily to triangular B-patches via blossoming, but getting from one B-patch to a collection of triangular patches that join together smoothly is not a simple task.

Exercises

1. What are the up and down recurrences for
   a. the pyramid algorithm for tensor product B-spline surfaces?
   b. the tetrahedral algorithm for B-patches?

2. Complete the analysis of the pyramid algorithm for tensor product B-spline surfaces by showing how to implement this algorithm when the degree in s is different from the degree in t.

3. Implement both the de Boor algorithm and the pyramid algorithm for tensor product B-spline surfaces. Which algorithm do you prefer? Why? Experiment with tensor product surfaces of different degrees. Determine how changing the location of the control points affects the shape of the surface.

4. Use the tetrahedral algorithm for evaluating B-patches to develop an algorithm for
   a. blossoming B-patches
   b. homogenizing B-patches
   c. differentiating B-patches

    d. converting B-patches to Bezier patches

    e. subdividing one B-patch into three B-patches

5. Consider a knot-net

$$(r_1,s_1),\ldots,(r_n,s_n),\ (t_1,u_1),\ldots,(t_n,u_n),\ (v_1,w_1),\ldots,(v_n,w_n).$$

Define the corresponding B-patch basis functions $\{b_{ijk}^n(s,t)\}$ by setting $b_{ijk}^n(s,t)$ = the sum of all paths from the point P_{ijk} at the base of the B-patch recurrence to the apex of the tetrahedron.

    a. Show that if $P(s,t)$ is any bivariate polynomial of degree n, then

$$P(s,t) =$$

$$\sum_{i+j+k=n} b_{ijk}^n(s,t)p\Big((r_1,s_1),\ldots,(r_i,s_i),(t_1,u_1),\ldots,(t_j,u_j),(v_1,w_1),\ldots,(v_k,w_k)\Big).$$

    b. Conclude that the blossom evaluated on the knot-net provides the dual functionals for the B-patch basis functions.

6. Given a homogenized knot-net

$$(r_1,s_1,t_1),\ldots,(r_n,s_n,t_n),\ (u_1,v_1,w_1),\ldots,(u_n,v_n,w_n),\ (x_1,y_1,z_1),\ldots,(x_n,y_n,z_n),$$

define the corresponding homogeneous L-patch basis functions $\{l_{ijk}^n(\rho,\sigma,\tau)\}$ by setting

$$L_{1i}(\rho,\sigma,\tau) = (r_1\rho + s_1\sigma + t_1\tau)\cdots(r_i\rho + s_i\sigma + t_i\tau)$$

$$L_{2j}(\rho,\sigma,\tau) = (u_1\rho + v_1\sigma + w_1\tau)\cdots(u_j\rho + v_j\sigma + w_j\tau)$$

$$L_{3k}(\rho,\sigma,\tau) = (x_1\rho + y_1\sigma + z_1\tau)\cdots(x_k\rho + y_k\sigma + z_k\tau)$$

$$l_{ijk}^n(\rho,\sigma,\tau) = L_{1i}(\rho,\sigma,\tau)L_{2j}(\rho,\sigma,\tau)L_{3k}(\rho,\sigma,\tau) \qquad i+j+k=n.$$

    a. Using the result of Exercise 5(a), show that

$$(sx + ty + uz)^n = \sum_{i+j+k=n} l_{ijk}^n(x,y,z)b_{ijk}^n(s,t,u).$$

    b. Using part (a) and the result of Section 6.5.1, Exercise 17, show that

$$[l_{pqr}^n(s,t,u),b_{ijk}^n(s,t,u)]_n = \delta_{ip}\delta_{jq}\delta_{kr}.$$

7. Two homogeneous polynomial bases $\{B_{ijk}^n(s,t,u)\}$ and $\{D_{ijk}^n(x,y,z)\}$ of degree n are said to be *dual* if they satisfy the identity in Exercise 6(a). That is, $\{B_{ijk}^n(s,t,u)\}$ and $\{D_{ijk}^n(x,y,z)$ are called *dual bases* if

$$(sx + ty + uz)^n = \sum_{i+j+k=n} D_{ijk}^n(x,y,z)B_{ijk}^n(s,t,u).$$

    a. Find the dual basis to each of the following homogeneous B-patch bases:

      i.   Bernstein basis

      ii. Monomial basis

b. Show that the dual to the homogeneous lineal Lagrange basis $L_{ijk}^n(x,y,z)$ with nodes $Q_{ijk} = (a_{ijk}, b_{ijk}, c_{ijk})$ is the B-patch power basis

$$P_{ijk}^n(s,t,u) = (a_{ijk}s + b_{ijk}t + c_{ijk}u)^n.$$

What is the knot-net for the B-patch power basis?

c. Show that if $\{B_{ijk}^n(x,y,z)\}$ and $\{D_{ijk}^n(s,t,u)\}$ are dual bases, then

$$[D_{pqr}^n(s,t,u), D_{ijk}^n(s,t,u)]_n = \delta_{ip}\delta_{jq}\delta_{kr}.$$

8. Consider the blossoming algorithm for B-patches derived in Exercise 4.

a. Show that the homogenized version of this B-patch blossoming recurrence remains valid if blossom values are replaced by homogeneous polynomials in the following fashion:

$$p\big((x_1,y_1,z_1),\ldots,(x_n,y_n,z_n)\big) \to (rx_1 + sy_1 + tz_1)\cdots(rx_n + sy_n + tz_n).$$

b. Recall the bracket operator defined in Section 6.5.1, Exercise 17. What do you get if you bracket each polynomial in this algorithm with a fixed homogeneous polynomial $P(r,s,t)$? (Hint: Compare to Section 6.2, Exercise 10.)

9. Consider a B-patch $P(s,t)$ of degree n with control points $\{P_{ijk}\}$ and knot-net

$$(r_1,s_1),\ldots,(r_n,s_n),\ (t_1,u_1),\ldots,(t_n,u_n),\ (v_1,w_1),\ldots,(v_n,w_n).$$

a. Show that the labels along the edges of the tetrahedral evaluation algorithm are nonnegative when

$$(s,t) \in \bigcap_{i+j+k\leq n+2} \Delta(r_i,s_i)(t_j,u_j)(v_k,w_k)$$

b. Conclude that the point $P(s,t)$ lies in the convex hull of the control points $\{P_{ijk}\}$ when

$$(s,t) \in \bigcap_{i+j+k\leq n+2} \Delta(r_i,s_i)(t_j,u_j)(v_k,w_k).$$

10. a. Consider a quadratic B-patch $P(s,t)$ with control points $\{P_{ijk}\}$ and knot-net $(r_1,s_1),(r_2,s_2),\ (t_1,u_1),(t_2,u_2),\ (v_1,w_1),(v_2,w_2)$. Show that the value of the surface and its first-order partial derivatives along the line determined by the points (t_1,u_1) and (v_1,w_1) is independent of the value of the control point P_{002} and the node (r_2,s_2).

b. Generalize the result in part (a) to B-patches $P(s,t)$ of arbitrary degree.

7.13 Summary

Our primary purpose has been to give you a sound foundation in the fundamentals of polynomial and spline interpolation and approximation. This goal, we trust, has now been realized. With this background in hand you are well prepared for graduate seminars on more advanced topics. We hope you enjoyed reading this material as much as we enjoyed writing it. This is almost, but not quite, the end of our story. In the next chapter we shall discuss pyramid algorithms for multisided surface patches.

We have covered a lot of material in this chapter. Below we summarize the different techniques we encountered for analyzing B-splines, collect in one place the standard properties of and algorithms for B-spline curves and surfaces, and identify a few special progressive and spline bases. Then we list for easy access a collection of useful identities for the B-spline basis functions harvested from the text and the exercises.

- *Tools for Analyzing B-Spline Curves and Surfaces*
  1. De Boor algorithm
     - Path arguments
     - Induction + recursion
  2. Knot insertion procedures
     - Convergence as knot spacing decreases to zero
     - Conversion to piecewise Bezier form
  3. Blossoming
     - Powerful change of basis method
     - Effective tool for analyzing derivative properties
  4. Divided difference
     - Identities
  5. Continuous convolution
     - Uniform B-splines only
- *Properties of B-Spline Curves and Surfaces*
  1. Piecewise polynomial
  2. Continuity of order $C^{n-\mu}$ at knots of multiplicity μ
  3. Local control
  4. Affine invariance
  5. Local convex hull
  6. Locally nondegenerate
  7. Interpolation at knots where the multiplicity μ is equal to the degree n
  8. Variation diminishing (curves only)
- *Algorithms for B-Spline Curves and Surfaces*
  1. Evaluation
  2. Differentiation
  3. Integration
  4. Blossoming

5. Homogenization
6. Knot insertion
   - Boehm's algorithm
   - Oslo algorithm
   - Factored knot insertion algorithm
   - Lane-Riesenfeld algorithm (uniform knots)
7. Conversion to piecewise Bezier form

■ *Special Progressive and Spline Bases*
   1. Bezier

   $$t_1 = \cdots = t_n = a$$

   $$t_{n+1} = \cdots = t_{2n} = b$$

   $$B_k^n(t) = \frac{\binom{n}{k}(t-a)^k(b-t)^{n-k}}{(b-a)^n} \qquad 0 \le k \le n$$

   2. Monomial

   $$t_1 = \cdots = t_n = a$$

   $$t_{n+1} = \cdots = t_{2n} = \delta$$

   $$M_k^n(t) = \binom{n}{k}(t-a)^k \qquad 0 \le k \le n$$

   3. Power

   $$t_1,\ldots,t_n,t_0,\ldots,t_{n-1}$$

   $$P_k^n(t) = \frac{(t-t_k)^n}{\prod_{j \ne k}(t_k - t_j)}, \qquad 0 \le k \le n$$

   4. Uniform

   $$t_k = k$$

   $$N_{0,n}(t) = \underbrace{\chi_{[0,1)}(t) * \cdots * \chi_{[0,1)}(t)}_{n+1 \ \ factors}$$

7.13.1 Identities for the B-Spline Basis Functions

1. *Compact support*

   $$Support\{N_{k,n}(t)\} = [t_k, t_{k+n+1}]$$

2. *Smoothness at the knots*

   τ has multiplicity μ in the sequence $t_k,\ldots,t_{k+n+1} \Rightarrow N_{k,n}(t)$ is $C^{n-\mu}$ at τ

3. *Interpolation at the knots*

   $$t_{j+1} = \cdots = t_{j+n} \Rightarrow N_{j,n}(t_{j+1}) = 1$$

4. *Evaluation at the knots*

$$N_{k,n}(t_{k+j} \mid t_k,...,t_{k+n+1}) = N_{k,n-1}(t_{k+j} \mid t_k,...,t_{k+j-1},t_{k+j+1},...,t_{k+n+1})$$

5. *Invariance under affine transformations of the knots*

$$N_{k,n}(at+b \mid at_k+b,...,at_{k+n+1}+b) = N_{k,n}(t \mid t_k,...,t_{k+n+1})$$

6. *Nonnegativity*

$$N_{k,n}(t) \geq 0$$

7. *Partition of unity*

$$\sum_k N_{k,n}(t) \equiv 1$$

8. *Recursion*

$$N_{k,0}(t) = 1 \qquad\qquad t_k \leq t < t_{k+1}$$

$$N_{k,n}(t) = \frac{t-t_k}{t_{k+n}-t_k} N_{k,n-1}(t) + \frac{t_{k+n+1}-t}{t_{k+n+1}-t_{k+1}} N_{k+1,n-1}(t)$$

$$N_{k,n}(t) = (t-t_k) \frac{N_{k,n-1}(t)}{Support} + (t_{k+n+1}-t) \frac{N_{k+1,n-1}(t)}{Support}$$

9. *Nonstandard recursion*

$$N_{k,n}(t \mid t_k,...,t_{k+n+1}) = \frac{\tau-t_k}{t_{k+n}-t_k} N_{k,n}(t \mid t_k,...,\tau,...,t_{k+n})$$

$$+ \frac{t_{k+n+1}-\tau}{t_{k+n+1}-t_{k+1}} N_{k+1,n}(t \mid t_{k+1},...,\tau,...,t_{k+n+1})$$

$$N_{k,n}(t \mid t_k,...,t_{k+n+1}) = \frac{t-t_{k+i}}{t_{k+j}-t_{k+i}} N_{k,n-1}(t \mid t_k,...,\hat{t}_{k+j},...,t_{k+n+1})$$

$$+ \frac{t_{k+j}-t}{t_{k+j}-t_{k+i}} N_{k,n-1}(t \mid t_k,...,\hat{t}_{k+i},...,t_{k+n+1})$$

10. *Differentiation*

$$\frac{dN_{k,n}(t)}{dt} = n \left(\frac{N_{k,n-1}(t)}{t_{k+n}-t_k} - \frac{N_{k+1,n-1}(t)}{t_{k+n+1}-t_{k+1}} \right)$$

$$\frac{dN_{k,n}(t)}{dt} = n \left(\frac{N_{k,n-1}(t)}{Support} - \frac{N_{k+1,n-1}(t)}{Support} \right)$$

11. *Nonstandard differentiation*

$$\frac{dN_{k,n}(t)}{dt} =$$

$$n\left(\frac{N_{k,n-1}(t \mid t_k,\ldots,\hat{t}_{k+j},\ldots,t_{k+n+1})}{t_{k+j} - t_{k+i}} - \frac{N_{k+1,n-1}(t \mid t_k,\ldots,\hat{t}_{k+i},\ldots,t_{k+n+1})}{t_{k+j} - t_{k+i}}\right)$$

12. *Recursion for the derivative*

$$\left(\frac{n-1}{n}\right)N'_{k,n}(t) = \left(\frac{t - t_k}{t_{k+n} - t_k}\right)N'_{k,n-1}(t) + \left(\frac{t_{k+n+1} - t}{t_{k+n+1} - t_{k+1}}\right)N'_{k+1,n-1}(t)$$

13. *Integration*

$$\int \frac{N_{k,n}(t \mid t_k,\ldots,t_{k+n+1})dt}{Support} = \frac{N_{k,n+1}(t \mid t_k,\ldots,t_{k+n+1},\delta)}{n+1}$$

$$\int_{Support} \frac{N_{k,n}(t)}{Support} dt = \frac{1}{n+1}$$

$$\int_{Support} \left\{\frac{N_{k,n}(t)}{Support}\right\}\left\{\frac{F^{(n+1)}(t)}{n!}\right\}dt = F[t_k,\ldots,t_{k+n+1}]$$

$$\int_{Support} \left\{\frac{N_{k,n}(t)}{Support}\right\}\left\{\frac{F^{(n+1)}(t)}{n!}\right\}dt = \int_{\Delta^n} F^{(n+1)}(\textstyle\sum_j v_j t_{k+j})dv_1\cdots dv_{n+1}$$

$$\int_{-\infty}^{\infty} \left\{\frac{N_{i,m}(t)}{Support}\right\}\left\{\frac{N_{j,n}(t)}{Support}\right\}dt$$

$$= (-1)^{m+1}\frac{m!\,n!}{(m+n)!}(y-x)_+^{m+n+1}[t_j,\ldots,t_{j+n+1}]_y[t_i,\ldots,t_{i+m+1}]_x$$

14. *Linear independence*

$$\textstyle\sum_k c_k N_{k,n}(t) = 0 \Leftrightarrow c_k = 0 \text{ for all } k$$

15. *Descartes' Law of Signs*

$$Sign\ Alternations\{\textstyle\sum_k c_k N_{k,n}(t)\} \leq Sign\ Alternations\{c_k\}$$

16. *Nodes*

$$t = \sum_k \left\{\frac{t_{k+1} + \cdots + t_{k+n}}{n}\right\}N_{k,n}(t)$$

17. *Representation of the monomials*

$$\binom{n}{j}t^j = \sum_k \left\{ \sum_\sigma t_{k+\sigma(1)} \cdots t_{k+\sigma(j)} \right\} N_{k,n}(t) \quad (\sigma = \text{permutation of } \{1,...,n\})$$

18. *Divided difference formula*

$$N_{k,n}(t) = (t_{k+n+1} - t_k)(x - t)_+^n [t_k,...,t_{k+n+1}]$$

$$\frac{N_{k,n}(t)}{Support} = (x - t)_+^n [t_k,...,t_{k+n+1}]$$

19. *Marsden identity*

$$(x - t)^n = \sum_k (t_{k+1} - t) \cdots (t_{k+n} - t) N_{k,n}(x)$$

20. *De Boor–Fix formula*

$$\sum_p \frac{(-1)^{n-p}}{n!} N_{j,n}^{(p)}(\tau) \psi_{k,n}^{(n-p)}(\tau) = 1 \qquad\qquad j = k$$

$$= 0 \qquad\qquad j \neq k$$

$$\psi_{k,n}(\tau) = (t_{k+1} - \tau) \cdots (t_{k+n} - \tau) \quad \tau \in (t_{k+1}, t_{k+n})$$

21. *Blossoming as dual functionals*

$$n_{k,n}(t_{j+1},...,t_{j+n}) = \delta_{jk}$$

$$S(t) = \sum_i s(t_{i+1},...,t_{i+n}) N_{i,n}(t)$$

22. *Knot insertion*

$$N_{k,n}(t \mid t_k,...,t_{k+n+1}) = \frac{\tau - t_k}{t_{k+n} - t_k} N_{k,n}(t \mid t_k,...,\tau,...,t_{k+n})$$

$$+ \frac{t_{k+n+1} - \tau}{t_{k+n+1} - t_{k+1}} N_{k+1,n}(t \mid t_{k+1},...,\tau,...,t_{k+n+1})$$

23. *Degree elevation*

$$N_{k,n}(t \mid t_k,...,t_{k+n+1}) = \frac{\sum_j N_{k,n+1}(t \mid t_k,...,t_{k+j},t_{k+j},...,t_{k+n+1})}{n+1}$$

24. *Partial derivatives with respect to the knots*

$$\frac{\partial N_k^n(t \mid t_k,...,t_j,...,t_{k+n+1})}{\partial t_j}$$

$$= \mu_j \left(\frac{N_{k+1}^n(t \mid t_{k+1},...,t_j,t_j,...,t_{k+n+1})}{t_{k+n+1} - t_{k+1}} - \frac{N_k^n(t \mid t_k,...,t_j,t_j,...,t_{k+n})}{t_{k+n} - t_k} \right)$$

$$\mu_j = \text{multiplicity } t_j$$

25. *Geometric characterization*

$$N_{0,n-1}(t \mid t_0,...,t_n) = \frac{(t_n - t_0)Vol_{n-1}(\sigma \in \Delta^n(P) \mid \sigma_1 = t)}{nVol_n(\Delta^n(P))}$$

$P_0,...,P_n$ affinely independent points with $P_{i1} = t_i$, $i = 0,...,n$

$\Delta^n(P) = n$-simplex with vertices $P_0,...,P_n$

26. *Special identities for uniform B-splines*

a. *Translation invariance*

$$N_{k,n}(t) = N_{0,n}(t-k)$$

b. *Recursion*

$$N_{k,n}(t) = \frac{t-k}{n}N_{k,n-1}(t) + \frac{k+n+1-t}{n}N_{k+1,n-1}(t)$$

c. *Differentiation*

$$\frac{dN_{k,n}(t)}{dt} = N_{k,n-1}(t) - N_{k+1,n-1}(t)$$

d. *Integration*

$$\int N_{k,n}(t)dt = N_{k,n+1}(t \mid k,...,k+n+1,\delta)$$

$$\int_{Support} N_{k,n}(t)dt = 1$$

e. *Continuous convolution*

$$N_{k,n}(t) = \int_0^1 N_{k,n-1}(t-x)dx$$

$$N_{0,n}(t) = \underbrace{\chi_{[0,1)}(t) * \cdots * \chi_{[0,1)}(t)}_{n+1 \;\; factors}$$

f. *Subdivision*

$$N_{0,n}(t) = \sum_i \frac{\binom{n+1}{i}}{2^n}N_{0,n}(2t-i)$$

CHAPTER 8

Pyramid Algorithms for Multisided Bezier Patches

In Chapter 5 we studied three-sided and four-sided Bezier patches built using dynamic programming procedures based upon three-term and four-term recurrence relations. In this chapter we are going to study Bezier patches with an arbitrary number of sides. Three-sided and four-sided patches are typically used for free-form design; multisided patches are commonly required when it is necessary to fill n-sided holes.

Our approach to multisided patches will be similar to our approach to three-sided and four-sided patches: we shall build dynamic programming procedures—pyramid algorithms—based upon special recurrence relations. By now we have a good deal of experience with generating useful recurrences. We have seen that such recurrences arise quite naturally in three settings: Lagrange interpolation, discrete convolution, and blossoming. We shall discover, however, in Section 8.3 that standard Lagrange interpolation procedures do not lend themselves to generating multisided patches. Therefore, discrete convolution and blossoming will be the two central motifs of this chapter. We shall see here as well that these two devices are necessarily interrelated.

To construct three-sided and four-sided Bezier patches, we need three-sided and four-sided arrays of control points and barycentric coordinate functions for the triangle and the rectangle. Similarly, to construct multisided Bezier patches we require multisided arrays of control points and barycentric coordinate functions for multisided polygonal domains. But what are multisided arrays of control points and how do we construct barycentric coordinates for multisided polygons? There is no single answer to either of these questions: different answers lead to different types of multisided Bezier schemes. Based on three different answers to these questions, we shall develop three different types of multisided Bezier patches: S-patches, C-patches, and toric Bezier patches. Three common threads tie these three schemes together: discrete convolution, Minkowski sum (a device we shall introduce in Section 8.2), and the general pyramid algorithm.

445

There is no simple answer to the question: which of these multisided schemes is best? Although special types of *S*-patches have been around for some time, their properties have yet to be fully explored. Moreover, toric Bezier patches are quite new, so it is premature to judge the relative merits of these schemes. We present these schemes here to elucidate their constructions, explain their basic properties, and clarify their interrelationships; we shall not try to pick a winner.

A key role in each of these constructions is played by barycentric coordinate functions, so we begin with a discussion of a generalization of barycentric coordinates to arbitrary convex polygons.

8.1   Barycentric Coordinates for Convex Polygons

The main properties of the barycentric coordinates $\beta_1, \beta_2, \beta_3$ for a triangle P with vertices P_1, P_2, P_3 are summarized in Section 1.2.3, Theorem 1.1. We recall these properties in Table 8.1 along with the analogous properties we want to hold for the barycentric coordinate functions $\beta_1, \ldots, \beta_n$ for a convex polygon Q with ordered vertices $Q_1, \ldots, Q_n$.

The first two properties of the barycentric coordinate functions for convex polygons are required because we want the multisided Bezier surfaces constructed from these functions to be affine invariant and to lie in the convex hulls of their control points. The third property will guarantee that the boundary curves of these surfaces are the Bezier curves determined by their boundary control points, and the fourth property ensures that these surfaces will interpolate their corner control points. The fourth property is also key later on in ensuring that the *S*-patch blossom evaluated at the vertices of the domain polygon provides the dual functionals for *S*-patches (see Section 8.4.4). The final property asserts that the functions describing these surfaces are not too complicated—that these surfaces are defined by rational expressions.

For triangles we discussed two approaches to barycentric coordinates in Section 1.2.3: normalized areas and normalized line equations. Consider the triangle in Figure 8.1, with vertices P_1, P_2, P_3 and edges defined by the equations $L_{23}(u,v) = 0$,

Table 8.1 Properties of barycentric coordinate functions for triangles and convex polygons.

Triangles	*Convex Polygons*
1. $\sum_{k=1}^{3} \beta_k \equiv 1$.	1. $\sum_{k=1}^{n} \beta_k \equiv 1$.
2. $\beta_k > 0$ in the interior of P.	2. $\beta_k > 0$ in the interior of Q.
3. $\beta_k = 0$ on the line $P_i P_{i+1}$, $k \neq i, i+1$.	3. $\beta_k = 0$ on the line $Q_i Q_{i+1}$, $k \neq i, i+1$.
4. $\beta_k(P_j) = 0 \qquad j \neq k$ $\qquad\quad = 1 \qquad j = k$.	4. $\beta_k(Q_j) = 0 \qquad j \neq k$ $\qquad\quad = 1 \qquad j = k$.
5. $\beta_1, \beta_2, \beta_3$ are linear functions.	5. $\beta_1, \ldots, \beta_n$ are rational functions.

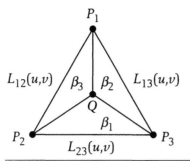

Figure 8.1 Barycentric coordinates for the triangle $\Delta P_1 P_2 P_3$.

$L_{13}(u,v) = 0$, $L_{12}(u,v) = 0$. The barycentric coordinates $\beta_1, \beta_2, \beta_3$ of a point Q with respect to the vertices P_1, P_2, P_3 are given by

$$\beta_k(Q) = \pm \frac{area(\Delta Q P_i P_j)}{area(\Delta P_1 P_2 P_3)} \qquad i \neq j \neq k, \tag{8.1}$$

or equivalently by

$$\beta_k(Q) = \frac{L_{ij}(Q)}{L_{ij}(P_k)} \qquad i \neq j \neq k. \tag{8.2}$$

We are now going to provide an explicit construction for the barycentric coordinate functions of a convex polygon by generalizing the construction of the barycentric coordinates for triangles given in Equation (8.2). An equivalent extension of Equation (8.1) from triangles to convex polygons is provided in Exercise 3.

Consider then a convex n-gon with ordered vertices $Q_1,...,Q_n$, and let $L_{i,i+1}(u,v)$, $i = 1,...,n$, be the equation of the line joining Q_i and Q_{i+1}, where $Q_{n+1} = Q_1$. (See, for example, the hexagon in Figure 8.2.)

Normalize each function $L_{i,i+1}(u,v)$ so that the normal of the line is pointing into the polygon. Then we can define barycentric coordinate functions $\beta_1,...,\beta_n$ with respect to the vertices $Q_1,...,Q_n$ by setting

$$\alpha_k(Q) = c_k \prod_{i,i+1 \neq k} L_{i,i+1}(Q) \tag{8.3}$$

$$\beta_k(Q) = \frac{\alpha_k(Q)}{\sum_j \alpha_j(Q)} \qquad k = 1,...,n. \tag{8.4}$$

Notice that, up to a constant multiple, α_k is the product of the edges $L_{i,i+1}(u,v)$ on which the vertex Q_k does not lie, and β_k is equal to α_k normalized by the sum of the α's. Hence, in general, β_k is a rational function of degree $n - 2$. The constants $c_k > 0$ are arbitrary; one canonical choice for these constants is presented in Exercise 4. Now in analogy with Theorem 1.1, we have the following results.

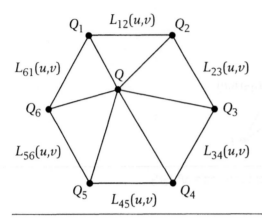

Figure 8.2 A hexagon with vertices Q_i and bounding lines $L_{i,i+1}(u,v)$, $i = 1,...,6$.

THEOREM
8.1

Existence of Barycentric Coordinates for Convex Polygons

Let $\beta_1,...,\beta_n$ be the barycentric coordinate functions defined in Equation (8.4) relative to the ordered vertices $Q_1,...,Q_n$ of a convex polygon. Then

1. $\sum_{k=1}^{n} \beta_k \equiv 1$.

2. $\beta_k > 0$ in the interior of the polygon.

3. $\beta_k = 0$ on the line $Q_i Q_{i+1}$, $k \neq i, i+1$.

4. $\beta_k(Q_j) = 0 \qquad j \neq k$

   $\qquad\qquad = 1 \qquad j = k$.

5. $\beta_1,...,\beta_n$ are rational functions in the rectangular coordinates u,v.

Proof
1. Property 1 is immediate from Equation (8.4).

2. Property 2 follows because $L_{i,i+1}(u,v)$ is chosen with its normal vector pointing into the polygon. Hence in the interior of the polygon $L_{i,i+1}(u,v) > 0$, so by Equations (8.3) and (8.4) $\alpha_k, \beta_k > 0$ in the interior of the polygon.

3. Property 3 is valid because by Equation (8.3) $L_{i,i+1}(u,v)$ is a factor of α_k whenever $k \neq i, i+1$. Therefore, $\alpha_k, \beta_k = 0$ on the line $Q_i Q_{i+1}$, whenever $k \neq i, i+1$.

4. Property 4 is a direct consequence of Property 1 and Property 3. By Property 3, $\beta_k(Q_j) = 0, j \neq k$. Therefore, by Property 1, $\beta_k(Q_k) = 1$.

5. Property 5 follows immediately from Equations (8.3) and (8.4).

Theorem 8.1 establishes the existence of barycentric coordinate functions for convex polygons with the properties listed in Table 8.1. Although typically we shall use the functions defined in Equation (8.4), we shall call any set of functions that satisfies these five properties barycentric coordinate functions for the convex polygon.

Exercises

1. Let $Q_1,\ldots,Q_n$ be the ordered vertices of a convex polygon, and let $\beta_1,\ldots,\beta_n$ be the functions defined by Equation (8.4). Show that β_i, β_{i+1} are rational linear functions along the line $Q_i Q_{i+1}$.

2. Let $P = (p_1, p_2)^T$, $Q = (q_1, q_2)^T$, $R = (u, v)^T$ be the vertices of a triangle. Show that

   a. $$2 \times area(\Delta PQR) = \pm \det \begin{pmatrix} p_1 & q_1 & u \\ p_2 & q_2 & v \\ 1 & 1 & 1 \end{pmatrix} = \pm \det \begin{pmatrix} P & Q & R \\ 1 & 1 & 1 \end{pmatrix}.$$

   b. $\det \begin{pmatrix} P & Q & R \\ 1 & 1 & 1 \end{pmatrix} = 0$ is the equation of the line joining P and Q.

3. Let $Q_1,\ldots,Q_n$ be the ordered vertices of a convex polygon. Using the results of Exercise 2, show that Equation (8.1) for the barycentric coordinate functions of a triangle can be generalized to barycentric coordinate functions $\beta_1,\ldots,\beta_n$ for a convex polygon by setting

   $$\alpha_k(Q) = c_k \prod_{i,i+1 \neq k} area(Q_i Q_{i+1} Q)$$

   $$\beta_k(Q) = \frac{\alpha_k(Q)}{\sum_j \alpha_j(Q)},$$

   where $c_k > 0$ are arbitrary constants.

4. Choose the constants c_k in Exercise 3 by setting $c_k = area(Q_{k-1} Q_k Q_{k+1})$. We are going to show that for this choice of constants

   $$L(Q) = \sum_{k=1}^{n} \beta_k(Q) L(Q_k)$$

   for every linear function L. Define $M(Q) = \sum_{k=1}^{n} \alpha_k(Q)\big(L(Q) - L(Q_k)\big)$, and let $P = (1-\lambda)Q_i + \lambda Q_{i+1}$.

   a. Show that $L(P) = (1-\lambda)L(Q_i) + \lambda L(Q_{i+1})$.

   b. Show that $M(P) =$

   $$\prod_{j \neq i-1, i, i+1} area(Q_j, Q_{j+1}, P) \left\{ \begin{array}{l} area(Q_{i-1} Q_i Q_{i+1}) area(Q_{i+1} Q_{i+2} P)\big(L(P) - L(Q_i)\big) \\ + area(Q_i Q_{i+1} Q_{i+2}) area(Q_{i-1} Q_i P)\big(L(P) - L(Q_{i+1})\big) \end{array} \right\}.$$

    c. $M(P) = 0$. (Hint: Use parts (a) and (b).)

    d. Conclude from Section 2.13, Proposition 2.12, that every edge of the domain polygon is a linear factor of M, and hence, since $degree(M) = n - 1$, that $M \equiv 0$.

    e. Show that $M \equiv 0 \Rightarrow L(Q) = \sum_{k=1}^{n} \beta_k(Q) L(Q_k)$.

    f. Conclude from part (e) that $Q = \sum_{k=1}^{n} \beta_k(Q) Q_k$. (Compare to Section 1.2.3, Exercise 4.)

5. Here we are going to show that barycentric coordinate functions for convex polygons with more than four sides cannot be polynomials. Suppose that $\beta_1, \dots \beta_n$ are polynomials, and that Q is a point of intersection between two nonadjacent edges of the convex polygon with ordered vertices $Q_1, \dots, Q_n$.

    a. Show that if $\beta_1, \dots, \beta_n$ satisfy Property 3 of Theorem 8.1, then $\beta_j(Q) = 0$, $j = 1, \dots, n$.

    b. Conclude that $\sum_k \beta_k \neq 1$, and hence that $\beta_1, \dots, \beta_n$ cannot be barycentric coordinate functions for the convex polygon with vertices $Q_1, \dots, Q_n$.

    c. Why do rational functions not suffer from the same problem?

    d. Explain why barycentric coordinates for rectangles can be polynomials.

6. Consider a convex polygon with ordered vertices $Q_1, \dots, Q_n$, and let $L_{i,i+1}(u,v)$, $i = 1, \dots, n$, be the equation of the line joining Q_i and Q_{i+1}, where $Q_{n+1} = Q_1$. Normalize each function $L_{i,i+1}(u,v)$ in the usual manner so that the normal of the line is pointing into the polygon. Define functions $\beta_1, \dots, \beta_n$ with respect to the vertices $Q_1, \dots, Q_n$ by setting

$$\alpha_k(Q) = c_k \prod_{i=1}^{n} L_{i,i+1}^{p_{i,k}}(Q) \qquad\qquad c_k > 0$$

$$\beta_k(Q) = \frac{\alpha_k(Q)}{\sum_j \alpha_j(Q)} \qquad\qquad k = 1, \dots, n \ .$$

Show that these functions $\beta_1, \dots, \beta_n$ satisfy the five properties of barycentric coordinate functions listed in Theorem 8.1 if and only if $p_{k-1,k} = p_{k,k} = 0$ for $k = 1, \dots, n$, and all the other $p_{i,k}$ are positive integers.

8.2   Polygonal Arrays

To construct a polygonal patch, we must begin with a polygonal array of control points. But what exactly is a polygonal array of control points? So far we have seen only two examples of polygonal arrays: rectangular arrays and triangular arrays. A $(d+1) \times (d+1)$ rectangular array of control points is a set of points $\{P_{ij}\}$, where

$0 \le i, j \le d$; an *order d* triangular array of control points is a collection of points $\{P_{ijk}\}$, where $i + j + k = d$. We are now going to generalize these notions to arbitrary polygonal arrays of points.

The key technique we shall use is the Minkowski sum. Let I and J be two arbitrary sets of p-tuples. The *Minkowski sum* of I and J is the set

$$I \oplus J = \{i + j \mid i \in I \text{ and } j \in J\},$$

where $i + j$ denotes the p-tuple formed by adding the coordinates of i and j. For example, if

$$I = \{(0,0),(1,0),(0,1),(1,1)\}$$
$$J = \{(0,0),(1,0),(0,1)\} ,$$

then

$$I \oplus J = \{(0,0),(1,0),(0,1),(1,1),(0,2),(1,2),(2,1),(2,0)\}$$

(see Figure 8.3). For future reference, we define

$$I^0 = \underbrace{(0,\dots,0)}_{p}$$

$$I^d = \underbrace{I \oplus \cdots \oplus I}_{d \ summands} \cdot$$

The Minkowski sum facilitates the indexing of discrete convolution. Let $A(t) = \{A_i(t) \mid i \in I\}$ and $B(t) = \{B_j(t) \mid j \in J\}$. Then the discrete convolution $A(t) \otimes B(t)$ is indexed by the Minkowski sum $I \oplus J$—that is, $A(t) \otimes B(t) = (A \otimes B)(t) = \{(A \otimes B)_k(t) \mid k \in I \oplus J\}$—since, by definition,

$$(A \otimes B)_k(t) = \sum_{i+j=k} A_i(t)B_j(t) .$$

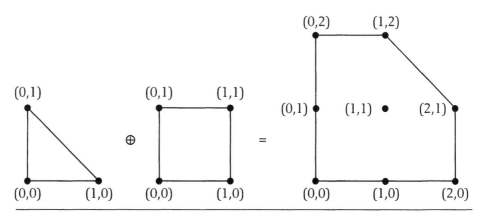

Figure 8.3 The Minkowski sum of a triangular array and a rectangular array is a pentagonal array.

Similarly,

$$A(t)^d = \underbrace{A(t) \otimes \cdots \otimes A(t)}_{d \ factors}$$

is indexed by I^d. We shall take advantage of this indexing for discrete convolution in Section 8.4, when we study S-patches.

To form an n-sided polygonal array, start by selecting an indexing set I of n distinct p-tuples. An n-sided polygonal array of depth d for the index set I is any set of points indexed by I^d.

For example, let

$$I_R = \{(0,0),(1,0),(1,1),(0,1)\}$$
$$I_T = \{(0,0,1),(1,0,0),(0,1,0)\}.$$

Then $I_R^d = \{(j,k) \mid 0 \le j,k \le d\}$ is the indexing set for a $(d+1) \times (d+1)$ rectangular array $\{P_{jk}\}$, and $I_T^d = \{i,j,k) \mid i+j+k=d\}$ is the indexing set for an order d triangular array $\{P_{ijk}\}$.

But there is more to a polygonal array than just the indexing of the points; we must also indicate how the points are related topologically in the array—that is, which points are adjacent to which points. For example, in the triangular array of order 3 in Figure 5.43(b), the point P_{003} is adjacent to the points P_{012} and P_{102}, but P_{003} is not adjacent to any of the other points in the array. To establish adjacency, we consider the indexing set I not just as a set, but as an ordered sequence of p-tuples $I = (i_1,\ldots,i_n)$. Points indexed by I inherit their adjacency from I. Thus points indexed by a set I with n elements lie, topologically, on the vertices of an n-gon. To close the polygon, we consider the first element of I to be adjacent to the last element of I.

Adjacency in I^d is defined recursively. Adding a fixed index in I to each index in I translates the polygon, so adjacent indices remain adjacent. Similarly, adjacent indices in I^{d-1} remain adjacent under translation. (For further details and definitions, see Exercise 6.)

An n-sided polygonal array of depth d is a set of points $\{P_\lambda\}$ in affine space indexed by a set I^d, where $I = (i_1,\ldots,i_n)$ is an ordered set of n distinct p-tuples. The points in the array $\{P_\lambda\}$ inherit their adjacency from the adjacency relation in I^d. That is, two points P_μ, P_ν in the array $\{P_\lambda\}$ are *adjacent* if ν is adjacent to μ in I^d. When we draw a polygonal array, we shall connect adjacent points with straight lines (see Figures 8.4 and 8.5). The rth *boundary* of the n-sided polygonal array $\{P_\lambda\}$ of depth d consists of the points indexed by $(d-k)i_r + ki_{r+1}$, $k = 0,\ldots,d$, where $i_{n+1} = i_1$. Figures 8.4 and 8.5 illustrate two distinct pentagonal arrays of depth 2.

Exercises

1. Draw triangular arrays of depths 2 and 3, where
   a. *I* consists of the vertices of the standard 2-simplex
   b. $I = \{0,1,2\}$

2. Draw rectangular arrays of depths 2 and 3, where

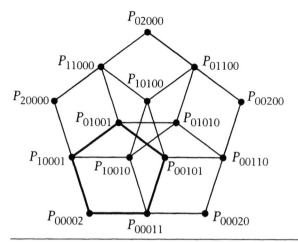

Figure 8.4 A pentagonal array of depth 2 with 15 points indexed by I^2, where I consists of the vertices of the standard 4-simplex—that is, I is the set of all 5-tuples with a 1 in one position and a 0 everywhere else. The outline of one of the five small pentagons indexed by the subset of I^2 generated by translating I by $(0,0,0,0,1) \in I$ is darkened.

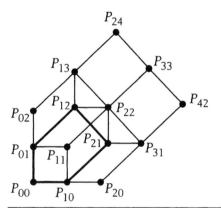

Figure 8.5 A pentagonal array of depth 2 with 14 points indexed by the set I^2, where $I = \{(0,0),(1,0),(2,1),(1,2),(0,1)\}$. The outline of one of the five small pentagons indexed by the subset of I^2 generated by translating I by $(0,0) \in I$ is darkened.

      a. I consists of the vertices of the standard 3-simplex

      b. $I = \{(0,0),(1,0),(1,1),(0,1)\}$

  3. Draw a pentagonal array of depth 3, where

      a. I consists of the vertices of the standard 4-simplex

      b. $I = \{(0,0),(1,0),(2,1),(1,2),(0,1)\}$

4. Draw hexagonal arrays of depths 2 and 3, where

   a. I consists of the vertices of the standard 5-simplex

   b. $I = \{(0,0),(1,0),(2,1),(2,2),(1,2),(0,1)\}$

5. Show that the number of points in a polygonal array of depth d indexed by the set I^d, where I consists of the vertices of the standard n-simplex, is $\binom{n+d}{d}$.

6. Show that the following two definitions of adjacency in I^d are equivalent: Let $i, j_1, \ldots, j_{d-1} \in I$, and let $\mu, \nu \in I^d$. Then μ is adjacent to ν if there are indices i_μ, i_ν such that

Definition 1	Definition 2
i. i_μ is adjacent to i_ν in I.	i. i_μ is adjacent to i_ν in I^{d-1}.
ii. $\mu = i_\mu + j_1 + \cdots + j_{d-1}$.	ii. $\mu = i_\mu + i$.
iii. $\nu = i_\nu + j_1 + \cdots + j_{d-1}$.	iii. $\nu = i_\nu + i$.

7. Show that if $I = \{i_1, \ldots, i_n\}$, then

   a. $I^d = \{k_1 i_1 + \cdots + k_n i_n \mid k_1 + \cdots + k_n = d\}$

   b. I^d is isomorphic to the set of equivalence classes $\{(k_1, \ldots, k_n) \mid k_1 + \cdots + k_n = d\}$, where two n-tuples $(k_1, \ldots, k_n)$ and $(h_1, \ldots, h_n)$ are members of the same equivalence class of I^d if
   $$k_1 i_1 + \cdots + k_n i_n = h_1 i_1 + \cdots + h_n i_n$$

8. Show that

   a. $(I_1 \oplus \cdots \oplus I_m)^d = I_1^d \oplus \cdots \oplus I_m^d$.

   b. $(I_1^{c_1} \oplus \cdots \oplus I_m^{c_m}) \oplus (I_1^{d_1} \oplus \cdots \oplus I_m^{d_m}) = I_1^{c_1+d_1} \oplus \cdots \oplus I_m^{c_m+d_m}$.

8.3   Neville's Pyramid Algorithm and Multisided Grids

The three main ingredients we needed in Chapter 2 to construct pyramid algorithms for three-sided and four-sided interpolating Lagrange patches were barycentric coordinate functions, arrays of control points, and specialized grids. In Section 8.1 we addressed the issue of barycentric coordinates for convex polygons, and in Section 8.2 we introduced polygonal arrays of control points. In this section we shall consider general n-sided grids.

Actually we have already encountered n-sided grids in Section 8.2. A polygonal array of points typically lives in three- or higher-dimensional space, but if we consider a polygonal array in two dimensions, then we have a polygonal grid. Figures 8.4 and 8.5 represent pentagonal grids in the plane. Straight lines in the grid connect points that are adjacent in the array. In general, three points in such an array are collinear only by accident. For example, in the pentagonal grid depicted in Figure 8.5, the points P_{20}, P_{31}, P_{42} need not be collinear.

Unfortunately, we cannot, in general, build pyramid algorithms to interpolate on polygonal grids where the polygons have more than four sides and the depth of the grid is greater than 1 even if many of the grid points are collinear. To see why, let us try to apply dynamic programming to the pentagonal grids in Figures 8.4 and 8.5.

To begin, we must first construct pentagonal interpolants for the subgrids of depth 1. This base case is easy. Let $\beta_k(u,v)$ be barycentric coordinates for the pentagon with vertices Q_k, $k = 1,...,5$, and let

$$B(u,v) = \beta_1(u,v)P_1 + \beta_2(u,v)P_2 + \beta_3(u,v)P_3 + \beta_4(u,v)P_4 + \beta_5(u,v)P_5 \ .$$

Then by Theorem 8.1

$$B(Q_k) = P_k \qquad k = 1,...,5.$$

Now to form the pentagonal interpolants $P(u,v)$ for the grids of depth 2, we try to proceed as we did for the three-sided and four-sided interpolants in Chapter 2. Let $P_1(u,v),...,P_5(u,v)$ be the five pentagonal interpolants of depth 1 emanating from the five corner vertices $Q_1,...,Q_5$ of the depth 2 pentagonal grid. Let $\beta_k(u,v)$ be the barycentric coordinates of the pentagon with vertices Q_k, $k = 1,...,5$, and let

$$P(u,v) = \beta_1(u,v)P_1(u,v) + \cdots + \beta_5(u,v)P_5(u,v).$$

This approach works for the grids of three and four sides in Chapter 2, but it fails for pentagonal grids. For grids with three or four sides, every node not on the boundary of the grid lies in each of the subgrids of one lower depth (see, for example, Figure 2.27). Therefore, each of the lower-depth interpolants certainly interpolates at each of the interior nodes of the grid. Since barycentric coordinates sum to one, the higher-depth surface $P(u,v)$ also interpolates the data at all the interior nodes. But for the pentagonal grids in Figures 8.4 and 8.5, the interior nodes do not lie on all the grids of depth 1. Therefore, the surface $P(u,v)$ will not interpolate the data at the interior nodes. The same problem arises for polygonal grids of arbitrary depth with an arbitrary number of sides.

What then can we do? Recall that triangular Bezier patches are simpler to construct than triangular Lagrange interpolants because in the triangular Bezier construction there is only one domain triangle. In the construction of the pyramid algorithm for triangular Lagrange interpolants, we change the domain triangle—that is, the barycentric coordinates—as we move from node to node and level to level. But in the construction of triangular Bezier patches, we use the same domain triangle and the same barycentric coordinates everywhere in the pyramid algorithm. Therefore, once we have a general notion of barycentric coordinates for convex polygons, this Bezier construction should work just as well for polygonal arrays of control points as it does for triangular arrays. Giving up on interpolation over polygonal grids, we should still be able to construct the analogues of triangular Bezier patches for arbitrary convex polygons. This idea actually does work, and the resulting multisided Bezier surfaces are called *S-patches*, so it is to these *S*-patches that we now turn our attention.

Exercises

1. Show that it is not, in general, possible to build a pyramid algorithm to interpolate over the quadrilateral grid in Figure 8.6. Explain what goes wrong with the standard dynamic programming construction in Chapter 2.

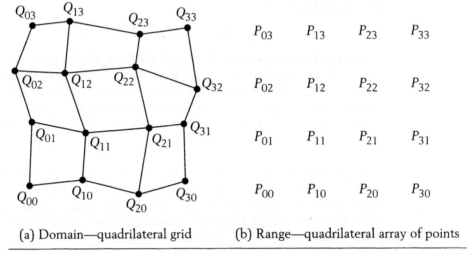

(a) Domain—quadrilateral grid   (b) Range—quadrilateral array of points

Figure 8.6 Data for a quadrilateral interpolant: (a) represents the nodes in the domain, and (b) represents the control points in the range.

2. a. Construct a pyramid algorithm to interpolate data over the quadrilateral grid in Figure 8.7, using barycentric coordinates for an arbitrary quadrilateral.

   b. Show that for the barycentric coordinate functions defined in Equation (8.4) this interpolant is a rational surface of degree $2n$.

3. Consider again the domain in Figure 8.7. This quadrilateral grid is defined by two sets of intersecting lines: $M_k(u,v)$ and $N_k(u,v)$, $k = 0,\ldots,n$, so that each node $Q_{ij} = M_i \cap N_j$, $i,j = 0,\ldots,n$. Let $Q_{ij} = (u_{ij}, v_{ij})$ and define

$$\Lambda_{ij}(u,v) = \left(\prod_{\alpha \neq i} \frac{M_\alpha(u,v)}{M_\alpha(u_{ij}, v_{ij})} \right) \left(\prod_{\beta \neq j} \frac{N_\beta(u,v)}{N_\beta(u_{ij}, v_{ij})} \right)$$

$$L_{ij}(u,v) = \frac{\Lambda_{ij}(u,v)}{\sum_{kl} \Lambda_{kl}(u,v)}$$

$$L(u,v) = \sum_{ij} L_{ij}(u,v) P_{ij}$$

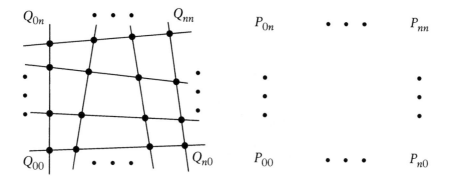

(a) Domain—quadrilateral grid   (b) Range—quadrilateral array of points

Figure 8.7 Data for a quadrilateral interpolant: (a) represents the nodes in the domain, and (b) represents the control points in the range. The nodes must lie on a special quadrilateral grid, but the control points may be in arbitrary positions. The surface $P(u,v)$ must interpolate the control points P_{ij} at the nodes Q_{ij}—that is, $P(Q_{ij}) = P_{ij}$.

    a. Show that the functions $L_{ij}(u,v)$ satisfy the cardinal conditions

$$L_{ij}(Q_{kl}) = L_{ij}(u_{kl}, v_{kl}) = 0 \quad (k,l) \neq (i,j)$$
$$= 1 \quad (k,l) = (i,j) \cdot$$

    b. Show that the surface $L(u,v)$ is a rational surface of degree $2n$ that interpolates the control points P_{ij} at the nodes Q_{ij}.

    c. Compare the interpolating surface $L(u,v)$ to the interpolating surface constructed in Exercise 2.

8.4 *S-Patches*

Given a polygonal array of control points, we can apply a pyramid algorithm to construct a polygonal surface patch just as we did for triangular Bezier patches. The only difference is that we must replace the barycentric coordinates for the domain triangle by barycentric coordinate functions for the domain polygon. Instead of a tetrahedral algorithm, this procedure generates a pyramid algorithm over a polygon, whose boundary at the base is formed by the boundary control points of the polygonal array. Such a surface is called an *S-patch*. Figure 8.8 illustrates the pyramid algorithm for a pentagonal *S*-patch for the pentagonal array of depth 2 in Figure 8.4.

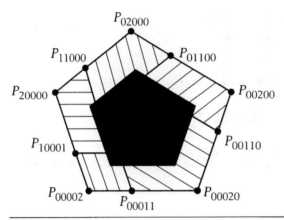

Figure 8.8 Schematic version of the pyramid algorithm of depth 2 for an S-patch over a pentagonal domain. Each pentagonal panel represents the computation of a point at its center calculated by multiplying the points at its vertices by the barycentric coordinates of the domain pentagon and adding the results. The five striped corner pentagons represent pentagonal patches of depth 1, and the dark central pentagon represents the pentagonal patch of depth 2. Interior control points are obscured by the panels. Compare to the pyramid algorithm for a triangular Bezier patch in Figure 5.44 .

By construction, S-patches have the following underlying framework:

- Domain—convex polygon
- Control points—polygonal array
- Blending functions—convolutions of generalized barycentric coordinates

Moreover, in analogy with triangular and tensor product Bezier patches, S-patches also have the associated properties and procedures listed below:

- Properties of S-patches
  - Affine invariance
  - Convex hull property
  - Boundary curves—Bezier curves determined by the boundary control points
  - Degree—rational parametric degree $d(n-2)$, where d is the depth of the control net and n is the number of sides of the domain polygon
  - Special cases—triangular and tensor product Bezier patches
- Procedures for S-patches
  - Evaluation algorithm—pyramid algorithm whose edges are labeled with generalized barycentric coordinates
  - Differentiation algorithm
  - Blossoming algorithm

We begin our investigation of *S*-patches with a formal presentation of the pyramid algorithm, along with properties and formulas directly related to this algorithm. We shall then go on to explore simplicial *S*-patches and to develop a general theory of differentiation and blossoming for arbitrary *S*-patches.

8.4.1 The Pyramid Algorithm and the *S*-Patch Blending Functions

To define the pyramid algorithm formally, we need

1. a convex domain polygon with ordered vertices $Q_1,\dots,Q_n$ and with barycentric coordinate functions $\beta_1(u,v),\dots,\beta_n(u,v)$
2. an *n*-sided polygonal array $\{P_\lambda\}$ of points in the range indexed by a set I^d, where $I = (i_1,\dots,i_n)$ is an ordered set of *n* distinct *p*-tuples

The *n*-sided *S*-patch $S(u,v)$ of depth *d* for the control points $\{P_\lambda\}$ is defined recursively by the following procedure.

Pyramid Algorithm for S-Patches

$$P_\lambda^0(u,v) = P_\lambda \qquad\qquad\qquad \lambda \in I^d$$

$$P_\rho^l(u,v) = \sum_{h=1}^{n} \beta_h(u,v) P_{\rho+i_h}^{l-1}(u,v) \qquad\qquad \rho \in I^{d-l} \qquad\qquad (8.5)$$

$$S(u,v) = P_0^d(u,v) \ .$$

Notice that the intermediate functions $P_\rho^l(u,v)$ that emerge during this computation are the *S*-patches of depth *l* for the control points $\{P_\mu\}$ indexed by $\rho \oplus I^l$. It is these functions $\{P_\rho^l(u,v)\}$ that are depicted by polygonal panels in Figure 8.8.

S-patches over convex polygonal domains are affine invariant and lie in the convex hull of their control points. These properties hold because generalized barycentric coordinate functions sum to one and are nonnegative in the interior and on the boundary of their domain polygon. Hence, by induction, the functions $\{P_\rho^l(u,v)\}$ appearing in the nodes at every level of the pyramid algorithm are affine invariant and lie in the convex hull of the control points.

The boundaries of an *S*-patch are the images of the boundaries of the domain polygon. We are now going to show that these boundary curves are actually the Bezier curves determined by the boundary control points of the control net. Moreover, when the domain parameters are restricted to a boundary of the domain polygon, restricting the pyramid algorithm to the corresponding lateral triangular face of the pyramid generates the de Casteljau algorithm for this bounding Bezier curve.

Let $Q_1,\dots,Q_n$ be the vertices of the domain polygon, and consider a specific bounding edge $Q_h Q_{h+1}$. All but two of the barycentric coordinate functions vanish along this boundary, since $\beta_g = 0$ on the edge $Q_h Q_{h+1}$, $g \neq h, h+1$. Therefore, along the boundary $Q_h Q_{h+1}$ only the corresponding boundary control points—the control points $P_{(d-k)i_h + k i_{h+1}}$, $k = 0,\dots,d$—contribute to the boundary of the surface. Moreover, since the barycentric coordinate functions sum to one, $\beta_{h+1} = 1 - \beta_h$ along the boundary Q_hQ_{h+1}. Hence along this boundary, the pyramid algorithm for the S-patch reduces to de Casteljau's triangle along the lateral face of the pyramid for the boundary control points, with the edges in the algorithm labeled by β_h and $1 - \beta_h$. It follows that the boundary curves of an S-patch of depth d are just the Bezier curves of degree d (reparametrized by setting $t = \beta_h$) determined by the boundary control points.

S-patches are not, in general, polynomial patches because the barycentric coordinates of arbitrary polygons are not polynomial functions. Instead, since the barycentric coordinates of an n-gon are rational functions of degree $n - 2$, S-patches are rational surfaces of parametric degree $d(n - 2)$, where n is the number of sides of the domain polygon and d is the depth of the control net, or equivalently the number of levels in the pyramid algorithm. (Here we use the barycentric coordinate functions constructed in Equation (8.4); different barycentric coordinate functions could give higher parametric degrees.) Nevertheless, although an S-patch is a rational surface, it follows by our preceding analysis that the boundaries of an S-patch are polynomial curves.

The rational blending functions $\{S_\lambda^d(u,v)$ for an S-patch of depth d can be computed by either the up recurrence or the down recurrence. In the up recurrence, we place a 1 in the position indexed by $\lambda \in I^d$ and a 0 everywhere else at the base of the pyramid and run the pyramid algorithm. The function that emerges at the apex of the pyramid is $S_\lambda^d(u,v)$. In the down recurrence we place a 1 at the apex of the pyramid, reverse all the arrows, and run the recurrence. The functions that emerge at the base of the pyramid are the blending functions $\{S_\lambda^d(u,v)\}$. The S-patch with control points $\{P_\lambda\}$ is given by

$$S(u,v) = \sum_{\lambda \in I^d} S_\lambda^d(u,v)P_\lambda .$$

Another simple and convenient way to construct the bivariate Bernstein basis functions is via discrete convolution (see Section 5.8.2). We can apply this convolution technique here as well to generate both explicit and recursive formulas for the S-patch blending functions.

A word first about notation. We are, of course, using the same indexing for our blending functions that we use for our control points. Let $\beta_1(u,v),\ldots,\beta_n(u,v)$ be the barycentric coordinates of the domain polygon. Since these barycentric coordinate functions are the blending functions for an S-patch of depth 1,

$$\beta_k(u,v) \equiv \beta_{i_k}(u,v) \qquad k = 1,\ldots,n.$$

Thus the index set I indexes the barycentric coordinate functions.

Adopting this convention, we now can define an array of blending functions $\{S_\lambda^d(u,v)\}$, $\lambda \in I^d$, using discrete convolution by setting

$$\{S_\lambda^d(u,v)\} = \underbrace{\{\beta_1(u,v),\ldots,\beta_n(u,v)\} \otimes \cdots \otimes \{\beta_1(u,v),\ldots,\beta_n(u,v)\}}_{d \; factors}. \tag{8.6}$$

Notice that there are exactly as many blending functions $\{S_\lambda^d(u,v)\}$ as there are control points $\{P_\rho\}$ in a polygonal array of depth d, since both sets are indexed by I^d. Now it follows by induction from Equation (8.6) that the functions $\{S_\lambda^d(u,v)\}$ satisfy the n-term recurrence

$$S_\lambda^d(u,v) = \beta_1(u,v)S_{\lambda-i_1}^{d-1}(u,v) + \cdots + \beta_n(u,v)S_{\lambda-i_n}^{d-1}(u,v), \tag{8.7}$$

where, by convention, $S_{\lambda-i_k}^{d-1}(u,v) = 0$ if $\lambda - i_k \notin I^{d-1}$.

Equation (8.7) is equivalent to the down recurrence. Hence the functions $\{S_\lambda^d(u,v)\}$ constructed from discrete convolution are indeed the blending functions for an *S*-patch of depth d whose domain is the convex polygon with ordered vertices $Q_1,...,Q_n$ and whose indexing set is I^d. Moreover, it follows from Equation (8.6) and the definition of discrete convolution that the blending functions for an *S*-patch are given explicitly by the formula

$$S_\lambda^d(u,v) = \sum_{k_1 i_1 + \cdots + k_n i_n = \lambda} \binom{d}{k_1 \cdots k_n} \beta_1^{k_1}(u,v) \cdots \beta_n^{k_n}(u,v), \tag{8.8}$$

Since *S*-patches are already rational surfaces we may as well introduce scalar weights $\{w_\lambda\}$ and define rational *S*-patches by setting

$$R(u,v) = \frac{\sum_{\lambda \in I^d} S_\lambda^d(u,v) w_\lambda P_\lambda}{\sum_{\rho \in I^d} w_\rho S_\rho^d(u,v)}. \tag{8.9}$$

There is also a pyramid algorithm for rational *S*-patches: simply replace the input $P_\lambda^0(u,v) = P_\lambda$ at the base of the pyramid by $P_\lambda^0(u,v) = (w_\lambda P_\lambda, w_\lambda)$ and divide the output of the algorithm by the weight. One advantage of rational *S*-patches is that we can ignore the common denominator in the barycentric coordinate functions $\beta_1(s,t),...,\beta_n(s,t)$, since this denominator cancels in Equation (8.9). The numerators $\alpha_1(s,t),...,\alpha_n(s,t)$ of the barycentric coordinate functions are polynomials, so the pyramid algorithm performs a polynomial computation in Grassmann space and then divides by the weight to get the rational surface $R(u,v)$ in affine space.

Exercises

1. Give three proofs that $\sum_{\lambda \in I^d} S_\lambda^d(u,v) \equiv 1$.

2. Give an example to show that the blending functions of an *S*-patch are not necessarily linearly independent.

3. Show that both triangular and tensor product Bezier patches are special cases of *S*-patches.

4. Let $I = \{0,1\}$. Show that

   a. $I^d = \{0,...,d\}$.

b. The pyramid algorithm for the points $\{P_i\}$, $i \in I^d$, is the de Casteljau algorithm for the Bezier curve with control points $\{P_i\}$.

5. Let $I = \{0,1,2\}$.

    a. Show that

        i. $I^d = \{0,\ldots,2d\}$.

        ii. The S-patch for the points $\{P_\lambda\}$ indexed by I^d is a polynomial patch of degree d.

    b. Implement the pyramid algorithm for points $\{P_\lambda\}$ indexed by I^d.

6. Implement the pyramid algorithm for points $\{P_\lambda\}$ indexed by I^d, where

    a. I consists of the vertices of the standard 4-simplex

    b. $I = \{(0,0),(1,0),(2,1),(1,2),(0,1)\}$

    Compare the pentagonal surfaces generated by these two indexing sets.

7. For $k = 1,\ldots,d$, let Q_k denote a convex polygon with ordered vertices $Q_{i_{k1}},\ldots,Q_{i_{km}}$ and barycentric coordinate functions $\beta_{i_{k1}}(u,v),\ldots,\beta_{i_{km}}(u,v)$ indexed by a set of p-tuples I_k. Generalize the pyramid algorithm for S-patches by replacing I^{d-l} with $I_1 \oplus \cdots \oplus I_{d-l}$.

    a. What is the domain of this patch?

    b. How must Equation (8.5) be modified to make the pyramid algorithm valid?

    c. Describe the blending functions for this patch.

8. Let $I = \{i_1,\ldots,i_n\}$ be the vertices of a planar polygon. Show that if we use the barycentric coordinate functions in Section 8.1, Exercise 4, then the corresponding S-patch of depth d with domain I and control points indexed by I^d reproduces linear functions on I^d.

9. Show that the boundaries of a rational S-patch are the rational Bezier curves determined by the mass-points along the boundary of the control net.

10. Let $\{S_\lambda^d(u,v)\}$ be a collection of S-patch blending functions, and let $\{w_\lambda\}$ be a collection of nonzero scalar weights. Define

$$R_\lambda^d(u,v) = \frac{w_\lambda S_\lambda^d(u,v)}{\sum\limits_{\mu \in I^d} w_\mu S_\mu^d(u,v)}, \quad \lambda \in I^d.$$

Show that these functions behave like rational S-patch basis functions. In particular, show that

    a. $\sum_{\lambda \in I^d} R_\lambda^d(u,v) \equiv 1$

    b. $R(u,v) = \sum_{\lambda \in I^d} R_\lambda^d(u,v)P_\lambda$

8.4.2 Simplicial *S*-Patches

A *simplicial S-patch* is an n-sided S-patch whose index set is $(\Delta^{n-1})^d$, where Δ^{n-1} consists of the n vertices $\{(1,0,\ldots,0),\ldots,(0,\ldots,0,1)\}$ of the standard $(n-1)$-simplex. Simplicial S-patches have the following two rather special properties:

    i.   Every n-sided S-patch of depth d can be represented as an n-sided simplicial S-patch of depth d.

    ii. Every polynomial surface of degree d can be represented as an n-sided simplicial S-patch of depth d, where the barycentric coordinate functions are chosen as in Section 8.1, Exercise 3, and normalized as in Section 8.1, Exercise 4.

To establish the first property, observe that

$$(\Delta^{n-1})^d = \{(k_1,\ldots,k_n) \mid \sum_{j=1}^{n} k_j = d\}.$$

Therefore, by Equation (8.8), the blending functions for a simplicial S-patch of depth d are given by

$$T_{k_1\cdots k_n}^d(u,v) = \binom{d}{k_1\cdots k_n}\beta_1^{k_1}(u,v)\cdots\beta_n^{k_n}(u,v). \tag{8.10}$$

Let $\{S_\lambda^d(u,v)\}$ denote the blending functions for an arbitrary n-sided S-patch of depth d with the same domain polygon, but with index set I^d, where $I = \{i_1,\ldots,i_n\}$. Then by Equations (8.8) and (8.10),

$$S_\lambda^d(u,v) = \sum_{k_1 i_1 + \cdots + k_n i_n = \lambda} \binom{d}{k_1\cdots k_n}\beta_1^{k_1}(u,v)\cdots\beta_n^{k_n}(u,v) = \sum_{k_1 i_1 + \cdots + k_n i_n = \lambda} T_{k_1\cdots k_n}^d(u,v).$$

Now suppose that $S(u,v)$ is an arbitrary n-sided S-patch of depth d with control points $\{P_\lambda\}$. Let $T(u,v)$ be the simplicial S-patch of depth d with control points $\{Q_{k_1\cdots k_n}\}$, where

$$Q_{k_1\cdots k_n} = P_{k_1 i_1 + \cdots + k_n i_n}.$$

Then

$$T(u,v) = \sum_{k_1 + \cdots + k_n = d} T_{k_1\cdots k_n}^d(u,v)Q_{k_1\cdots k_n} = \sum_{\lambda \in I^d} \{\sum_{k_1 i_1 + \cdots + k_n i_n = \lambda} T_{k_1\cdots k_n}^d(u,v)\} P_\lambda$$

$$= \sum_{\lambda \in I^d} S_\lambda^d(u,v)P_\lambda = S(u,v) \ .$$

Hence the S-patch of depth d with control points $\{P_\lambda\}$ is equivalent to the simplicial S-patch of depth d with control points $\{Q_{k_1\cdots k_n}\}$. Thus by equating specific control points in the control net, we can represent any arbitrary S-patch by a simplicial S-patch of the same depth.

Moreover, every polynomial surface $P(u,v)$ of degree d can be represented as an n-sided simplicial S-patch of depth d. To understand how, let $Q_1,\ldots,Q_n$ be the vertices of a convex polygon Q with the barycentric coordinate functions $\beta_1,\ldots,\beta_n$ defined in Section 8.1, Exercise 3, and normalized as in Section 8.1, Exercise 4. By Section 8.1, Exercise 4, these barycentric coordinate functions reproduce linear functions, so, in particular,

$$(u,v) = \beta_1(u,v)Q_1 + \cdots + \beta_n(u,v)Q_n.$$

Let

$$\Gamma^n = \{(u_1,\ldots,u_n) \mid u_1,\ldots,u_n \geq 0 \;\; \text{and} \;\; \sum_{k=1}^{n} u_k = 1\},$$

and define the affine map $\pi : \Gamma^n \to Q$ by setting $\pi(u_1,\ldots,u_n) = u_1 Q_1 + \cdots + u_n Q_n$. Then

$$P(u,v) = P\big(\beta_1(u,v)Q_1 + \cdots + \beta_n(u,v)Q_n\big) = (P \circ \pi)(\beta_1(u,v),\ldots,\beta_n(u,v)).$$

But $(P \circ \pi)(u_1,\ldots,u_n)$ is a polynomial map of degree d in n variables, and, just as in the bivariate setting, the Bernstein polynomials

$$B^d_{k_1 \cdots k_n}(u_1,\ldots,u_n) = \binom{d}{k_1 \cdots k_n} u_1^{k_1} \cdots u_n^{k_n}, \quad k_1 + \cdots + k_n = d,$$

form a basis for these polynomials over Γ^n (see Exercise 2). Hence there are points $P_{k_1 \cdots k_n}$ such that

$$(P \circ \pi)(u_1,\ldots,u_n) = \sum_{k_1 + \cdots + k_n = d} B^d_{k_1 \cdots k_n}(u_1,\ldots,u_n) P_{k_1 \cdots k_n}. \tag{8.11}$$

It follows that $P(u,v)$

$$= (P \circ \pi)(\beta_1(u,v),\ldots,\beta_n(u,v))$$

$$= \sum_{k_1 + \cdots + k_n = d} B^d_{k_1 \cdots k_n}(\beta_1(u,v),\ldots,\beta_n(u,v)) P_{k_1 \cdots k_n} = \sum_{k_1 + \cdots + k_n = d} T^d_{k_1 \cdots k_n}(u,v) P_{k_1 \cdots k_n}.$$

Therefore, $P(u,v)$ can be represented as an S-patch of depth d with control points $P_{k_1 \cdots k_n}$.

We can exploit Equation (8.11) to find the S-patch control points $P_{k_1 \cdots k_n}$ of the polynomial surface $P(u,v)$ by evaluating the blossom of P at the vertices of the domain polygon. Since π is degree 1, the blossom of $P \circ \pi$ is

$$p \circ \pi^d = p \circ (\underbrace{\pi \times \cdots \times \pi}_{d \text{ factors}}).$$

Let $E_1,\ldots,E_n$ be the vertices of Δ^{n-1}. Then by the dual functional property of the d-variate blossom

$$P_{k_1 \cdots k_n} = (p \circ \pi^d)(\underbrace{E_1,\ldots,E_1}_{k_1},\ldots,\underbrace{E_n,\ldots,E_n}_{k_n}) = p(\underbrace{\pi(E_1),\ldots,\pi(E_1)}_{k_1},\ldots,\underbrace{\pi(E_n),\ldots,\pi(E_n)}_{k_n}).$$

But $\pi(E_k) = Q_k, \; k = 1,\ldots,n$. Hence

$$P_{k_1 \cdots k_n} = p(\underbrace{Q_1,\ldots,Q_1}_{k_1},\ldots,\underbrace{Q_n,\ldots,Q_n}_{k_n}). \tag{8.12}$$

Exercises

1. Show that the blending functions $\{T^d_{k_1 \cdots k_n}(u,v)\}$ of a simplicial *S*-patch are composites of the multivariate Bernstein basis functions defined by

$$\{B^d_{k_1 \cdots k_n}(u_1,\ldots,u_n)\} = \underbrace{\{u_1,\ldots,u_n\} \otimes \cdots \otimes \{u_1,\ldots,u_n\}}_{d \; factors},$$

where u_j has the index $(0,\ldots,0,\underset{j}{1},0,\ldots,0)$, and the functions $u_j = \beta_j(u,v)$.

2. Prove that the Bernstein polynomials

$$B^d_{k_1 \cdots k_n}(u_1,\ldots,u_n) = \left(\begin{smallmatrix} d \\ k_1 \cdots k_n \end{smallmatrix}\right) u_1^{k_1} \cdots u_n^{k_n}, \quad k_1 + \cdots + k_n = d,$$

form a basis for the polynomials of degree d in n variables over Γ^n. (Hint: Mimic the trick in Section 5.4 for converting the Bernstein basis into the monomial basis.)

3. Consider a tensor product Bezier surface of bidegree d with control points $\{P_{lm}\}$. Show that the control points $\{Q_{k_1 \cdots k_4}\}$ of the equivalent simplicial *S*-patch are given by $Q_{k_1 \cdots k_4} = P_{k_2+k_4, k_3+k_4}$, $k_1 + k_2 + k_3 + k_4 = d$.

4. Let $P(u,v) = (u,v)$ be the identity map, and let $Q_1,\ldots,Q_n$ be the vertices of a convex polygon. Using Equation (8.12), show that the *S*-patch control points for $P(u,v)$ are given by

$$P_{k_1 \cdots k_n} = \frac{k_1 Q_1 + \cdots + k_n Q_n}{d}.$$

5. Here we develop a depth elevation algorithm for simplicial *S*-patches.

   a. Show that

   i. $\beta_j(u,v) T^d_{k_1 \cdots k_n}(u,v) = \dfrac{k_j + 1}{d+1} T^{d+1}_{k_1 \cdots k_j+1 \cdots k_n}(u,v).$

   ii. $T^d_{k_1 \cdots k_n}(u,v) = \sum_j \dfrac{k_j + 1}{d+1} T^{d+1}_{k_1 \cdots k_j+1 \cdots k_n}(u,v)$

   b. Let $\{P^d_{k_1 \cdots k_n}\}$ be the control points for a simplicial *S*-patch $T(u,v)$ of depth d, and let $R(u,v)$ be the simplicial *S*-patch of depth $d+1$ with control points

   $$R^{d+1}_{h_1 \cdots h_n} = \sum_{j=1}^{n} \frac{h_j}{d+1} P^d_{h_1 \cdots h_j-1 \cdots h_n}.$$

   Show that $R(u,v) = S(u,v)$.

(Compare to Section 5.8.2, Exercise 9.)

6. In this exercise you will use the generating function for the blending functions of a simplicial S-patch to derive identities for these S-patch blending functions. Define

$$T^d(u,v,x_1,\ldots,x_n) = \left(\beta_1(u,v)x_1 + \cdots + \beta_n(u,v)x_n\right)^d.$$

a. Show that

$$T^d(u,v,x_1,\ldots,x_n) = \sum_{\lambda \in (\Delta^{n-1})^d} T_\lambda^d(u,v)x_1^{\lambda_1} \cdots x_n^{\lambda_n}.$$

b. Using part (a), derive the following identities:

i.
$$\sum_{\lambda \in (\Delta^{n-1})^d} T_\lambda^d(u,v) \equiv 1$$

ii.
$$\sum_{\lambda \in (\Delta^{n-1})^d} \lambda_k \, T_\lambda^d(u,v) = d\,\beta_k(u,v) \qquad k = 1,\ldots,n$$

iii.
$$\sum_{\lambda \in (\Delta^{n-1})^d} \lambda_{k_1} \cdots \lambda_{k_j} \, T_\lambda^d(u,v) = \frac{d!}{(d-j)!}\,\beta_{k_1}(u,v)\cdots\beta_{k_j}(u,v)$$

iv.
$$\sum_{\lambda \in (\Delta^{n-1})^d} \frac{\lambda_{k_1}! \cdots \lambda_{k_j}!}{(\lambda_{k_1}-i_1)! \cdots (\lambda_{k_j}-i_j)!} T_\lambda^d(u,v) =$$

$$\frac{d!}{(d-i_1-\cdots-i_j)!}\,\beta_{k_1}^{i_1}(u,v)\cdots\beta_{k_j}^{i_j}(u,v)$$

(Hint: Differentiate the generating function and evaluate at $x_1 = \cdots = x_n = 1$.)

8.4.3   **Differentiating S-Patches**

It is easy to find the first-order partial derivatives of an S-patch. Recall from Equation (8.6) that the S-patch blending functions $S_\lambda^d(u,v)$ can be expressed in terms of discrete convolutions of the barycentric coordinate functions:

$$\{S_\lambda^d(u,v)\} = \underbrace{\{\beta_1(u,v),\ldots,\beta_n(u,v)\} \otimes \cdots \otimes \{\beta_1(u,v),\ldots,\beta_n(u,v)\}}_{d \ \ factors}.$$

Hence, by Equation (5.21),

$$\left\{\frac{\partial S_\lambda^d(u,v)}{\partial u}\right\} = d\left\{\frac{\partial \beta_1(u,v)}{\partial u},\ldots,\frac{\partial \beta_n(u,v)}{\partial u}\right\} \otimes \underbrace{\{\beta_1(u,v),\ldots,\beta_n(u,v)\} \otimes \cdots \otimes \{\beta_1(u,v),\ldots,\beta_n(u,v)\}}_{d-1 \ \ factors}$$

$$\left\{\frac{\partial S_\lambda^d(u,v)}{\partial v}\right\} = d\left\{\frac{\partial \beta_1(u,v)}{\partial v},\ldots,\frac{\partial \beta_n(u,v)}{\partial v}\right\} \otimes \underbrace{\{\beta_1(u,v),\ldots,\beta_n(u,v)\} \otimes \cdots \otimes \{\beta_1(u,v),\ldots,\beta_n(u,v)\}}_{d-1 \ \ factors}$$

or equivalently,

$$\frac{\partial S_\lambda^d(u,v)}{\partial u} = d\left(\frac{\partial \beta_1(u,v)}{\partial u} S_{\lambda-i_1}^{d-1}(u,v) + \cdots + \frac{\partial \beta_n(u,v)}{\partial u} S_{\lambda-i_n}^{d-1}(u,v) \right)$$

$$\frac{\partial S_\lambda^d(u,v)}{\partial v} = d\left(\frac{\partial \beta_1(u,v)}{\partial v} S_{\lambda-i_1}^{d-1}(u,v) + \cdots + \frac{\partial \beta_n(u,v)}{\partial v} S_{\lambda-i_n}^{d-1}(u,v) \right).$$

Therefore, the first-order partial derivatives of the *S*-patch

$$S(u,v) = \sum_{\lambda \in I^d} S_\lambda^d(u,v) P_\lambda$$

are given by

$$\frac{\partial S(u,v)}{\partial u} = d \sum_{\rho \in I^{d-1}} S_\rho^{d-1}(u,v) \left(\sum_{i \in I} \frac{\partial \beta_i(u,v)}{\partial u} P_{\rho+i} \right)$$

$$\frac{\partial S(u,v)}{\partial v} = d \sum_{\rho \in I^{d-1}} S_\rho^{d-1}(u,v) \left(\sum_{i \in I} \frac{\partial \beta_i(u,v)}{\partial v} P_{\rho+i} \right).$$

Algorithmically, these formulas say that to find a first-order partial derivative of an *S*-patch of depth *d*, we need only take the first-order partial derivative of the bottom level of the pyramid algorithm, then run the algorithm, and multiply the result by *d*. In fact, it follows by our convolution formulas that we could, if we choose, take the first-order partial derivative of any level of the pyramid algorithm, then run the algorithm, and multiply the result by *d* (see Exercise 2).

Along any boundary of an *S*-patch most of the functions $\{S_\rho^{d-1}(u,v)\}$ vanish because most of the barycentric coordinate functions are zero along the boundary. In fact, along the boundary corresponding to the edge $Q_h Q_{h+1}$ of the domain polygon, $\beta_g = 0$, $g \neq h, h+1$. Let ∂I_h^{d-1} denote the indices $(d-1-k)i_h + ki_{h+1}$, $k = 0,...,$ $d-1$—that is, the indices corresponding to a boundary of an *S*-patch of depth $d-1$. Then, by Equation (8.6), along the boundary corresponding to the edge $Q_h Q_{h+1}$

$$S_\rho^{d-1}(u,v) = 0, \quad \rho \notin \partial I_h^{d-1}.$$

Therefore, along this boundary

$$\frac{\partial S(u,v)}{\partial u} = d \sum_{\rho \in \partial I_h^{d-1}} S_\rho^{d-1}(u,v) \left(\sum_{i \in I} \frac{\partial \beta_i(u,v)}{\partial u} P_{\rho+i} \right)$$

$$\frac{\partial S(u,v)}{\partial v} = d \sum_{\rho \in \partial I_h^{d-1}} S_\rho^{d-1}(u,v) \left(\sum_{i \in I} \frac{\partial \beta_i(u,v)}{\partial v} P_{\rho+i} \right).$$

It follows then that only the control points indexed by the elements of $\partial I_h^{d-1} \oplus I$ affect the first-order partial derivatives along, or the directional derivatives across, this boundary.

Higher-order partial derivatives are not much more difficult to compute. To simplify our notation, let

$$\beta_I(u,v) = \{\beta_1(u,v),\ldots,\beta_n(u,v)\}$$

$$\frac{\partial\beta_I(u,v)}{\partial u} = \left\{\frac{\partial\beta_1(u,v)}{\partial u},\ldots,\frac{\partial\beta_n(u,v)}{\partial u}\right\}$$

and so on for higher-order derivatives. Then since

$$\left\{\frac{\partial S_\lambda^d(u,v)}{\partial u}\right\} = d\frac{\partial\beta_I(u,v)}{\partial u} \otimes \underbrace{\beta_I(u,v) \otimes \cdots \otimes \beta_I(u,v)}_{d-1\ \ factors},$$

it follows by Equation (5.21) that

$$\left\{\frac{\partial^2 S_\lambda^d(u,v)}{\partial u^2}\right\} = d\frac{\partial^2\beta_I(u,v)}{\partial u^2} \otimes \underbrace{\beta_I(u,v) \otimes \cdots \otimes \beta_I(u,v)}_{d-1\ \ factors}$$

$$+ d(d-1)\frac{\partial\beta_I(u,v)}{\partial u} \otimes \frac{\partial\beta_I(u,v)}{\partial u} \otimes \underbrace{\{\beta_1(u,v),\ldots,\beta_n(u,v)\} \otimes \cdots \otimes \{\beta_1(u,v),\ldots,\beta_n(u,v)\}}_{d-2\ \ factors},$$

and similar results hold for the other second-order partial derivatives of the blending functions.

Again this formula has an interesting algorithmic interpretation for differentiating S-patches. To find a second-order partial derivative of an S-patch of depth d, we can proceed in the following fashion:

1. Take the second-order partial derivative of one level of the pyramid algorithm, then run the algorithm, and multiply the result by d.

2. Take the first-order partial derivative of two different levels of the pyramid algorithm, then run the algorithm, and multiply the result by $d(d-1)$.

3. Add the results of (1) and (2).

For higher-order partial derivatives, there are similar formulas and similar algorithms, involving higher-order derivatives as well as derivatives of more and more levels of the pyramid algorithm (see Exercise 4). Since such formulas are rarely necessary in practice, and since, in any event, these formulas can be derived from the product rule, we shall not pursue this topic further here.

Exercises

1. Show that the normal of an S-patch of depth d along the boundary corresponding to the edge $Q_h Q_{h+1}$ of the domain polygon depends only on the control points indexed by $\partial I_h^{d-1} + I$.

2. Show that to find the first-order partial derivatives of an *S*-patch of depth *d*, we could, if we choose, take the first-order partial derivative of any level of the pyramid algorithm, then run the algorithm, and multiply the result by *d*. Explain why in some cases it might be better to take the derivative of the last level of the algorithm, instead of the first level.

3. Consider an *S*-patch whose domain polygon has ordered vertices $Q_1,...,Q_n$.

   a. Show that only the control points indexed by the elements of $\partial I_h^{d-2} \oplus I^2$ affect the second-order partial derivatives of the *S*-patch along the boundary $Q_h Q_{h+1}$.

   b. Generalize the result in part (a) to higher-order partial derivatives.

4. Develop an algorithm for finding the third-order partial derivatives of an *S*-patch. How many different pyramids must you compute? What are the normalizations for each of these pyramids?

8.4.4 Blossoming *S*-Patches

We know how to define the blossom for a polynomial surface, but an arbitrary *S*-patch is not a polynomial surface. How then should we define the blossom of an *S*-patch? Since *S*-patches are generalizations of triangular Bezier patches, we will take our cue from blossoming for bivariate Bernstein polynomials.

One way to blossom a triangular Bezier patch is to start with de Casteljau's tetrahedral algorithm and replace the parameters (u,v) by a different parameter pair (u_k,v_k) on each level of the algorithm. Let's try the same device with *S*-patches of depth *d*. On the *k*th level of the pyramid algorithm for an *S*-patch $S(u,v)$, replace the parameters (u,v) by the parameter pair (u_k,v_k). The effect is to replace the values of the barycentric coordinate functions $\beta_1(u,v),...,\beta_n(u,v)$ on the *k*th level of the algorithm by the values $\beta_1(u_k,v_k),...,\beta_n(u_k,v_k)$. The function that emerges at the apex of the pyramid we shall call the *blossom* of the *S*-patch $S(u,v)$, and we shall denote this blossom by $s((u_1,v_1),...,(u_d,v_d))$.

The blossom $s((u_1,v_1),...,(u_d,v_d))$ of an *S*-patch $S(u,v)$ of depth *d* with domain polygon $Q_1,...,Q_n$ and index set $I = (i_1,...,i_n)$ has the following properties:

i. *Symmetry*

$$s((u_1,v_1),...,(u_d,v_d)) = s((u_{\sigma(1)},v_{\sigma(1)}),...,(u_{\sigma(d)},v_{\sigma(d)}))$$

ii. *Diagonal*

$$s((u,v),...,(u,v)) = S(u,v)$$

iii. *Dual functional*

$$s\left(\underbrace{Q_1,...,Q_1}_{k_1},...,\underbrace{Q_n,...,Q_n}_{k_n}\right) = P_{k_1 i_1 + \cdots + k_n i_n}$$

Notice that the multiaffine property no longer holds, in general, for the blossom of an S-patch because the functions $\beta_1(u_k,v_k),\dots,\beta_n(u_k,v_k)$ are not linear functions for arbitrary S-patches. The multiaffine property will hold only when $n = 3$, or when $n = 4$ and the domain is a rectangle with sides parallel to the coordinate axes. We can, however, prove the validity of the other properties of the blossom of an S-patch.

The diagonal property is easiest to establish. Since we blossom the pyramid algorithm by replacing the pair (u,v) by the pairs (u_k,v_k), reversing the process and replacing the pairs (u_k,v_k) by the pair (u,v) certainly retrieves the original surface.

The dual functional property can be established by the following argument. At the jth vertex of the domain polygon, all the barycentric coordinates except the jth one vanish—that is,

$$\beta_h(Q_j) = 0 \qquad h \ne j$$

$$= 1 \qquad h = j.$$

Hence if we evaluate the pyramid algorithm for the blossom at

$$\underbrace{Q_1,\dots,Q_1}_{k_1},\dots,\underbrace{Q_n,\dots,Q_n}_{k_n},$$

then on the first k_1 levels $\beta_1 = 1$ and the remaining $\beta_h = 0$; on the next k_2 levels $\beta_2 = 1$ and the remaining $\beta_h = 0$ ...; until on the last k_n levels $\beta_n = 1$ and the remaining $\beta_h = 0$. Thus tracing through the pyramid algorithm, only the control point with index $k_1i_1 + \dots + k_ni_n$ survives to the apex of the pyramid.

The symmetry property requires a bit more work. To establish this property, we shall take a different, alternative approach to blossoming for S-patches by developing formulas for the blossoms of the blending functions.

We can easily blossom the blending functions $\{S_\lambda^d(u,v)\}$. Recall from Equation (8.6) that

$$\{S_\lambda^d(u,v)\} = \underbrace{\{\beta_1(u,v),\dots,\beta_n(u,v)\} \otimes \dots \otimes \{\beta_1(u,v),\dots,\beta_n(u,v)\}}_{d \;\; factors}.$$

To blossom the blending functions, we need only replace the pair (u,v) by a different pair (u_k,v_k) in each factor of the convolution. Thus

$$\left\{s_\lambda^d\big((u_1,v_1),\dots,(u_d,v_d)\big)\right\}$$
$$= \underbrace{\{\beta_1(u_1,v_1),\dots,\beta_n(u_1,v_1)\} \otimes \dots \otimes \{\beta_1(u_d,v_d),\dots,\beta_n(u_d,v_d)\}}_{d \;\; factors}. \qquad (8.13)$$

This substitution is equivalent to replacing the parameters (u,v) by the parameter pair (u_k,v_k) on the kth level of the down recurrence for $\{S_\lambda^d(u,v)\}$, since, by induction, Equation (8.13) is equivalent to

$$s_\lambda^d\big((u_1,v_1),\dots,(u_d,v_d)\big) = \beta_1(u_d,v_d)s_{\lambda-i_1}^{d-1}\big((u_1,v_1),\dots,(u_{d-1},v_{d-1})\big) +$$
$$\dots + \beta_n(u_d,v_d)s_{\lambda-i_n}^{d-1}\big((u_1,v_1),\dots,(u_{d-1},v_{d-1})\big),$$

with the usual convention that $s_{\lambda - i_k}^{d-1}(u,v) = 0$ if $\lambda - i_k \notin I^{d-1}$.

We can now establish all the properties of the blossom of an *S*-patch directly from Equation (8.13) for the blossom of the blending functions. As before the diagonal property is easy to prove. Moreover, since convolution is commutative, it follows immediately from Equation (8.13) that blossoming is symmetric in the parameter pairs $(u_1,v_1),\ldots,(u_d,v_d)$.

To establish the dual functional property for the blending functions, substitute

$$\underbrace{Q_1,\ldots,Q_1}_{k_1},\ldots,\underbrace{Q_n,\ldots,Q_n}_{k_n}$$

for the parameters $(u_1,v_1),\ldots,(u_d,v_d)$ in Equation (8.13). Then

$$\left\{ s_\lambda^d(\underbrace{Q_1,\ldots,Q_1}_{k_1},\ldots,\underbrace{Q_n,\ldots,Q_n}_{k_n}) \right\} = \underbrace{\{\beta_1(Q_1),\ldots,\beta_n(Q_1)\}^{k_1} \otimes \cdots \otimes \{\beta_1(Q_n),\ldots,\beta_n(Q_n)\}^{k_n}}_{d \; factors},$$

where the power k_j means repeat the corresponding factor k_j times. But

$$\beta_k(Q_j) = 0 \qquad k \neq j$$
$$= 1 \qquad k = j,$$

so

$$s_\lambda^d(\underbrace{Q_1,\ldots,Q_1}_{k_1},\ldots,\underbrace{Q_n,\ldots,Q_n}_{k_n}) = 0 \qquad k_1 i_1 + \cdots + k_n i_n \neq \lambda$$
$$= 1 \qquad k_1 i_1 + \cdots + k_n i_n = \lambda .$$

Since

$$S(u,v) = \sum_{\lambda \in I^d} S_\lambda^d(u,v) P_\lambda,$$

it follows by linearity that

$$s\left(\underbrace{Q_1,\ldots,Q_1}_{k_1},\ldots,\underbrace{Q_n,\ldots,Q_n}_{k_n} \right) = P_{k_1 i_1 + \cdots + k_n i_n} .$$

Is this approach to blossoming useful? Computations with the standard blossom are made simple by the multiaffine property. This property is not available for the blossom of an *S*-patch, so actual computations are difficult to perform. Even though we have a dual functional property for *S*-patches, it is unclear how to exploit this property to generate change of basis algorithms for *S*-patches as we did for Bezier patches using blossoming.

One way to look at the problem is that there are too many terms in the factors of the convolution in Equation (8.13); n functions cannot be both linear and independent. But if each of these factors were itself to factor into products of simpler convolutions, then the computations would simplify. This serendipity is exactly what happens for tensor product surfaces, where there are four bilinear barycentric coordinate functions, but these four functions factor into the convolution of two sets of linear functions. So it is such blending functions, generated by convolutions that factor into linear factors, that we would like to study. The corresponding surfaces are called *C*-patches. But before we can investigate *C*-patches, we first need to generalize the pyramid algorithm so that it is valid for surfaces other than *S*-patches.

Exercises

1. Consider a tensor product Bezier surface $S(u,v)$ of bidegree d with control points $\{P_{lm}\}$. Show that for any integer r satisfying $0 \le r \le \min(k,l)$

$$P_{lm} = s\left(\underbrace{(0,0)}_{d-l-m+r}, \underbrace{(1,0),}_{l-r} \underbrace{(1,0),}_{m-r} \underbrace{(1,1)}_{r} \right).$$

2. Show that for an arbitrary *S*-patch, different choices of $k_1,\dots,k_n$ in the expression

$$s\left(\underbrace{Q_1,\dots,Q_1}_{k_1},\dots,\underbrace{Q_n,\dots,Q_n}_{k_n} \right)$$

can generate the same control point $\{P_\lambda\}$, but that for simplicial *S*-patches the choice of $k_1,\dots,k_n$ is unique for each control point $\{P_\lambda\}$.

3. Let $I = \{0,1,2\}$.

   a. Show that the blossom for a three-sided *S*-patch with indexing set I^d is multiaffine.

   b. Conclude that the blossom for a three-sided *S*-patch with indexing set I^d is the standard blossom for bivariate polynomials of degree d.

   c. Use the result in part (b) to develop an algorithm to convert three-sided *S*-patches with indexing set I^d to triangular Bezier form.

4. Consider the *S*-patch in Section 8.4.1, Exercise 7, generated by replacing I^{d-l} with $I_1 \oplus \cdots \oplus I_{d-l}$, where $I_1,\dots,I_d$ are d sets of p-tuples.

   a. What is the blossom of this *S*-patch?

   b. What is the dual functional property for this *S*-patch?

5. a. Explain how to blossom a rational *S*-patch.

   b. What is the dual functional property for rational *S*-patches?

6. Suppose that $S(u,v)$ is an n-sided *S*-patch of depth d with blending functions

$$\{S_\lambda^d(u,v)\} = \underbrace{\{\beta_{i_1}(u,v),\dots,\beta_{i_n}(u,v)\} \otimes \cdots \otimes \{\beta_{i_1}(u,v),\dots,\beta_{i_n}(u,v)\}}_{d \ \ factors}.$$

Let $\hat{S}(u_{i_1},\ldots,u_{i_n})$ be the polynomial generated by replacing the barycentric coordinate functions $\beta_{i_1}(u,v),\ldots,\beta_{i_n}(u,v)$ with the parameters $u_{i_1},\ldots,u_{i_n}$ in the pyramid algorithm for $S(u,v)$, and let

$$\{\hat{S}^d_{\lambda}(u_{i_1},\ldots,u_{i_n})\} = \underbrace{\{u_{i_1},\ldots,u_{i_n}\} \otimes \cdots \otimes \{u_{i_1},\ldots,u_{i_n}\}}_{d \ \text{factors}}.$$

Then $\hat{S}(u_{i_1},\ldots,u_{i_n})$ and $\hat{S}^d_{\lambda}(u_{i_1},\ldots,u_{i_n})$ are polynomials of degree d in n variables with polynomial blossoms

$$\hat{s}\big((u_{11},\ldots,u_{1n}),\ldots,(u_{d1},\ldots,u_{dn})\big) \text{ and } \hat{s}^d_{\lambda}\big((u_{11},\ldots,u_{1n}),\ldots,(u_{d1},\ldots,u_{dn})\big).$$

Show that

a. $s^d_{\lambda}\big((u_1,v_1),\ldots,(u_d,v_d)\big) =$

$$\hat{s}^d_{\lambda}\Big((\beta_{i_1}(u_1,v_1),\ldots,\beta_{i_n}(u_1,v_1)),\ldots,(\beta_{i_1}(u_d,v_d),\ldots,\beta_{i_n}(u_d,v_d))\Big)$$

b. $s\big((u_1,v_1),\ldots,(u_d,v_d)\big) =$

$$\hat{s}\Big((\beta_{i_1}(u_1,v_1),\ldots,\beta_{i_n}(u_1,v_1)),\ldots,(\beta_{i_1}(u_d,v_d),\ldots,\beta_{i_n}(u_d,v_d))\Big)$$

7. Let $S(u,v)$ be a simplicial n-sided S-patch of depth d with domain polygon $Q_1,\ldots,Q_n$ and control points $\{P^d_{k_1\cdots k_n}\}$.

a. Show that $s\left(\underbrace{Q_1,\ldots,Q_1}_{k_1},\ldots,\underbrace{Q_n,\ldots,Q_n}_{k_n}\right) = P^d_{k_1\cdots k_n}.$

b. Let $R(u,v)$ be the simplicial S-patch of depth $d+1$ with control points

$$R^{d+1}_{h_1\cdots h_n} = \sum_{j=1}^{n} \frac{h_j}{d+1} P^d_{h_1\cdots h_j-1\cdots h_n}.$$

Using part (a), Exercise 6(b), and the multivariate version of Proposition 6.5, show that $R(u,v) = S(u,v)$. (Compare to Section 6.5.1, Exercise 14.)

8.5 **Pyramid Patches and the General Pyramid Algorithm**

Consider again the pyramid algorithm in Equation (8.5), where we use the set I to index the barycentric coordinate functions.

The General Pyramid Algorithm

$$1. \quad P_\lambda^0(u,v) = P_\lambda \qquad\qquad\qquad\qquad \lambda \in I^d$$

$$2. \quad P_\rho^l(u,v) = \sum_{i \in I} \beta_i(u,v) P_{\rho+i}^{l-1}(u,v) \qquad\qquad \rho \in I^{d-l}$$

$$3. \quad P(u,v) = P_0^d(u,v) \ . \tag{8.14}$$

The key step is clearly the recursion in step 2. What do we really need to make this step a valid recurrence?

If we want to guarantee that the surface patch $P(u,v)$ is affine invariant and lies in the convex hull of the control points $\{P_\lambda\}$, then the functions $\{\beta_i(u,v)\}_{i \in I}$ must sum to one and be nonnegative inside the domain polygon. This constraint is all we require. For S-patches, the number of elements in I is required to match the number of vertices in the domain polygon. But the recurrence in Equation (8.14) will work even without this constraint.

A *pyramid patch*, then, or more simply a *P-patch*, is a surface $P(u,v)$ defined by the general pyramid algorithm in Equation (8.14), where the functions $\{\beta_i(u,v)\}_{i \in I}$ sum to one and are nonnegative inside some domain polygon. In analogy with S-patches, we shall call the functions $\{\beta_i(u,v)\}_{i \in I}$ the *barycentric coordinate functions* for the P-patch. Clearly every S-patch is a P-patch, but P-patches represent other surfaces as well. Since the barycentric coordinate functions $\beta_{i_1}(u,v),\dots,\beta_{i_n}(u,v)$ for the P-patch need no longer be barycentric coordinates for the domain polygon $Q_1,\dots,Q_m$—in fact, in general, $n \geq m$—we may give up the conditions

$$\beta_{i_k} = 0 \text{ on the line } Q_i Q_{i+1}, \ k \neq i, i+1.$$

$$\beta_{i_k}(Q_j) = 0 \qquad j \neq k$$

$$\qquad\qquad = 1 \qquad j = k \ .$$

Eliminating these constraints will cause us to lose some control over the boundaries of the P-patch, but this loss is the price we pay for generality. Nevertheless, when the barycentric coordinate functions $\{\beta_i(u,v)\}_{i \in I}$ are well chosen, P-patches have other compensating features. In Section 8.6, we shall see that if we select the functions $\{\beta_i(u,v)\}_{i \in I}$ in an appropriate fashion, there are more efficient recursive evaluation algorithms for P-patches than for S-patches. In this case there is also a novel blossoming procedure that provides the dual functionals for P-patches. In Section 8.7, we shall show how to define the barycentric coordinate functions $\{\beta_i(u,v)\}_{i \in I}$ to regain control over the boundaries of the patch.

Most of the formulas we developed for S-patches extend readily to P-patches. The blending functions $\{P_\lambda^d(u,v)\}$ for a P-patch of depth d can be computed either by the up recurrence or by the down recurrence. The P-patch with control points $\{P_\lambda\}$ is given by

$$P(u,v) = \sum_{\lambda \in I^d} P_\lambda^d(u,v) P_\lambda \ .$$

By introducing scalar weights $\{w_\lambda\}$, we can define rational P-patches by setting

$$R(u,v) = \frac{\sum\limits_{\lambda \in I^d} P_\lambda^d(u,v) w_\lambda P_\lambda}{\sum\limits_{\rho \in I^d} w_\rho P_\rho^d(u,v)}.$$

There is also a pyramid algorithm for rational P-patches: simply replace the input $P_\lambda^0(u,v) = P_\lambda$ at the base of the pyramid by $P_\lambda^0(u,v) = (w_\lambda P_\lambda, w_\lambda)$ and divide the output of the algorithm by the weight. Notice that for rational P-patches the barycentric coordinate functions need not even sum to one, since the blending in Equation (8.14) occurs in Grassmann space, not in affine space.

We can compute the blending functions $\{P_\lambda^d(u,v)\}$ by discrete convolution. Let $I = \{i_1, \ldots, i_n\}$. Then

$$\{P_\lambda^d(u,v)\} = \underbrace{\{\beta_{i_1}(u,v), \ldots, \beta_{i_n}(u,v)\} \otimes \cdots \otimes \{\beta_{i_1}(u,v), \ldots, \beta_{i_n}(u,v)\}}_{d \text{ factors}}. \tag{8.15}$$

It follows by induction from Equation (8.15) that the functions $\{P_\lambda^d(u,v)\}$ satisfy the n-term recurrence

$$P_\lambda^d(u,v) = \beta_{i_1}(u,v) P_{\lambda-i_1}^{d-1}(u,v) + \cdots + \beta_{i_n}(u,v) P_{\lambda-i_n}^{d-1}(u,v), \tag{8.16}$$

with the usual convention that $P_{\lambda-i_k}^{d-1}(u,v) = 0$ if $\lambda - i_k \notin I^{d-1}$.

Equation (8.16) is equivalent to the down recurrence for P-patches. Hence the functions $\{P_\lambda^d(u,v)\}$ constructed from discrete convolution are indeed the blending functions for the P-patch of depth d whose indexing set is I^d. Moreover, it follows from the definition of discrete convolution that the blending functions for a P-patch are given explicitly by the formula

$$P_\lambda^d(u,v) = \sum_{k_1 i_1 + \cdots + k_n i_n = \lambda} \binom{d}{k_1 \cdots k_n} \beta_{i_1}^{k_1}(u,v) \cdots \beta_{i_n}^{k_n}(u,v). \tag{8.17}$$

All of the preceding formulas for P-patches are already familiar to us from S-patches, but there are other properties of S-patches that do not extend so readily to P-patches.

The boundaries of a P-patch are the images of the boundaries of its domain polygon. However, unlike an S-patch, the boundaries of a P-patch are not, in general, the Bezier curves generated by the control points indexed by $(d-k)i_r + ki_{r+1}$, $k = 0, \ldots, d$. In fact, since the number of elements in I need no longer match the number of vertices in the domain polygon, we should no longer expect points indexed in this fashion to have any special relation to a boundary of the patch. This lack of control over the boundary curves is one thing we lose when we generalize from S-patches to P-patches. We shall see how to overcome this problem in Section 8.7, where we discuss toric Bezier patches.

What about blossoming for P-patches? We can certainly try to blossom a P-patch just like we blossom an S-patch, by replacing the parameters (u,v) by a different parameter pair (u_k, v_k) on each level of the pyramid algorithm. This function would be symmetric and satisfy the diagonal property just like the blossom of an S-patch, but, in general, the dual functional property would no longer hold because the dual functional property is a consequence of the identity

$$\beta_{i_k}(Q_j) = 0 \qquad j \neq k$$
$$= 1 \qquad j = k ,$$

which is no longer valid for arbitrary P-patches. We shall see, however, in the next section that if we choose the functions $\beta_{i_1}(u,v), \ldots, \beta_{i_n}(u,v)$ carefully, then there is an alternative blossoming procedure that does provide the dual functionals for P-patches, and this blossoming for P-patches is simpler than the blossoming for S-patches with the same domain and depth.

Exercises

1. For $k = 1, \ldots, d$, let Q_k denote a convex polygon and let $\beta_{i_{k1}}(u,v), \ldots, \beta_{i_{km}}(u,v)$ be a collection of barycentric coordinate functions for Q_k indexed by a set of p-tuples I_k. Generalize the pyramid algorithm for P-patches by replacing I^{d-l} with $I_1 \oplus \cdots \oplus I_{d-l}$.

   a. What is the domain of this patch?

   b. How must Equation (8.14) be modified to make the pyramid algorithm valid?

   c. Describe the blending functions for this patch.

   (Compare to Section 8.4.1, Exercise 7.)

2. Show that the partial derivative with respect to u or v of a pyramid patch of depth d can be computed by differentiating any one level of the pyramid algorithm and multiplying the result by d. Explain how to extend this result to second-order partial derivatives.

8.6   C-Patches

C-patches are convolutions of S-patches—or more accurately a C-patch is a P-patch whose barycentric coordinate functions are generated by convolution from the barycentric coordinate functions of several S-patches. The C in C-patch stands for *convolution*.

Since we are going to use several S-patches at the same time, we will indicate the kth S-patch by the subscript k. To fix our notation, let $(Q_{k1}, \ldots, Q_{kn_k})$ be the ordered vertices of the domain polygon Q_k with barycentric coordinate functions $\beta_{k1}(u,v), \ldots, \beta_{kn_k}(u,v)$, and let $\{S_{\lambda_k}^{d_k}(u,v)\}$ be the corresponding S-patch blending functions indexed by the set of p-tuples $I_k^{d_k}$, where $I_k = (i_{k1}, \ldots, i_{kn_k})$. Recall that $\beta_{kj}(u,v) \equiv \beta_{i_{kj}}(u,v)$, and let $\beta_{I_k}(u,v) = (\beta_{i_{k1}}(u,v), \ldots, \beta_{i_{kn_k}}(u,v))$. Then with this notation

$$\{S_{\lambda_k}^{d_k}(u,v)\} = \underbrace{\beta_{I_k}(u,v) \otimes \cdots \otimes \beta_{I_k}(u,v)}_{d_k \ \ factors} \equiv \beta_{I_k}^{d_k}(u,v).$$

To construct *C*-patch barycentric coordinate functions from *S*-patch barycentric coordinate functions, let $I = I_1 \oplus \cdots \oplus I_m$ and define

$$\beta_I(u,v) = \beta_{I_1}(u,v) \otimes \cdots \otimes \beta_{I_m}(u,v). \tag{8.18}$$

Since the functions $\beta_{I_k}(u,v)$ are barycentric coordinates, these functions sum to one; therefore, the functions $\beta_I(u,v)$ also sum to one (see Exercise 1). Moreover, the functions $\beta_I(u,v)$ are nonnegative over the domain polygon $Q = Q_1 \cap \cdots \cap Q_m$. Now let $d = (d_1, \ldots, d_m)$. Then the *C*-patch blending functions of depth d are defined by

$$\{C_\lambda^d(u,v)\} = \beta_I^d(u,v) = \beta_{I_1}^{d_1}(u,v) \otimes \cdots \otimes \beta_{I_m}^{d_m}(u,v) = \{S_{\lambda_1}^{d_1}(u,v)\} \otimes \cdots \otimes \{S_{\lambda_m}^{d_m}(u,v)\},$$

where $\lambda \in I^d \equiv I_1^{d_1} \oplus \cdots \oplus I_m^{d_m}$. Notice that if $d_1 = \cdots = d_m$, then the *C*-patch of depth d is a *P*-patch of depth d_1 with barycentric coordinate functions $\beta_I(u,v)$.

The preceding construction is rather abstract, so before proceeding any further let us take a look at some concrete examples.

EXAMPLE 8.1

Tensor Product Bezier Patches

Recall that the blending functions for a tensor product Bezier surface can be generated by convolving the blending functions for Bezier curves. Thus for a tensor product Bezier patch, the analogue of Equation (8.18) is

$$\{(1-u)(1-v), u(1-v), (1-u)v, uv\} = \{(1-u), u\} \otimes \{(1-v), v\}$$
$$\{\ \ (0,0), \quad (1,0), \quad (0,1), \quad (1,1)\ \ \} = \{(0,0),(1,0)\} \oplus \{\ (0,0),(0,1)\},$$

where beneath each term we have indicated the indexing assigned to each factor. The *S*-patch blending functions $S_{\lambda_1 0}^{d_1}(u,v), S_{0\lambda_2}^{d_2}(u,v)$

are univariate Bernstein basis functions, since

$$\{S_{\lambda_1 0}^{d_1}(u,v)\} = \underbrace{\{(1-u), u\} \otimes \cdots \otimes \{(1-u), u\}}_{d_1} = \{B_{\lambda_1}^{d_1}(u)\}$$

$$\{S_{0\lambda_2}^{d_2}(u,v)\} = \underbrace{\{(1-v), v\} \otimes \cdots \otimes \{(1-v), v\}}_{d_2} = \{B_{\lambda_2}^{d_2}(v)\},$$

and the C-patch blending functions $C^d_{\lambda_1\lambda_2}(u,v)$ are the tensor product Bernstein basis functions

$$\{C^d_{\lambda_1\lambda_2}(u,v)\} = \{S^{d_1}_{\lambda_1 0}(u,v)\} \otimes \{S^{d_2}_{0\lambda_2}(u,v)\} = \{B^{d_1}_{\lambda_1}(u)B^{d_2}_{\lambda_2}(v)\}.$$

If $d_1 = d_2$, then the C-patch corresponding to the blending functions

$$\{C^d_{\lambda_1\lambda_2}(u,v)\}$$

is also an S-patch because the functions $\{(1-u)(1-v), u(1-v), (1-u)v, uv\}$ are barycentric coordinates for the unit square (see Figure 8.9).

EXAMPLE
8.2

A Pentagonal C-Patch

Consider the barycentric coordinate functions

$$\left(\beta_{ij}(u,v)\right) = ((2-u)/2, u/2) \otimes ((2-v)/2, v/2) \otimes ((3-u-v)/3, u/3, v/3)$$

$$I \quad = \quad ((0,0),(1,0)) \quad \oplus \quad ((0,0),(0,1)) \quad \oplus \quad ((0,0),(1,0),(0,1)) \quad (8.19)$$

where beneath each term on the right-hand side we have indicated the indexing assigned to the factor. By construction

$$\{S^{d_1}_{\lambda_1,0}(u,v)\} = \underbrace{\{(2-u)/2, u/2\} \otimes \cdots \otimes \{(2-u)/2, u/2\}}_{d_1} = \{B^{d_1}_{\lambda_1,0}(u/2)\}$$

$$\{S^{d_2}_{0,\lambda_2}(u,v)\} = \underbrace{\{(2-v)/2, v/2\} \otimes \cdots \otimes \{(2-v)/2, v/2\}}_{d_2} = \{B^{d_2}_{0,\lambda_2}(v/2)\}$$

are univariate Bernstein basis functions, and

$$\{S^{d_3}_{\lambda_3,\lambda_4}(u,v)\} = \underbrace{\{(3-u-v)/3, u/3, v/3\} \otimes \cdots \otimes \{(3-u-v)/3, u/3, v/3\}}_{d_3}$$

$$= \{B^{d_3}_{\lambda_3,\lambda_4}(u/3, v/3)\}$$

are bivariate Bernstein basis functions. Hence

$$\{C^d_\lambda(u,v)\} = \{B^{d_1}_{\lambda_1,0}(u/2)\} \otimes \{B^{d_2}_{0,\lambda_2}(v/2)\} \otimes \{B^{d_3}_{\lambda_3,\lambda_4}(u/3, v/3)\}.$$

The indexing set I for the functions $\{\beta_{ij}(u,v)\}$ is the Minkowski sum of the indexing sets of the three factors (Figure 8.10), and the domain polygon for the C-patch is the intersection of the domains of the three S-patches (Figure 8.11). Notice that, even if $d_1 = d_2 = d_3$, since there are eight functions $\{\beta_{ij}(u,v)\}$, the corresponding C-patches are not S-patches for the domain pentagon, which has only five barycentric coordinate functions.

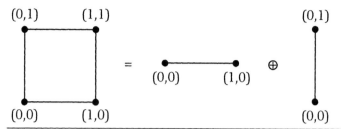

Figure 8.9 The unit square is the Minkowski sum of a horizontal line and a vertical line. Therefore, a tensor product Bezier patch is a *C*-patch generated by convolving the univariate Bernstein basis functions for the two lines.

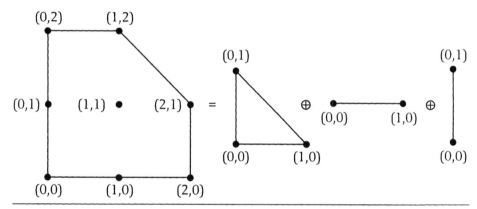

Figure 8.10 The indexing set *I* for the functions $(\beta_{ij}(u,v))$ in Example 8.2 is the set of lattice points in the pentagon on the left, which is the Minkowski sum of the lattice points of the triangle and the two lines on the right.

EXAMPLE 8.3

A Hexagonal C-Patch

Consider the functions

$$\left(\beta_{ij}(u,v)\right) = ((2-u)/2, u/2) \otimes ((2-v)/2, v/2) \otimes ((1+u-v)/2, (1-u+v)/2)$$

$$I \quad = \quad ((0,0),(1,0)) \quad \oplus \quad ((0,0),(0,1)) \quad \oplus \quad ((0,0),(1,1)) \quad (8.20)$$

where beneath each term on the right-hand side we have indicated the indexing assigned to the factor. Here the *C*-patch blending functions are given by

$$\{C_\lambda^d(u,v)\} = \{B_{\lambda_1,0}^{d_1}(u/2)\} \otimes \{B_{0,\lambda_2}^{d_2}(v/2)\} \otimes \{B_{\lambda_3,\lambda_4}^{d_3}((1-u+v)/2)\}.$$

The indexing set *I* for the barycentric coordinate functions along with the domain hexagon for the *C*-patch are illustrated in Figure 8.12. Again even if $d_1 = d_2 = d_3$, since there are seven functions $\{\beta_{ij}(u,v)\}$, the corresponding *C*-patches are not *S*-patches for the domain hexagon, which has only six barycentric coordinate functions.

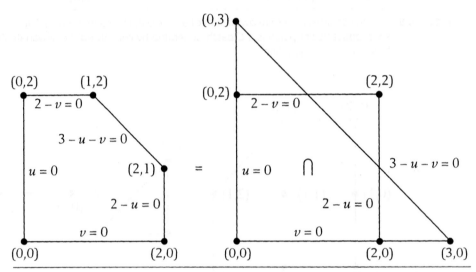

Figure 8.11 The domain for the *C*-patch in Example 8.2 is the pentagon on the left, which is the intersection of the triangle and the square bounded by the four lines on the right.

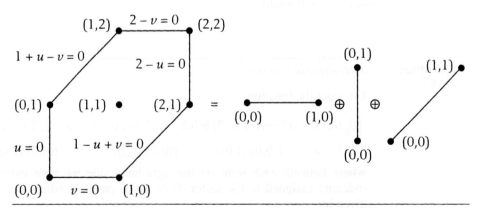

Figure 8.12 The indexing set *I* for the barycentric coordinate functions $(\beta_{ij}(u,v))$ in Example 8.3 is the set of lattice points in the hexagon on the left, which is the Minkowski sum of the lattice points of the three lines on the right. The domain for the *C*-patch is the hexagon on the left, which is bounded by the three pairs of lines that bound the domains for the barycentric coordinate functions of the three *S*-patches (unbounded rectangles) that generate the *C*-patch.

The domains of the *S*-patches in the preceding three examples were carefully chosen so that their intersection would produce interesting domains for the corresponding *C*-patches. The indexing on their barycentric coordinate functions, however, is somewhat arbitrary and was chosen so that the index sets for the *C*-patches would coincide with lattice points inside their domains. Nevertheless, this indexing will be important in Section 8.7, when we study toric Bezier patches.

C-patches are special kinds of *P*-patches when $d_1 = \cdots = d_m$, so the general pyramid algorithm in Equation (8.14) is a recursive evaluation algorithm for these *C*-patches. Recall, however, that tensor product Bezier patches have two recursive evaluation algorithms: a bilinear pyramid algorithm (Figures 5.41 and 5.42) and a two-tier de Casteljau algorithm (Figure 5.40). The bilinear pyramid algorithm is the *S*-patch evaluation algorithm for tensor product control nets. But where does the two-tier evaluation algorithm come from?

The two-tier de Casteljau algorithm arises because the bilinear barycentric coordinates for the square factor into linear factors via discrete convolution:

$$\{(1-u)(1-v), u(1-v), (1-u)v, uv\} = \{(1-u), u\} \otimes \{(1-v), v\}.$$

Hence the tensor product blending functions also factor in a similar fashion:

$$\left\{ B_{i,j}^{d_1, d_2}(u,v) \right\} = \{(1-u), u\}^{d_1} \otimes \{(1-v), v\}^{d_2} = \left\{ B_i^{d_1}(u) \right\} \otimes \left\{ B_j^{d_2}(v) \right\},$$

where powers denote repeated convolution. Therefore, we can replace bilinear blends in (u,v) by linear blends in u followed by linear blends in v. This procedure is exactly the two-tier de Casteljau evaluation algorithm for tensor product Bezier patches. Since the barycentric coordinate functions for a *C*-patch factor via discrete convolution, *C*-patches also have a multitier recursive evaluation algorithm.

Multitier Evaluation Algorithm for C-Patches

Initialize $P_\lambda(u,v) = P_\lambda$ for $\lambda \in I^d = I_1^{d_1} \oplus \cdots \oplus I_m^{d_m}$.

For $k = 1, \ldots, m$

  For each $\xi \in I_{k+1}^{d_{k+1}} \oplus \cdots \oplus I_m^{d_m}$:

$$P_{\xi+\eta}^0(u,v) = P_{\xi+\eta}(u,v) \qquad\qquad \eta \in I_k^{d_k}$$

$$P_{\xi+\rho}^l(u,v) = \sum_{h=1}^{n_k} \beta_{i_{kh}}(u,v) P_{\xi+\rho+i_{kh}}^{l-1}(u,v) \qquad\qquad \rho \in I_k^{d_k - l} \qquad (8.21)$$

$$P_\xi(u,v) = P_\xi^{d_k}(u,v)$$

Set $C(u,v) = P_0^{d_m}(u,v)$.

The input to this algorithm is a collection of control points $\{P_\lambda\}$ indexed by

$$I^d = I_1^{d_1} \oplus \cdots \oplus I_m^{d_m}.$$

The output is a point on the C-patch with blending functions $\{C_\lambda^d(u,v)\}$. Each inner loop of this algorithm is the S-patch pyramid algorithm for the points $\{P_{\xi+\eta}(u,v)\}$, where ξ is a fixed value in

$$I_{k+1}^{d_{k+1}} \oplus \cdots \oplus I_m^{d_m}$$

and η varies over $I_k^{d_k}$, so the output $P_\xi(u,v)$ of each inner loop is the S-patch for the points $\{P_{\xi+\eta}(u,v)\}$. Since, by assumption, the C-patch basis functions are convolutions of the S-patch basis functions, this algorithm generates the C-patch for the control points $\{P_\lambda\}$.

The formal proof that the multitier evaluation algorithm generates the C-patch proceeds by induction on m. Suppose that $m = 2$. On the first pass, the algorithm computes the values

$$P_{\lambda_2}(u,v) = \sum_{\lambda_1 \in I_1^{d_1}} S_{\lambda_1}^{d_1}(u,v)P_{\lambda_1+\lambda_2} \qquad \lambda_2 \in I_2^{d_2}.$$

On the second pass, the algorithm computes the surface

$$
\begin{aligned}
C(u,v) &= \sum_{\lambda_2 \in I_2^{d_2}} S_{\lambda_2}^{d_2}(u,v)P_{\lambda_2}(u,v) \\
&= \sum_{\lambda_2 \in I_2^{d_2}} S_{\lambda_2}^{d_2}(u,v) \sum_{\lambda_1 \in I_1^{d_1}} S_{\lambda_1}^{d_1}(u,v)P_{\lambda_1+\lambda_2} \\
&= \sum_{\lambda_1+\lambda_2 \in I_1^{d_1} \oplus I_2^{d_2}} C_{\lambda_1+\lambda_2}^d(u,v)P_{\lambda_1+\lambda_2} ,
\end{aligned}
$$

which is indeed the C-patch for the points $\{P_\lambda\}$, $\lambda \in I^d = I_1^{d_1} \oplus I_2^{d_2}$. It follows then by induction on m that the multitier algorithm is an evaluation algorithm for C-patches (Exercise 8).

We observed in Section 5.8.1 that the two-tier de Casteljau evaluation algorithm for tensor product Bezier patches is a good deal more efficient than the bilinear pyramid algorithm. Similarly, for C-patches the multitier evaluation algorithm is much more efficient than the general P-patch pyramid algorithm. For example, consider the pentagonal C-patch of Example 8.2 for $d_1 = d_2 = d_3$. The standard P-patch pyramid algorithm for this pentagonal patch is an eight-term recurrence over a five-sided domain. We can now replace this eight-term recurrence by two sets of two-term Bernstein recurrences and one set of three-term Bernstein recurrences because

$$
\begin{aligned}
\{C_\lambda^d(u,v)\} &= \left(\beta_{ij}(u,v)\right)^d \\
&= ((2-u)/2,u/2)^{d_1} \otimes ((2-v)/2,v/2)^{d_1} \otimes ((3-u-v)/3,u/3,v/3)^{d_1},
\end{aligned}
$$

where powers denote repeated convolution. So this pentagonal C-patch has an alternative evaluation algorithm consisting sequentially of two tiers of univariate de Casteljau algorithms and one tier of the bivariate tetrahedral algorithm.

Blossoming also works quite nicely for *C*-patches. Recall that simply replacing the parameters (u,v) by a different parameter pair (u_k,v_k) on each level of the *P*-patch pyramid algorithm does not, in general, give rise to the dual functionals for *P*-patches. But for *C*-patches, something wonderful happens: there is a new way to blossom.

We can blossom a tensor product patch from the two-tier de Casteljau evaluation algorithm by blossoming the lower tiers in the *u* parameter and the upper tier in the *v* parameter. Let's try an analogous tactic with the multitier evaluation algorithm for *C*-patches. That is, blossom the multitier algorithm for *C*-patches by blossoming the *S*-patch pyramid algorithms for the different tiers independently. The effect would be to replace the values of the barycentric coordinate functions $\beta_{k1}(u,v),\ldots,\beta_{kn_k}(u,v)$ on the *j*th level where they appear in the corresponding *S*-patch algorithm by the values $\beta_{k1}(u_{kj},v_{kj}),\ldots,\beta_{kn_k}(u_{kj},v_{kj})$. We call the function generated in this fashion the *blossom* of the *C*-patch $C(u,v)$ and denote this blossom by

$$c\big((u_{11},v_{11}),\ldots,(u_{1d_1},v_{1d_1}),\ldots,(u_{m1},v_{m1}),\ldots,(u_{md_m},v_{md_m})\big).$$

Equivalently, we can blossom the basis functions of the *C*-patch

$$\{C_\lambda^d(u,v)\} = \beta_{I_1}^{d_1}(u,v) \otimes \cdots \otimes \beta_{I_m}^{d_m}(u,v) = \{S_{\lambda_1}^{d_1}(u,v)\} \otimes \cdots \otimes \{S_{\lambda_m}^{d_m}(u,v)\}$$

by replacing $\beta_{k1}(u,v),\ldots,\beta_{kn_k}(u,v)$ by $\beta_{k1}(u_{kj},v_{kj}),\ldots,\beta_{kn_k}(u_{kj},v_{kj})$ in *j*th factor of $\beta_{I_k}^{d_k}(u,v)$. Evidently then

$$\left\{ c_\lambda^d\big((u_{11},v_{11}),\ldots,(u_{1d_1},v_{1d_1}),\ldots,(u_{m1},v_{m1}),\ldots,(u_{md_m},v_{md_m})\big)\right\}$$

$$= \left\{ s_{\lambda_1}^{d_1}\big((u_{11},v_{11}),\ldots,(u_{1d_1},v_{1d_1})\big)\right\} \otimes \cdots \otimes \left\{ s_{\lambda_m}^{d_m}\big((u_{m1},v_{m1}),\ldots,(u_{md_m},v_{md_m})\big)\right\}.$$

Thus we blossom the *C*-patch blending functions by blossoming, independently, the *S*-patch blending functions that define them.

The blossom $c\big((u_{11},v_{11}),\ldots,(u_{1d_1},v_{1d_1}),\ldots,(u_{m1},v_{m1}),\ldots,(u_{md_m},v_{md_m})\big)$ of a *C*-patch $C(u,v)$ of depth *d* has the following properties:

i. *Multisymmetry*

$$c\big((u_{11},v_{11}),\ldots,(u_{1d_1},v_{1d_1}),\ldots,(u_{m1},v_{m1}),\ldots,(u_{md_m},v_{md_m})\big)$$

$$= c\big((u_{1\sigma_1(1)},v_{1\sigma_1(1)}),\ldots,(u_{1\sigma_1(d_1)},v_{1\sigma_1(d_1)}),\ldots,$$

$$(u_{m\sigma_m(1)},v_{m\sigma_m(1)}),\ldots,(u_{m\sigma_m(d_m)},v_{m\sigma_m(d_m)})\big)$$

ii. *Diagonal*

$$c\big((u,v),\ldots,(u,v)\big) = C(u,v)$$

iii. *Dual functional*

$$
c\left(\underbrace{Q_{11},\dots,Q_{11}}_{\gamma_{11}},\dots,\underbrace{Q_{1n_1},\dots,Q_{1n_1}}_{\gamma_{1n_1}},\dots,\underbrace{Q_{m1},\dots,Q_{m1}}_{\gamma_{m1}},\dots,\underbrace{Q_{mn_m},\dots,Q_{mn_m}}_{\gamma_{mn_m}}\right) = P_\lambda,
$$

where $\lambda = \lambda_1 + \cdots + \lambda_m$ and $\lambda_k = \gamma_{k1}i_{k1} + \cdots + \gamma_{kn_k}i_{n_k}$.

The diagonal property is immediate from the definition of the blossom, and the multisymmetry property follows from the symmetry of the blossom for S-patches. The dual functional property is also a consequence of the dual functional property for S-patches because

$$
\left\{c_\lambda^d\left(\underbrace{Q_{11},\dots,Q_{11}}_{\gamma_{11}},\dots,\underbrace{Q_{1n_1},\dots,Q_{1n_1}}_{\gamma_{1n_1}},\dots,\underbrace{Q_{m1},\dots,Q_{m1}}_{\gamma_{m1}},\dots,\underbrace{Q_{mn_m},\dots,Q_{mn_m}}_{\gamma_{mn_m}}\right)\right\}
$$

$$
= \left\{s_{\lambda_1}^{d_1}\left(\underbrace{Q_{11},\dots,Q_{11}}_{\gamma_{11}},\dots,\underbrace{Q_{1n_1},\dots,Q_{1n_1}}_{\gamma_{1n_1}}\right)\right\} \otimes \cdots \otimes \left\{s_{\lambda_m}^{d_m}\left(\underbrace{Q_{m1},\dots,Q_{m1}}_{\gamma_{m1}},\dots,\underbrace{Q_{mn_m},\dots,Q_{mn_m}}_{\gamma_{mn_m}}\right)\right\},
$$

so

$$
c_\lambda^d\left(\underbrace{Q_{11},\dots,Q_{11}}_{\gamma_{11}},\dots,\underbrace{Q_{1n_1},\dots,Q_{1n_1}}_{\gamma_{1n_1}},\dots,\underbrace{Q_{m1},\dots,Q_{m1}}_{\gamma_{m1}},\dots,\underbrace{Q_{mn_m},\dots,Q_{mn_m}}_{\gamma_{mn_m}}\right) = 1
$$

if and only if $\lambda = \lambda_1 + \cdots + \lambda_m$ where $\lambda_k = \gamma_{k1}i_{k1} + \cdots + \gamma_{kn_k}i_{n_k}$.

In general, the blossom of a C-patch, like the blossom of an S-patch, is not multiaffine, but if any factor $\beta_{I_k}(u,v)$ consists entirely of linear functions, then the blossom of the C-patch will be multiaffine in the corresponding blossom parameters because each of these parameters in the blossom will appear only to the first power. Thus if the C-patch is the convolution of S-patches that represent Bezier curves or triangular Bezier patches, the blossom of the C-patch will be multiaffine in every parameter pair. For example, the blossom of the pentagonal patch in Example 8.2 is multiaffine in all its parameters and so too is the blossom of the hexagonal patch in Example 8.3.

To summarize: two good things happen for C-patches: there is a multitier evaluation algorithm, and the dual functionals can be constructed by blossoming this algorithm. Moreover, if all the S-patches in the convolution for the C-patch are standard Bezier patches (or Bezier curves), then the blossom is multiaffine in all its parameters.

In fact, even if the S-patch barycentric coordinate functions are not normalized to sum to one, as long as the blending functions for the C-patch sum to one, all the results in this section including the multitier evaluation algorithm remain valid. Blossoming the multitier evaluation algorithm still provides the dual functionals for the *C*-patch. The proofs are exactly the same (see Exercise 11). We shall make use of these observations in Section 8.7.11, when we blossom toric Bezier *C*-patches.

As usual, we can construct rational *C*-patches by introducing scalar weights $\{w_\lambda\}$ and setting

$$R(u,v) = \frac{\sum_{\lambda \in I^d} C_\lambda^d(u,v) w_\lambda P_\lambda}{\sum_{\rho \in I^d} w_\rho C_\rho^d(u,v)}.$$

There is also a multitier evaluation algorithm for rational *C*-patches: simply replace the input $P_\lambda^0(u,v) = P_\lambda$ at the base of the algorithm by $P_\lambda^0(u,v) = (w_\lambda P_\lambda, w_\lambda)$ and divide the output of the algorithm by the weight. Since the blending in Equation (8.21) now occurs in Grassmann space, the blending functions for a rational *C*-patch need not even sum to one. Therefore, we can generate the barycentric coordinate functions for a rational *C*-patch by convolving the unnormalized barycentric coordinate functions of rational *S*-patches. Blossoming still provides the dual functionals for rational *C*-patches constructed in this fashion (see Exercise 10).

To get a better feel for the relative advantages and disadvantages of *S*-patches and *C*-patches, we close with a comparison (Table 8.2) of the hexagonal *S*-patch and hexagonal *C*-patch, where $d_1 = d_2 = d_3$, for the domain in Example 8.3.

From this comparison we see that while *C*-patches have many fine properties, they have one embarrassing deficiency: the boundary of a *C*-patch is not easy to determine. This lack of control over the boundary makes it hard to use arbitrary *C*-patches in practice to fill *n*-sided holes. Toric Bezier patches overcome this difficulty, so it is to these patches that we next turn our attention. Nevertheless, in the study of toric Bezier patches we shall need the technology of *C*-patches, since the methods applied here are required to effectively blossom toric Bezier schemes (see Section 8.7.11).

Table 8.2 Comparison of hexagonal *S*-patches and hexagonal *C*-patches.

Property	*Hexagonal S-Patch*	*Hexagonal C-Patch*
Type	rational	polynomial
Parametric degree	$4 \times depth$	$3 \times depth$
Number of barycentric coordinate functions	6	7
Evaluation	pyramid algorithm	multitier algorithm
Blossom	symmetric, not multiaffine	multisymmetric, multiaffine
Boundaries	Bezier curves determined by boundary control points	unknown

Exercises

1. Let $A(u,v) = \{A_i(u,v) \mid i \in I\}$, $B(u,v) = \{B_j(u,v) \mid j \in J\}$, and $C(u,v) = A(u,v) \otimes B(u,v)$. Show that

   a. $\displaystyle\sum_{k \in I \oplus J} C_k(u,v) = \{\sum_{i \in I} A_i(u,v)\}\{\sum_{j \in J} B_j(u,v)\}$

   b. $\displaystyle\sum_{i \in I} A_i(u,v) = \sum_{j \in J} B_j(u,v) \equiv 1 \Rightarrow \sum_{k \in I \oplus J} C_k(u,v) \equiv 1$

2. Implement both the pyramid algorithm and the multitier evaluation algorithm for the pentagonal C-patch in Example 8.2 with $d_1 = d_2 = d_3$. Which algorithm do you prefer? Why? Experiment with pentagonal C-patches of different depths. Determine how changing the location of the control points affects the shape of the surface.

3. Show that a surface whose barycentric coordinate functions are generated by convolution from the barycentric coordinate functions of several C-patches is a C-patch.

4. Compute the blossom of the basis functions for the C-patches constructed in Examples 8.2 and 8.3 with $d_1 = d_2 = d_3 = 2$.

5. Consider the functions defined by

   $$\left(\beta_{ij}(u,v)\right) = ((1+u-v)/2,(1-u+v)/2) \otimes ((3-u-v)/3,u/3,v/3)$$

   $$I \quad = \quad ((0,0),(1,1)) \qquad \oplus \qquad ((0,0),(1,0),(0,1))$$

   where beneath each term on the right-hand side we have indicated the indexing assigned to the factor.

   a. How many functions are in the set $\{\beta_{ij}(u,v)\}$? What is their degree?

   b. Compute explicit expressions for the functions $\{\beta_{ij}(u,v)\}$.

   c. Show that the functions $\{\beta_{ij}(u,v)\}$ generate a pentagonal C-patch.

   d. Compute the blossom of the basis functions for this C-patch of depth $d = (2,2)$.

   e. How does this pentagonal patch differ from the pentagonal patch in Example 8.2?

6. Consider the functions defined by

   $$\left(\beta_{ij}(u,v)\right) = ((1+u-v)/3,(2-u)/3,v/3) \otimes ((1-u+v)/3,u/3,(2-v)/3)$$

   $$I \quad = \quad ((0,0),(1,0),(0,1)) \qquad \oplus \qquad ((0,0),(0,1),(1,1))$$

   where beneath each term on the right-hand side we have indicated the indexing assigned to the factor.

   a. How many functions are in the set $\{\beta_{ij}(u,v)\}$? What is their degree?

   b. Compute explicit expressions for the functions $\{\beta_{ij}(u,v)\}$.

c. Show that the functions $\{\beta_{ij}(u,v)\}$ generate a hexagonal *C*-patch.

d. Compute the blossom of the basis functions for this *C*-patch of depth $d = (2,2)$.

e. How does this hexagonal patch differ from the hexagonal patch in Example 8.3?

7. Consider the hexagonal domain in Figure 8.12. Which algorithm is more efficient:

a. the pyramid algorithm for the *S*-patch with this hexagonal domain or

b. the multitier algorithm for the *C*-patch with this hexagonal domain?

Justify your answer.

8. Complete the proof of the validity of the multitier evaluation algorithm for *C*-patches by completing the inductive argument.

9. Prove that the multitier evaluation algorithm is valid for rational *C*-patches.

10. Prove that blossoming the multitier evaluation algorithm provides the dual functionals for rational *C*-patches.

11. Let $\beta_I(u,v) = \beta_{I_1}(u,v) \otimes \cdots \otimes \beta_{I_m}(u,v)$ be the barycentric coordinate functions for a *C*-patch. Suppose that the functions in the set $\{C_\alpha^d(u,v)\}$ sum to one, and that the functions in the sets $\beta_{I_k}(u,v)$ satisfy

$$\beta_{kj}(Q_{kl}) = 0 \qquad j \neq l$$
$$= 1 \qquad j = l$$

but are not normalized to sum to one. Show that

a. The multitier evaluation algorithm is still valid.

b. Blossoming the multitier evaluation algorithm still provides the dual functionals for the *C*-patch.

12. Suppose that $I = I_1 \oplus I_2$ and that $\beta_I(u,v) = \beta_{I_1}(u,v) \otimes \beta_{I_2}(u,v)$. Let $d = (d_1,d_2)$ with $d_2 \geq d_1$. Consider the *C*-patch with control points $\{P_\lambda\}, \lambda \in I^d = I_1^{d_1} \oplus I_2^{d_2}$.

a. Show that this *C*-patch has the following evaluation algorithm:

i. For each $\xi \in I_2^{d_2-d_1}$, run the standard pyramid algorithm with the initial data

$$P_{\lambda_1}^0(u,v) = P_{\lambda_1+\xi}, \quad \lambda_1 \in I^{d_1}$$

and barycentric coordinate functions $\{\beta_I(u,v)\}$. The output is a collection of values

$$P_\xi(u,v), \quad \xi \in I_2^{d_2-d_1}.$$

ii.   Run the standard pyramid algorithm with the initial data

$$P_\xi^0(u,v) = P_\xi(u,v), \quad \xi \in I_2^{d_2-d_1}$$

computed in step (i) and barycentric coordinate functions $\{\beta_{I_2}(u,v)\}$.

b. Generalize the result in part (a) to $I \equiv I_1 \oplus \cdots \oplus I_m$ and $d = (d_1,\ldots,d_m)$.

13. Show that the first-order partial derivative with respect to u or v of a C-patch of depth $d = (d_1,\ldots,d_m)$ can be computed by summing the following m terms. For each $k = 1,\ldots,m$, take the first-order partial derivative with respect to u or v of any one level in the kth tier of the multitier evaluation algorithm, then run the entire evaluation algorithm, and multiply the result by d_k. Explain how to apply similar techniques to calculate the second-order partial derivatives of a C-patch.

8.7   Toric Bezier Patches

Toric Bezier patches are pyramid patches defined by special indexing sets and special barycentric coordinate functions. These indexing sets and barycentric coordinate functions are chosen to overcome some of the deficiencies we have observed in arbitrary P-patches and in general C-patches.

Blossoming the pyramid algorithm does not provide the dual functionals for arbitrary pyramid patches. C-patches overcome this deficiency by blossoming the multitier evaluation algorithm, but C-patches have two other serious shortcomings. First, as we observed in the previous section, the boundary of a C-patch is not easy to determine. More serious, but perhaps more subtle, we do not even know what the boundary control points are for a C-patch.

Consider again the C-patches in Examples 8.2 and 8.3. In both cases, the points in the indexing set lie inside or on the boundary of the domain polygon. We have carefully chosen the indexing sets for the S-patches that generate these C-patches to force this concurrence. This juxtaposition creates the illusion that the boundary control points of the C-patch are the points whose indices lie on the boundaries of the domain polygon. But, in general, there is no relationship at all between the domain polygon of a C-patch and its indexing set. We could easily have chosen a different indexing for the S-patches, without altering the domain polygon of the C-patch. In fact, the domain of a C-patch need not have the same shape as the convex hull of its indexing set. So what exactly are the boundary control points of a C-patch?

When the index set of a C-patch consists of points in the plane, there are generally two distinct polygons associated with the C-patch: the domain polygon and the convex hull of the index set. The domain polygon specifies the geometry of the patch; the index set describes the topology of the control structure. If these two sets were to coincide, we might be better able to control the boundary of the patch.

For toric Bezier patches these two sets do coincide. So we can talk about the boundary control points of a toric Bezier patch. Toric Bezier patches overcome both of the problems we encountered with C-patches: their domain coincides with the

convex hull of their indexing set, and their boundaries are the Bezier curves determined by their boundary control points.

Moreover, unlike general pyramid patches, blossoming the pyramid algorithm
provides the dual functionals for most of the blending functions of a toric Bezier
patch. In addition, many toric Bezier patches are also C-patches, so these patches
inherit the multitier evaluation algorithm and the blossoming procedure generic to
C-patches.

To define toric Bezier patches, we need to go back to where we began this chapter and to reexamine what we mean both by polygonal arrays of control points and
by barycentric coordinate functions. We begin, then, with a discussion of lattice
polygons and barycentric coordinates for lattice polygons. We will then apply these
new kinds of indexing sets and new types of barycentric coordinate functions to
build multisided toric Bezier patches based on techniques already familiar to us from
S-patches and C-patches.

Exercise

1. Give an example to show that the domain of a C-patch indexed by an array
   of 2-tuples need not have the same shape as the convex hull of its indexing
   set.

8.7.1 Lattice Polygons

What is a polygonal array of points? We have asked this question once before, in
Section 8.2. There our answer was that a polygonal array of points is any collection
of points indexed by a set of the form I^d, where I is an ordered set of n distinct p-
tuples. This definition was certainly useful from the perspective of S-patches. But
look again at Figures 8.10 and 8.12. The points in these figures surely look like they
form pentagonal and hexagonal arrays. Yet these configurations do not conform with
the definition of pentagonal and hexagonal arrays in Section 8.2.

These arrangements of points appear polygonal because their convex hulls are
polygons. So evidently there is another possible definition of a polygonal array: An
array is *polygonal* if it is indexed by a finite set of points in the plane. The shape of
the array is the shape of the convex hull of the indexing set. A boundary of the array
consists of the points in the array corresponding to the indices on a boundary of the
convex hull of the index set. If we adopt these alternative definitions, then the control
points for the C-patches in Examples 8.2 and 8.3 form pentagonal and hexagonal
arrays.

To construct barycentric coordinates for these indexing sets, we cannot, as it
happens, choose just any finite collection of points in the plane for our indexing set.
Instead we must restrict ourselves to lattice polygons. A *lattice polygon* is the intersection of the convex hull of a set of points having integer coordinates with the lattice $\mathbf{Z} \times \mathbf{Z}$. Equivalently, a lattice polygon consists of all the points with integer
coordinates inside or on the boundary of a convex polygon whose vertices have inte-
ger coordinates. The convex hull of a lattice polygon I is often called the *Newton*

polygon of *I*. We shall see in Section 8.7.2 that the elements of the index set are used as exponents to construct barycentric coordinate functions. So we choose a lattice polygon as our index set to ensure that these exponents are always integers.

The arrays of points in Figures 8.10 and 8.12 are lattice polygons, and their Newton polygons are the domains for the corresponding *C*-patches in Examples 8.2 and 8.3. Other important examples are easy to construct. Let *i*,*j*,*d* denote nonnegative integers. Then

$$I_R^d = \{(i,j) \mid 0 \le i, j \le d\}$$

is a lattice rectangle and is the indexing set for a $(d + 1) \times (d + 1)$ rectangular array $\{P_{ij}\}$—that is, for the control points of a tensor product surface. Similarly,

$$I_T^d = \{(i,j) \mid 0 \le i + j \le d\}$$

is a lattice triangle and is the indexing set for an order *d* triangular array $\{P_{ij}\}$—that is, for the control points of a triangular patch.

An array of control points for a toric Bezier patch of depth *d* is a collection of points $\{P_\lambda\}$ indexed by a set I^d, where *I* is a lattice polygon. The boundary points of the array $\{P_\lambda\}$ are the points in the array indexed by the points of I^d lying on the boundary of the Newton polygon of I^d. We shall see in Section 8.7.3 that for toric Bezier patches the Newton polygon of *I* is the domain polygon, so for toric Bezier patches the domain polygon and the indexing set—the geometry of the patch and the combinatorial structure of the control net—are intimately related. It turns out as well that the boundary curves of a toric Bezier patch are the Bezier curves determined by the boundary control points. But before we can define precisely what we mean by a toric Bezier patch, we need to introduce barycentric coordinate functions for lattice polygons.

Exercises

1. Show that the only polygonal arrays of depth *d* that are also lattice polygons are I_R^d and I_T^d.

2. Let *E* and *F* be two sets of points in the lattice $\mathbf{Z} \times \mathbf{Z}$ lying on distinct line segments in the *xy*-plane. Show that $E \oplus F$ consists of the points in the lattice $\mathbf{Z} \times \mathbf{Z}$ lying in a parallelogram with sides equal and parallel to the line segments containing *E* and *F*.

3. Let $L(u,v) \equiv au + bv + c = 0$ be a boundary of the Newton polygon of a lattice polygon *I*. Show that $L^d(u,v) \equiv au + bv + dc = 0$ is a boundary of the Newton polygon of I^d.

4. Let $Q_1,\ldots,Q_n$ be the vertices of the Newton polygon of the lattice polygon *I*. Show that

   a. The points
   $$\underbrace{Q_1 \oplus \cdots \oplus Q_1}_{d \text{ summands}},\ldots,\underbrace{Q_n \oplus \cdots \oplus Q_n}_{d \text{ summands}}$$
   are the vertices of the Newton polygon of I^d.

b. The boundaries of the Newton polygon of I^d are the d-fold Minkowski sums of the corresponding boundaries of the Newton polygon of I.

c. The boundaries of the Newton polygon of I^d are parallel to the corresponding boundaries of the Newton polygon of I.

d. $Area(\text{Newton polygon of } I^d) = d^2 \times Area(\text{Newton polygon of } I)$.

8.7.2 Barycentric Coordinates for Lattice Polygons

To build toric Bezier patches, we are going to construct barycentric coordinate functions $\{\beta_{ij}\}_{(i,j)\in I}$ for lattice polygons I that have properties similar to the barycentric coordinates $\{\beta_1,...,\beta_n\}$ associated with the vertices of convex polygons Q. The main properties of the barycentric coordinate functions $\{\beta_1,...,\beta_n\}$ for a convex polygon Q with ordered vertices $Q_1,...,Q_n$ are described in Table 8.1. We reproduce these properties in Table 8.3; alongside we list the analogous properties we want to hold for the barycentric coordinate functions $\{\beta_{ij}\}_{(i,j)\in I}$ of a lattice polygon I whose Newton polygon has vertices $Q_1,...,Q_n$.

The first two properties of the barycentric coordinate functions for lattice polygons are required because we want the toric Bezier surfaces defined by these functions to be affine invariant and to lie in the convex hulls of their control points. The third property guarantees that the boundary curves of these surfaces are determined only by their boundary control points, and the fourth property ensures that these surfaces interpolate their corner control points. The fourth property is also key in ensuring that the blossom of a toric Bezier patch evaluated at the vertices of the domain polygon provides at least some of the dual functionals for toric Bezier patches (see Section 8.7.10). This property is crucial as well for the construction of toric S-patches (see Section 8.7.6). The final property asserts that the functions describing these surfaces are not too complicated—that toric Bezier surfaces are defined by rational expressions.

Table 8.3 Properties of barycentric coordinates for convex polygons and lattice polygons.

Convex Polygons	Lattice Polygons
1. $\sum_{k=1}^{n} \beta_k \equiv 1.$	1. $\sum_{(i,j)\in I} \beta_{ij} \equiv 1.$
2. $\beta_k > 0$ in the interior of Q.	2. $\beta_{ij} > 0$ in the interior of Convex Hull (I)
3. $\beta_k = 0$ on the line Q_iQ_{i+1}, $k \neq i, i+1$.	3. $\beta_{ij} = 0$ on the line Q_kQ_{k+1}, if and only if $(i,j) \notin Q_kQ_{k+1}$.
4. $\beta_k(Q_j) = 0 \qquad j \neq k$ $ = 1 \qquad j = k$.	4. $\beta_{ij}(Q_k) = 0 \qquad (i,j) \neq Q_k$ $\phantom{\beta_{ij}(Q_k)} = 1 \qquad (i,j) = Q_k$.
5. $\beta_1,...,\beta_n$ are rational functions.	5. $\{\beta_{ij}\}$ are rational functions.

Just as we gave an explicit construction for the barycentric coordinate functions of a convex polygon Q, we can provide an explicit construction for the barycentric coordinate functions of a lattice polygon I. Let $Q_1,\ldots,Q_n$ be the vertices of the Newton polygon of I, and let $L_k(u,v) \equiv a_k u + b_k v + c_k = 0$, $k = 1,\ldots,n$, be the equation of the kth boundary line $Q_k Q_{k+1}$. Normalize these equations so that the normal vector (a_k, b_k) of the line $L_k(u,v)$ satisfies the following two constraints:

i. (a_k, b_k) points into the Newton polygon of I.

ii. (a_k, b_k) is the shortest vector in this direction with integer coordinates.

The first condition will guarantee that our barycentric coordinate functions are positive inside the domain polygon; the second condition will ensure that below we deal only with integer exponents and that these exponents are of the lowest possible size. Notice that we can always enforce the second condition because the vertices of the Newton polygon of I have integer coordinates.

By the third property in Table 8.3, a barycentric coordinate function must vanish on each boundary to which its index does not belong. The easiest way to construct a rational function $\beta(u,v)$ that vanishes on the kth boundary $Q_k Q_{k+1}$ of the Newton polygon of I is to make $L_k(u,v)$ a factor of the numerator of $\beta(u,v)$. We took advantage of precisely this observation in Section 8.1 to construct barycentric coordinates for convex polygons. Thus the numerators of the barycentric coordinate functions $\{\beta_{ij}\}_{(i,j)\in I}$ for a lattice polygon I are all going to be products of powers of the functions $L_k(u,v)$. In particular, define

$$\alpha_{ij}(u,v) = c_{ij}\{L_1(u,v)\}^{L_1(i,j)}\cdots\{L_n(u,v)\}^{L_n(i,j)} \tag{8.22}$$

$$\beta_{ij}(u,v) = \frac{\alpha_{ij}(u,v)}{\sum\limits_{(k,l)\in I} \alpha_{kl}(u,v)}. \tag{8.23}$$

The constants $c_{ij} > 0$ are arbitrary normalizing constants that will be chosen later (see Sections 8.7.4, 8.7.5, 8.7.6, and 8.7.10 and Examples 8.4 and 8.5) to guarantee that certain desirable formulas are satisfied.

THEOREM
8.2

Properties of Barycentric Coordinates for Lattice Polygons

Let $\{\beta_{ij}(u,v)\}$ be the functions defined by Equation (8.23) for the lattice polygon I whose Newton polygon has vertices $Q_1,\ldots,Q_n$. Then:

1. $\sum\limits_{(i,j)\in I} \beta_{ij} \equiv 1$.

2. $\beta_{ij} > 0$ in the interior of the Newton polygon of I.

3. $\beta_{ij} = 0$ on the boundary $Q_k Q_{k+1}$, if and only if $(i,j) \notin Q_k Q_{k+1}$.

4. $\beta_{ij}(Q_k) = 0 \qquad (i,j) \neq Q_k$
    $\qquad\quad = 1 \qquad (i,j) = Q_k$.

5. $\{\beta_{ij}\}$ are rational functions.

Proof   Property 1 is immediate from Equation (8.23).

Property 2 follows because $L_k(u,v)$ is chosen with its normal pointing into the Newton polygon of I. Hence in the interior of the Newton polygon $L_k(u,v) > 0$, so by Equations (8.22) and (8.23) $\alpha_{ij}(u,v)$, $\beta_{ij}(u,v)$ in the interior of the Newton polygon of I.

Property 3 is valid for the following reason. If (i,j) does not lie on the kth boundary Q_kQ_{k+1}, then the exponent $L_k(i,j) \neq 0$, so $L_k(u,v)$ is a factor of $\alpha_{ij}(u,v)$. Hence $\alpha_{ij}(u,v)$ and $\beta_{ij}(u,v)$ vanish on Q_kQ_{k+1}. If, however, (i,j) does lie on the kth boundary Q_kQ_{k+1}, then the exponent $L_k(i,j) = 0$, so the factor $L_k(u,v)$ disappears from Equation (8.22) and $\alpha_{ij}(u,v)$ does not vanish on Q_kQ_{k+1}. Hence $\beta_{ij}(u,v)$ does not vanish on Q_kQ_{k+1}.

Property 4 is immediate from Properties 1 and 3. In fact, notice that at a vertex (i,j) of the Newton polygon of I, the function $\alpha_{ij}(u,v)$ is just the product of powers of the edges of the polygon not passing through the vertex (i,j). Thus the numerator $\alpha_{ij}(u,v)$ of $\beta_{ij}(u,v)$ is very similar to the numerator $\alpha_k(u,v)$ for the corresponding barycentric coordinate function $\beta_k(u,v)$ at the vertex (i,j) defined in Equation (8.3). (For further comparisons between the barycentric coordinate functions at the vertices of a lattice polygon and the barycentric coordinate functions of the corresponding convex polygon, see Section 8.7.6.)

Property 5 follows immediately from Equations (8.22) and (8.23) because we have chosen the coefficients (a_k, b_k) to be integers.

EXAMPLE
8.4

Lattice Squares

Consider the lattice square

$$I_R^d = \{(i,j) \mid 0 \leq i, j \leq d\}.$$

The equations of the boundaries of the Newton polygon of I_R^d are the four lines: $u = 0$, $v = 0$, $d - u = 0$, $d - v = 0$. Let

$$c_{ij} = \binom{d}{i}\binom{d}{j}.$$

Then

$$\beta_{ij}(u,v) =$$

$$\frac{\binom{d}{i}\binom{d}{j}u^i(d-u)^{d-i}v^j(d-v)^{d-j}}{\sum\limits_{k,l=0}^{d}\binom{d}{k}\binom{d}{l}u^k(d-u)^{d-k}v^l(d-v)^{d-l}} = \frac{\binom{d}{i}\binom{d}{j}u^i(d-u)^{d-i}v^j(d-v)^{d-j}}{d^{2d}},$$

which are the standard tensor product Bernstein basis functions for the square with side d.

EXAMPLE
8.5

Lattice Triangles

Consider the lattice triangle

$$I_T^d = \{(i,j) \mid 0 \le i + j \le d\}.$$

The equations of the boundaries of the Newton polygon of I_T^d are the lines $u = 0$, $v = 0$, $d - u - v = 0$. Let

$$c_{ij} = \binom{d}{i \; j \; d-i-j}.$$

Then

$$\beta_{ij}(u,v) =$$

$$\frac{\binom{d}{i \; j \; d-i-j} u^i v^j (d-u-v)^{d-i-j}}{\sum_{0 \le k+l \le d} \binom{d}{k \; l \; d-k-l} u^k v^l (d-u-v)^{d-k-l}} = \frac{\binom{d}{i \; j \; d-i-j} u^i v^j (d-u-v)^{d-i-j}}{d^d},$$

which are the standard bivariate Bernstein basis functions for the isosceles right triangle with side d.

Thus barycentric coordinates for lattice polygons are generalizations of the bivariate Bernstein basis functions. We shall see shortly that surfaces defined using barycentric coordinate functions for lattice polygons are multisided generalizations of standard Bezier patches.

Exercises

1. Let I be the lattice polygon in Figure 8.10 or Figure 8.12. For each $(i,j) \in I$, define

   $\alpha_{ij}(u,v) = $ product of the boundary lines on which (i,j) does not lie

   $$\beta_{ij}(u,v) = \frac{\alpha_{ij}(u,v)}{\sum_{(k,l) \in I} \alpha_{kl}(u,v)}.$$

   a. Show that these functions $\{\beta_{ij}(u,v)\}$ satisfy all the conditions of Theorem 8.2.

   b. Show that these functions are not the same as the functions defined by Equations (8.22) and (8.23).

2. Let I be a lattice polygon, and let $L_k(u,v)$ be a boundary of the Newton polygon of I. Show that there exist integers (a_k, b_k) such that (a_k, b_k) is normal to $L_k(u,v)$.

3. Compute the barycentric coordinate functions for the lattice polygons in Figures 8.10 and 8.12.

4. Suppose that the number of lattice points is the same along each edge of a lattice polygon I. Let $L_k(u,v) \equiv a_k u + b_k v + c_k = 0$, $k = 1,...,n$, be the equation of the kth boundary of the Newton polygon of I, normalized so that (a_k,b_k) is the shortest normal vector of the line $L_k(u,v) = 0$ with integer coordinates pointing into the convex hull of I. Show that

a. $\displaystyle\sum_{k=1}^{n} a_k = \sum_{k=1}^{n} b_k = 0$.

b. The numerators of the barycentric coordinate functions all have the same total degree.

5. Let $Q_1,...,Q_n$ be the vertices of the Newton polygon of a lattice polygon I, and let $L_k(u,v) \equiv a_k u + b_k v + c_k = 0$, $k = 1,...,n$, be the equation of the kth boundary line $Q_k Q_{k+1}$, where the normal (a_k,b_k) points into the Newton polygon of I and (a_k,b_k) is the shortest vector in this direction with integer coordinates. Let R_k be the lattice point on the line $Q_k Q_{k+1}$ closest to Q_k, and let $T \in I$. Show that

a. $L_k(P) = Det\begin{pmatrix} Q_k & R_k & P \\ 1 & 1 & 1 \end{pmatrix} = 2\,Area(\Delta Q_k R_k P)$

b. $\alpha_T(P) =$

   $\qquad 2^n c_{ij} \{Area(\Delta Q_1 R_1 P)\}^{2\,Area(\Delta Q_1 R_1 T)} \cdots \{Area(\Delta Q_n R_n P)\}^{2\,Area(\Delta Q_n R_n T)}$

(Compare to Section 8.1, Exercise 3.)

6. Here we are going to show that barycentric coordinate functions for a lattice polygon I whose Newton polygon has more than four sides cannot be polynomials. Suppose that $\{\beta_{ij}\}_{(i,j)\in I}$ are polynomials, and that P is a point of intersection between two nonadjacent edges of the Newton polygon of I.

a. Show that if $\{\beta_{ij}\}_{(i,j)\in I}$ satisfy Property 3 of Theorem 8.2, then $\beta_{ij}(P) = 0$ for all $(i,j) \in I$.

b. Conclude that $\sum_{ij} \beta_{ij} \neq 1$, and hence that $\{\beta_{ij}\}_{(i,j)\in I}$ cannot be barycentric coordinate functions for the lattice polygon I.

c. Why do rational functions not suffer from the same problem?

d. Explain why barycentric coordinates for lattice rectangles can be polynomials. (Compare to Section 8.1, Exercise 5.)

8.7.3 The Pyramid Algorithm for Toric Bezier Patches

A *toric Bezier patch of depth d* is a pyramid patch of depth d, whose indexing array I is a lattice polygon and whose barycentric coordinate functions are the barycentric coordinates of I (see Figure 8.13).

Let us unwind what this definition really means. We start with a set of control points $\{P_\lambda\}$ indexed by a set I^d, where I is a lattice polygon. To compute points on the corresponding toric Bezier patch of depth d, we run the pyramid algorithm, where the barycentric coordinate functions $\{\beta_{ij}(u,v)\}$ are the barycentric coordinates

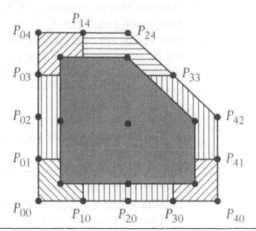

Figure 8.13 The pyramid algorithm for a pentagonal toric Bezier patch of depth 2. The eight overlapping striped pentagons at the base represent pentagonal toric Bezier patches of depth 1—one of these patches is completely hidden by the dark pentagon—and the dark central pentagon represents the pentagonal toric Bezier patch of depth 2. Each pentagonal panel represents the result of multiplying the points in the corresponding pentagonal array by the barycentric coordinates of the lattice pentagon in Figure 8.10 and adding the results. Interior control points are obscured by the dark panel. Compare to the pyramid algorithm for a pentagonal S-patch in Figure 8.8.

for the lattice polygon I. Notice that the domain of a toric Bezier patch of depth d is the Newton polygon of I, not the Newton polygon of I^d.

The Pyramid Algorithm for Toric Bezier Patches

$$
\begin{aligned}
&1. \quad P_\lambda^0(u,v) = P_\lambda & &\lambda \in I^d \\
&2. \quad P_\gamma^l(u,v) = \sum_{\rho \in I} \beta_\rho(u,v) P_{\gamma+\rho}^{l-1}(u,v) & &\gamma \in I^{d-l} & &(8.24) \\
&3. \quad B(u,v) = P_0^d(u,v) \; .
\end{aligned}
$$

By Equation (8.15) the blending functions $\{B_\lambda^d(u,v)\}_{\lambda \in I^d}$ for a toric Bezier patch of depth d can be computed by convolving the barycentric coordinate functions. That is,

$$
\{B_\lambda^d(u,v)\} = \underbrace{\beta_I(u,v) \otimes \cdots \otimes \beta_I(u,v)}_{d \; factors}, \tag{8.25}
$$

where $\beta_I(u,v) = \{\beta_{ij}(u,v)\}_{(i,j)\in I}$. Moreover, by Equation (8.17), if $I = \{i_1,j_1),\ldots,(i_n,j_n)\}$, then we also have an explicit formula for the blending functions:

$$
B_\lambda^d(u,v) = \sum_{k_1(i_1,j_1)+\cdots+k_n(i_n,j_n)=\lambda} \binom{d}{k_1\cdots k_n} \beta_{i_1 j_1}^{k_1}(u,v) \cdots \beta_{i_n j_n}^{k_n}(u,v). \tag{8.26}
$$

EXAMPLE
8.6

Tensor Product Bezier Patches

Consider the lattice rectangle $I_R = \{(0,0),(1,0),(1,1),(0,1)\}$. The equations of the boundaries of the Newton polygon of I_R are the four lines $u = 0$, $v = 0$, $1 - u = 0$, $1 - v = 0$. Hence the barycentric coordinate functions of I_R are just the barycentric coordinates of the unit square: $(1-u)(1-v)$, $u(1-v)$, $(1-u)v$, uv. Therefore, by Equation (8.25), the blending functions for the corresponding toric Bezier patch of depth d are the functions

$$\{B_\lambda^d(u,v)\} =$$

$$\underbrace{\{(1-u)(1-v),u(1-v),(1-u)v,uv\} \otimes \cdots \otimes \{(1-u)(1-v),u(1-v),(1-u)v,uv\}}_{d \ factors},$$

which are the blending functions for a tensor product Bezier patch of bidegree d.

EXAMPLE
8.7

Triangular Bezier Patches

Consider the lattice triangle $I_T = \{(0,0),(1,0),(0,1)\}$. The equations of the boundaries of the Newton polygon of I_T are the three lines $1 - u - v = 0$, $u = 0$, $v = 0$. Hence the barycentric coordinate functions of I_T are just the barycentric coordinates of the standard triangle: $(1 - u - v)$, u, v. Therefore, by Equation (8.26), the blending functions for the corresponding toric Bezier patch of depth d are the functions

$$B_{ij}^d(u,v) = \binom{d}{i \ j \ d-i-j}u^i v^j (1-u-v)^{d-i-j}, \quad 0 \le i + j \le n,$$

which are the blending functions for a triangular Bezier patch of degree d.

EXAMPLE
8.8

Pentagonal Toric Bezier Patches

Consider the lattice pentagon in Figure 8.10. The equations of the boundaries of the Newton polygon are the five lines $u = 0$, $v = 0$, $2 - u = 0$, $2 - v = 0$, $3 - u - v = 0$. Thus, by Equation (8.22), the numerators of the eight barycentric coordinate functions are

$$\alpha_{00}(u,v) = (2-u)^2(2-v)^2(3-u-v)^3 \qquad \alpha_{11}(u,v) = uv(2-u)(2-v)(3-u-v)$$

$$\alpha_{01}(u,v) = v(2-u)^2(2-v)(3-u-v)^2 \qquad \alpha_{10}(u,v) = u(2-u)(2-v)^2(3-u-v)^2$$

$$\alpha_{20}(u,v) = u^2(2-v)^2(3-u-v) \qquad \alpha_{02}(u,v) = v^2(2-u)^2(3-u-v)$$

$$\alpha_{12}(u,v) = uv^2(2-u) \qquad\qquad\qquad \alpha_{21}(u,v) = u^2v(2-v)$$

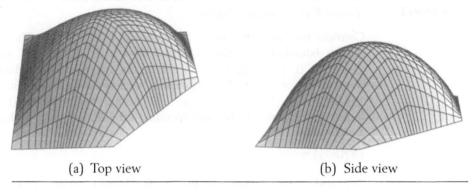

(a) Top view   (b) Side view

Figure 8.14 A pentagonal toric Bezier patch of depth $d = 1$.

Figure 8.14 shows an example of a pentagonal toric Bezier patch of depth $d = 1$ generated using these eight barycentric coordinate functions.

Toric Bezier patches of depth d have the following underlying structure:

- Domain—Newton polygon of a lattice polygon I
- Control points—indexed by a power I^d of the lattice polygon I
- Blending functions—convolutions of barycentric coordinates for the lattice polygon I

We shall show in subsequent sections that, in common with triangular and tensor product Bezier patches, toric Bezier patches also share the following associated properties and procedures. Notice that affine invariance and the convex hull property follow immediately from the first two properties of barycentric coordinates for lattice polygons listed in Theorem 8.2.

- Properties of toric Bezier patches
    - Affine invariance
    - Convex hull property
    - Nondegenerate
    - Boundary curves—Bezier curves determined by the boundary control points
    - Implicit Degree—$d^2 \times 2Area$(Newton polygon of I)
    - Special cases—Triangular and tensor product Bezier patches
- Procedures for toric Bezier patches
    - Evaluation algorithm—pyramid algorithm whose edges are labeled with the barycentric coordinate functions for the lattice polygon I
    - Differentiation algorithm

- Blossoming algorithm
- Subdivision algorithm
- Depth elevation algorithm

Toric Bezier patches are, in general, rational surfaces because the barycentric coordinate functions for a lattice polygon are, in general, rational functions. Since toric Bezier patches are already rational surfaces, we may as well introduce scalar weights $\{w_\lambda\}$ and define rational toric Bezier patches using mass-points in Grassmann space instead of affine points in affine space.

A *rational toric Bezier patch of depth d* is a rational pyramid patch of depth d, whose indexing array I is a lattice polygon and whose barycentric coordinate functions are the barycentric coordinates of I. Thus, for a rational toric Bezier patch, the input to the pyramid algorithm is a collection of mass-points $P_\lambda^0(u,v) = (w_\lambda P_\lambda, w_\lambda)$, $\lambda \in I^d$, and the output of the algorithm must be divided by the weight. In terms of the blending functions $\{B_\lambda^d(u,v)\}$ a rational toric Bezier patch $R(u,v)$ is given by

$$R(u,v) = \frac{\sum\limits_{\lambda \in I^d} B_\lambda^d(u,v) w_\lambda P_\lambda}{\sum\limits_{\gamma \in I^d} w_\gamma B_\gamma^d(u,v)}. \tag{8.27}$$

Notice that for a rational toric Bezier patch the denominator

$$\sum_{(k,l) \in I} \alpha_{kl}(u,v)$$

of the barycentric coordinate functions $\beta_{ij}(u,v)$ appears to the same power in the numerator and denominator of $R(u,v)$. Therefore, we can cancel these factors. Thus for rational toric Bezier patches, we can use the numerators $\alpha_{ij}(u,v)$ in place of the barycentric coordinate functions $\beta_{ij}(u,v)$—that is, we do not need to normalize the barycentric coordinate functions to sum to one, since the division in Equation (8.27) performs the necessary normalization automatically. Hence, we can replace the barycentric coordinate functions $\beta_{ij}(u,v)$ by the numerators $\alpha_{ij}(u,v)$ in the pyramid algorithm for a rational toric Bezier patch.

Exercises

1. Consider the lattice line $I = \{(0,0),(1,0)\}$. Describe the toric Bezier patches of depth d corresponding to this lattice line.

2. Consider the lattice rectangle $I_R^n = \{(i,j) \mid 0 \le i, j \le n\}$.

   a. Describe the toric Bezier patches of depth 1 corresponding to this lattice rectangle.

   b. Describe the toric Bezier patches of depth d corresponding to this lattice rectangle.

3. Consider the lattice triangle $I_T^n = \{(i,j) \mid 0 \le i + j \le n\}$.

   a. Describe the toric Bezier patches of depth 1 corresponding to this lattice triangle.

   b. Describe the toric Bezier patches of depth d corresponding to this lattice triangle.

4. Compute the blending functions for the pentagonal and hexagonal toric Bezier patches of depth 2 defined by the lattice polygons in Figures 8.10 and 8.12.

5. Implement the pyramid algorithm for the pentagonal toric Bezier patches defined by the lattice pentagon in Figure 8.10. Experiment with pentagonal patches of different depths. Determine how changing the location of the control points affects the shape of the surface.

6. Implement the pyramid algorithm for the hexagonal toric Bezier patches defined by the lattice hexagon in Figure 8.12. Experiment with hexagonal patches of different depths. Determine how changing the location of the control points affects the shape of the surface.

7. Generalize the pyramid algorithm for toric Bezier patches by replacing I^{d-l} with $I_1 \oplus \cdots \oplus I_{d-l}$, where $I_1, \ldots, I_d$ are lattice polygons.

   a. What is the domain of the patch?

   b. How must Equation (8.24) be altered to make the pyramid algorithm valid?

   c. Describe the blending functions for this patch. (Compare to Section 8.5, Exercise 1.)

8. Show that the only toric Bezier patches that are also S-patches are

   a. Bezier curves of degree 1

   b. triangular Bezier patches of degree 1

   c. tensor product patches of bidegree 1

8.7.4 The Boundaries of a Toric Bezier Patch

The boundaries of a toric Bezier patch are the Bezier curves determined by their boundary control points. This result holds provided only that the constant coefficients c_{ij} for the barycentric coordinate functions corresponding to lattice points along the boundary of the Newton polygon of the lattice polygon are properly chosen. We are going to prove this assertion shortly below. Thus, unlike general pyramid patches, in toric Bezier patches we maintain control over the boundary of the patch. This control is one of the main advantages of toric Bezier patches over arbitrary pyramid patches.

To analyze the boundaries, consider a toric Bezier patch $B(u,v)$ with control points $\{P_\lambda\}$ indexed by a lattice polygon I with barycentric coordinate functions $\{\beta_{ij}\}_{(i,j) \in I}$. Let $Q_1, \ldots, Q_m$ be the vertices of the Newton polygon of I. Then by Theorem 8.2,

$$\beta_{ij} = 0 \text{ on the boundary } Q_k Q_{k+1}, \text{ if } (i,j) \notin Q_k Q_{k+1}.$$

Thus the kth boundary of $B(u,v)$ is completely determined by the points of $\{P_\lambda\}$ indexed by the values of I along the boundary Q_kQ_{k+1}, since the coefficients $\beta_{ij}(u,v)$ of all the other control points vanish along this boundary. Hence the boundary curves of a toric Bezier patch of depth $d = 1$ are completely determined by the boundary control points of the patch.

To show that these boundary curves are actually Bezier curves, let us focus on a specific boundary Q_lQ_{l+1}. Suppose that $R_0,\dots,R_n$ are the points in the lattice polygon I along the boundary Q_lQ_{l+1}, and denote by $\beta_0,\dots,\beta_n$ the barycentric coordinate functions of I corresponding to the points $R_0,\dots,R_n$. We are going to apply a reparametrization to show that when restricted to the boundary Q_lQ_{l+1}, these barycentric coordinate functions are univariate Bernstein polynomials.

Let $L_k(u,v) \equiv a_ku + b_kv + c_k = 0$ be the equation of the kth boundary line Q_kQ_{k+1}, $k = 1,\dots,m$, of the Newton polygon of I. Then by Equations (8.22) and (8.23), when (u,v) are restricted to the boundary Q_lQ_{l+1}

$$\alpha_h(u,v) = c_h\{L_1(u,v)\}^{L_1(R_h)}\cdots\{L_m(u,v)\}^{L_m(R_h)} \tag{8.28}$$

$$\beta_h(u,v) = \frac{\alpha_h(u,v)}{\sum\limits_{g=0}^{n}\alpha_g(u,v)} \qquad h = 0,\dots,n \cdot \tag{8.29}$$

But since $R_0,\dots,R_n$ lie along the edge of a lattice polygon, there are integers p,q,r,s such that

$$R_h = (p + rh, q + sh) \qquad h = 0,\dots,n \cdot$$

Therefore, on the boundary Q_lQ_{l+1}

$$\alpha_h(u,v) = c_h\{L_1(u,v)\}^{a_1(p+rh)+b_1(q+sh)+c_1}\cdots\{L_m(u,v)\}^{a_m(p+rh)+b_m(q+sh)+c_m}.$$

We can now split $\alpha_h(u,v)$ into two factors:

1. $C = \{L_1(u,v)\}^{a_1p+b_1q+c_1}\cdots\{L_m(u,v)\}^{a_mp+b_mq+c_m}$

2. $D_h = c_h\left\{\{L_1(u,v)\}^{a_1r+b_1s}\cdots\{L_m(u,v)\}^{a_mr+b_ms}\right\}^h.$

The first factor is independent of h, and hence is common to $\alpha_h(u,v)$ for all $h = 0,\dots,n$. Therefore, this factor cancels in the expression for $\beta_h(u,v)$ in Equation (8.29). We are free to choose c_h in any manner we like, so let $c_h = \binom{n}{h}$. Now apply the reparametrization

$$\tau = \{L_1(u,v)\}^{a_1r+b_1s}\cdots\{L_m(u,v)\}^{a_mr+b_ms}.$$

Then along the boundary Q_lQ_{l+1}

$$\beta_h(u,v) = \frac{\binom{n}{h}\tau^h}{\sum\limits_{g=0}^{n}\binom{n}{g}\tau^g}. \tag{8.30}$$

Substituting $\tau = t/(1-t)$ into Equation (8.30) and multiplying the numerator and denominator by $(1-t)^n$ yields

$$\beta_h(u,v) = \frac{\binom{n}{h}t^h(1-t)^{n-h}}{\sum\limits_{g=0}^{n}\binom{n}{g}t^g(1-t)^{n-g}} = \binom{n}{h}t^h(1-t)^{n-h} = B_h^n(t).$$

Thus, when restricted to the boundary Q_lQ_{l+1}, the barycentric coordinate functions of the lattice polygon I corresponding to the lattice points $R_0,\ldots,R_n$ along this boundary reduce to the univariate Bernstein basis functions. Therefore, the corresponding boundary curve of $B(u,v)$ is the Bezier curve determined by the control points of $\{P_\lambda\}$ indexed by the points of I lying along the boundary Q_lQ_{l+1}. Notice that the boundary curve is polynomial even though the surface is rational.

All that work was just for the base case of a toric Bezier patch—that is, for a toric Bezier patch of depth $d = 1$. What about the boundaries of a toric Bezier patch of depth $d > 1$? Here we can reason as follows. By Equation (8.25) the blending functions $\{B_\lambda^d(u,v)\}$ can be computed from the barycentric coordinate functions $\beta_I(u,v) = \{\beta_{ij}\}_{(i,j)\in I}$ by the discrete convolution formula

$$\{B_\lambda^d(u,v)\} = \underbrace{\beta_I(u,v) \otimes \cdots \otimes \beta_I(u,v)}_{d\ \ factors}.$$

But the boundaries of the Newton polygon of I^d are just the d-fold Minkowski sums of the corresponding boundaries of the Newton polygon of I (see Section 8.7.1, Exercise 4). Therefore, when restricted to the kth boundary of I, the blending functions $\{B_\lambda^d(u,v)\}$ where λ lies along the kth boundary of the Newton polygon of I^d, are just d-fold convolutions of the barycentric coordinate functions corresponding to the lattice points along the kth boundary of the Newton polygon of I. But we have just proved that under a reparametrization when restricted to the kth boundary of the Newton polygon of I, these barycentric coordinate functions are univariate Bernstein basis functions. Since the convolution of Bernstein bases are Bernstein bases of higher degree, the blending functions of $\{B_\lambda^d(u,v)\}$ where λ lies along the kth boundary of the Newton polygon of I^d, when restricted to the kth boundary of the Newton polygon of I, are univariate Bernstein basis functions. Hence the boundary curves of a toric Bezier patch of depth $d > 1$ are the Bezier curves determined by their boundary control points.

Exercises

1. Show that the boundaries of a rational toric Bezier patch are the rational Bezier curves determined by the mass-points along their boundaries.

2. Consider a boundary of a toric Bezier patch.

    a. Show that along a patch boundary, the pyramid algorithm reduces to the triangular computation along the lateral face of the pyramid whose base consists of the corresponding boundary control points.

    b. Explain how this triangular computation is related to the de Casteljau algorithm for a Bezier curve with the same control points.

8.7.5 The Monomial and Bernstein Representations of a Toric Bezier Patch

What is the parametric degree of a toric Bezier patch? Since a toric Bezier patch is a pyramid patch, the degree of the patch depends on the degree of the barycentric coordinate functions of the patch. By Equation (8.23), the degree of the barycentric coordinate functions $\{\beta_{ij}\}_{(i,j)\in I}$ depends, in turn, on the degree of the functions $\{\alpha_{ij}\}_{(i,j)\in I}$. Let $L_k(u,v) \equiv a_k u + b_k v + c_k = 0$, $k = 1,...,p$, be the equations of the boundaries of the Newton polygon of I. Then by Equation (8.22)

$$degree\{\alpha_{ij}(u,v)\} = \sum_{k=1}^{p} L_k(i,j).$$

Therefore, naively, for a toric Bezier patch of depth $d = 1$,

$$parametric\ degree = max_{(i,j)\in I}\{\sum_{k=1}^{p} L_k(i,j)\},$$

and for patches of depth d this degree must be multiplied by d.

The preceding analysis drastically overestimates the parametric degree of a toric Bezier patch. The functions $\{\alpha_{ij}\}_{(i,j)\in I}$ actually have many common factors, and these factors cancel in the expression in Equation (8.23) for the barycentric coordinate functions $\{\beta_{ij}\}_{(i,j)\in I}$. The truth is that after this cancellation and a simple change of variables the only powers of the parameters that appear in the barycentric coordinate functions are the values $(i,j) \in I$—that is, the powers indexed by the lattice polygon of the patch.

The techniques used in the previous section to analyze the boundaries of a toric Bezier patch can be employed here as well to find simple monomial and Bernstein representations for the barycentric coordinate functions of a toric Bezier patch. Recall that by Equation (8.22)

$$\alpha_{ij}(u,v) = c_{ij}\{L_1(u,v)\}^{L_1(i,j)} \cdots \{L_p(u,v)\}^{L_p(i,j)}$$

$$= c_{ij}\{L_1(u,v)\}^{a_1 i + b_1 j + c_1} \cdots \{L_p(u,v)\}^{a_p i + b_p j + c_p}. \tag{8.31}$$

Thus we can split $\alpha_{ij}(u,v)$ into three factors:

1. $C = \{L_1(u,v)\}^{c_1} \cdots \{L_p(u,v)\}^{c_p}$

2. $D_i = \left\{\{L_1(u,v)\}^{a_1} \cdots \{L_p(u,v)\}^{a_p}\right\}^i$

3. $E_j = \left\{\{L_1(u,v)\}^{b_1} \cdots \{L_p(u,v)\}^{b_p}\right\}^j$.

The first factor is independent of i,j, and hence is common to $\alpha_{ij}(u,v)$ for all $(i,j) \in I$. Therefore, this factor cancels in the expression for $\beta_{ij}(u,v)$ in Equation (8.23). Now let

$$\sigma = \{L_1(u,v)\}^{a_1} \cdots \{L_p(u,v)\}^{a_p}$$

$$\tau = \{L_1(u,v)\}^{b_1} \cdots \{L_p(u,v)\}^{b_p} .$$

Substituting these parameters into Equation (8.31) yields

$$\alpha_{ij}(u,v) = c_{ij}\sigma^i \tau^j \qquad (i,j) \in I \, . \tag{8.32}$$

Hence the barycentric coordinate functions of a toric Bezier patch with lattice polygon I can be represented in rational form by monomials indexed by the set I. The control points and weights of the patch are unchanged. To find the blending functions for a patch of depth d, we just convolve the barycentric coordinate functions. Thus, the blending functions of a toric Bezier patch of depth d with lattice polygon I can be represented in rational form by monomials indexed by the set I^d.

With just a bit more work, we can also use Equation (8.32) to find both tensor product and triangular Bezier representations for a toric Bezier patch. We will illustrate the technique for the tensor product basis and leave the triangular case as an exercise (see Exercise 2). Let $m = max\{i\}$ and $n = max\{j\}$ for all $(i,j) \in I$. Substitute $\sigma = s/(1-s)$ and $\tau = t/(1-t)$ into Equation (8.32), and multiply the numerator and denominator by $(1-s)^m(1-t)^n$ to obtain

$$\alpha_{ij}(u,v) = \frac{c_{ij}s^i(1-s)^{m-i}t^j(1-t)^{n-j}}{(1-s)^m(1-t)^n} = \frac{B_i^m(s)B_j^n(t)}{(1-s)^m(1-t)^n},$$

where we have chosen $c_{ij} = \binom{m}{i}\binom{n}{j}$. Again the denominators $(1-s)^m(1-t)^n$ are common to $\alpha_{ij}(u,v)$ for all $(i,j) \in I$. Therefore, this factor cancels in the expression for $\beta_{ij}(u,v)$ in Equation (8.23). Hence, the barycentric coordinate functions for a toric Bezier patch with lattice polygon I can be represented in rational form by bivariate Bernstein basis functions of bidegree (m,n) indexed by the set I. Consequently, the blending functions of a toric Bezier patch with lattice polygon I of depth d can be represented in rational form by bivariate Bernstein basis functions of bidegree (md,nd) indexed by the set I^d.

The implicit degree of a rational surface whose exponents lie in a lattice polygon I is known to be $2 \times Area$(Newton polygon of I). But by Section 8.7.1, Exercise 4(d),

$$Area(\text{Newton polygon of } I^d) = d^2 \times Area(\text{Newton polygon of } I).$$

Therefore, it follows from the preceding analysis that the implicit degree of a toric Bezier patch of depth d with lattice polygon I is $d^2 \times 2Area$(Newton polygon of I).

Exercises

1. Show that the blending functions of a toric Bezier patch are linearly independent. Conclude that toric Bezier patches are nondegenerate—that is, that these surfaces never collapse to a single point unless all their control points are located at that point.

2. Show that the barycentric coordinate functions of a toric Bezier patch with lattice polygon I can be represented in rational form by triangular Bernstein basis functions of total degree $n = max\{i + j\}$, $(i,j) \in I$, indexed by the set I. Conclude that the blending functions of a toric Bezier patch of depth d with lattice polygon I can be represented in rational form by triangular Bernstein basis functions of total degree nd indexed by the set I^d. (Hint: Consider the change of variables $\sigma = s/(1 - s - t)$ and $\tau = t/(1 - s - t)$.)

3. Consider the barycentric coordinate functions of a rational toric Bezier patch with lattice polygon I. Show that these functions can be represented in rational form by

   a. the monomials indexed by the set I

   b. the bivariate Bernstein basis functions of bidegree (m,n) indexed by the set I, where $m = max\{i\}$ and $n = max\{j\}$ for all $(i, j) \in I$

   c. the bivariate Bernstein basis functions of total degree n indexed by the set I, where $n = max\{i + j\}$ for all $(i, j) \in I$

4. Suppose that J is a lattice polygon generated by rotating and translating another lattice polygon I. Show that the toric Bezier patches built using the lattice polygon J are identical to the toric Bezier patches built using the lattice polygon I.

5. Let I be a lattice polygon. Show that

$$2 \times Area(\text{convex hull } I) = 2 \times (\text{number of interior points } I)$$
$$+ (\text{number of boundary points } I) - 2.$$

(Hint: Split I into two lattice polygons and apply induction.)

6. Let I be a lattice polygon, and let $L_k(u,v) = 0$, $k = 1,...,p$, be the equations of the boundaries of the Newton polygon of I. For all $(i,j) \in I$, define

$$\alpha_{ij}(x_1,...,x_p) = c_{ij} x_1^{L_1(i,j)} \cdots x_p^{L_p(i,j)}$$

$$\beta_{ij}(x_1,...,x_p) = \frac{\alpha_{ij}(x_1,...,x_p)}{\sum_{(k,l)\in I} \alpha_{kl}(x_1,...,x_p)}.$$

Show that if we replace the barycentric coordinate functions $\beta_{ij}(u,v)$ for the lattice polygon I with the functions $\beta_{ij}(x_1,...,x_p)$, we generate the same toric Bezier patch for $x_1,...,x_p \geq 0$. (Hint: Consider a change of variables.)

8.7.6 Toric S-Patches

If we use the barycentric coordinate functions for a convex polygon constructed in Section 8.1, then the parametric degree of an n-sided S-patch of depth d is $d(n - 2)$.

But what is the implicit degree? In general, the implicit degree of a rational surface is given by the formula

implicit degree = (parametric degree)^2 – number of base points (with multiplicity),

where a *base point* is a parameter value where the blending functions evaluate to $0/0$.

The blending functions for S-patches have base points wherever nonadjacent sides of the domain polygon intersect because by Equations (8.3) and (8.4) the numerators and denominators of all the barycentric coordinate functions vanish at these points. For an *n*-sided S-patch there are $\binom{n}{2} - n = n(n-3)/2$ such base points, and for an S-patch of depth *d* these base points each have multiplicity d^2. Therefore, for a generic *n*-sided S-patch of depth *d*,

$$\text{implicit degree} = d^2(n-2)^2 - d^2 n(n-3)/2 = d^2(n^2 - 5n + 8)/2.$$

For example, for a three-sided S-patch of depth *d* (i.e., a Bezier patch of degree *d*), the implicit degree is d^2; for a four-sided S-patch of depth *d* (e.g, a tensor product Bezier patch of bidegree *d*), the implicit degree is $2d^2$. For a five-sided S-patch of depth *d*, the implicit degree is $4d^2$, and for a six-sided S-patch of depth *d*, the implicit degree is $7d^2$.

By contrast, we observed in Section 8.7.5 that the implicit degree of a toric Bezier patch of depth *d* with lattice polygon *I* is $d^2 \times 2Area(\text{Newton polygon of } I)$. For three-sided and four-sided toric schemes, we get the same degrees as for three-sided and four-sided S-patches, since three-sided and four-sided toric Bezier patches are equivalent to triangular and tensor product Bezier patches. But consider the lattice pentagon $I = \{(0,0),(1,0),(2,1),(1,2),(0,1),(1,1)\}$ depicted in Figure 8.5. For this lattice pentagon $2 \times Area(\text{Newton polygon of } I) = 5$. Hence for the corresponding toric Bezier patches of depth *d*, the implicit degree is $5d^2$, which is clearly larger than $4d^2$, the implicit degree of a five-sided S-patch of depth *d*. On the other hand, consider the lattice hexagon in Figure 8.12. Here $2 \times Area(\text{Newton polygon of } I) = 6$. Hence the corresponding toric Bezier patches of depth *d* have implicit degree $6d^2$, which is clearly smaller than $7d^2$, the implicit degree of a six-sided S-patch of depth *d*. So sometimes S-patches have lower implicit degree, sometimes toric Bezier patches have lower implicit degree, and sometimes both schemes have the same implicit degree.

We can, however, always generate *n*-sided S-patches of depth *d* with the same implicit degree as an *n*-sided toric Bezier patch of depth *d* by choosing a different set of barycentric coordinate functions. Suppose that the vertices $Q_1,...,Q_n$ of a convex polygon Q have integer coordinates in the plane. Let $\beta_1,...,\beta_n$ be the barycentric coordinate functions at $Q_1,...,Q_n$ for the lattice polygon *I* whose vertices are at $Q_1,...,Q_n$, and set the constant coefficients of the $\alpha_{ij}(u,v)$ in Equation (8.22) to zero for all indices $(i,j) \in I$ not at the vertices of *I*. By Theorem 8.2 these functions $\beta_1,...,\beta_n$ satisfy all the properties of barycentric coordinate functions for the convex polygon Q. Explicitly if $L_k(u,v) \equiv a_k u + b_k v + c_k = 0$, $k = 1,...,n$, is the equation of the *k*th boundary line $Q_k Q_{k+1}$, normalized in the usual fashion (see Section 8.7.2), then

$$\alpha_k(u,v) = e_k \{L_1(u,v)\}^{L_1(Q_k)} \cdots \{L_n(u,v)\}^{L_n(Q_k)} \tag{8.33}$$

$$\beta_k(u,v) = \frac{\alpha_k(u,v)}{\displaystyle\sum_{l=1}^{n}\alpha_l(u,v)} \qquad k = 1,\ldots,n. \tag{8.34}$$

Contrast this construction with the standard construction of barycentric coordinates for the convex polygon Q given in Equations (8.3) and (8.4):

$$\alpha_k(u,v) = \frac{e_k L_1(u,v)\cdots L_n(u,v)}{L_{k-1}(u,v)L_k(u,v)} \tag{8.35}$$

$$\beta_k(u,v) = \frac{\alpha_k(u,v)}{\displaystyle\sum_{l=1}^{n}\alpha_l(u,v)} \qquad k = 1,\ldots,n. \tag{8.36}$$

Notice that in both sets of equations—Equations (8.33) and (8.35)—the lines $L_{k-1}(u,v), L_k(u,v)$ are suppressed in $\alpha_k(u,v)$. In Equation (8.33), these lines disappear because their exponents $L_{k-1}(Q_k) = L_k(Q_k) = 0$; in Equation (8.35), these lines are canceled by the denominator.

At first glance Equations (8.33) and (8.34) look more complicated and of higher parametric degree than Equations (8.35) and (8.36). But proceeding as in Section 8.7.5, we can simplify the expressions for $\beta_1,\ldots,\beta_n$ in Equation (8.34). First remove the common factor

$$C = \{L_1(u,v)\}^{c_1}\cdots\{L_n(u,v)\}^{c_n}$$

from the functions $\alpha_k(u,v)$ and then apply the change of variables

$$\sigma = \{L_1(u,v)\}^{a_1}\cdots\{L_n(u,v)\}^{a_n}$$

$$\tau = \{L_1(u,v)\}^{b_1}\cdots\{L_n(u,v)\}^{b_n}.$$

After performing these operations, we are left with

$$\alpha_k(u,v) = e_k\sigma^{k_1}\tau^{k_2} \qquad Q_k = (k_1,k_2),$$

so the only powers that appear in the barycentric coordinate functions are the coordinates of the vertices of the polygon Q. Hence the implicit degree of the corresponding S-patch of depth d is $d^2 \times 2Area(Q)$, the same degree as the toric Bezier patch of depth d whose lattice polygon consists of all the points with integer coordinates inside or on the boundary of Q.

We can use the barycentric coordinate functions in Equation (8.34) to construct an n-sided S-patch, without necessarily using the polygon Q to index the polygonal array of control points. Any indexing set I of n distinct p-tuples will still work. For example, we can use the barycentric coordinate functions generated by the lattice hexagon depicted in Figure 8.12 together with the simplicial indexing set $\Delta_6^d = \{k_1,\ldots,k_6)\mid k_1 +\cdots+k_6 = d\}$ to generate hexagonal S-patches of depth d. This approach keeps the degree of the S-patch low and still allows arbitrary hexagonal arrays of control points. We call the barycentric coordinate functions in Equation (8.34) *toric barycentric coordinate functions* for the convex polygon Q, and we call an S-patch that uses toric barycentric coordinate functions a *toric S-patch*.

Exercises

1. Show that the blending functions of an *S*-patch generated by the toric barycentric coordinates are linearly independent. Conclude that *S*-patches generated by the toric barycentric coordinate functions are nondegenerate—that is, that these surfaces never collapse to a single point unless all their control points are located at that point. Compare this result to Section 8.4.1, Exercise 2.

2. Consider the pentagon with vertices $Q = \{(0,0),(1,0),(2,1),(1,2),(0,1)\}$.

   a. Implement the pyramid algorithm for control points $\{P_\lambda\}$ indexed by I^d, where I consists of the vertices of the standard 4-simplex, using

      i. the standard barycentric coordinate functions for the pentagon with vertices Q

      ii. the toric barycentric coordinate functions for the pentagon with vertices Q

   b. What are the implicit degrees of these *S*-patches of depth *d*?

   c. Compare the shapes of the pentagonal surfaces generated by these two sets of barycentric coordinates.

3. Consider the hexagon Q depicted in Figure 8.12.

   a. Implement the pyramid algorithm for control points $\{P_\lambda\}$ indexed by I^d, where I consists of the vertices of the standard 5-simplex, using

      i. the standard barycentric coordinate functions for the hexagon Q

      ii. the toric barycentric coordinate functions for the hexagon Q

   b. What are the implicit degrees of these *S*-patches of depth *d*?

   c. Compare the shapes of the hexagonal surfaces generated by these two sets of barycentric coordinates.

4. What is the implicit degree of

   a. a seven-sided *S*-patch of depth *d*?

   b. an eight-sided *S*-patch of depth *d*?

   c. the lowest-degree seven-sided toric Bezier patch of depth *d*?

   d. the lowest-degree eight-sided toric Bezier patch of depth *d*?

8.7.7 Subdividing Toric Bezier Patches into Tensor Product Bezier Patches

In Section 8.7.5 we showed how to extend a lattice polygon to a lattice rectangle and then perform a change of variables to convert any toric Bezier patch to rational monomial or rational tensor product Bezier form. By applying this technique, we could treat an *n*-sided toric Bezier patch as a rational tensor product Bezier patch and apply the subdivision algorithm that we already know for tensor product patches. This approach, however, would subdivide an *n*-sided toric patch into four tensor product Bezier patches without regard for the *n*-sided structure of the patch. Our goal here is to develop an alternative subdivision procedure for *n*-sided patches that respects the *n*-sided structure of the patch.

Every n-sided toric Bezier patch can be split into n rational tensor product Bezier patches. This subdivision can be performed by placing a tensor product patch at each corner of the toric patch, so that the tensor product patches all meet at a common point and patches at adjacent vertices join along smooth curves on the toric surface (see Figure 8.15). This subdivision procedure can be applied to reduce the analysis of toric Bezier patches—for example, rendering and intersection algorithms—to the analysis of standard rational tensor product Bezier patches.

How is this done? Consider first a rational tensor product Bezier patch

$$P(u,v) = \frac{\sum\limits_{i=0}^{m}\sum\limits_{j=0}^{n} w_{ij}P_{ij}B_i^m(u)B_j^n(v)}{\sum\limits_{i=0}^{m}\sum\limits_{j=0}^{n} w_{ij}B_i^m(u)B_j^n(v)} \qquad 0 \le u,v \le 1.$$

If we subdivide this patch at $u = 1/2$ and $v = 1/2$ by the method in Section 5.8.1, then we split the patch into four subpatches as illustrated in Figure 8.16.

But there is another more direct way to find the control points of each of these subpatches. Let us focus for now on the subpatch at the corner P_{00}. To find the control points of this subpatch, divide the numerator and denominator of $P(u,v)$ by $(1-u)^m(1-v)^n$ and then apply the change of variables

$$s = \frac{u}{1-u} \qquad t = \frac{v}{1-v}$$

to obtain the reparametrized patch

$$P*(s,t) = \frac{\sum\limits_{i=0}^{m}\sum\limits_{j=0}^{n} w_{ij}P_{ij}\binom{m}{i}\binom{n}{j}s^i t^j}{\sum\limits_{i=0}^{m}\sum\limits_{j=0}^{n} w_{ij}\binom{m}{i}\binom{n}{j}s^i t^j}.$$

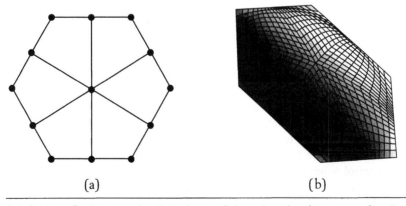

(a)	(b)

Figure 8.15 Subdivision of a hexagonal toric Bezier patch into six rational tensor product Bezier patches: (a) schematic diagram and (b) actual surface patch.

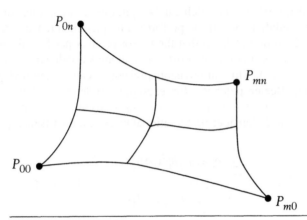

Figure 8.16 Schematic diagram of a rational tensor product Bezier patch subdivided into four rational tensor product Bezier patches.

Point for point, $P(u,v)$ and $P^*(s,t)$ represent the same surface, but the domain of $P(u,v)$ is $0 \leq u,v \leq 1$ whereas the corresponding domain of $P^*(s,t)$ is $0 \leq s,t < \infty$. Suppose, however, that we restrict the domain of $P^*(s,t)$ to $0 \leq s,t \leq 1$. Within this restricted domain, $P^*(s,t)$ is precisely the subpatch of $P(u,v)$ at the corner P_{00}, since $0 \leq s,t \leq 1$ if and only if $0 \leq u,v \leq 1/2$. But $0 \leq s,t \leq 1$ is the standard tensor product domain. So to find the Bezier control points of the subpatch at P_{00}, we need only convert the expression for $P^*(s,t)$ from the monomial basis to the tensor product Bernstein basis. We can easily perform this change of basis using any one of the standard change of basis algorithms such as blossoming for converting from monomial to Bezier form. The control points for the subpatches at the other three corners can be found in a similar fashion (see Exercise 1).

To recapitulate what we have just done: At each corner of a tensor product Bezier patch, we have a rectangular array of control points. To find the subpatches at each corner, we perform:

·1. a change of variables from tensor product Bernstein to monomial form

2. a change of basis from monomial to tensor product Bernstein form

We would like to proceed in a similar fashion to subdivide an arbitrary toric Bezier patch into a collection of rational tensor product Bezier patches. But one problem we immediately encounter is that for arbitrary toric Bezier patches we do not have a rectangular array of control points at each corner of the patch (see Figure 8.17). So we shall need to begin by extending the lattice polygon for the toric patch to a rectangular tensor product lattice at each corner of the patch. When we extend the lattice, we may introduce additional lattice points; the corresponding additional control points are simply set to zero. We can then proceed to subdivide the toric Bezier patch into a collection of rational tensor product Bezier patches just as we were able to subdivide rational tensor product Bezier patches.

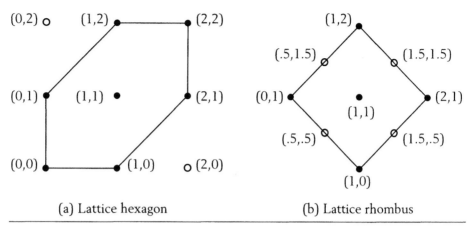

(a) Lattice hexagon   (b) Lattice rhombus

Figure 8.17 Extending the lattice at a corner for (a) a lattice hexagon and (b) a lattice rhombus. Notice that for the lattice rhombus we not only need to add points to the lattice polygon, but we must also add points to the lattice $\mathbf{Z} \times \mathbf{Z}$.

To summarize: the subdivision algorithm for toric Bezier patches has the following four steps:

1. Extend the lattice polygon at each corner of the patch to a lattice rectangle.
2. Fill in the additional control points with zeros.
3. Perform a change of variables to convert to monomial form.
4. Perform a change of basis from monomial to tensor product Bernstein form.

Now we shall explain in detail how to execute each of these steps.

To begin, we need to construct for each vertex of the toric patch a rectangular array of control points indexed by a lattice rectangle anchored at the corresponding vertex of the domain polygon. These new lattice rectangles must contain the original lattice polygon to ensure that we can still represent all the control points of the original patch by indices in the new lattices. Thus to prepare the way for subdivision, we are going to construct at each vertex Q_k of the lattice polygon I for the toric patch a new lattice Λ_k that contains the lattice $\mathbf{Z} \times \mathbf{Z}$.

Let R_k be the nearest point in I to Q_k along the boundary $Q_{k-1}Q_k$, and let $S_k \in I$ be the nearest point in I to Q_k along the boundary Q_kQ_{k+1}. Define

$$D_k = 2 \times Area(\Delta R_k Q_k S_k)$$

$$E_{k1} = \frac{(R_k - Q_k)}{D_k}$$

$$E_{k2} = \frac{(S_k - Q_k)}{D_k} .$$

Then starting from the point Q_k and adding integer multiples of E_{k1}, E_{k2} forms a new planar lattice Λ_k. We are now going to show that the lattice Λ_k contains the lattice $\mathbf{Z} \times \mathbf{Z}$—that is, Λ_k is a *lattice extension* of $\mathbf{Z} \times \mathbf{Z}$.

LEMMA   *The lattice Λ_k contains the lattice $\mathbf{Z} \times \mathbf{Z}$.*
8.3

Proof   To prove that Λ_k contains $\mathbf{Z} \times \mathbf{Z}$, we shall first show that for every point R in the plane

$$R = Q_k + L_k(R)E_{k1} + L_{k-1}(R)E_{k2},$$

where $L_j(u,v)$, $j = 1,\ldots,p$, denote the boundaries of the Newton polygon of I. Since E_{k1}, E_{k2} are parallel to different edges of the Newton polygon of I, certainly for every point R in the plane there are real numbers λ, μ such that

$$R = Q_k + \lambda E_{k1} + \mu E_{k2}.$$

But because $Q_k, R_k, S_k \in I$, there must exist integers h_{ij}, $i, j = 1,2$, such that

$$R_k - Q_k = (h_{11}, h_{12})$$
$$S_k - Q_k = (h_{21}, h_{22}). \tag{8.37}$$

Moreover,

$$D_k = \det\begin{pmatrix} h_{11} & h_{12} \\ h_{21} & h_{22} \end{pmatrix} = h_{11}h_{22} - h_{12}h_{21}.$$

Applying L_{k-1} to both sides of the equation $R = Q_k + \lambda E_{k1} + \mu E_{k2}$ and recalling that $Q_k + \lambda E_{k1}$ lies on L_{k-1} yields

$$L_{k-1}(R) = \mu(N_{k-1} \bullet E_{k2}),$$

where N_{k-1} denotes the normal of L_{k-1}. But by Equation (8.37)

$$N_{k-1} = (-h_{12}, h_{11}).$$

Hence

$$N_{k-1} \bullet E_{k2} = \frac{N_{k-1} \bullet (S_k - Q_k)}{D_k} = \frac{h_{11}h_{22} - h_{12}h_{21}}{D_k} = 1.$$

Therefore, $\mu = L_{k-1}(R)$. A similar argument shows that $\lambda = L_k(R)$. Thus for every point R in the plane

$$R = Q_k + L_k(R)E_{k1} + L_{k-1}(R)E_{k2}. \tag{8.38}$$

It follows immediately that Λ_k is a lattice extension of $\mathbf{Z} \times \mathbf{Z}$, since if $R \in \mathbf{Z} \times \mathbf{Z}$, then $L_{k-1}(R), L_k(R)$ are integers.

To extend the original lattice polygon I to a lattice rectangle I_k^*, we now take all lattice points in the new lattice Λ_k with $0 \le i \le \max_{R \in I}\{L_{k-1}(R)\}$ and $0 \le j \le \max_{R \in I}\{L_k(R)\}$. By Equation (8.38) this lattice rectangle I_k^* contains the original lattice polygon I. Control points corresponding to lattice points in I_k^* not in the original lattice polygon I are simply set to zero.

Next, we need to perform a change of variables to represent the toric Bezier patch in monomial form. We proceed exactly as in Section 8.7.5, but we replace the original lattice polygon I by the new lattice rectangle I_k^*. Using the expression in Equation (8.38) for R, we can rewrite $\alpha_R(u,v)$ as

$$\alpha_R(u,v) = c_R\{L_1(u,v)\}^{L_1(R)}\cdots\{L_p(u,v)\}^{L_p(R)} \tag{8.39}$$

$$= c_R\{L_1(u,v)\}^{L_1(Q_k)+L_k(R)(N_1\bullet E_{k1})+L_{k-1}(R)(N_1\bullet E_{k2})}$$

$$\cdots\{L_p(u,v)\}^{L_p(Q_k)+L_k(R)(N_p\bullet E_{k1})+L_{k-1}(R)(N_p\bullet E_{k2})}$$

Thus, as in Section 8.7.5, we can split $\alpha_R(u,v)$ into three factors:

1. $C = \{L_1(u,v)\}^{L_1(Q_k)}\cdots\{L_p(u,v)\}^{L_p(Q_k)}$

2. $D_k = \left\{\{L_1(u,v)\}^{(N_1\bullet E_{k1})}\cdots\{L_p(u,v)\}^{(N_p\bullet E_{k1})}\right\}^{L_k(R)}$

3. $E_{k-1} = \left\{\{L_1(u,v)\}^{(N_1\bullet E_{k2})}\cdots\{L_p(u,v)\}^{(N_p\bullet E_{k2})}\right\}^{L_{k-1}(R)}$

The first factor is independent of R, and hence is common to $\alpha_R(u,v)$ for all $R \in I$. Therefore, this factor cancels in the expression for $\beta_R(u,v)$ in Equation (8.23). Let

$$t = \{L_1(u,v)\}^{(N_1\bullet E_{k1})}\cdots\{L_p(u,v)\}^{(N_p\bullet E_{k1})}$$

$$s = \{L_1(u,v)\}^{(N_1\bullet E_{k2})}\cdots\{L_p(u,v)\}^{(N_p\bullet E_{k2})}. \tag{8.40}$$

Substituting these parameters into Equation (8.39) and discarding C yields

$$\alpha_R(u,v) = c_R s^{L_{k-1}(R)}t^{L_k(R)} \qquad R \in I. \tag{8.41}$$

Thus $\alpha_R(u,v)$ and hence too $\beta_R(u,v)$ is now represented in monomial form.

Consider the image on the toric patch of the unit square $\{(s,t)\mid 0 \le s,t \le 1\}$. Along the boundary $s = 0$, $\alpha_R(u,v) \ne 0$ if and only if R lies on the boundary L_{k-1}. Therefore, as in Section 8.7.4, the curve corresponding to $s = 0$ is the Bezier curve whose control points are indexed by the lattice points along the boundary L_{k-1}—that is, the $(k-1)$st boundary of the toric Bezier patch. (If $L_k(R_{k-1}) \ne 1$, then reparametrize by setting $u = t^{L_k(R_{k-1})}$.) In particular, if $P_0,\ldots,P_m$ are the control points corresponding to the lattice points along L_{k-1}, then in monomial form the boundary curve $s = 0$ is

$$P(t) = \sum_{h=0}^{m} \frac{\binom{m}{h}t^h}{\sum_{g=0}^{m}\binom{m}{g}t^g}P_h.$$

When $s = t = 0$, $\alpha_R(u,v) \neq 0$ if and only if $R = Q_k$, so this curve starts at P_{Q_k}, which is the initial point along the $(k-1)$st boundary of the toric Bezier patch. Moreover, when $s = 0$, $t = 1$, we arrive at the point

$$P(1) = \sum_{h=0}^{m} \frac{\binom{m}{h}}{2^m} P_h,$$

which is the point corresponding to the parameter value $1/2$ for the Bezier curve with control points $P_0,\ldots,P_m$. Similarly, the curve corresponding to $t = 0$ is the segment of the kth boundary of the toric Bezier patch extending from the initial point to the midpoint of the boundary. Finally notice that the boundary $s = 1$ is generated by the functions $\alpha_R(u,v) = c_R t^{L_k(R)}$, so this boundary is identical to the boundary corresponding to $t = 1$ for the rectangular patch located at Q_{k+1}. Hence these patches join along smooth curves on the toric surface as indicated in Figure 8.15.

The unit square $\{(s,t) \mid 0 \leq s,t \leq 1\}$ is the domain of a tensor product patch. Since $R = Q_k + L_k(R)E_{k1} + L_{k-1}(R)E_{k2}$, the index $R \in I$ corresponds to the index $\left(L_{k-1}(R), L_k(R)\right)$ in the new lattice Λ_k anchored at Q_k. Thus the monomial coefficients T_{ij} for this tensor product patch are given by

$$\begin{aligned} T_{ij} &= P_R & i = L_{k-1}(R) \text{ and } j = L_k(R) \\ &= 0 & \text{otherwise,} \end{aligned}$$

where $0 \leq i \leq \max_{R \in I}\{L_{k-1}(R)\}$ and $0 \leq j \leq \max_{R \in I}\{L_k(R)\}$. With this indexing the toric Bezier surface defined by the control points $\{P_R\}$ and the tensor product patch defined by the monomial coefficients $\{T_{ij}\}$ are identical by the change of variables in Equation (8.40).

Equation (8.41) represents the barycentric coordinate functions of the tensor product surface in monomial form. Thus the control points $\{T_{ij}\}$ are the monomial coefficients for the tensor product patch. To find the corresponding Bezier control points, we can simply apply any one of the standard change of basis algorithms such as blossoming to convert from monomial to tensor product Bezier form.

The preceding analysis applies to toric Bezier patches of depth $d = 1$. But the same approach to subdivision works as well for toric Bezier patches of depth $d > 1$ because the blending functions for these patches are generated by convolution from the blending functions of patches of depth $d = 1$. Hence we can simply replace the boundaries $L_k(u,v) = a_k u + b_k v + c_k$ by $L_k^d(u,v) = a_k u + b_k v + dc_k$, $k = 1,\ldots,p$ and proceed with the same analysis. Thus the monomial form in Equation (8.40) for the numerators of the barycentric coordinate functions extends by convolution to the blending functions of toric Bezier patches of arbitrary depth (see Exercise 4).

Exercises

1. Consider a rational tensor product Bezier patch $P(u,v)$ of bidegree (m,n) with control points $(w_{ij}P_{ij}, w_{ij})$. Show how to apply a change of variables and a change of basis to find the control points for the subpatches induced by subdivision at $(u,v) = (.5,.5)$ at the corners P_{m0}, P_{0n}, P_{mn}.

2. Implement both the standard subdivision algorithm in Section 5.8.1 and the toric subdivision algorithm for rational tensor product Bezier surfaces. Which algorithm do you prefer? Why?

3. Show that in the subdivision of a toric Bezier patch into rational tensor product Bezier patches, the tensor product patches all share a common point. Find an explicit expression for this common point.

4. Develop an algorithm for subdividing n-sided toric Bezier patches of arbitrary depth d into n rational tensor product Bezier patches.

5. Implement the subdivision algorithm for hexagonal toric Bezier patches whose lattice hexagon is illustrated in Figure 8.12. That is, implement the algorithm to subdivide these hexagonal toric Bezier patches into six rational tensor product Bezier patches. Then apply the subdivision algorithm to render and to intersect these hexagonal toric Bezier patches.

6. Consider a rational triangular Bezier patch $P(u,v)$ of total degree n with control points P_{ijk}.

   a. Show that the subdivision technique for toric Bezier patches applied at the corner P_{00n} is equivalent to the following procedure:

      i. Divide numerator and denominator by $(1-u-v)^n$.

      ii. Apply the change of variables: $s = u/(1-u-v)$, $t = v/(1-u-v)$.

      iii. Convert from monomial to tensor product Bernstein form.

   b. Use the subdivision technique for toric Bezier patches to subdivide a rational Bezier triangle into three rational tensor product Bezier patches.

7. Consider a rational triangular Bezier patch

$$P(u,v) = \frac{\sum_{i+j+k=n} w_{ijk} P_{ijk} B_{ijk}^n(u,v)}{\sum_{i+j+k=n} w_{ijk} B_{ijk}^n(u)} \qquad 0 \le u+v \le 1.$$

Divide the numerator and denominator of $P(u,v)$ by $(2u+2v-1)^n$ and then apply the change of variables

$$s = \frac{u}{2u+2v-1} \qquad t = \frac{v}{2u+2v-1}.$$

a. Show that the reparametrized surface is given by

$$P*(s,t) = \frac{\sum_{i+j+k=n} (-1)^{n-i-j} w_{ijk} P_{ijk} B_{ijk}^n(s,t)}{\sum_{i+j+k=n} (-1)^{n-i-j} w_{ijk} B_{ijk}^n(s,t)}.$$

b. Show that the change of variables $(u,v) \to (s,t)$ maps the triangle with vertices $(0,1),(1,0),(1/3,1/3)$ to the triangle with vertices $(0,1),(1,0),(1,1)$.

   c. Use parts (a) and (b) to show that $P*(s,t)$ restricted to the domain triangle with vertices $(0,1),(1,0),(1,1)$ is point for point the same surface as $P(u,v)$ restricted to the domain triangle with vertices $(0,1),(1,0),(1/3,1/3)$.

   d. Find the control points of $P*(s,t)$ relative to the Bernstein basis over the domain triangle with vertices $(0,1),(1,0),(1,1)$ by performing a change of basis.

   e. Using the change of variables $\sigma = 1-s$, $\tau = 1-t$, show that the control points in part (d) are the same as the control points relative to the Bernstein basis $\{B_{ijk}^n(\sigma,\tau)\}$ over the standard unit triangle with vertices at $(0,0),(1,0),(0,1)$ of the subpatch of the original patch restricted to the domain triangle with vertices $(1/3,1/3),(0,1),(1,1)$. Conclude that the change of basis in part (d) plays the role of a subdivision algorithm at the parameter values $(u,v) = (1/3,1/3)$ for a particular subpatch of the original triangular patch.

   f. Apply blossoming to show that the standard subdivision algorithm at $(u,v) = (1/3,1/3)$ and the change of basis in part (d) are exactly the same procedure.

8. Explain how to modify the subdivision technique for toric Bezier patches presented in the text so that the rational tensor product Bezier patches inserted at the corners are replaced by rational triangular Bezier patches. Show, however, that these triangular Bezier patches do not necessarily cover the original toric Bezier patch.

9. Let $P(u,v)$ be an n-sided toric Bezier patch with control points P_{ij} defined over a lattice polygon I whose boundaries are $L_1(u,v),...,L_n(u,v)$. Define

$$P*(x_1,...,x_n) = \frac{\sum_{(i,j)\in I} c_{ij}P_{ij}x_1^{L_1(i,j)}...x_n^{L_n(i,j)}}{\sum_{(i,j)\in I} c_{ij}x_1^{L_1(i,j)}...x_n^{L_n(i,j)}}.$$

   a. Show that the patches $P*(1,...,1,x_i,x_{i+1},1,...,1)$, $0 \le x_i,x_{i+1} \le 1$, are exactly the patches constructed in the text that subdivide the surface $P(u,v)$ into n rational tensor product Bezier patches.

   b. Show that these patches all meet at the point $P*(1,...,1)$.

   c. Show that adjacent patches share a common edge.

10. Let $P(u,v)$ be an n-sided toric Bezier patch with control points P_{ij} defined over a lattice polygon I whose boundaries are $L_1(u,v),...,L_n(u,v)$. Define $P*(x_1,...,x_n)$ as in Exercise 9.

   a. Show that the patches $P*(1,...,1,x_i,x_{i+1},1,...,1)$, $x_i x_{i+1} \ge 0$, $x_i + x_{i+1} \le 1$, subdivide the surface $P(u,v)$ into n rational triangular Bezier patches.

   b. Show that these patches all meet at the point $P*(1,...,1)$.

   c. Show that adjacent patches share a common edge.

8.7.8   Depth Elevation for Toric Bezier Patches

Every toric Bezier patch of depth d can be represented as a toric Bezier patch of depth $d + 1$ with the same lattice polygon. To prove this assertion, we will first show how to express the blending functions $\{B_\gamma^d(u,v)\}$ in terms of the blending functions $\{B_\gamma^{d+1}(u,v)\}$.

We begin by deriving another explicit expression for the toric Bezier blending functions $\{B_\gamma^d(u,v)\}$. Suppose that $I = \{\rho_1,...,\rho_m\}$ is a lattice polygon with barycentric coordinate functions $\{\beta_{\rho_1},...,\beta_{\rho_m}\}$. Let $L_j(u,v) \equiv a_j u + b_j v + c_j = 0, j = 1,...,n$ be the equation of the jth boundary of the Newton polygon of I. Then

$$L_j^d(u,v) \equiv a_j u + b_j v + dc_j = 0, \quad j = 1,...,n,$$

is the equation of the jth boundary of the Newton polygon of I^d (see Section 8.7.1, Exercise 3). By Equation (8.26)

$$B_\lambda^d(u,v) = \Sigma_{k_1\rho_1+\cdots+k_m\rho_m=\lambda} \, (_{k_1\cdots k_m}^{\quad d}) \beta_{\rho_1}^{k_1}(u,v)\cdots\beta_{\rho_m}^{k_m}(u,v).$$

Moreover, by Equations (8.22) and (8.23),

$$\beta_\rho(u,v) = \frac{c_\rho\{L_1(u,v)\}^{L_1(\rho)}\cdots\{L_n(u,v)\}^{L_n(\rho)}}{D(u,v)} \qquad \rho \in I$$

$$D(u,v) = \sum_{\sigma\in I}\alpha_\sigma(u,v).$$

Substituting these formulas for $\{\beta_{\rho_h}\}$ into the expression for $\{B_\gamma^d(u,v)\}$ yields exponents of the form

$$k_1 L_j(\rho_1) + \cdots + k_m L_j(\rho_m) = L_j^d(\lambda)$$

because $k_1\rho_1 + \cdots + k_m\rho_m = \lambda$ and $k_1 + \cdots + k_m = d$. Hence

$$B_\lambda^d(u,v) = \frac{\Sigma_{k_1\rho_1+\cdots+k_m\rho_m=\lambda} \, (_{k_1\cdots k_m}^{\quad d}) c_{\rho_1}^{k_1}\cdots c_{\rho_m}^{k_m}\{L_1(u,v)\}^{L_1^d(\lambda)}\cdots\{L_n(u,v)\}^{L_n^d(\lambda)}}{\{D(u,v)\}^d}.$$

Therefore, since the factor $\{L_1(u,v)\}^{L_1^d(\lambda)}\cdots\{L_n(u,v)\}^{L_n^d(\lambda)}$ appears in every term,

$$B_\lambda^d(u,v) = \frac{c_\lambda^d\{L_1(u,v)\}^{L_1^d(\lambda)}\cdots\{L_n(u,v)\}^{L_n^d(\lambda)}}{\{D(u,v)\}^d} \tag{8.42}$$

$$c_\lambda^d = \Sigma_{k_1\rho_1+\cdots+k_m\rho_m=\lambda} \, (_{k_1\cdots k_m}^{\quad d}) c_{\rho_1}^{k_1}\cdots c_{\rho_m}^{k_m}.$$

To express the blending functions $\{B_\gamma^d(u,v)\}$ in terms of the blending functions $\{B_\gamma^{d+1}(u,v)\}$, we shall multiply the expression for $B_\gamma^d(u,v)$ in Equation (8.42) by each of the barycentric coordinate functions of I and add the results. To proceed, observe that by Equation (8.42)

$$\beta_\rho(u,v)B_\lambda^d(u,v) = \frac{c_\rho c_\lambda^d \{L_1(u,v)\}^{L_1^d(\lambda)+L_1(\rho)} \cdots \{L_n(u,v)\}^{L_n^d(\lambda)+L_n(\rho)}}{\{D(u,v)\}^{d+1}}$$

$$= \frac{c_\rho c_\lambda^d \{L_1(u,v)\}^{L_1^{d+1}(\lambda+\rho)} \cdots \{L_n(u,v)\}^{L_n^{d+1}(\lambda+\rho)}}{\{D(u,v)\}^{d+1}}.$$

Therefore,

$$\beta_\rho(u,v)B_\lambda^d(u,v) = \frac{c_\rho c_\lambda^d}{c_{\lambda+\rho}^{d+1}} B_{\lambda+\rho}^{d+1}(u,v).$$

Summing both sides of this equation over all $\rho \in I$ and recalling that the barycentric coordinate functions sum to one yields

$$B_\lambda^d(u,v) = \sum_{\rho \in I} \frac{c_\rho c_\lambda^d}{c_{\lambda+\rho}^{d+1}} B_{\lambda+\rho}^{d+1}(u,v). \tag{8.43}$$

Now consider a toric Bezier patch $B_\lambda^d(u,v)$ of depth d with control points $\{P_\lambda^d\}$ and let $B^{d+1}(u,v)$ be the toric Bezier patch of depth $d+1$ with control points $\{P_\gamma^{d+1}\}$ defined by

$$P_\gamma^{d+1} = \sum_{\rho \in I} \frac{c_\rho c_{\gamma-\rho}^d}{c_\gamma^{d+1}} P_{\gamma-\rho}^d, \tag{8.44}$$

where $c_{\gamma-\rho}^d = 0$ if $\gamma - \rho \notin I^d$. Then

$$B^{d+1}(u,v) = \sum_{\gamma \in I^{d+1}} B_\gamma^{d+1}(u,v) P_\gamma^{d+1} = \sum_{\gamma \in I^{d+1}} B_\gamma^{d+1}(u,v) \sum_{\rho \in I} \frac{c_\rho c_{\gamma-\rho}^d}{c_\gamma^{d+1}} P_{\gamma-\rho}^d$$

$$= \sum_{\lambda \in I^d} \left(\sum_{\rho \in I} \frac{c_\rho c_\lambda^d}{c_{\lambda+\rho}^{d+1}} B_{\lambda+\rho}^{d+1}(u,v) \right) P_\lambda^d = \sum_{\lambda \in I^d} B_\lambda^d(u,v) P_\lambda^d$$

$$= B^d(u,v),$$

so the surface $B^{d+1}(u,v)$ is the depth-elevated form of the surface $B^d(u,v)$. Therefore, Equation (8.44) is the depth elevation formula for toric Bezier patches.

Exercise

1. Show that for triangular Bezier patches, the depth elevation formula in Equation (8.44) reduces to the standard three-term degree elevation formula given in Section 5.8.2, Exercise 9.

8.7.9 Differentiating Toric Bezier Patches

We can differentiate toric Bezier patches exactly in the same way that we can differentiate S-patches because the blending functions $\{B_\lambda^d(u,v)\}$ for a toric Bezier patch of depth d can be expressed in terms of discrete d-fold convolutions of the barycentric coordinate functions $\beta_I(u,v) = \{\beta_\rho(u,v)\}_{\rho \in I}$ of the lattice polygon I. That is, by Equation (8.25),

$$\{B_\lambda^d(u,v)\} = \underbrace{\beta_I(u,v) \otimes \cdots \otimes \beta_I(u,v)}_{d \ factors}.$$

Hence, the derivative formulas we derived for S-patches in Section 8.4.3 readily extend to toric Bezier patches.

For example, first-order partial derivatives are easy to compute. Indeed by Equation (5.21),

$$\left\{ \frac{\partial B_\lambda^d(u,v)}{\partial u} \right\} = d \left\{ \frac{\partial \beta_I(u,v)}{\partial u} \right\} \otimes \underbrace{\beta_I(u,v) \otimes \cdots \otimes \beta_I(u,v)}_{d-1 \ factors}$$

or equivalently,

$$\frac{\partial B_\lambda^d(u,v)}{\partial u} = d \sum_{\rho \in I} \frac{\partial \beta_\rho(u,v)}{\partial u} B_{\lambda-\rho}^{d-1}(u,v).$$

Therefore, the first-order partial derivative with respect to u of the toric Bezier patch

$$B(u,v) = \sum_{\lambda \in I^d} B_\lambda^d(u,v) P_\lambda$$

is given by

$$\frac{\partial B(u,v)}{\partial u} = d \sum_{\gamma \in I^{d-1}} B_\gamma^{d-1}(u,v) \left(\sum_{\rho \in I} \frac{\partial \beta_\rho(u,v)}{\partial u} P_{\gamma+\rho} \right).$$

An analogous formula holds, of course, for the first-order partial derivative with respect to v.

Algorithmically, these formulas say that to find the first-order partial derivative of a toric Bezier patch of depth d, we need only take the first-order partial derivative of the barycentric coordinate functions on the bottom level of the pyramid algorithm, then run the algorithm, and multiply the result by d. As with S-patches, it follows by our convolution formulas that we could, if we choose, take the first-order partial derivative of the barycentric coordinate functions on any level of the pyramid algorithm, then run the algorithm, and multiply the result by d.

Along any boundary of a toric Bezier patch most of the blending functions $\{(B_\gamma^{d-1}(u,v)\}$ vanish because most of the barycentric coordinate functions are zero along the boundary. In fact, let $L_k(u,v)$ denote the kth boundary of the Newton polygon of I, and let ∂I_k denote the indices of I lying on $L_k(u,v)$. Then along $L_k(u,v) = 0$,

$\beta_\rho(u,v) = 0$, $\rho \notin \partial I_k$. Now let ∂I_k^{d-1} denote the indices along the kth boundary of the Newton polygon of I^{d-1}. Then, by Equation (8.25), along the boundary $L_k(u,v)$

$$B_\gamma^{d-1}(u,v) = 0, \quad \gamma \notin \partial I_k^{d-1}.$$

Therefore, along $L_k(u,v)$

$$\frac{\partial B(u,v)}{\partial u} = d \sum_{\gamma \in \partial I_k^{d-1}} B_\gamma^{d-1}(u,v) \left(\sum_{\rho \in I} \frac{\partial \beta_\rho(u,v)}{\partial u} P_{\gamma+\rho} \right).$$

Again an analogous formula holds for the partial derivative with respect to v along $L_k(u,v)$. It follows then that only the control points indexed by the elements of $\partial I_k^{d-1} \oplus I$ affect the first-order partial derivatives along, or the directional derivatives across, the kth boundary of the patch.

Higher-order partial derivatives are not much more difficult to compute. For example,

$$\left\{ \frac{\partial^2 B_\lambda^d(u,v)}{\partial u^2} \right\} = d \frac{\partial^2 \beta_I(u,v)}{\partial u^2} \otimes \underbrace{\beta_I(u,v) \otimes \cdots \otimes \beta_I(u,v)}_{d-1 \ factors}$$

$$+ d(d-1) \frac{\partial \beta_I(u,v)}{\partial u} \otimes \frac{\partial \beta_I(u,v)}{\partial u} \otimes \underbrace{\beta_I(u,v) \otimes \cdots \otimes \beta_I(u,v)}_{d-2 \ factors},$$

and similar results hold for the other second-order partial derivatives of the blending functions. Again this formula has the same algorithmic interpretation that we found when differentiating S-patches. To find a second-order partial derivative of a toric Bezier patch of depth d:

1. Take the second-order partial derivative of one level of the pyramid algorithm, then run the algorithm, and multiply the result by d
2. Take the first-order partial derivative of two different levels of the pyramid algorithm, then run the algorithm, and multiply the result by $d(d-1)$
3. Add the results of 1 and 2

Higher-order partial derivatives of toric Bezier patches can be computed with similar algorithms.

Exercises

1. Show that the normal along the kth boundary of a toric Bezier patch of depth d with lattice polygon I depends only on the control points indexed by $\partial I_k^{d-1} \oplus I$. (Compare to Section 8.4.3, Exercise 1.)

2. Show that to find the first-order partial derivatives of a toric Bezier patch of depth d, we could, if we choose, take the first-order partial derivative of any level of the pyramid algorithm, then run the algorithm, and multiply the result by d. Explain why in some cases it might be better to take the derivative of the last level of the algorithm instead of the first level. (Compare to Section 8.4.3, Exercise 2.)

3. Consider a toric Bezier patch of depth d with lattice polygon I.

   a. Show that only the control points indexed by the elements of

   $$\partial I_k^{d-2} \oplus I^2$$

   affect the second-order partial derivatives of the patch along the kth boundary.

   b. Generalize the result in part (a) to higher-order partial derivatives.

   (Compare to Section 8.4.3, Exercise 3.)

4. Develop an algorithm for finding the third-order partial derivatives of a toric Bezier patch of depth d. How many different pyramids must you compute? What are the normalizations for each of these pyramids? (Compare to Section 8.4.3, Exercise 4.)

8.7.10   Blossoming Toric Bezier Patches

We could try to blossom a toric Bezier patch $B(u,v)$ of depth d just as we blossom an S-patch of depth d, by replacing the parameters (u,v) by a different parameter pair (u_k,v_k) on each level of the pyramid algorithm. Let us denote this function by $b((u_1,v_1),...,(u_d,v_d))$. This function would certainly be symmetric and satisfy the diagonal property, just like the blossom of an S-patch. But there is a problem with the dual functional property. What we would like to have happen is that each control point of the toric Bezier patch should be given by the blossom of the patch evaluated at some collection of vertices of the patch. Unfortunately, this cannot always occur, even for toric Bezier patches of depth $d = 1$ because there are more control points than there are vertices. There are just not enough vertices to go around.

Nevertheless, we do get a partial dual functional property. Let $\beta_I(u,v) = \{\beta_{ij}(u,v)\}$ be the barycentric coordinate functions for the lattice polygon I whose Newton polygon has vertices $Q_1,...,Q_n$. By Equation (8.25)

$$\{B_\lambda^d(u,v)\} = \underbrace{\beta_I(u,v) \otimes \cdots \otimes \beta_I(u,v)}_{d \ \text{factors}},$$

and by Theorem 8.2

$$\beta_{ij}(Q_k) = 0 \qquad (i,j) \neq Q_k$$
$$= 1 \qquad (i,j) = Q_k.$$

Hence for $k_1 + \cdots + k_n = d$,

$$
\left\{ b_\lambda^d (\underbrace{Q_1,\ldots,Q_1}_{k_1},\ldots,\underbrace{Q_n,\ldots,Q_n}_{k_n}) \right\} = \{\beta_I(Q_1)\}^{k_1} \otimes \cdots \otimes \{\beta_I(Q_n)\}^{k_n},
$$

so

$$
b_\lambda^d (\underbrace{Q_1,\ldots,Q_1}_{k_1},\ldots,\underbrace{Q_n,\ldots,Q_n}_{k_n}) = 1 \qquad \lambda = k_1 Q_1 \oplus \cdots \oplus k_n Q_n
$$

$$
= 0 \qquad \text{otherwise.}
$$

Now let $\{P_\lambda\}$ be the control points of $B(u,v)$. Since

$$
B(u,v) = \sum_{\lambda \in I^d} B_\lambda^d (u,v) P_\lambda,
$$

it follows by linearity that

$$
b\left(\underbrace{Q_1,\ldots,Q_1}_{k_1},\ldots,\underbrace{Q_n,\ldots,Q_n}_{k_n} \right) = P_{k_1 Q_1 \oplus \cdots \oplus k_n Q_n}.
$$

So if we can express the index of a control point as a d-fold Minkowski sum of the vertices of I, then we can compute the control point by evaluating the function $b\big((u_1,v_1),\ldots,(u_d,v_d)\big)$ at the corresponding vertices of I. Unfortunately, in general, not all indices in I^d, and hence not all control points in $\{P_\lambda\}$ can be expressed in this manner (see Exercise 1).

What is to be done? We need either more sets of vertices or more levels in the pyramid algorithm. To overcome our predicament, our strategy is going to be to try to express a toric Bezier patch as a C-patch—that is, to write the barycentric coordinate functions of a toric Bezier patch as convolutions of the barycentric coordinate functions of a collection of S-patches. If we are successful, then we can apply the blossoming procedure that we already know works for C-patches to compute the dual functionals for the toric Bezier patch. Therefore, we turn our attention next to toric Bezier C-patches.

Exercises

1. Let I denote the lattice pentagon depicted in Figure 8.10. Show that for every $d \geq 1$ there are indices in I^d that cannot be expressed as the d-fold Minkowski sum of the vertices of I. Conclude that blossoming the pyramid algorithm does not provide all the dual functionals for these pentagonal lattices for any depth.

2. Let I denote the lattice hexagon depicted in Figure 8.12. Show that every index of I^d for $d \geq 2$ can be expressed as the d-fold Minkowski sum of vertices in I. Conclude that blossoming the pyramid algorithm provides the dual functionals for these lattice hexagons of depth $d \geq 2$.

3. Suppose that $B(u,v)$ is a toric Bezier patch of depth d with barycentric coordinate functions $\{\beta_{\rho_1}(u,v),...,\beta_{\rho_m}(u,v)\}$ and blending functions

$$\{B_\lambda^d(u,v)\} = \underbrace{\{\beta_{\rho_1}(u,v),...,\beta_{\rho_m}(u,v)\} \otimes \cdots \otimes \{\beta_{\rho_1}(u,v),...,\beta_{\rho_m}(u,v)\}}_{d \ \ factors}.$$

Let $\hat{B}(u_{\rho_1},...,u_{\rho_m})$ be the polynomial generated by replacing the barycentric coordinate functions $\beta_{\rho_1}(u,v),...,\beta_{\rho_m}(u,v)$ with the parameters $u_{\rho_1},...,u_{\rho_m}$ in the pyramid algorithm for $B(u,v)$, and let

$$\{\hat{B}_\lambda^d(u_{\rho_1},...,u_{\rho_m})\} = \underbrace{\{u_{\rho_1},...,u_{\rho_m}\} \otimes \cdots \otimes \{u_{\rho_1},...,u_{\rho_m}\}}_{d \ \ factors}.$$

Then $\hat{B}(u_{\rho_1},...,u_{\rho_m})$ and $\hat{B}_\lambda^d(u_{\rho_1},...,u_{\rho_m})$ are polynomials of degree d in m variables with polynomial blossoms

$$\hat{b}\big((u_{11},...,u_{1m}),...,(u_{d1},...,u_{dm})\big) \ \ \text{and} \ \ \hat{b}_\lambda^d\big((u_{11},...,u_{1m}),...,(u_{d1},...,u_{dm})\big).$$

Show that

a. $b_\lambda^d\big((u_1,v_1),...,(u_d,v_d)\big) =$

$$\hat{b}_\lambda^d\Big((\beta_{\rho_1}(u_1,v_1),...,\beta_{\rho_m}(u_1,v_1)),...,(\beta_{\rho_1}(u_d,v_d),...,\beta_{\rho_m}(u_d,v_d))\Big)$$

b. $b\big((u_1,v_1),...,(u_d,v_d)\big) =$

$$\hat{b}\Big((\beta_{\rho_1}(u_1,v_1),...,\beta_{\rho_m}(u_1,v_1)),...,(\beta_{\rho_1}(u_d,v_d),...,\beta_{\rho_m}(u_d,v_d))\Big)$$

(Compare to Section 8.4.4, Exercise 6.)

8.7.11   Toric Bezier C-Patches

A toric Bezier C-patch is a toric Bezier patch that is also a C-patch. To determine if the functions $\alpha_I(u,v) = \{\alpha_{i,j}(u,v)\}_{(i,j)\in I}$ for a lattice polygon I define a C-patch, we must seek

1. a decomposition

$$I = I_1 \oplus \cdots \oplus I_m$$

2. a distribution

$$\alpha_I(u,v) = \alpha_{I_1}(u,v) \otimes \cdots \otimes \alpha_{I_m}(u,v).$$

A *decomposition* means that I can be expressed as a Minkowski sum of lattice polygons $I_1,...,I_m$. For an arbitrary lattice polygon I a decomposition does not always exist, but a decomposition of I does exist for many interesting lattice polygons. Indeed, for some lattice polygons, there are several different decompositions (see Exercise 1). Suppose that a decomposition of I does exist. By Equation (8.22) the functions $\alpha_I(u,v)$ are products of powers of the boundaries of the Newton polygon of I. Therefore, a *distribution* means that products of the lines bounding the Newton polygon of I can be apportioned to the nodes of the lattice polygons $I_1,\ldots,I_m$ in the decomposition so that

- the functions at the nodes are the numerators of barycentric coordinate functions for convex polygons $P_1,\ldots,P_m$;
- convolving these products gives exactly the functions $\alpha_I(u,v)$ defined in Equation (8.22)—the numerators of the barycentric coordinate functions for the lattice polygon I.

Although a decomposition of I need not exist, when there is a decomposition of I into lines (two nodes) and triangles (three nodes), there is always an associated distribution.

For example, consider the lattice hexagon in Figure 8.12. Evidently this lattice polygon can be decomposed into the Minkowski sum of the three linear lattices. Now we want to distribute products of the lines bounding the hexagon to the nodes of these linear lattices so that the functions $\alpha_I(u,v)$ for the lattice hexagon result from convolving these products. Figure 8.18 illustrates such a distribution (see Exercise 1).

We are still missing the polygons—the line segments—associated with these three linear lattices. (Recall that for S-patches, the polygon is not necessarily the convex hull of the indexing set.) So far with each linear array I_k in Figure 8.18, there are associated two quadratic functions, F_{k1}, F_{k2}. We want these functions to represent barycentric coordinate functions for a line segment. Thus for each array I_k, we need to find two points Q_{k1}, Q_{k2} so that

$$F_{kl}(Q_{kh}) = 0 \qquad l \neq h$$
$$= 1 \qquad l = h \ . \tag{8.45}$$

Actually it is enough to find points Q_{k1}, Q_{k2} such that

$$F_{kl}(Q_{kh}) = 0 \qquad l \neq h$$
$$\neq 1 \qquad l = h,$$

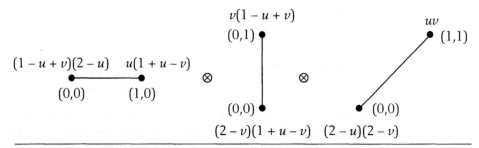

Figure 8.18 A distribution for the lattice hexagon in Figure 8.12. It is straightforward to verify directly that the functions $\alpha_{ij}(u,v)$ for the lattice hexagon are given by convolving the functions at the nodes of these three linear lattices.

since we can always multiply F_{kl} by a constant to force $F_{kl}(Q_{kl}) = 1$. We have the freedom to multiply by constants because we have the freedom to choose the constant coefficients in the functions $\alpha_l(u,v)$. In our example we can take the point Q_{k1} to be the intersection of the lines that factor F_{k2} and the point Q_{k2} to be the intersection of the lines that factor F_{k1}. Notice that these points are vertices of the Newton polygon of the original lattice hexagon.

With this distribution, we have essentially represented the barycentric coordinates for the hexagonal patch as a convolution of the barycentric coordinates of three linear S-patches. We say essentially because we have represented the functions $\alpha_l(u,v)$ by convolution instead of the barycentric coordinate functions $\beta_l(u,v)$. This problem is easily overcome; since there are three factors in the convolution, simply divide the function at each node by the factor

$$\{ \sum_{(k,l) \in I} \alpha_{kl}(u,v) \}^{1/3}.$$

(For rational toric Bezier patches, we can ignore this normalization entirely.) Notice, by the way, that even with this normalization, the functions labeling the nodes for each line do not sum to one. It turns out that this does not matter; we care only that the barycentric coordinate functions $\beta_l(u,v)$ for the lattice hexagon sum to one and that Equation (8.45) is satisfied so that we know where to evaluate the blossom to obtain the dual functionals for the C-patch (see Section 8.6, Exercise 11).

The domain of a C-patch is the intersection of the domains of the corresponding S-patches. For each of the S-patches whose index set is depicted in Figure 8.18, the domain is actually bounded by a quadrilateral—by the four lines contained in the factors at the two nodes. (The domain is bounded by four lines in uv-space even though the polygon itself is a line segment because the barycentric coordinate functions are nonnegative inside a quadrilateral.) But because we have a distribution, an apportioning of the products of the lines bounding the hexagon to the vertices of the lines in the decomposition, the intersection of these three quadrilaterals is exactly the hexagon. Whenever we have a distribution, the lines in the factors at the vertices of the S-patches—that is, the lines that factor the barycentric coordinate functions of the S-patches—are exactly the lines that bound the toric Bezier patch. Therefore, the intersection of the domains of the S-patches in the decomposition will always be the domain of the corresponding toric Bezier patch.

For the lattice hexagon, $I = I_1 \oplus I_2 \oplus I_3$ and

$$\beta_I(u,v) = \beta_{I_1}(u,v) \otimes \beta_{I_2}(u,v) \otimes \beta_{I_3}(u,v)\},$$

so the corresponding toric Bezier patches are C-patches. Let $d = (d_1, d_2, d_3)$. Then there is a three-tier evaluation algorithm for the hexagonal toric Bezier patches indexed by $I^d = I^{d_1} \oplus I^{d_2} \oplus I^{d_3}$. Moreover, the mechanism for blossoming C-patches applies without any modification to these hexagonal patches. Therefore, we can blossom the hexagonal toric Bezier patches indexed by I^d simply by blossoming each tier of the three-tier evaluation algorithm independently. Thus we have succeeded in constructing a blossom for these hexagonal toric Bezier patches that is tri-symmetric, reduces to the original patch along the diagonal, and satisfies the dual functional property. This blossom, however, is not multiaffine, since the functions in the distribution are not linear.

We have illustrated our approach with a particular example, but the details are much the same for all toric Bezier patches with lattice polygons that have a decomposition into a Minkowski sum of lines and triangles. The only missing detail is how the distribution is accomplished. We shall explain precisely how to perform the distribution in Theorem 8.6, but first we need some preparatory lemmas.

LEMMA 8.4

Suppose that $I = I_1 \oplus \cdots \oplus I_m$ is a decomposition of I. Let

$$L(u,v) \equiv au + bv + c = 0$$

be a bounding line of the Newton polygon of I, and let $N = (a,b)$ be a normal vector of $L(u,v)$ pointing into the Newton polygon of I. For each index set I_j, let Q_j be a node of I_j at which N points into the Newton polygon of I_j, and let $Q = Q_1 \oplus \cdots \oplus Q_m$. Then $L(Q) = 0$—that is, Q lies on L.

Proof Since N is an inward-pointing normal, $L(R) \geq 0$ for all $R \in I$. Let $R_j \neq Q_j$ be another node of I_j. By assumption, N points into the Newton polygon of I_j at Q_j, so

$$N \bullet (R_j - Q_j) \geq 0$$

or equivalently

$$N \bullet R_j \geq N \bullet Q_j.$$

Hence for all $R \in I_1 \oplus \cdots \oplus I_m = I$, we have

$$N \bullet R \geq N \bullet Q, \text{ so } L(R) \geq L(Q) \geq 0.$$

Therefore, $L(Q)$ is the minimum value of L on I. But L is zero on indices of I that lie on L, and positive for all other indices in I; hence $L(Q) = 0$.

LEMMA 8.5

Suppose that $I = I_1 \oplus \cdots \oplus I_m$ is a decomposition of I. Let

$$L(u,v) \equiv au + bv + c = 0$$

be a bounding line of the Newton polygon of I, and let $N = (a,b)$ be a normal vector of $L(u,v)$ pointing into the Newton polygon of I. For each index set I_j, let Q_j be a node of I_j at which N points into the Newton polygon of I_j, and let $L_j(u,v) \equiv au + bv + c_j = 0$ be the line through Q_j parallel to $L(u,v)$. If $R = R_1 \oplus \cdots \oplus R_m$, then $L(R) = L_1(R_1) + \cdots + L_m(R_m)$.

Proof   Let $c^* = c_1 + \cdots + c_m$ and let $L^*(u,v) = au + bv + c^*$. If $R = R_1 \oplus \cdots \oplus R_m$, then $L_1(R_1) + \cdots + L_m(R_m) = L^*(R)$. We shall now show that $L^* = L$. Certainly the line $L^*(u,v) = 0$ is parallel to the line $L(u,v) = 0$. Let $Q = Q_1 \oplus \cdots \oplus Q_m$. By Lemma 8.4, $L(Q) = 0$. But, by assumption,

$$L_j(Q_j) = 0,$$

so $L^*(Q) = L_1(Q_1) + \cdots + L_m(Q_m) = 0$. Hence Q must lie on both L^* and L. But L^* is parallel to L. Therefore, $L^* = L$.

THEOREM
8.6

Let I be a lattice polygon, and let $L_k(u,v) = 0$, $k = 1,\ldots,n$, be the equation of the kth boundary line of the Newton polygon of I with normal vector N_k pointing into the Newton polygon. Suppose that $I = I_1 \oplus \cdots \oplus I_m$ is a decomposition of I. For each index set I_j, let Q_{kj} be the node of I_j at which N_k points into the Newton polygon of I_j, and let $L_{kj}(u,v) = 0$ be the line through Q_{kj} parallel to $L_k(u,v) = 0$. Then the distribution associated to each node R_j of I_j is given by

$$\alpha_{R_j}(u,v) = \{L_1(u,v)\}^{L_{1j}(R_j)} \cdots \{L_n(u,v)\}^{L_{nj}(R_j)}.$$

Proof   Let $R = R_1 \oplus \cdots \oplus R_m$, where $R_j \in I_j$. By Lemma 8.5,

$$L_{k1}(R_1) + \cdots + L_{km}(R_m) = L_k(R).$$

Therefore,

$\alpha_{R_1}(u,v) \ldots \alpha_{R_m}(u,v)$

$$= \{L_1(u,v)\}^{L_{11}(R_1) + \cdots + L_{1m}(R_m)} \cdots \{L_n(u,v)\}^{L_{n1}(R_1) + \cdots + L_{nm}(R_m)}$$

$$= \{L_1(u,v)\}^{L_1(R)} \cdots \{L_n(u,v)\}^{L_n(R)}$$

$$= \alpha_R(u,v) .$$

Thus we have successfully apportioned products of the lines bounding the Newton polygon of I to the nodes of the lattice polygons $I_1,\ldots,I_m$ in the decomposition so that convolving these products gives exactly the functions $\alpha_R(u,v)$ defined in Equation (8.22)—the numerators of the barycentric coordinate functions for the lattice polygon I.

Exercises

1. Let I be the lattice hexagon in Figure 8.12.

   a. Verify that Figure 8.18 represents a distribution for I.

   b. Show that $I = I_1 \oplus I_2$, where $I_1 = \{(0,0),(1,0),(0,1)\}$ and $I_2 = \{(0,0),(0,1),(1,1)\}$, is another decomposition for I.

c. Find a distribution for the decomposition in part (b).

2. Consider the lattice pentagon $I_1 = \{(0,0),(1,0),(0,1),(1,1),(1,2),(2,1)\}$.

   a. Show that $I = I_1 \oplus I_2$, where $I_1 = \{(0,0),(1,0),(0,1)\}$ and $I_2 = \{(0,0),(1,1)\}$.

   b. Find a distribution for this decomposition.

3. Find a distribution for the decomposition of the lattice pentagon in Figure 8.10.

4. Explain why it is not necessary to normalize the distribution in Figure 8.18 for a rational hexagonal Bezier patch.

5. For the hexagonal toric Bezier patches whose lattice hexagon is illustrated in Figure 8.12, there are three evaluation algorithms:

   i.   The pyramid algorithm

   ii. A two-tier evaluation algorithm based on the decomposition in Exercise 1

   iii. A three-tier evaluation algorithm based on the decomposition in Figure 8.18

   Implement all three of these evaluation algorithms. Which algorithm do you prefer? Why? Experiment with toric hexagonal Bezier patches of different depths. Determine how changing the location of the control points affects the shape of the surface.

6. Here we develop a multidepth elevation formula for toric Bezier C-patches whose lattice polygon I has a decomposition $I = I_1 \oplus \cdots \oplus I_m$ into lines and triangles. Let $d = (d_1,\ldots,d_m)$ and suppose that

$$\alpha_I^d(u,v) = \alpha_{I_1}^{d_1}(u,v) \otimes \cdots \otimes \alpha_{I_m}^{d_m}(u,v)$$

is a distribution. Let $L_k(u,v) \equiv a_k u + b_k v + c_k = 0$, $k = 1,\ldots,n$, be the equation of the kth boundary line of the Newton polygon of I, and, adopting the notation of Theorem 8.6, define

$$L_{kj}(u,v) = a_k u + b_k v + c_{k,j} \qquad\qquad j = 1,\ldots,m\ ,$$

$$L_k^d(u,v) = a_k u + b_k v + c_{k,1} d_1 + \cdots + c_{k,m} d_m \qquad k = 1,\ldots,n\ .$$

a. Show that there are constants c_ρ^d, c_ξ such that if $\xi \in I_j$

   i. $\alpha_\rho^d(u,v) = c_\rho^d L_1(u,v)^{L_1^d(\rho)} \cdots L_n(u,v)^{L_n^d(\rho)}$

   ii. $\alpha_\xi(u,v)\alpha_\rho^d(u,v) = c_\xi c_\rho^d \{L_1(u,v)\}^{L_1^d(\rho)+L_{1j}(\xi)} \cdots \{L_n(u,v)\}^{L_n^d(\rho)+L_{nj}(\xi)}$

b. Now let e_j be the m-tuple with a 1 in the jth position and a zero everywhere else. Using part (a) show that

$$\left\{\sum_{\xi \in I_j} \alpha_\xi(u,v)\right\} \alpha_\rho^d(u,v) = \sum_{\xi \in I_j} \frac{c_\xi c_\rho^d}{c_{\rho+\xi}^{d+e_j}} \alpha_{\rho+\xi}^{d+e_j}(u,v).$$

c. Using the fact that $\{\alpha_{I_j}(u,v)\} \otimes \{\alpha_{I^d}^d(u,v)\} = \{\alpha_{I^{d+e_j}}^{d+e_j}(u,v)\}$, show that

$$\sum_{\xi \in I_j} \alpha_\xi(u,v) \sum_{\lambda \in I^d} \alpha_\lambda^d(u,v) = \sum_{\tau \in I^{d+e_j}} \alpha_\tau^{d+e_j}(u,v).$$

d. Dividing the result in part (b) by the result in part (c), conclude that

$$B_\rho^d(u,v) = \sum_{\xi \in I_j} \frac{c_\xi c_\rho^d}{c_{\rho+\xi}^{d+e_j}} B_{\rho+\xi}^{d+e_j}(u,v).$$

e. Applying the result of part (d), show that

$$P_\gamma^{d+e_j} = \sum_{\xi \in I_j} \frac{c_\xi c_{\gamma-\xi}^d}{c_\gamma^{d+e_j}} P_{\gamma-\xi}^d, \text{ where } c_{\gamma-\xi}^d = 0 \text{ if } \gamma - \xi \notin I^d,$$

is the depth elevation formula in the direction e_j. That is, using part (d), show that the toric Bezier surface of multidepth d with control points $\{P_\lambda^d\}$ is identical to the toric Bezier surface of multidepth $d + e_j$ with control points $\{P_\lambda^{d+e_j}\}$.

8.8 Summary

Multisided Bezier patches have many different formulations: S-patches, C-patches, and toric Bezier patches are the three most important paradigms. Each of these schemes is an example of a pyramid patch with a polygonal domain, but with a different type of indexing set and a different collection of barycentric coordinate functions. We review these differences in Table 8.4. Below we summarize the properties and algorithms that these multisided schemes share with the standard three-sided and four-sided Bezier patches.

Table 8.4 Framework for constructing multisided Bezier patches.

Patch	Index Set	Domain	Barycentric Coordinates
S-patch	p-tuples	convex polygon	convex polygon
C-patch	Minkowski sums	intersection polygon	convolutions
Toric Bezier	lattice polygon	Newton polygon	lattice polygon

- Properties of multisided Bezier patches
  1. Rational
  2. Affine invariant
  3. Lie in the convex hull of their control points
  4. Boundaries are Bezier curves determined by boundary control points (toric Bezier patches and S-patches, but not C-patches)
  5. Nondegenerate (toric Bezier patches and toric S-patches)
- Algorithms for multisided Bezier patches
  1. Pyramid evaluation algorithm
  2. Multitier evaluation algorithm (C-patches)
  3. Differentiation
  4. Depth elevation (toric Bezier patches and simplicial S-patches)
  5. Blossoming
  6. Subdivision (toric Bezier patches)

These pyramid patches do not by any means exhaust the possibilities for multi-sided surfaces. Other approaches to multisided patches include multivariate B-splines and subdivision surfaces. Each of these schemes could, in itself, be the subject for a separate book, so we have not attempted to cover these topics here. Our hope is that the tools you have learned throughout this text will give you an entry into these and other topics as the need arises. So, for now, we end our choreography here. We trust you have enjoyed the dance.

Index

About the Author

Ron Goldman has been a professor of computer science at Rice University since 1990. Previously he worked in industry, solving problems in computer graphics, geometric modeling, and computer-aided design. He served as a mathematician at Manufacturing Data Systems Inc., where he helped implement one of the first industrial solid-modeling systems. Later he was employed as a senior design engineer at Ford Motor Company, enhancing the capabilities of their corporate graphics and computer-aided design software. From Ford, Dr. Goldman moved to Control Data Corporation as a principal consultant for the development group devoted to computer-aided design and manufacturing. He left Control Data Corporation in 1987 to become an associate professor of computer science at the University of Waterloo in Ontario, Canada, before joining the faculty at Rice.

Dr. Goldman's current research interests lie in the mathematical representation, manipulation, and analysis of shape, using computers. He is particularly interested in algorithms for polynomial and piecewise polynomial curves and surfaces and has investigated both parametrically and implicitly represented geometry. His current work includes research in computer-aided geometric design, solid modeling, computer graphics, computer algebra, elimination theory, and splines, and he has published over 100 research articles on these and related topics. He also lectures extensively at national and international conferences devoted to these subjects.

Dr. Goldman holds a B.S. in mathematics from M.I.T. and an M.A. and a Ph.D. in mathematics from Johns Hopkins University. He is an associate editor of *Computer-Aided Design* and *Computer-Aided Geometric Design* and a co-organizer of the Dagstuhl meeting on geometric modeling.

Printed and bound by CPI Group (UK) Ltd, Croydon, CR0 4YY

03/10/2024

01040344-0003